MBL Lectures in Biology
Volume 2

TIME, SPACE, AND PATTERN IN EMBRYONIC DEVELOPMENT

MBL LECTURES IN BIOLOGY

Volume 1
The Origins of Life and Evolution
Harlyn O. Halvorson and K.E. Van Holde, *Editors*

Volume 2
Time, Space, and Pattern in Embryonic Development
William R. Jeffery and Rudolf A. Raff, *Editors*

TIME, SPACE, AND PATTERN IN EMBRYONIC DEVELOPMENT

Editors

William R. Jeffery

Department of Zoology
The University of Texas at Austin
Austin, Texas

Rudolf A. Raff

Department of Biology
Indiana University
Bloomington, Indiana

ALAN R. LISS, INC. • NEW YORK

Address all Inquiries to the Publisher
Alan R. Liss, Inc., 150 Fifth Avenue, New York, NY 10011

Printed in the United States of America.

Library of Congress Cataloging in Publication Data
Main entry under title:
Time, space, and pattern in embryonic development.
(MBL lectures in biology ; v. 2)
Includes index.
1. Embryology--Addresses, essays, lectures.
2. Ontogeny--Addresses, essays, lectures. 3. Sea urchin embryo--Addresses, essays, lectures.
I. Jeffery, William R. II. Raff, Rudolf A. III. Series.
[DNLM: 1. Embryology--Congresses. W1 MB999 v.2 / QS 604 T583 1982]
QL971.T55 1983 591.3′3 83-5393
ISBN 0-8451-2201-0

Contents

Contributors

Lynne M. Angerer [101] Department of Biology, University of Rochester, Rochester, NY 14627

Robert C. Angerer [101] Department of Biology, University of Rochester, Rochester, NY 14627

Pierre-André Bédard [29] Department of Biology, McGill University, Montreal, Quebec H3A 1B1, Canada

Jean Bennett [1] Laboratory of Human Reproduction and Reproductive Biology, Harvard Medical School, Boston, MA 02115

Steven D. Black [261] Department of Molecular Biology, University of California, Berkeley, CA 94720

Bruce P. Brandhorst [29] Department of Biology, McGill University, Montreal, Quebec H3A 1B1, Canada

Arthur M. Bruskin [87] Department of Biology, Indiana University, Bloomington, IN 47405

Gail W. Cannon [157] Department of Zoology, Duke University, Durham, NC 27706

Clifford D. Carpenter [87] Department of Biology, Indiana University, Bloomington, IN 47405

Clarissa M. Cheney [49] Department of Biology, Johns Hopkins University, Baltimore, MD 21218

M.R. Dohmen [197] Zoological Laboratory, University of Utrecht, Padualaan 8, 3508 TB Utrecht, The Netherlands

Elizabeth D. Eldon [87] Department of Biology, Indiana University, Bloomington, IN 47405

Rachel D. Fink [157] Department of Zoology, Duke University, Durham, NC 27706

Gary Freeman [171] Department of Zoology, University of Texas at Austin, Austin, TX 78712

John C. Gerhart [261] Department of Molecular Biology, University of California, Berkeley, CA 94720

Robert L. Gimlich [261] Department of Molecular Biology, University of California, Berkeley, CA 94720

Paul R. Gross [xiii] Marine Biological Laboratory, Woods Hole, MA 02543

Michael Alan Harkey [131] Departments of Zoology and Microbiology, University of Washington, Seattle, WA 98195

Tim Hunt [49] Department of Biochemistry, University of Cambridge, Cambridge CB2 1QW, England

William R. Jeffery [ix, 241] Department of Zoology, University of Texas at Austin, Austin, TX 78712

The number in brackets is the opening page number of the contributor's article.

Martin H. Johnson [287] Department of Anatomy, University of Cambridge, Cambridge CB2 3DY, England

Klaus Kalthoff [313] Department of Zoology, University of Texas at Austin, Austin, TX 78712

Thomas C. Kaufman [365] Department of Biology, Indiana University, Bloomington, IN 47405

William H. Klein [87] Department of Biology, Indiana University, Bloomington, IN 47405

John S. Laufer [221] Department of Molecular, Cellular, and Developmental Biology, University of Colorado, Boulder, CO 80309

Anthony P. Mahowald [349] Department of Developmental Genetics and Anatomy, Case Western Reserve University, Cleveland, OH 44106

Richard B. Marchase [157] Department of Anatomy, Duke University, Durham, NC 27706

Daniel Mazia [1] Hopkins Marine Station, Stanford University, Pacific Grove, CA 93950

David R. McClay [157] Department of Zoology, Duke University, Durham, NC 27706

Hester P.M. Pratt [287] Department of Anatomy, University of Cambridge, Cambridge CB2 3DY, England

Rudolf A. Raff [ix, 65] Department of Biology, Indiana University, Bloomington, IN 47405

Eric T. Rosenthal [49] Departments of Cell and Developmental Biology and Anatomy, Harvard Medical School, Boston, MA 02115

Joan V. Ruderman [49] Departments of Cell and Developmental Biology and Anatomy, Harvard Medical School, Boston, MA 02115

Stanley R. Scharf [261] Department of Molecular Biology, University of California, Berkeley, CA 94720

Lisa M. Spain [87] Department of Biology, Indiana University, Bloomington, IN 47405

Susan Strome [221] Department of Molecular, Cellular, and Developmental Biology, University of Colorado, Boulder, CO 80309

Terese R. Tansey [49] Departments of Cell and Developmental Biology and Anatomy, Harvard Medical School, Boston, MA 02115

Frank Tufaro [29] Department of Biology, McGill University, Montreal, Quebec H3A 1B1, Canada

Angela L. Tyner [87] Department of Biology, Indiana University, Bloomington, IN 47405

Gary M. Wessel [157] Department of Anatomy, Duke University, Durham, NC 27706

William B. Wood [221] Department of Molecular, Cellular, and Developmental Biology, University of Colorado, Boulder, CO 80309

Preface

Embryology appears to be unique among modern experimental disciplines in biology in not only still possessing but still celebrating its ancient unsolved problems. One of the most important as well as one of the oldest of these is the origin of complex morphology from an apparently formless egg. The first mechanistic hypotheses directed toward the solution of this problem were framed in the 18th century and suggested that the egg contained a miniature, preformed "homunculus". Development was thus seen as the unfolding and growth of an already existing structure, and the miracle of epigenesis, or form arising spontaneously, was avoided. Unfortunately, this idea could not be maintained in the face of the careful embryological investigations of the late 18th and early 19th centuries. There was no homunculus, and epigenesis appeared to be inescapable.

As has been pointed out by Ernst Mayr in **The Growth of Biological Thought**, strikingly opposed viewpoints in biology are often ultimately resolved in a synthesis. In the case of preformation vs. epigenesis, structure does appear de novo, but it is also true that preformed directions for the generation of structures are present in the egg. No one today, of course, has any difficulty recognizing those instructions as those of the DNA genome present in the egg pronucleus. There is, however, a source of preformed information in eggs distinct from and complementary to genomic DNA. It was recognized in the late 19th and early 20th centuries from studies conducted at the Marine Biological Laboratory (MBL) in Woods Hole and elsewhere that the cytoplasm of eggs and early cleavage stage embryos was not isotropic. Instead, it was found that informational molecules or determinants, localized in specific regions of the cytoplasm, were crucial for correct differentiation of the blastomeres containing them.

The relationship of localized cytoplasmic determinants to the action of the nuclear genome in development posed a conundrum. For example, F.R. Lillie, in a lecture delivered at MBL in the summer of 1927, envisioned embryonic development to be entirely directed by selectively localized cytoplasmic information because all cells contain the same genome. Our current view that localized determinants act by influencing nuclear gene expression in particular regions of the embryo was first explicitly stated by T.H. Morgan in 1934 in his book **Embryology and Genetics.**

The chapters in this book are derived from lectures which constituted the central theme of the Embryology course at the MBL in the summer of 1982.

The book presents the current state of investigation on cytoplasmically localized information in embryos, its spatial distribution, possible molecular nature, time of expression, and relationship to the function of genes involved in pattern formation. Although many of the topics considered here have long histories, it is apparent that recent investigations are at last beginning to produce significant new insights.

Localization phenomena are dynamic processes linked to the cleavage cycle and to other temporal control systems in the embryo. Investigations with the embryos of spiralians, nematodes, insects, and in nonmosaic mammalian embryos reveal a considerable evolutionary flexibility in the management of determinative processes during early development. In some instances, such as the germ-line determinants of insects, determinants are already positioned in the region of the egg in which they will ultimately function during oogenesis. Regionalized information systems already established in the egg may determine the cleavage patterns during early development. However, in most embryos, determinants reach their final positions via a set of progressive cytoplasmic movements as cleavage proceeds. In mammalian embryos, preformed determinants appear not to exist or at least to play no demonstrably significant roles in early decisions. Instead, interactions between cells and relative location of cells in the cleaving embryo provide the information for determinative decisions.

The chemical identities of localized determinants are still poorly understood. However, recent advances in nucleic acid technology, notably the use of cloned DNA sequences to probe for specific mRNAs in fractionated embryos or in sectioned embryos in situ, have made it possible to map the spatial and temporal distributions of particular mRNAs. The use of monoclonal antibodies makes it possible to launch a similar search for potential protein determinants. Some rather surprising results have emerged from these studies. Both maternal mRNAs and proteins exhibit discrete spatial localization patterns. Perhaps of equal importance is that the localization and translational expression of mRNAs are under temporal controls. The finding that molecular events such as the release of physically sequestered maternal mRNAs is controlled by timing mechanisms is particularly exciting because there is ample evidence that the functioning of developmental determinants is tied to the cleavage clock. For example, the timing of micromere formation in sea urchin embryos is specified by the cleavage clock whereas the site of micromere formation is specified by information localized in the vegetal portion of the egg. In some cases, the cleavage clock can be uncoupled from cleavage. Thus, if cleavage is delayed by one cycle relative to the clock, sea urchin embryos can be made to produce micromeres at the correct clock time but at the third instead of the fourth cleavage.

A final conceptual and experimental advance stems from the developmental-genetic analysis of pattern formation in *Drosophila*. The initial pattern (i.e., anterior-posterior, dorsal-ventral) is determined by genes active during oogenesis. Such maternally established patterns are interpreted and elaborated by genes active in the embryo. These include genes whose action is spatially regulated by maternal patterning elements to establish the number, location, and polarity of segments. Genes revealed by homoeotic mutations establish segment identities. Cloning of these genes is beginning to provide a molecular approach to the action of genes so far accessible only by genetic and developmental analysis.

The classic problem of the origin of form in developing embryos remains, but efforts such as those presented in this volume not only have revealed much about the temporal and spatial functions of information systems present in the early embryo but also have allowed the problem to be cast in new, experimentally accessible terms. The problem we face (in evolution as well as in embryology) is one of accounting for how the activity of genes is converted into morphology. Development presents us with a sort of black box in which the crucial processes occur. Clearly there is more to it than the relatively well-understood processes of transcription and translation or the self-assembly of proteins into higher-order structures. The genes that act during oogenesis to establish pattern dictate the subsequent spatial and temporal events related to gene expression. Current methodologies are begining to allow us to document and study such events. The relationship between localized or temporally specified mRNAs and morphogenesis is still obscure, but the existence and involvement of a complex cytoskeletal architecture in mRNA localization in eggs and embryos suggest a crucial role for cell structure in the expression of genes in development. The complementary use of genetic, molecular, and cell biological techniques to attack the problems posed here is illustrated by the chapters in this volume. We hope that this book proves as exciting to readers as it has been to us. Perhaps the best mark of the book's success will be if it helps to stimulate research which will quickly render it obsolete.

No project of this kind grows in isolation. We have profited from the encouragement and help of many of our colleagues and students. In particular, we wish to thank Paul Gross for providing the resources of the MBL which made possible the rich lecture series that underlies this book. We also thank Joan Howard and many others of the MBL staff for their resourcefulness and cheerful help in making the Embryology course work smoothly, and Alan R. Liss and Paulette Cohen, who provided some of the initial suggestions for this project. Finally, we are grateful to our students and colleagues in the Embryology course for the intellectual stimulation and good companionship they provided throughout the summer, and to our coauthors in this volume for their superb lecture presentations and for their cooperation in writing and editing these chapters.

Rudolf A. Raff
William R. Jeffery

Foreword

Among the research interests of the founders of the Marine Biological Laboratory in 1888, embryology was second to comparative morphology and physiology. Yet within a very short time—five years, more or less—certain issues of embryology came to dominate discussion and investigation during the intense and productive summer sessions of the Laboratory. This was in no small measure a consequence of the personalities and intellectual strengths of C.O. Whitman, the first Director of the Laboratory, and E.B. Wilson, followed by such younger colleagues and students as E.G. Conklin, T.H. Morgan, and F.R. Lillie, who was to succeed Whitman, after the latter's brilliant tenure as Director of the MBL.

A formal course in embryology was established in 1893 in response to considerable pressure from many quarters for such an effort. Its main virtue would be to bring together in one place, for the benefit of serious students and younger investigators, a faculty of quality not duplicable in any single university, with an abundance of biological material uniquely suited to the studies of interest. The establishment of the Embryology course followed by five years the founding of the Laboratory and of its first course offering, Invertebrate Zoology. (Marine Botany was first given in 1890, and General Physiology—the first such course in the world—in 1892.)

The Embryology course, whose 1982 lectures are the chapters of this volume, has thus been offered essentially without hiatus since its founding. Its 89-year history is surely as long and distinguished a record as that of any science-teaching activity in the world. Those who are acquainted with scientific pedagogy, however, as well as with the history of science, will find in this some contradiction.

Most of us are aware that a science course offered unchanged even over half a decade is likely to be obsolete in most of its parts; lecture notes a decade old, however nicely retyped, are useful for little more than nostalgia. How, then, can the MBL's Embryology course have survived, in its primary organization (as it has), for the better part of a century, and yet have been continually "distinguished"?

The answer is that it has always been a kind of sculptor's armature, cleaned of dry, adherent clay every winter, and remodeled by the faculty and the leading thinkers of the field every summer, according to the highest current canon. That is, indeed, true of all the MBL summer courses, so that their titles are sometimes a less-than-explicit guide to their contents. Aside from this little problem of titles, the system has worked well. Its product has been an educational program that is evolutionary in the best sense: a scientific training in which there is continuity with change, the useful changes stabilized by selection pressure. Ideas, techniques, arguments that do not work well, or prove to be of low relevance to the central problems, fail to reproduce themselves. Important ideas and methods, issues that demand resolution independently of immediate changes of style or technique, reproduce themselves and are themselves subject to further evolutionary change.

The issues toward which the 1982 lectures were directed, and these chapters based upon them, are a splendid example of the process. At the base of all is a transcendent issue and problem, one of the first such problems to be identified with the MBL and with American biology quite specifically: the physical basis of determination in early animal development.

Out of it came some of the finest biological research of the late 19th and early 20th centuries, much of it done or inspired by the founders of the MBL. They did not solve the problem, but they did reformulate it eventually in such terms as to make it honestly investigable. That was a powerful advance over the natural-philosophical notions of the driving forces of embryogenesis that had become dominant in Europe during the preceding thirty years.

Yet the chapters of this volume deal with methods and fundamental concepts that would have been unimaginable to the MBL's founders—methods and concepts that are in fact remote from those that dominated "experimental embryology" and "developmental genetics" as recently as the 1940s and 1950s. I can imagine no more poignant lesson in the excitement, the aesthetic excellence of good biological research, and at the same time its awful transience, than to read these chapters side-by-side with lectures, papers, and books produced by MBL embryologists just a few decades ago. The work must indeed be its own reward: that, too, we try to teach students at the MBL.

The first line of entirely original, internationally influential research at the new seaside laboratory in Woods Hole was a kind of investigation dubbed "cell lineage." Exploiting energetically the lead provided by Van Beneden and others in Europe during the 1870s and 1880s, Director Whitman, in powerful work begun as his doctoral dissertation of 1878, established the existence of a precocious bilateral symmetry and of organ-specific regions in the egg of the leech *Clepsine*. E.B. Wilson did likewise for the clamworm *Nereis*, and that was followed, among others (but none so elegantly), by the beautiful studies of Conklin on predetermination as manifested in cell-lineages of the *Crepidula* embryo.

These investigations and their offspring appeared unequivocally to refute antecedent notions of phylogeny and germ-layer homology as the *specific* processes driving early embryogenesis; and, more importantly, to disprove the proposal of Weismann that early nuclear (i.e., genetic) differentiation—a direct consequence of cleavage— is the agency of cytoplasmic divergence among the early blastomeres. The groundwork was provided for what was to become a transcendent idea: that the cytoplasm of an egg, or of a zygote, is regionally inhomogeneous, and that it is to that extent possessed of information not implicit simply in the (average) chemistry of the "protoplasm."

Nor was the experimental evidence limited to the "mosaic" eggs of molluscs, worms, and tunicates: it was not long before such simple but powerful experiments as those of Morgan on centrifuged sea urchin eggs demonstrated that even "regulative" embryos begin development with an axis of polarity, at least, preformed.

There were then, as there are today, tensions among groups of investigators of different disciplinary background. The rising and brilliant group with allegiance to the General Physiology course and to the leaders of that field, most notably Jacques Loeb, found (not entirely without justice) some tendency toward vitalism, or at least a smell of it, in talk of "organ forming substances," or later, of "morphogenetic determinants"; in the idea that the "protoplasmic organization" contains a blueprint, or program, for the three-dimensional reality that is the larva of an invertebrate animal.

On the other hand, the embryologists, closer to the biological material and with an unshakeable sense of the spontaneity and *directedness* of cleavage and morphogenetic movements, saw in the notion of "isotropy" of the egg, as put forward originally by Loeb and others, an example of simplistic, premature emphasis upon chemistry and physics, to whose primacy the physiologists were committed on philosophical grounds.

Driesch's success in separating the early blastomeres of sea urchin embryos was a great boost for the argument of "isotropy." It seemed to refute the possibility that *any* predetermination of morphologic outcome can exist, at least up to the four-cell stage. (At the same time, Driesch's result was a further blow to Weismann's proposal of nuclear differentiation and divergence.) It remained for Wilson and Morgan, and for two entire generations of their followers at the MBL and elsewhere, to show that there are *not* two perfectly separable and distinct kinds of development, "mosaic" and "regulative"; that the sea urchin egg is *not* isotropic; and that the molluscan and annelid eggs are *not* perfectly and finally predetermined to the last feature of larval form and function.

Simultaneous with the rising conviction that to determine when and how *genes* function in development (in a discipline that came to be known as "developmental genetics") is an indispensable step in the analysis of embryogenesis, there was established the certainty that all or nearly all animal embryos have their cleavages guided and their emerging form determined by cytoplasmic elements, independent of the nuclear genes. It was a convincing argument of E.B. Wilson's monumental textbook (third edition in 1925) that while the "morphogenetic determinants" may not be in their final locations at the very start of development, and may indeed be redistributed in various ways after cleavage is underway, most if not all early development is directed by them, in the sense that the fundamental symmetries and the developmental potencies of blastomeres result from their presence or absence in a particular place.

The contemporary era (which I suspect is nearly at an end, to be followed by an even more extraordinary one of *directed* transformation of embryos) began with the mutual interaction of molecular genetics—itself an offspring of formal genetics and microbial physiology—and experimental embryology. One can set the time as the middle 1960s. There was no longer any question about whether or not genes have something to do with development, and if they do, whether or not it is early. It was by that date clear that genes have *everything* to do with development, in the broadest sense; because, among other reasons, at every step of the way the last step of gene expression—protein synthesis—must occur. Inevitably, in the new interface between molecular genetics and embryology, the old issue, settled in principle but certainly not in chemistry, came once again to the fore.

What is a "morphogenetic determinant"—in chemical terms? Where is it located? How does it work? How is its information content related to the program explicit in the nuclear genome? What relationship exists, if any, between it and the emergence of *form*—between it, in short, and the spontaneity and directedness that so astonished the first embryologists who followed cell lineages; and that so enchant today's high-school student, set down before a microscope with a clock, a culture of newly fertilized marine invertebrate eggs, and, perhaps, a sandwich and a book?

This is an old issue, as indicated, but it is one to which every trick of four trades—descriptive embryology, biochemistry, molecular genetics, molecular cell biology—has been applied with profit. That, in fact, is what this book of lectures is all about. I leave it to the reader to judge the quality of the work and the amount of progress made, should he or she be familiar with the way things were just a few years ago. I say only that there *are* morphogenetic determinants in the cytoplasms of early embryos, of nearly all kinds. The only likely exceptions are those of birds and, as implied in the Preface to this volume, mammals. I accept even that qualification with suspicion: it seems to me to be stretching homology to equate all the early blastomeres of a mammalian zygote (as opposed to the first cells in the inner cell mass), or the products of incomplete cleavage in the avian blastoderm, with early blastomeres of an annelid, or echinoderm, or a tunicate.

There is, moreover, an enormous amount of information stored in the cytoplasm of an egg, over and above that to be found generally in the cytoplasm of a somatic cell or a late embryonic cell. Some of it is in the form of RNA transcribed during oogenesis, and some of *that* is messenger RNA, responsible for directing a large part of the protein synthesis during early development. Some of the RNA is apparently *not* mature messenger RNA, but is rather a complex set of macromolecules of still unknown function, containing interspersed repetitive sequences found otherwise only in the nuclear RNA of somatic cells. As Eric Davidson and his colleagues have shown, these transcripts of genes are not likely to be accidentally present in the egg; they are a specific subset of the nuclear sequences, and probably have a specific information function.

There are also cytoplasmic proteins, some of them apparently released from the germinal vesicle at the time of its breakdown, that direct or facilitate morphogenetic processes in later development. And there appear to be, remarkably, architecturally-differentiated domains of the egg cytoplasm, as evidenced by differential organization of the molecular cytoskeleton, that match the domains of those morphogenetic "plasms" so painstakingly traced out by the first investigators of cell lineage.

From the point of view of molecular biology, in fact, the uncleaved egg contains an information-rich, asymmetrically-distributed *secondary genome*; relatively long-lived by the time-scale of early development; some of whose ultimate products (e.g., chromosomal proteins) almost certainly influence subsequent gene expression in the cells to which they are sequestered by cleavage.

In a sense, if and when these last assertions are known—and generally accepted—to be true, the long quest upon which the MBL's first Director embarked will have been completed (by his descendants, of course). But as is always the case in basic science, the end is just a beginning. Of what it is a beginning will undoubtedly be a major interest of next summer's Embryology course at the MBL; but it is also announced, explicitly or by implication, in the inspiring chapters that follow.

Paul R. Gross
Woods Hole, Massachusetts
April 1983

Time, Space, and Pattern in Embryonic Development, pages 1–27

Fusion of Sea Urchin Eggs

Jean Bennett and Daniel Mazia

Department of Zoology, University of California at Berkeley, Berkeley, California 94720

Little is known about the contributions of different parts of the egg to the more advanced embryo or adult organism. This is in part due to the fact that few species exist in whose eggs there are natural markers which can be followed over the course of development. From studies following the fates of those egg markers that do exist (such as the yellow pigment in certain ascidian eggs [Morgan, 1927], the red pigment band in eggs of one species of sea urchin, *Paracentrotus lividus* [Morgan, 1927], and the gray crescent in the frog egg [Ancel and Vintemberger, 1948]), we know that defined regions of the egg cytoplasm can give rise to particular structures during the course of embryogenesis. However, we still know very little about the interactions between cytoplasmic domains *within* the eggs themselves. In order to investigate what types of information might be transmitted or shared between different realms of the egg, we devised a method of fusing eggs together. We chose to work with two types of eggs which have numerous gross morphological (and biochemical) differences: fertilized and unfertilized sea urchin eggs. By using donor eggs from sea urchins of different species *(Strongylocentrotus purpuratus* and *Lytechinus pictus),* we can recognize the two cytoplasms by light microscopy. With this system, we attempt to examine the structural integrity of the cytoplasmic, surface, and nuclear components of the egg.

Jean Bennett's present address is Harvard Medical School, % Registrar, 25 Shattuck Street, Boston, MA 02115.

Daniel Mazia's present address is Hopkins Marine Station, Stanford University, Pacific Grove, CA 93950.

FUSION OF SEA URCHIN EGGS

Descriptions of fused sea urchin eggs appearing spontaneously in a normal population of unfused eggs exist in the literature. Fused eggs have been observed either at the end of the spawning season or after treatments designed to remove the fertilization envelope [Driesch, 1900; Bennett, personal observations]. Such spontaneous fusions are very infrequent and it is not possible to predict when or whether they will occur.

The experimental induction of the fusion of sea urchin eggs has also been reported. Several different methods have been used. These are: (1) to treat the eggs with Ca-free seawater after removal of the fertilization membrane [Driesch, 1900], (2) to treat eggs with alkaline seawater and then to centrifuge them [Driesch, 1900; Bierens de Haan, 1913a,b], (3) to place eggs in hypotonic solutions of NaCl in seawater [Goldfarb, 1913], (4) to centrifuge eggs together in capillary tubes [Tyler, 1935, 1942], and (5) to apply an inhomogeneous alternating electric field to eggs in a nonconductive medium [Richter et al., 1982]. Although a few of the eggs treated with these various methods were described as being fused, only in the last situation were the two separate plasma membranes really united rather than just tightly apposed.

Recently we developed a new technique which furnishes fair numbers of fused eggs and permits the fusions of unfertilized and fertilized eggs of different species [Bennett and Mazia, 1981a,b]. This technique takes advantage of the ability of positively charged polymers to induce adhesion of the negatively charged eggs to one another. When then immersed in a medium conducive to fusion, many of the adherent eggs merge [Bennett and Mazia, 1981a]. In this chapter, we will demonstrate what happens when two cells with very different and very well-studied morphological and biochemical characteristics (i.e., fertilized and unfertilized sea urchin eggs) are fused. We will discuss the questions: (1) Is the autonomy of one egg maintained when it is fused to another? (2) Is the fused egg intermediate in its characteristics or does it have characteristics of just one or the other original eggs? (3) Is the unfertilized portion fertilized upon fusion? and (4) What happens at mitosis?

SEA URCHIN EGGS—BACKGROUND

At the time of spawning, unfertilized sea urchin eggs have completed metaphase II of meiosis. They are in a relatively quiescent G_0 state, awaiting the arrival of the sperm or some parthenogenic agent for entry into the cell cycle [Giudice, 1973; Czihak, 1975]. Immediately upon fertilization or parthenogenic activation, the eggs enter the cell cycle.

Fertilized and unfertilized sea urchin eggs provide two very different morphological cell types—an advantage in egg fusion studies, since it is easy to identify the partners in the fusion. The main morphological differences stem from the cortical reaction. The unfertilized egg is covered with short, stubby microvilli, as shown in Figure 1 [Eddy and Shapiro, 1976; Guidice, 1973]. Immediately after fusion of the sperm and egg, the cortical reaction is propagated around the egg. The cortical granules underneath the egg surface fuse to the plasma membrane and release their contents to the outside environment. The release of cortical granule material causes the vitelline envelope to balloon out from the egg and it becomes the "fertilization envelope." Meanwhile, partially as a result of the cortical reaction, but also due to activation of the egg itself, the short, stubby microvilli present on the unfertilized surface are transformed. The surface of the fertilized (or activated) egg (Fig. 2) becomes covered with long, slender microvilli [Eddy and Shapiro, 1976; Kidd and Mazia, 1980; Schroeder, 1979].

Fig. 1. Unfertilized *S. purpuratus* egg as observed with scanning electron microscopy (SEM). Note the short stubby microvilli. ×3,400.

The Bearing of the Cortical Reaction on the Egg Fusion Procedure

The successful fusion of sea urchin eggs calls for the solution of three problems: (1) the removal of materials peripheral to the plasma membrane; (2) the agglutination of the eggs; and (3) provision of a suitable environment for fusion [Bennett and Mazia, 1981a].

Removal of peripheral materials. There are several barriers present on the surfaces of fertilized and unfertilized eggs which interfere with fusion of plasma membranes. In the case of unfertilized eggs, the jelly layer and the vitelline envelope must be removed. For fusion of fertilized eggs, not only must the jelly layer be removed, but also the hardened fertilization envelope and the hyaline material released after the cortical reaction. These materials can be removed by incubating the eggs in artificial seawater containing proteolytic enzymes and reagents which reduce thiol groups [Bennett and Mazia, 1981a].

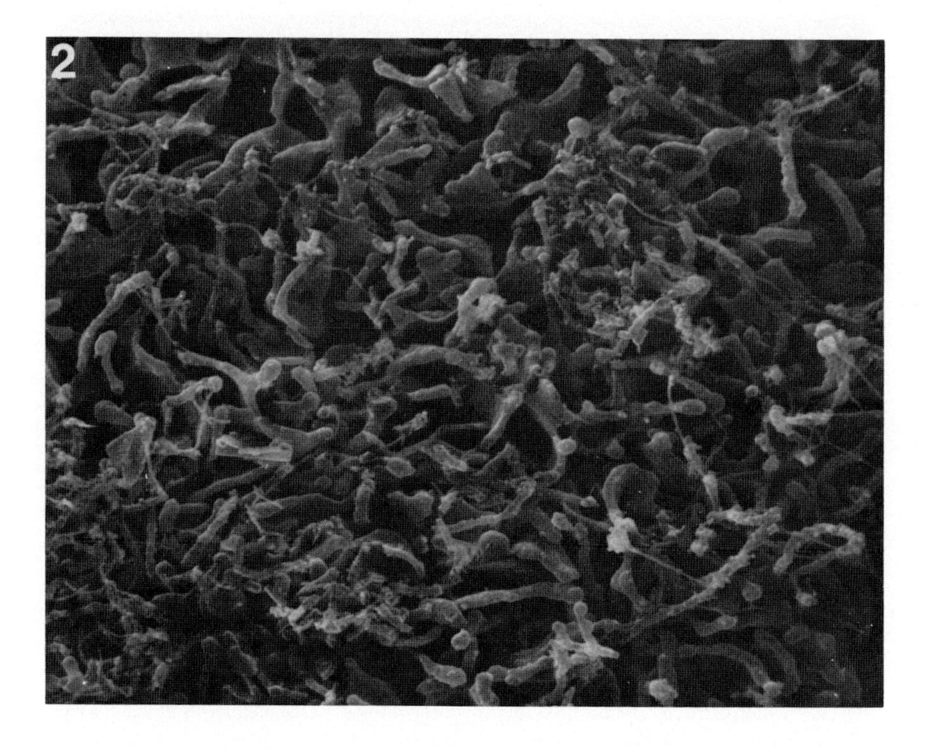

Fig. 2. Fertilized *S. purpuratus* egg as observed with SEM. Fixed 65 minutes after fertilization. Note long, slender microvilli. ×3,400.

The agglutination of the eggs. Since the membranes of most cells are negatively charged, positively charged polymers can be used to induce adhesion to various substrates [Maciera-Coelho and Aurameas, 1972; Mazia et al., 1975]. We make use of this principle to induce adhesion of eggs to one another with the polycation poly(Arg). The advantage of poly(Arg) is that its charge can be removed at the end of the procedure by the addition of arginase.

Satisfaction of the conditions for fusion. The microvilli which cover the surfaces of sea urchin eggs (Figs. 1, 2) prove to be obstacles to fusion. In our standard procedure, the surfaces of both fertilized and unfertilized eggs are made smooth and no microvilli are observed when the eggs are incubated in the hypotonic conditions of the fusion medium, SW/F (SW/F is 75% seawater). The hypotonic medium containing poly(Arg) seems to swell and expand the microvilli to such an extent that they are no longer barriers to fusion. Other treatments which render the egg smooth can substitute for the hypotonic medium [Bennett and Mazia, 1981a].

The fusion of sea urchin eggs is also enhanced by increasing the concentration of calcium in our fusion medium (25 mM) from that normally found in seawater (10 mM).

FUSION PROCESS

Recognition of Fused Cells

Once the above three conditions for fusion are met, fertilized and unfertilized eggs can be fused together. Most fusions take place between unfertilized eggs. In the field of fused and unfused (internal control) eggs portrayed in Figure 3, one can see that the fused eggs have a diameter approximately 1.4× that of unfused eggs, as can be predicted from geometric principles [Bennett, 1980]. Two nuclei or one large, fused nucleus is also visible in some of the fused eggs. If the fusion conditions are slightly modified, one can induce a fusion reaction in a whole chain of eggs (Fig. 4). However, such fusion products are difficult to "harness" for culture because of their fragility—presumably because substances peripheral to the plasma membranes have been removed.

We believe that once the conditions for fusion have been satisfied, any type of cell can be fused by our protocol. In fact, we have had success in fusing several types of differentiated cells to one another and to sea urchin eggs. For example, in Figure 5 one can see a "supersperm" formed by the fusion of three separate spermatozoa. A more immediately practical fusion-produced organism is that formed by the fusion of a plant protoplast to an animal egg. A science fiction strain of "planimals" is indeed a possibility, for we have succeeded in inserting chromosomes of the broadbean, *Vicia faba,* into sea urchin eggs (Fig. 6).

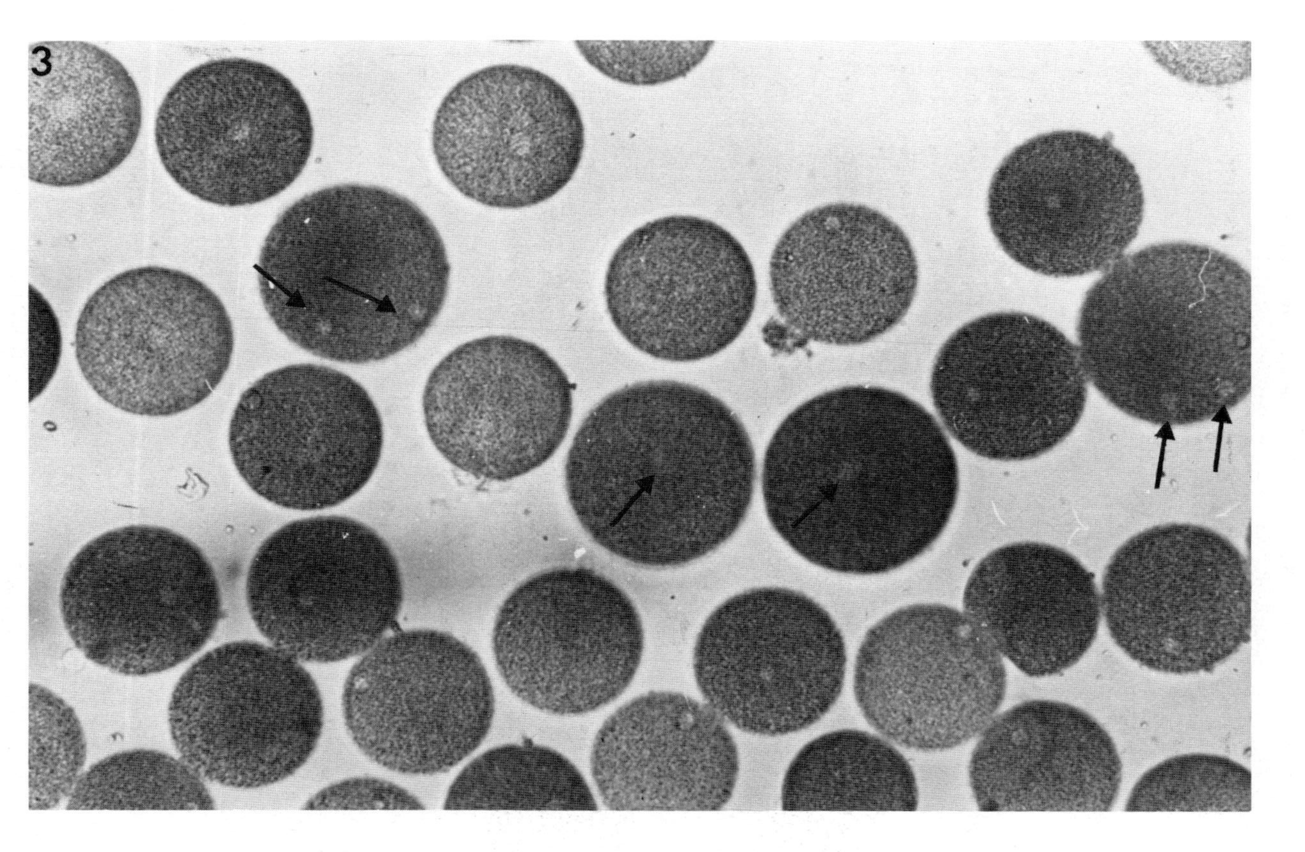

Fig. 3. A field of unfertilized *S. purpuratus* eggs, some of which are fused. The fused eggs have two nuclei or one large, fused nucleus and a diameter ×1.4 that of unfused eggs. Arrows point to nuclei in fused eggs. ×68.

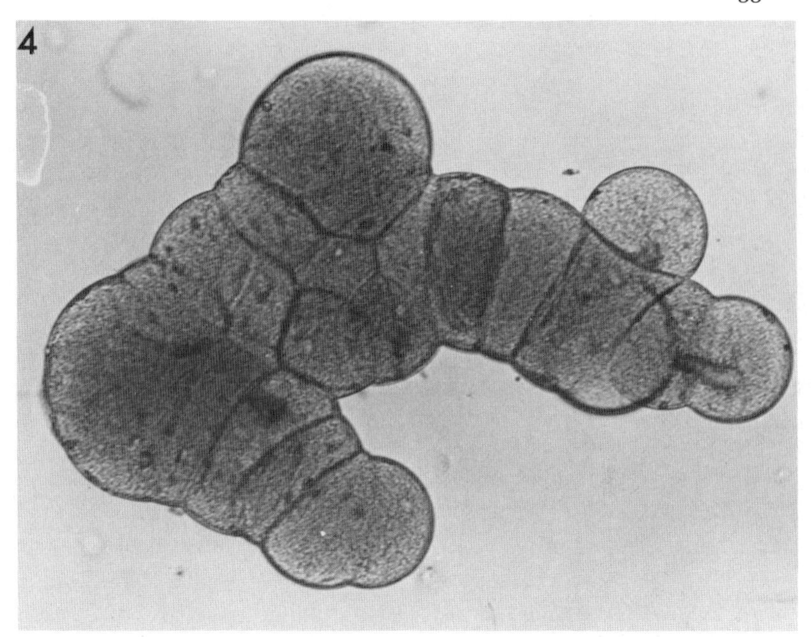

Fig. 4. Partial syncytium composed of unfertilized *L. pictus* eggs; 15 minutes after poly(Arg) treatment. ×50.

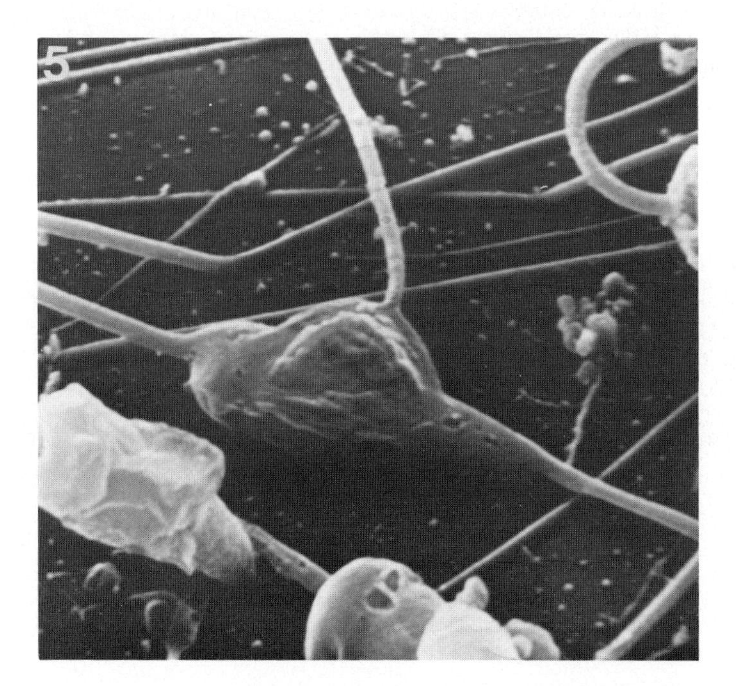

Fig. 5. Three *S. purpuratus* spermatozoa have fused after poly(Arg) treatment [Bennett, 1980]. Note that the fusions have occurred in a "head-to-head" manner. ×13,330.

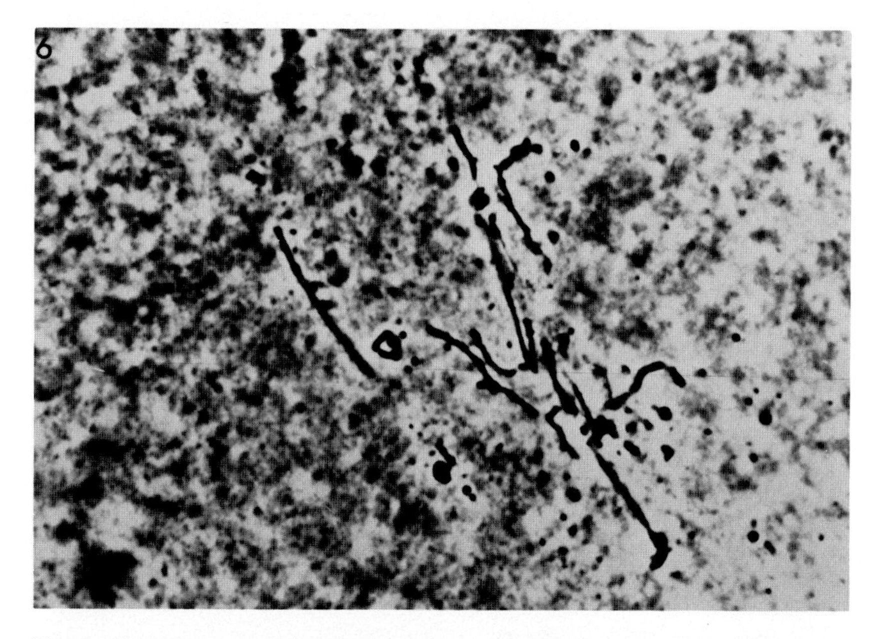

Fig. 6. Plant chromosomes in animal cytoplasm. Unfertilized *S. purpuratus* egg fused with *V. faba* protoplasts are sometimes found to contain *V. faba* chromosomes [Bennett, 1980]. *V. faba* chromosomes are distinguishable from sea urchin chromosomes on the basis of size and number. Sea urchin chromosomes, of which 2n = 36, are very short and of similar length [Makino, 1951]. In contrast, as seen here, the 2n number of *V. faba* is 12, of which two are extremely long and the remaining ten are almost as long [Chooi, 1971]. ×40.

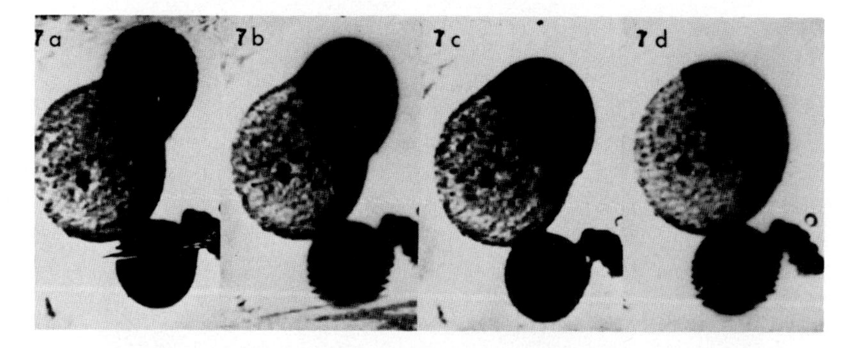

Fig. 7. Unfertilized *L. pictus* egg in process of fusing with unfertilized *S. purpuratus egg*. Unfused *S. purpuratus* egg is also present in the sequence. Sequence begins 20 minutes after start of poly(Arg) treatment. Photos are taken at five-minute intervals (beginning with a) from a video recording. Thus, the time span encompassed in a–d is 15 minutes. ×50.

Identification of Fusing Eggs

It is not difficult to identify fusing eggs. As shown in Figure 7, two spherical eggs merge progressively to form a single spherical cell which can be identified by its size. In the fusion of eggs of different species whose cytoplasms are easily distinguished, the two cytoplasms occupy distinct regions within the newly fused eggs (Figs. 7, 9) [Bennett and Mazia, 1981a].

MIXING OF CYTOPLASMIC YOLK PARTICLES

An advantage of fusing eggs of different species is that the donor cytoplasms can be identified in the fused egg by looking for the species-specific cytoplasmic yolk particles observable with the light microscope. The cytoplasm of eggs of *S. purpuratus* appears dark and relatively opaque; that of *L. pictus* is light and relatively transparent. These appearances reflect optical properties of yolk particles, and strictly speaking, the following descriptions of cytoplasmic mixing refer to the movement of yolk particles. It is possible that macromolecules and ions on the one hand, and large organelles and cytoskeletal components on the other, have quite different rates of mixing.

The two cytoplasms do not mix in response to the events of fusion themselves. When unfertilized eggs are fused to unfertilized eggs, the initial sharp separation of the cytoplasms is followed by slow mixing (Fig. 8). The yolk granules do not always spread uniformly. In Figure 8b, for example,

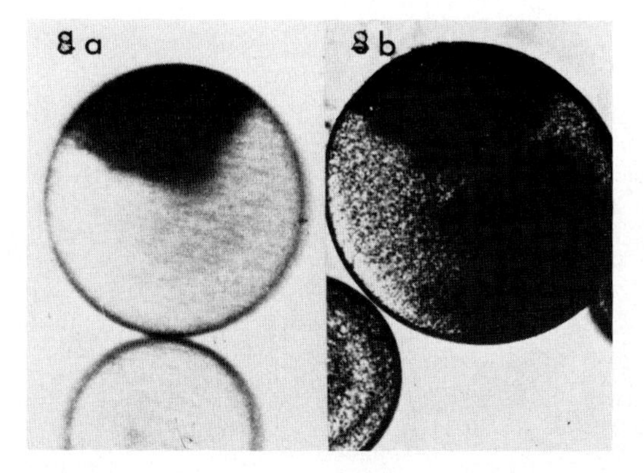

Fig. 8. a. Interspecies hybrid composed of three unfertilized eggs: two *L. pictus* eggs and one *S. purpuratus* egg. Observed 25 minutes after incubation in poly(Arg). Note the initial separation of the "dark" *S. purpuratus* and "transparent" *L. pictus* cytoplasms. b. Same hybrid as seen in Figure 3, but flattened for observation 1.5 hours after incubation in poly(Arg)-containing medium. Note the diffusion of the "dark" *S. purpuratus* cytoplasm. ×68.

the dark cytoplasm of *S. purpuratus* moves away from its original location, but seems to spread along a linear path. However, even though the mixing may appear to be nonrandom at first, the yolk granules become uniformly distributed after about three to six hours.

If fertilized eggs are fused with fertilized eggs immediately after fertilization, the cytoplasms in the fused eggs remain distinct (Fig. 9). However, there is some mixing of the cytoplasms in the region surrounding the mitotic apparatus at the time the nuclei break down for the first mitosis. Cleavage interferes with the completion of the mixing. The net result is that the cytoplasms remain distinct through the following cleavages.

When unfertilized eggs are fused with fertilized eggs, the appearance of the cytoplasm is much the same as when fertilized eggs are fused with fertilized eggs. The yolk particles stay separate up until the time of mitosis, when there is some mixing around the mitotic apparatus. Mitosis of the fused fertilized and unfertilized hybrid gives rise to one furrow and cleavage into two daughter cells. The furrow can form in various planes relative to the boundary between the two cytoplasms (Fig. 10). The dark *S. purpuratus* cytoplasm can be evenly distributed between the two daughter blastomeres (Fig. 10a) or one blastomere can receive most of the dark cytoplasm (Fig. 10b) [Bennett and Mazia, 1981a].

In older experiments the overall structure of the cytoplasm was expressed in terms of viscosity. Our observations of the mixing of the cytoplasmic yolk

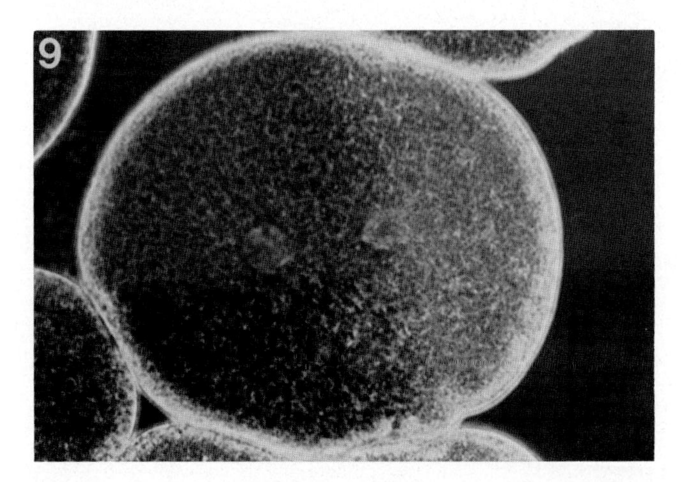

Fig. 9. Interspecies hybrid composed of two fertilized eggs, flattened for observation of nuclei and cytoplasms. *S. purpuratus* eggs were fertilized 25 minutes in advance of the *L. pictus* eggs. Hybrids are observed 65 minutes after poly(Arg) treatment. Note separation of the "dark" *S. purpuratus* and "transparent" *L. pictus* cytoplasms. Also note separations of the two nuclei. ×100.

particles of eggs of two species after fusion sustain that work. Unfertilized eggs were judged to have a relatively low viscosity by various physical criteria, including changes in surface tension as measured by compression [Yoneda et al., 1978; Ikeda et al., 1976; Hiramoto, 1976; Hiramoto, 1969a,b] and changes in the force necessary to move a magnetic particle embedded in

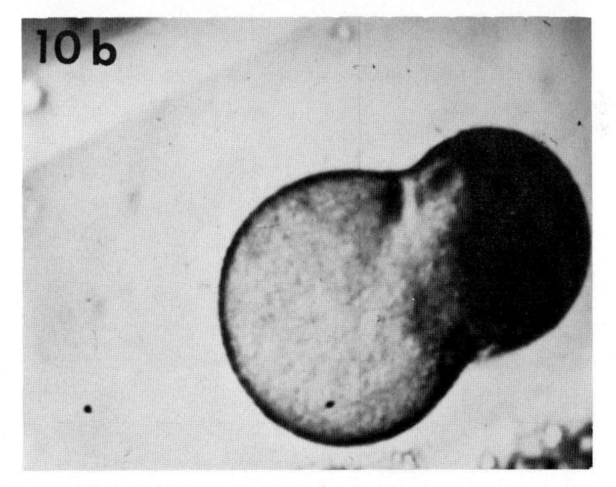

Fig. 10. Interspecific hybrids composed of one fertilized *S. purpuratus* and one unfertilized *L. pictus* egg which have undergone first cleavage. a. Note the even distribution of the "dark" *S. purpuratus* cytoplasmic yolk granules in the two daughter blastomeres. Photo taken from a video recording 3.5 hours after fertilization. Note the presence of an unfertilized *L. pictus* egg—an internal control. b. Note that one blastomere is receiving most of the dark *S. purpuratus* cytoplasm. Photo taken from a video recording three hours after fertilization. ×75.

the protoplasm [Hiramoto, 1974]. Following fertilization, these same measurements indicate that there is an increase in viscosity, or a gelation [Hiramoto, 1969b]. In accord with this work, we find that yolk particles eventually mix completely when unfertilized eggs are fused with other unfertilized eggs. We also observe that when two fertilized eggs are fused the cytoplasmic yolk particles do not mix. Similarly, in the fusion of unfertilized with fertilized eggs, the cytoplasms remain in separate regions, presumably because the cytoplasm contributed by the fertilized eggs retains its greater rigidity and cohesion. At the time of mitosis the mobility of the yolk particles increases and they begin to spread around the region of the mitotic apparatus. This is consistent with Hiramoto's [Hiramoto, 1969b] observation that the cytoplasm surrounding the mitotic apparatus has a relatively low viscosity. The fact that the period of mitosis is so brief limits the extent of mixing. Thus, even by the time of the second mitosis, when, in the hybrids of unfertilized and fertilized eggs, cytokinesis finally takes place, one can still see the segregation of the two types of yolk granules. Although we can interpret the observations on cytoplasmic mixing in terms of overall viscosity, their relationship to cytoskeletal structure remains a mystery. Perhaps the fundamental alterations in the rigidity of the cortex at fertilization are due to the new appearance of assembled microfilaments following sperm-egg fusion.

One wonders whether the information which transforms cytoskeletal structure at fertilization might be transferable, for instance, from a fertilized egg when it is fused with an unfertilized egg. There is an elegant assay for this query: the examination of the mixing of cell surfaces of fused eggs.

MIXING OF CELL SURFACES

The surfaces of fertilized and unfertilized sea urchin eggs are easily distinguished by the presence or absence of long microvilli (Figs. 1, 2) and of undischarged cortical granules [Eddy and Shapiro, 1976; Kidd and Mazia, 1980; Schroeder, 1979; Guidice, 1973]. We have examined these features on the surfaces of the fused fertilized and unfertilized eggs and followed any changes in distribution as the eggs progress through the mitotic cycle.

In order to interpret the results, it is necessary to be aware of the effects of the fusion medium itself on the appearances of fertilized and unfertilized eggs. As described above, the surfaces of both unfertilized and fertilized eggs become smooth in our hypotonic fusion medium, SW/ F. The surfaces of unfertilized eggs remain smooth after the fusion treatment, as shown in Figure 11. One no longer sees the short, stubby microvilli which cover the surfaces of normal unfertilized eggs, either on the fused unfertilized eggs or on the unfused eggs which have gone through the fusion procedure. Fertilized

Fig. 11. Two unfertilized *S. purpuratus* eggs fused and prepared for SEM. Fixed 60 minutes after completion of poly(Arg) treatment. Note the lack of microvilli. ×1,400.

eggs, however, regain their microvilli within 20 minutes after they have been returned to normal seawater. Figure 12 shows the surface of an egg formed by the fusion of two fertilized eggs; the appearance of eggs which have gone through the fusion procedure but have failed to fuse is the same. The microvilli are neither as long nor as numerous as those which cover a normal fertilized egg such as the one portrayed in Figure 13 [Bennett and Mazia, 1981b].

When fertilized eggs are fused with unfertilized eggs, a distinct boundary remains between the respective partners, as is shown in Figure 14. The delineation remains sharp until the time of the first mitosis, at which point the whole egg becomes covered with microvilli (Fig. 15). Before that time, there is no blurring of the boundary between the two regions [Bennett and Mazia, 1981b].

The changes in appearance of the outer surface of the sea urchin egg are paralleled by behavior of the cortical granules on the inner surface. Normally a layer of these granules lies immediately under the plasma membrane of the mature unfertilized egg [Eddy and Shapiro, 1976; Kidd and Mazia, 1980; Schroeder, 1979; Guidice, 1973]. The granules are easily observed with the

Fig. 12. A hybrid composed of two fertilized *S. purpuratus* eggs, 65 minutes after fertilization. Note long microvilli. These are not as long as those on untreated fertilized eggs at mitosis (see Fig. 13). ×2,800.

light microscope after eggs have been fixed with ethanol/acetic acid, as can be seen in Figure 16a. The surfaces of fertilized eggs are devoid of cortical granules (Fig. 16b).

When unfertilized eggs are fixed with fertilized eggs, one can distinguish the unfertilized and fertilized portions of the surface by the presence or absence of cortical granules. Again, as in the distribution of microvilli, the two regions of the cortex remain distinct during the period up to the time of the first mitosis (Fig. 16c). There is no sign of diffusion or mixing, no spread

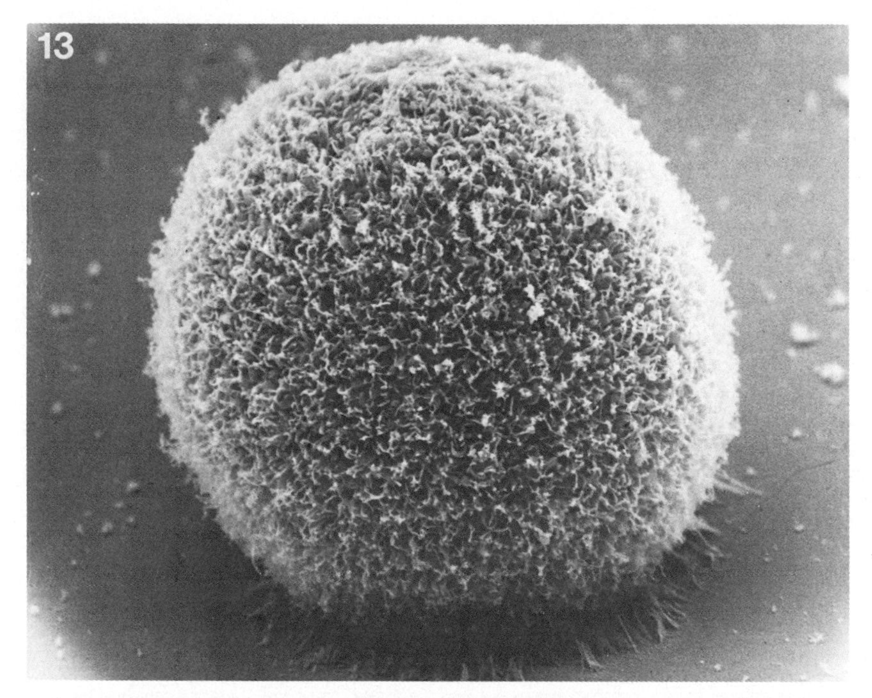

Fig. 13. A nonpoly(Arg)-treated fertilized *S. purpuratus* egg, 65 minutes after fertilization. Note the long and densely packed microvilli. ×1,500.

nor dilution of the cortical granules of the unfertilized surface nor of the microvilli of the fertilized surface [Bennett and Mazia, 1981b]. Since these changes are presumed to be propagated from the point of entry of the spermatozoon in normal fertilization, it must be concluded that the events responsible for the propagation have been completed by the time when unfertilized and fertilized eggs fuse in our procedure. The observations agree with the prevailing view that the early surface events are set off by a transient increase in Ca^{+2} ions.

At the time of mitosis, the boundaries disappear. The whole surface is covered by microvilli and the cortical granules have disappeared from the portion of the surface contributed by the unfertilized egg.

The fact that there is neither spreading nor diffusion of the features of the fertilized egg surface during the time between fusion and mitosis indicates that some event associated with the onset of mitosis causes the discharge of cortical granules and the formation of microvilli on the unfertilized portion of the surface. Normally, these are events which take place at the time of

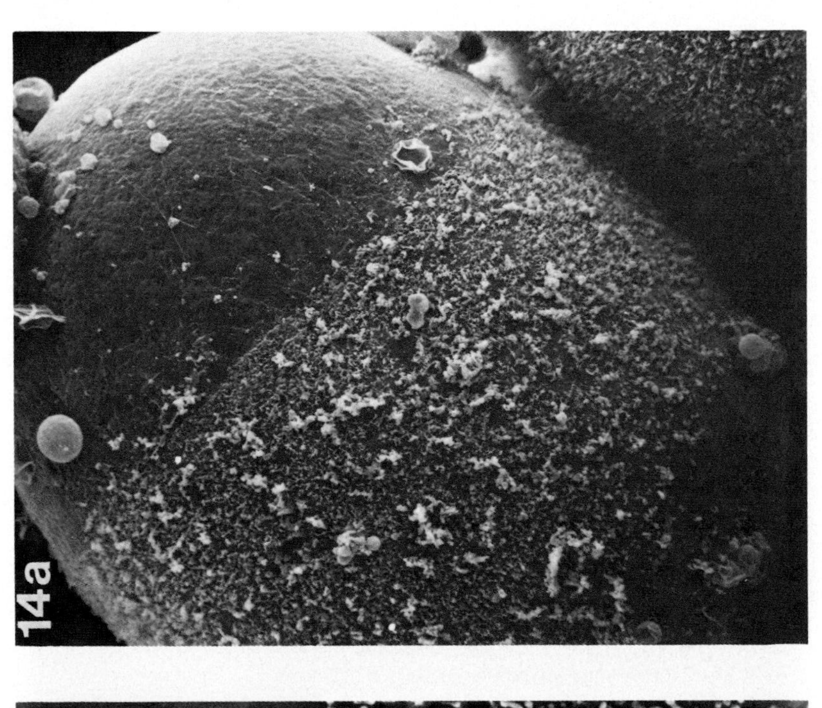

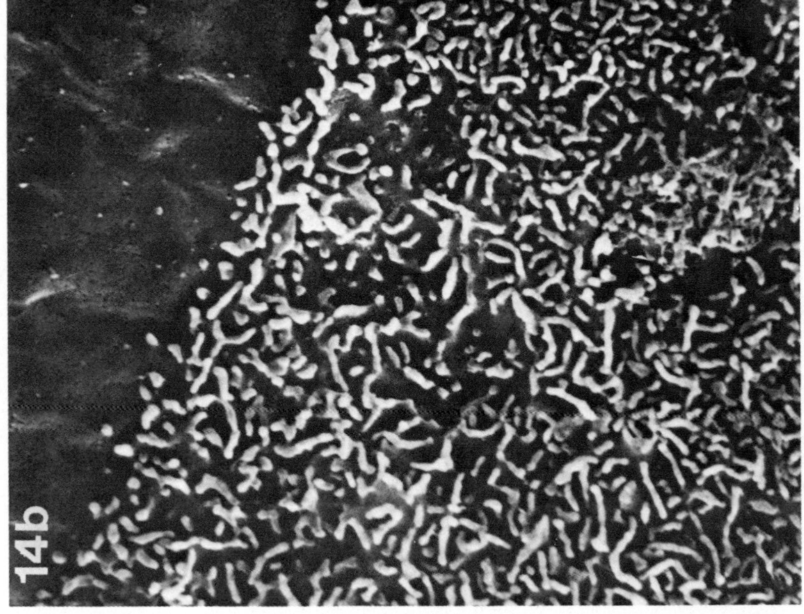

Fig. 14.

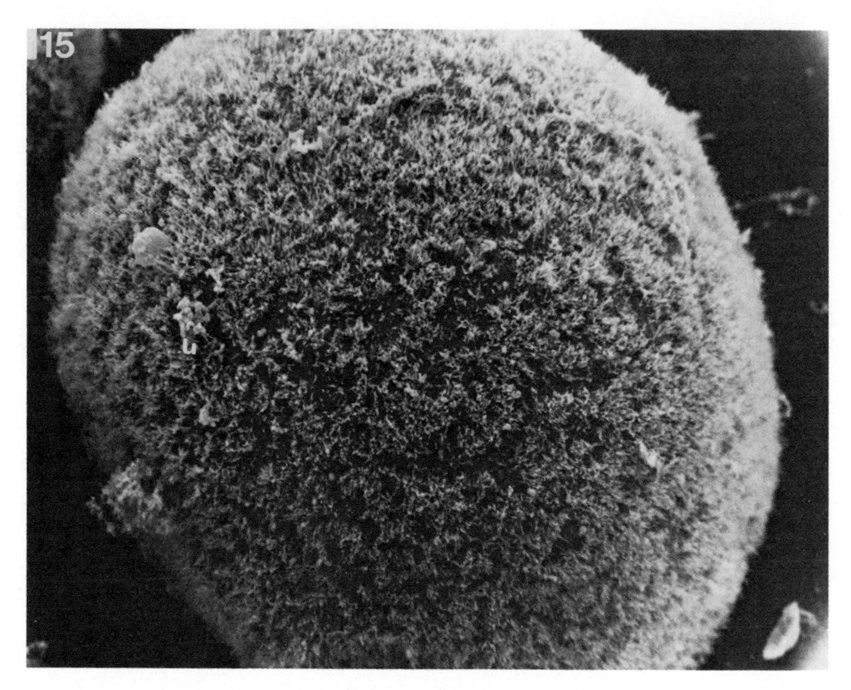

Fig. 15. An interspecies hybrid composed of fused fertilized and unfertilized eggs. Observed at the time of prophase of mitosis. Egg is covered with microvilli. However, these microvilli are still not as densely packed as those on a nonpoly(Arg)-treated fertilized egg at the same stage of the cycle (see Fig. 13). ×1,500.

fertilization. It is striking that at the time the boundaries formed by microvilli and cortical granules disappear in fused fertilized and unfertilized eggs there is also mixing of cytoplasmic yolk particles (see above). Our findings invite speculation that the fertilized egg repeats, at the time of mitosis, some underlying changes which are known to occur at fertilization. (In the fused eggs, these changes would be transmitted to the unfertilized region). One such event, a transient elevation of the Ca^{+2} ion concentration, is a candidate for the cause of the discharge of the cortical granules at fertilization [Stein-

---

Fig. 14. a. Fused fertilized and unfertilized eggs fixed and prepared for SEM 25 minutes after fertilization and the beginning of the fusion treatment. Note the boundary marked by the merging of the extended microvilli with the smooth membrane. b. Higher magnification of the interspecies fused egg composed of fertilized and unfertilized eggs seen in Figure 3. a, ×868; b, ×7,140.

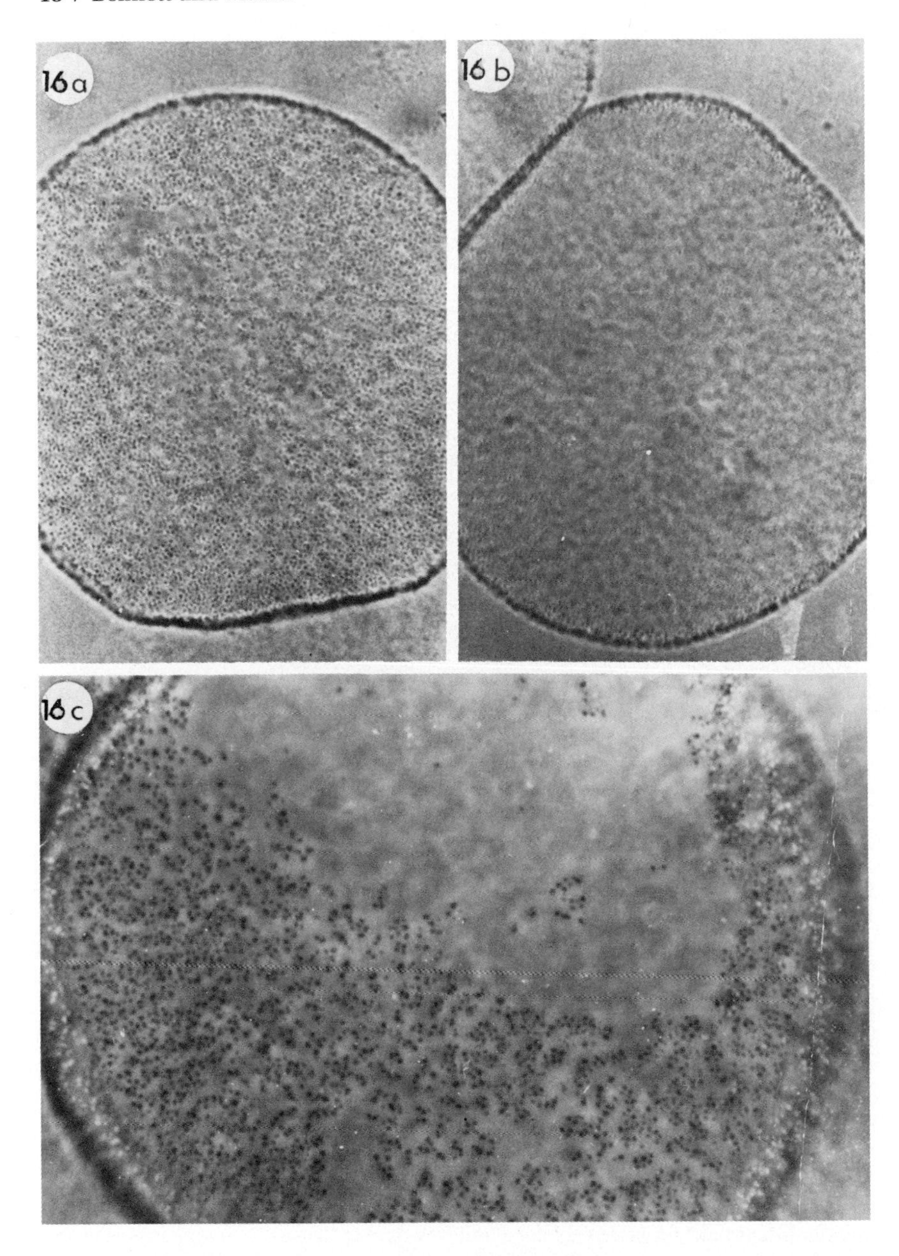

Fig. 16. Unfertilized (a) and fertilized (b) *S. purpuratus* eggs prepared for observation of cortical granules. Fused fertilized and unfertilized egg (c) prepared for observation of cortical granules 35 minutes after fusion. Note the absence of cortical granules in one half (the "fertilized" portion) of the fused egg. a,b, ×121; c, ×242.

hardt et al., 1977, 1978; Steinhardt and Epel, 1974]. The release of Ca^{+2} and the ensuing elevation of pH may also play roles in the elongation of rigid microvilli at fertilization [Begg and Rebhun, 1979, 1980].

INTERACTIONS OF NUCLEI AND MITOTIC FIGURES

We have seen that after the fusion of a fertilized egg with another egg, the two portions of the hybrid egg maintain morphological characteristics they held before fusion: the cell cytoplasms do not mix, as judged by the lack of mixing of cytoplasmic yolk particles, and neither do the cortical granules nor the microvilli. Here we ask whether, despite the apparent structural boundaries between the two portions of the egg, the two nuclei physically interact.

After the fusion of two unfertilized eggs, the two nuclei may or may not merge into one large nucleus (Fig. 3). They merge if they are close together at the time of fusion. In neither case do the nuclei enter mitosis; no breakdown of the nuclear envelope is seen and no condensed chromosomes are found in fixed material, even when observed several hours later. This is an important point for the interpretation of other experiments; it says that the various treatments involved in the fusion technique do not themselves activate the unfertilized eggs of the species we have studied.

When fertilized eggs are fused with fertilized eggs, the two nuclei remain distinct throughout the cell cycle (Fig. 9). This observation is consistent with evidence of the relative rigidity of the two cytoplasms as seen by the lack of diffusion of yolk particles in these hybrids [Bennett and Mazia, 1981a].

At mitosis, the chromosomes condense and mitotic apparatuses form simultaneously in the cytoplasm(s) of fused fertilized eggs. This is to be expected since, in general, mitotic synchrony is observed among hybrid somatic cells of other species [Lau et al., 1977]. Here, not only do we find mitotic synchrony, but we also find synchrony of the substages of mitosis, even when eggs are fused which had been fertilized at different times. The two sets of chromosomes undergo events of (Fig. 17a) prophase, (Fig. 17b) metaphase, and (Fig. 17c) anaphase at the same time. It is difficult to test for the phenomenon of synchrony of each of the substages of mitosis in somatic cell hybrids. Most somatic cells are asynchronous and spend a low percentage of their generation time in the stage of mitosis.

As one might predict from the fact that neither nuclei nor chromosomes merge in fused fertilized eggs, neither do the mitotic poles. One assay for interaction of poles is the observation of equatorial chromosomes in between two sets of poles. We find no chromosomes aligned in this manner. One would expect that if the two poles were close enough together in the same cytoplasm they would be capable of interaction. Interactions between two neighboring monopoles to form a bipolar mitotic apparatus have been ob-

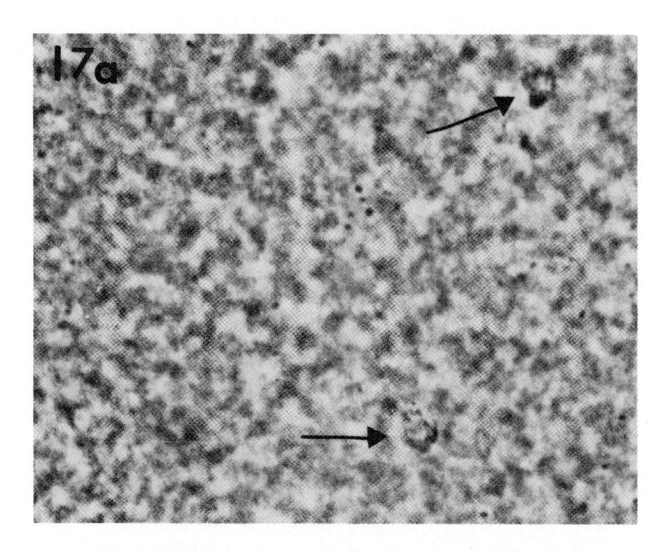

Fig. 17. Mitotic figures and chromosomes observed at the first mitosis of fused *S. purpuratus* eggs. a. Arrows point to two sets of chromosomes in prophase. ×256. b. Two metaphase mitotic figures. ×129. c. Two anaphase mitotic figures. ×91.

served [Sluder, 1978]. The observations can be summarized in the statement that we observe two bipolar spindles rather than one tetrapolar figure, which would be easy to recognize.

The fact that the nuclei and the mitotic figures in fused fertilized eggs do not physically interact is consistent with the evidence that indicates that the two cytoplasms are relatively rigid. It seems that even though the cytoplasm becomes more fluid at mitosis, the two mitotic figures are still unable to come together at this time.

The independence of the mitotic figures in fused fertilized eggs is also evident in their abilities to set up separate cleavage furrows. It is believed that early in mitosis the egg cortex receives the signal to furrow from the mitotic apparatus [Hiramoto, 1956; Tilney and Marsland, 1969; Asnes and Schroeder, 1979]. In the fused fertilized eggs we generally observe the formation of two separate cleavage furrows, which then give rise to four daughter cells (Fig. 18). Occasionally, the two spindles and their poles are so close to one side of the egg that this side receives the "furrow signal" in advance of the far side (Fig. 19). In such cases it usually takes longer to organize a cleavage plane in the fused eggs than in the nonfused eggs.

The remaining divisions keep pace with those of the unfused fertilized eggs, however (Fig. 18). The four cells derived from the unfused fertilized

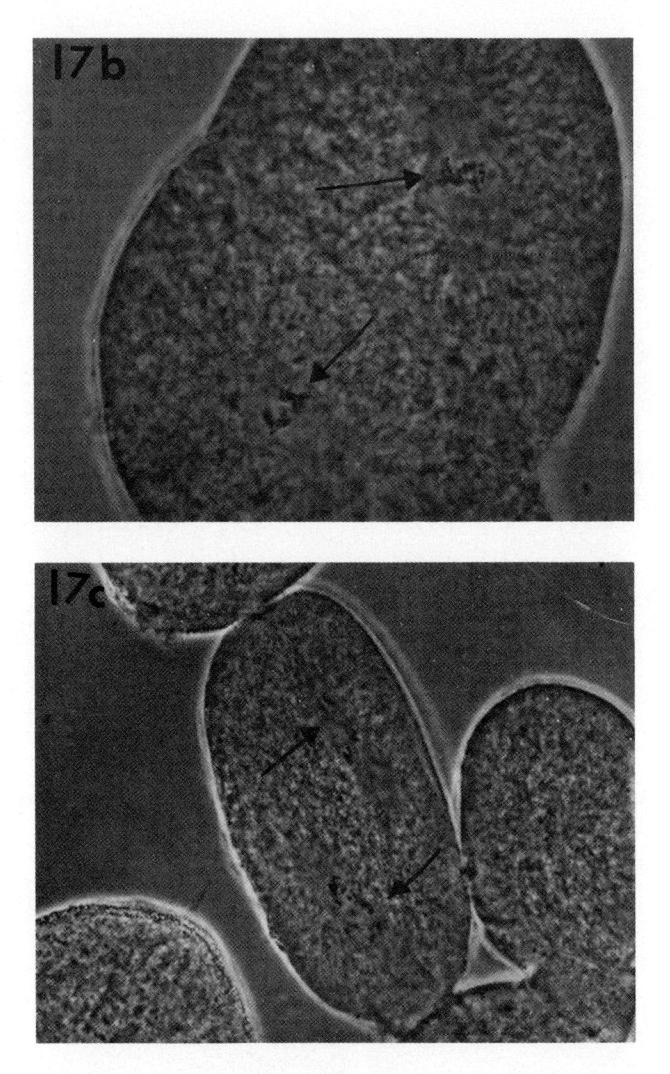

Fig. 17. (Continued)

egg begin to cleave to form eight cells while the four cells of the fused fertilized egg also cleave to form eight cells. When the unfused fertilized egg reaches the 16-cell stage, the hybrid egg also cleaves to form 16 cells (Fig. 18f). Likewise, blastula formation (Fig. 18g) and the swimming embryo stage (Fig. 18h) occur at the same times in fused and unfused fertilized eggs.

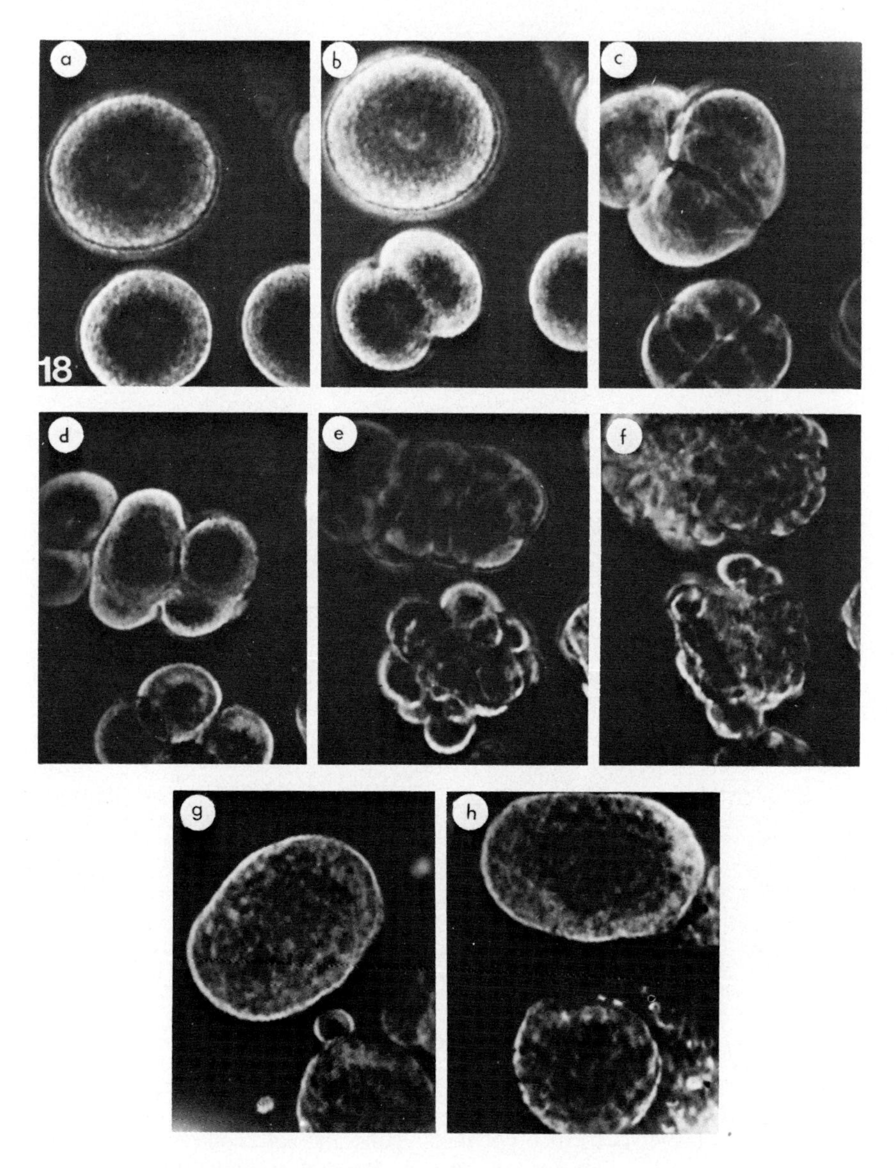

Fig. 18. Embryological development in fused and unfused fertilized *S. purpuratus* eggs. Note the delay of first cleavage of the fused egg in comparison to that of the unfused egg. Taken from a time-lapse video recording. ×40.

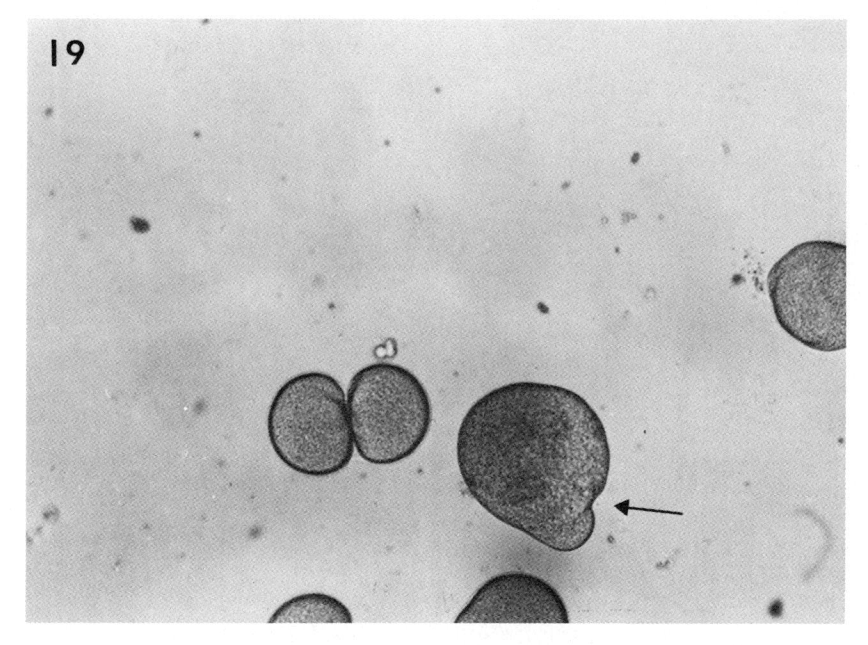

Fig. 19. A fused fertilized *S. purpuratus* egg in the process of cytokinesis. Arrow points to the furrow. Also note that the unfused egg (internal control) has completed first division. ×70.

Note that at least until the morula stage, the cells derived from the fused egg are large compared to those from the unfused egg. Even the micromeres, which are formed at this stage, are comparatively large in the fused hybrid.

Following the fusion of fertilized eggs with unfertilized eggs, both nuclei—the zygote nucleus and the unfertilized egg nucleus—are distinct. However, unlike the nuclei in the fused fertilized eggs, these nuclei merge with each other when the fertilized portion begins to enter mitosis. The unfertilized egg nucleus moves from its peripheral location in the "unfertilized" cytoplasm to the center of the cell, where it meets the zygote nucleus. Here, the two nuclei coalesce.

About 30 minutes after the zygote nucleus and the "unfertilized" nucleus have merged, a cortical contraction, initiating from the fertilized portion of the egg, spreads over the "unfertilized" surface of the egg. This contraction usually causes some distortion of the spherical egg, such as that seen in Figure 20b. Occasionally such contractions are so violent that the egg lyses. The eggs are quite prone to lysis since all of the protective materials peripheral to the plasma membranes had been removed for the fusion procedure.

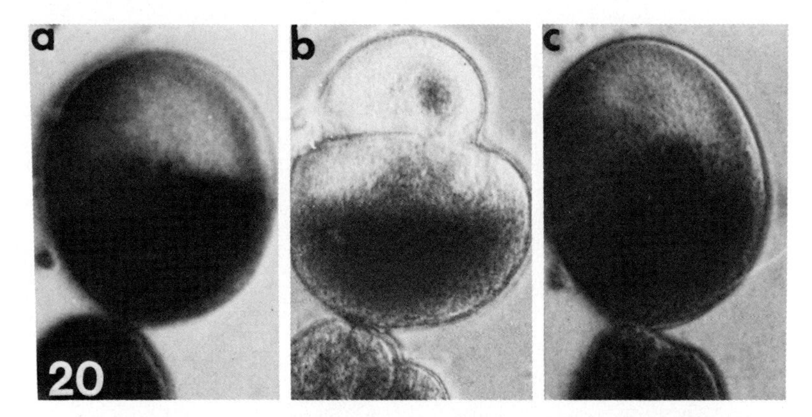

Fig. 20. Cortical contraction at mitosis in an interspecies hybrid composed of fused fertilized and unfertilized eggs. a, Forty-five minutes after fertilization and initiation of fusion-inducing procedure. b, Ninety minutes after. c, One hundred minutes after. ×91.

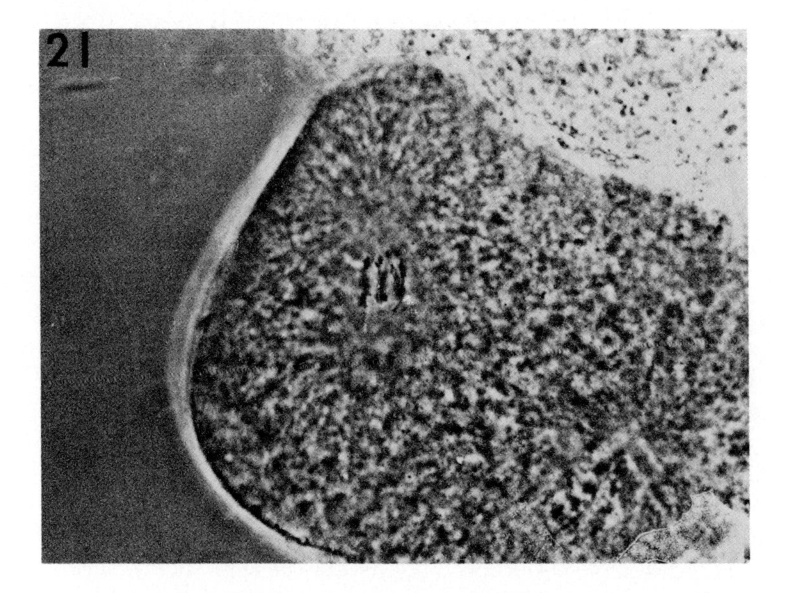

Fig. 21. Bipolar mitotic figure at the first mitosis of a fertilized-unfertilized interspecies hybrid. Note that all of the chromosomes present in this hybrid egg are crowded onto the one spindle. Fixed with Carnoy's fixative 120 minutes after fertilization. ×129.

However, if the egg survives the cortical contractions, it soon returns to its original spherical shape, as seen in Figure 20c.

At the same time the cortical contraction occurs, the hybrid nuclear envelope breaks down for mitosis and a bipolar spindle is formed. As is seen in Figure 21, this spindle contains all of the chromosomes present in the hybrid cell. The hybrid then undergoes cytokinesis and two daughter cells are formed (as described above).

The contractile wave at the first mitosis in fused fertilized and unfertilized eggs may be related to the other changes in the cytoplasmic, surface, and cortical morphologies we observe at this time. It is not clear what causes the contractions and/or morphological changes, but other evidence indicates that such events occur normally at both fertilization and mitosis. Kirschner et al. [1980] have observed that periodic surface contraction waves occur at fertilization and just prior to cell cleavage in fertilized amphibian eggs. These contractions also occur just prior to the equivalent time of mitosis in activated nucleate and anucleate fragments and at fertilization in the sea urchin egg [Schatten, 1979]. Although no such wavelike surface contraction is seen in the sea urchin egg at mitosis, an effect like bubbling is thought to be the equivalent (Mazia, personal communication).

Changes in surface morphology have also been observed in eggs at fertilization and at mitosis. As described above, microvillar elongation occurs at fertilization. At mitosis the microvilli become even longer [Burgess and Schroeder, 1979]. Perhaps the extension of microvilli over the unfertilized portion of unfertilized and fertilized fused eggs at this time is due to this same phenomenon.

CONCLUSIONS

One could have expected that when fused, two cells (eggs) would become one and all subunits would attain random distribution. We have demonstrated that this is not the case. In fact, each donor cell (egg) may maintain its original integrity—the cytoplasmic yolk particles stay in place, the microvilli and cortical granules do not diffuse, and the nuclei do not interact. One wonders what other components are unable to mix (cell organelles, different cytoplasmic domains, calcium ions, hydrogen ions (pH), precursors or products of DNA, RNA, protein syntheses). The findings are quite in accord with the emerging image of the cell as a body structured by its fibrillar system and the association of organelles with that system [Schliwa et al., 1981]. The present observations say that cell structure is not in a state of rapid turnover, which might lead to the mixing of two cytoplasms.

We have also demonstrated that the "unfertilized" portion of fused fertilized and unfertilized eggs undergoes changes equivalent to fertilization when the fertilized portion enters mitosis. It is of particular interest to determine whether cytoplasmic activation-associated events (DNA and protein synthesis, elevation of cytoplasmic pH) begin in the "unfertilized" portion of the fused egg at or before this time.

REFERENCES

Ancel P, Vintemberger P (1948): Recherches sur le determinisme de la symetrie bilaterale dans l'oeuf des amphibiens. Bull Biol [Suppl] 21:1–182.

Asnes CF, Schroeder T (1979): Cell Cleavage. Exp Cell Res 122:327–338.

Begg DA, Rebhun LI (1979): pH regulates the polymerization of actin in the sea urchin egg cortex. J Cell Biol 83:241–248.

Begg DA, Rebhun LI (1980): Microvillar elongation in the absence of actin filament bundle formation. J Cell Biol 87:226.

Bennett J (1980): "Fusion of Sea Urchin Eggs." PhD thesis, University of California, Berkeley.

Bennett J, Mazia D (1981a): Interspecific fusion of sea urchin eggs. Surface events and cytoplasmic mixing. Exp Cell Res 131:197–207.

Bennett J, Mazia D (1981b): Fusion of fertilized and unfertilized sea urchin eggs. Maintenance of cell surface integrity. Exp Cell Res 134:494–498.

Bierens de Haan JA (1913a): Uber homogene and heterogene keimverschmelzungen bei Echindiden. Arch Entwicklungsmech Mech 36:474.

Bierens de Haan JA (1913b): Uber die Entwicklung heterogener Verschmelzungen bei Echiniden. Arch Entwicklungs Mech 37:420.

Chooi W (1971): Variation in nuclear DNA content in the genus *Vicia.* Genetics 68:195–211.

Czihak G (1975): "The Sea Urchin Embryo: Biochemistry and Morphogenesis." Berlin: Springer-Verlag.

Driesch H (1900): Studien uber das regulationsvermogen der Organismen 4. Die verschmelzung der individualitat bei Echinidenkeimen. Arch Entwiklungs Mech 10:411–434.

Eddy EM, Shapiro BM (1976): Changes in topography of the sea urchin egg after fertilization. J Cell Biol 71:35–48.

Giudice G (1973): "Developmental Biology of the Sea Urchin Embryo." New York: Academic Press.

Goldfarb AJ (1913): Studies in the production of grafted embryos. Biol Bull 24:73.

Hiramoto Y (1956): Cell division without mitotic apparatus in sea urchin eggs. Exp Cell Res 11:630–636.

Hiramoto Y (1969a): Mechanical properties of the protoplasm of the sea urchin egg. I. Unfertilized egg. Exp Cell Res 56:201–208.

Hiramoto Y (1969b): Mechanical properties of the protoplasm of the sea urchin egg. II. Fertilized Egg. Exp Cell Res 56:209–218.

Hiramoto Y (1974): Mechanical properties of the surface of the sea urchin egg at fertilization and during cleavage. Exp Cell Res. 89:320–326.

Hiromato Y (1976): Mechanical properties of sea urchin eggs. III. Visco-elasticity of the cell surface. Dev Growth Differ 18:377–386.

Ikeda M, Nemoto S, Yoneda M (1976): Periodic changes in the content of protein-bound sulfhydryl groups and tension at the surface of starfish oocytes in correlation with the meiotic division cycle. Dev Growth Differ 18:221–225.

Kidd P, Mazia D (1980): The ultrastructure of surface layers isolated from fertilized and chemically stimulated sea urchin eggs. J Ultrastruct Res 70:58–69.

Kirschner M, Gerhart J, Hara K, Ubbels GA (1980): "The Cell Surface, Mediator of Developmental Processes." In Subtelney S, Wessells N (eds): "Proceedings of the 38th Symposium of the Society for Developmental Biology." New York: Academic Press.

Lau Y, Brown RL, Arrighi FR (1977): Induction of premature chromosome condensation in CHO cells fused with polyethylene glycol. Exp Cell Res 110:57–61.

Maceira-Coelho A, Aurameas S (1972): Modulation of cell behavior in vitro by the substration fibroblastic and leukemic mouse cell lines. Proc NH Acad Sci 69:2469–2473.

Makino S (1951): "An Atlas of the Chromosome Numbers in Animals." Ames, Iowa: Iowa State College Press.

Mazia S, Schatten G, Sale W (1975): Premature chromosome condensation and cell cycle analysis. J Cell Physiol 91:131–141.

Richter H-P, Scheurich P, Zimmerman U (1982): Electric field-induced fusion of sea urchin eggs. Dev Growth Differ 23(5):479–486.

Schatten G (1979): Pronuclear movement and fusion at fertilization. J Cell Biol 83:198.

Schliwa M, Van Blerkom J, Porter KR (1981): Stabilization of the cytoplasmic ground substance in detergent-opened cells and a structural and biochemical analysis of its composition. Proc Natl Acad Sci USA 78(7):4329–4333.

Schroeder TE (1979): Surface area change at fertilization: Resorption of the mosaic membrane. Dev Biol 70:306–326.

Sluder G (1978): The reproduction of mitotic centers: New information on an old experiment. In Dirkson ER, Prescott DM, Fox CF (eds): "ICN-UCLA Symposium on Molecular and Cellular Biology." New York: Academic Press, 12:563–569.

Steinhardt R, Epel D (1974): Activation of sea urchin eggs by a calcium ionophore. Proc Natl Acad Sci USA 71:1915–1919.

Steinhardt R, Shen S, Zucker R (1978): Direct evidence for ionic messengers in the two phases of metabolic derepression at fertilization in the sea urchin egg. In Dirkson ER, Prescott DM, Fox CF (eds): "ICN-UCLA Symposium on Molecular and Cellular Biology." New York: Academic Press, 12:415–424.

Steinhardt R, Zucker R, Schatten G (1977): Intracellular calcium release at fertilization in the sea urchin egg. Dev Bio 58:185–196.

Tilney L, Marsland D (1969): A fine structural analysis of cleavage induction and furrowing in the eggs of *Arbacia punctulata.* J Cell Biol 42:170–184.

Tyler A (1935): On the energetics of differentiation. A comparison of the rates of development of giant and of normal sea-urchin embryos. Biol Bull 68:451–460.

Tyler A (1942): Developmental processes and energetics. Q Rev Biol 17:339–353.

Yoneda M (1973): Tension at the surface of sea urchin eggs on the basis of "liquid drop" concept. Adv Biophys 4:153–190.

Yoneda M, Ikeda M, Wahitani S (1978): Periodic change in the tension at the surface of activated non-nucleate fragments of sea urchin eggs. Dev Growth Differ 20:329–336.

Time, Space, and Pattern in Embryonic Development, pages 29–47

Patterns of Protein Metabolism and the Role of Maternal RNA in Sea Urchin Embryos

Bruce P. Brandhorst, Pierre-André Bédard, and Frank Tufaro

Department of Biology, McGill University, Montreal, Quebec H3A 1B1, Canada (B.P.B., P.-A.B., F.T.), and Marine Biological Laboratory, Woods Hole, Massachusetts 02543 (B.P.B.)

Proteins are the final products in the flow of genetic information through cells. Consequently many investigations of the regulation of gene expression in developing systems have concerned protein metabolism. Investigations of protein synthesis in sea urchin eggs and embryos began shortly after radiolabeled precursors became available [Hultin, 1950]. Experiments in which eggs were preloaded with ^{35}S-methionine demonstrated that fertilization leads to a rapid increase in the rate of incorporation of the soluble pool of amino acids into proteins [Nakano and Monroy, 1958]. The rate of protein synthesis increases by ten to 20-fold within 20 minutes after fertilization [Epel, 1967; Humphreys, 1969] and continues to increase throughout much of embryonic development [Goustin and Wilt, 1981]. The mass of protein per embryo remains essentially constant until the pluteus begins to feed [Fry and Gross, 1970]. The rate of synthesis of proteins in blastulae of *Strongylocentrotus purpuratus* of about 500 pg per hour per embryo compares with a mass of about 40 ng protein per embryo, indicating that substantial turnover of proteins occurs during embryonic development.

Investigations of types of proteins synthesized have progressed as methods of fractionation and separation have developed. Measurements by nucleic acid hybridization techniques indicate that there are about 10^4 different, mostly rare, messenger RNA molecules translated in a sea urchin embryo [Galau et al., 1974]. Consequently, the many investigations of populations of proteins of embryos separated by electrophoresis in one dimension or by column chromatography have not been very informative because of limited resolution. Exceptions are a variety of elegant investigations of histone gene expression based on the unusual properties of these proteins [Ruderman and Gross, 1974; Newrock et al., 1978; Childs et al., 1979]; histone metabolism is beyond the scope of the present review.

A major advance in investigations of individual proteins came with the introduction of high-resolution two-dimensional (2-D) electrophoresis, which separates proteins on the basis of size and isoelectric point [O'Farrell, 1975]. In this review we summarize the behavior of about 1,000 individual polypeptides of sea urchin embryos and discuss the levels of regulation of their expression. We describe recent evidence that maternal mRNA may persist throughout embryonic development and code for proteins whose synthesis is developmentally regulated. Some of these maternal mRNAs may be involved in cytoplasmic localization and determination phenomena.

RECRUITMENT OF STORED MATERNAL RNA INTO POLYSOMES UPON FERTILIZATION

Evidence for Stored Maternal mRNA

The rate of protein synthesis following fertilization increases normally even when RNA synthesis is inhibited by actinomycin D [Gross and Cousineau, 1963; Gross et al., 1964; Greenhouse et al., 1971]. Protein synthesis also increases in response to parthenogenic activation of enucleate merogones [Brachet et al., 1963; Denny and Tyler, 1964]. These observations led to the hypothesis that eggs contain a store of maternal mRNA utilized after fertilization. Convincing evidence that stored mRNA is rapidly recruited into polysomes after fertilization came from two types of investigations. Humphreys [1969, 1971] demonstrated that the increased rate of protein synthesis is largely due to an increased amount of mRNA being translated after fertilization and that this mass of mRNA entering polysomes cannot be accounted for by the synthesis of new mRNA. The maternal mRNA residing in subribosomal fractions of sea urchin eggs includes translatable transcripts for proteins actively synthesized following fertilization such as the tubulins and histones [Raff et al., 1972; Skoultchi and Gross, 1973; Gross et al., 1973]; these transcripts are eventually found mostly in polysomes after fertilization.

The development of embryos is rather normal until the hatching blastula stage, when synthesis of new mRNA is inhibited by actinomycin D [Gross et al., 1964], and the pattern of protein synthesis is not detectably altered in early embryos by actinomycin D treatment [Terman and Gross, 1965; Terman, 1970]. Interspecies hybrid embryos of echinoids tend to show only maternal characteristics before hatching [for review, see Davidson, 1976; Tufaro and Brandhorst, 1982]. These observations led to the interpretation that most protein synthesis in the first several hours of embryonic development is dependent on the translation of maternal mRNA stored in the egg and recruited into polysomes during the first few hours after fertilization, but that RNA transcribed from the embryonic genome replaces the maternal mRNA in polysomes around the time of hatching.

Patterns of Protein Synthesis in Oocytes, Eggs, and Zygotes

Does the population of stored maternal mRNA include transcripts coding for specialized proteins whose synthesis is not required by eggs but is needed for embryonic development? This question was addressed by incubating unfertilized eggs, zygotes, and early embryos with ^{35}S-methionine [Brandhorst, 1976; Tufaro and Brandhorst, 1979]. Labeled proteins were anlayzed by 2-D electrophoresis and autoradiography. As shown in Figure 1, the patterns of the nearly 1,000 polypeptides detected are nearly identical for eggs and early embryos, indicating that the relative rates of synthesis of nearly all of these proteins have remained the same in spite of the large increase in the amount of mRNA being translated.

Most or all of the mRNA being translated in an unfertilized egg is newly synthesized and unstable [Brandhorst, 1980]. A much larger quantity of mRNA is synthesized and stored during oogenesis, and, unlike RNA synthesized in eggs, is stable and becomes available for translation only after fertilization. Thus the egg contains two populations of RNA coding for the same spectrum of detectable proteins which have very different functional and metabolic properties. Other than their times of synthesis, it is not yet clear what the distinctive properties of these two populations of mRNA are. The possible mechanisms of storage and recruitment of maternal mRNAs have been reviewed recently [Raff, 1980; Raff and Showman, 1983]. Maternal histone mRNA is not recruited into polysomes as rapidly as other stored RNAs [Wells et al., 1981]. It is possible that other maternal mRNA sequences are not recruited until much later in development (see discussion below).

In several other organisms fertilization is accompanied by distinct changes in the patterns of protein synthesis [e.g., Rosenthal et al., 1980; Braude et al., 1979], but in these organisms meiotic maturation normally follows fertilization. Sea urchin eggs are stored in, and shed from, ovaries after the completion of meiosis. Oocytes of another echinoderm, the starfish *Asterias,* can be collected before maturation and induced to complete meiosis by the addition of the natural hormone 1-methyladenine [Kanatani and Shirai, 1967; Schuetz and Biggers, 1967]. The breakdown of the germinal vesicle is accompanied by a rapid increase in the rate of protein synthesis [Houk and Epel, 1974], increased polyadenylation of RNA [Jefferey, 1977], and changes in the patterns of protein synthesis [Rosenthal et al., 1982]. Fertilization does not stimulate any further changes. The changes in the rate and pattern of protein synthesis are translationally mediated, being dependent on the recruit-

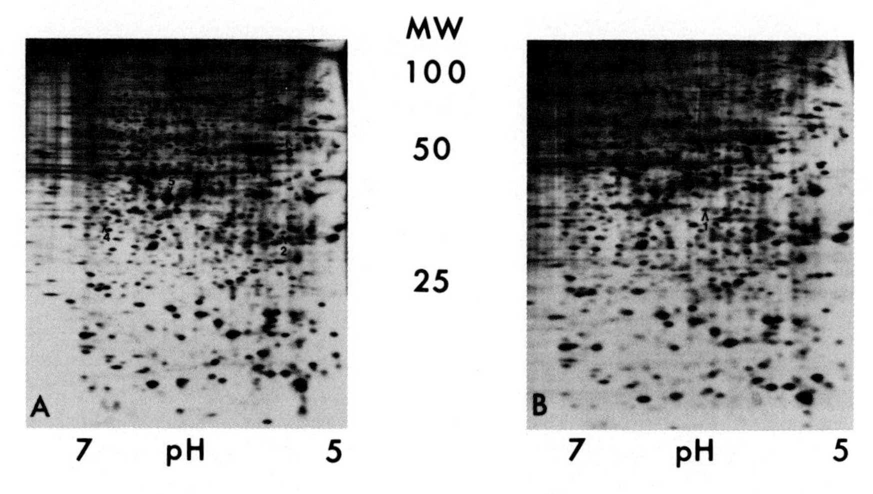

Fig. 1. Comparison of proteins synthesized by unfertilized eggs and 16-cell embryos. Proteins labeled with ^{35}S-methionine were extracted and separated by two-dimensional (2-D) electrophoresis. Autoradiographs show 997 spots. Spot 1 was observed only in embryos, while spots 2–4 were observed only in eggs. Spot 5 was observed in these eggs and embryos but not observed in several other batches of eggs and embryos. A, eggs; B, embryos. Reproduced from Tufaro and Brandhorst [1979].

ment into polysomes of RNA already present in the oocyte and translatable in cell-free systems. Such changes appear to occur in a variety of oocytes upon maturation [Rosenthal et al., 1982]. We predict that sea urchin oocytes would show a similar transition in pattern of protein synthesis upon maturation. The mass mobilization of a population of stored mRNA is a simple mechanism by which the rate and/or pattern of protein synthesis can be rapidly altered by a specific trigger, such as fertilization or hormonal induction of maturation.

CHANGES IN PATTERNS OF PROTEIN SYNTHESIS DURING EMBRYONIC DEVELOPMENT

While there are a few changes in the patterns of protein synthesis during the first several hours of development, substantial changes occur in later embryos [Brandhorst, 1976; Bédard and Brandhorst, 1983]. Figure 2 shows a comparison of proteins synthesized in zygotes, mesenchyme blastulae, and plutei of *S. purpuratus*. Two types of changes are pointed out: quantitative changes in which the relative rate of synthesis of the polypeptide has changed detectably and qualitative changes in which synthesis of the polypeptide was

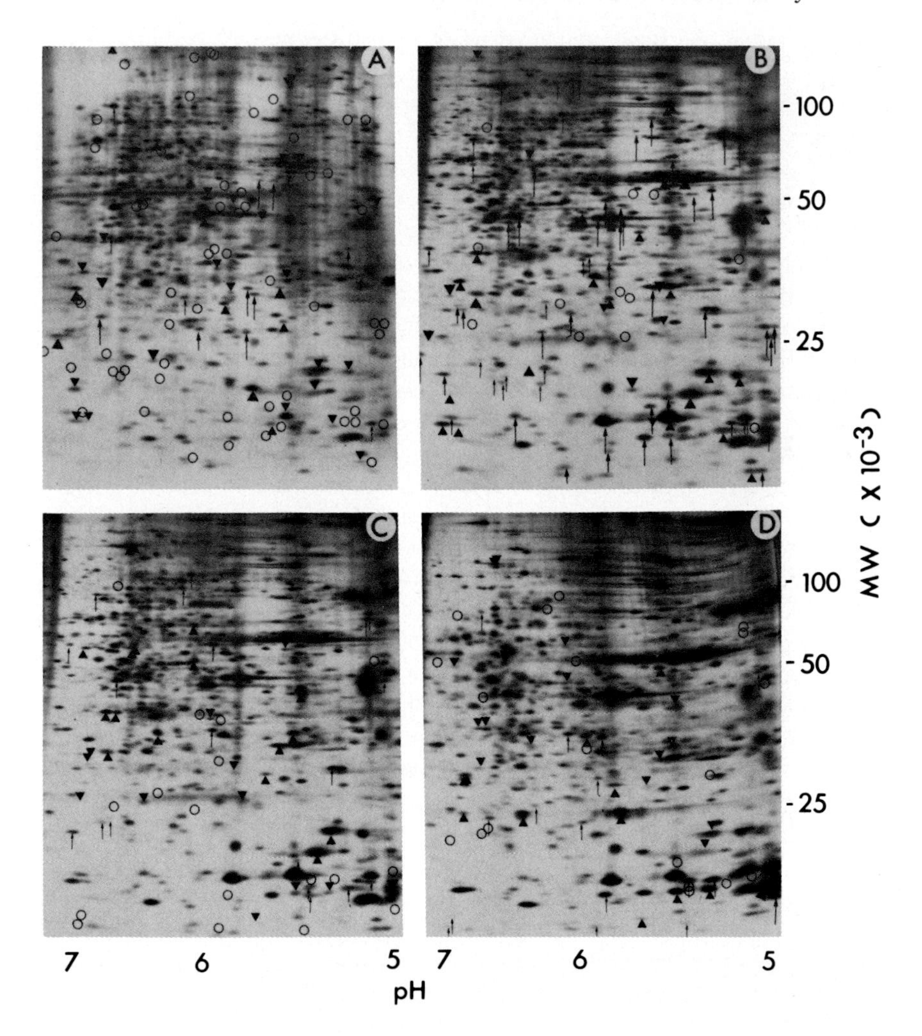

Fig. 2. Comparison of proteins synthesized by early embryos, mesenchyme blastulae, and plutei. Embryos were incubated with ^{35}S-methionine for 1 hour beginning 1.5, 24, and 72 hours after fertilization. Proteins were extracted and separated by 2-D electrophoresis: Autoradiography was for 1.8×10^{10} disintegrations applied to the first dimension. The 2-cell stage embryo (A) is compared with the late mesenchyme blastula (B), and the late mesenchyme blastula (C) is compared with the pluteus (D); panels B and C are the same autoradiograph. Arrows point to spots detected at one stage but not at the other stage compared; circles indicate the corresponding area of the missing spot on the gel for the stage compared. Triangles pointing up indicate that the relative intensity of labeling of that spot is greater at that stage while triangles pointing down indicate that the relative intensity of that spot is less at that stage.

not detected at one stage compared using standard exposure conditions; the definition of qualitative changes is necessarily arbitrary. There is an increase in the number of polypeptides detectably synthesized during development, and more increase than decrease in relative rates of synthesis. The extent of the changes was estimated by using multiple autoradiographic exposure times [Bédard and Brandhorst, 1983]. During development the relative intensity of labeling of about 20% of the nearly 900 spots analyzed changes. Of these about half change by at least tenfold; only about 1% of the spots change by at least 100-fold. The changes in synthesis of most polypeptides are greater before gastrulation, but many of the polypeptides undergoing changes during early development continue to change during late development. Only a few polypeptides show transient, reversible changes, all increasing and then declining. All but one of the proteins changing by 100-fold or more during development increase in relative rate of synthesis. These proteins are synthesized at low or undetectable levels in eggs but increase to become major synthetic products at later embryonic stages. They include an actin variant comigrating with mammalian β-actin [Durica and Crain, 1982] and a set of small acidic proteins highly enriched in the ectoderm [Bruskin et al., 1982].

The developmental timing of changes in protein synthesis was defined by pulse labeling proteins for 1–2 hours at various times throughout development [Bédard and Brandhorst, 1983]. Figure 3 shows the number of qualitative changes which have occurred during each labeling interval compared to the previous interval as detected by autoradiography of 2-D gels. Many changes occur at the time of hatching and during mesenchyme formation, while few changes occur during gastrulation. Quantitative changes have similar timing. The abrupt changes in the synthesis of many proteins at the time of hatching suggests that expression of some or all of these proteins is coordinately regulated.

Could this regulation be at the level of transcription? Comparison by 2-D electrophoresis of the products of cell-free translation of cytoplasmic RNA with proteins synthesized in vivo demonstrates that (for the approximately 50% of the spots which match) [see Brandhorst et al., 1979] most translatable mRNAs are only detectably present when their corresponding proteins are actually synthesized in the embryo (our unpublished observations). The development of embryos reared in the presence of actinomycin D is blocked at the time of hatching [Gross et al., 1964]. A switch from the synthesis of early to late histone variants occurs at about the time of hatching and is mediated by a replacement of early histone mRNAs by newly synthesized late histone mRNAs [Newrock et al., 1978; Grunstein, 1978; Hieter et al., 1979; Childs et al., 1979]. It is thus likely that an important transition in embryonic development occurs around the time of hatching and that it requires synthesis of RNA. However, it is not yet clear that all the mRNAs

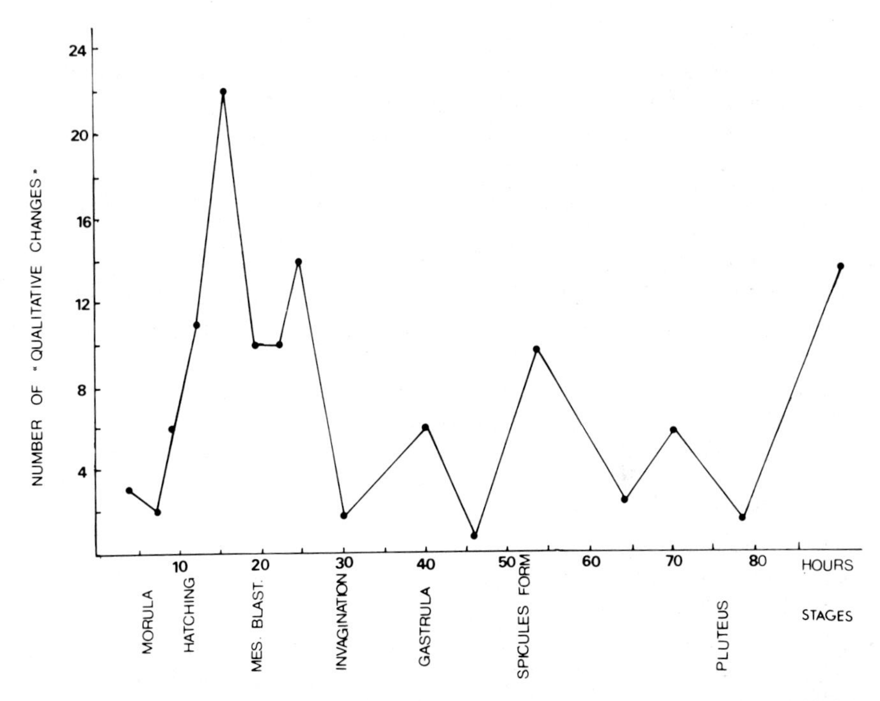

Fig. 3. Qualitative changes in protein synthesis during embryonic development. Embryos were labeled with ^{35}S-methionine for 1–2 hours at various times throughout embryonic development. Proteins were extracted, separated by 2-D electrophoresis, and analyzed by autoradiography for exposures of 1.8×10^{10} disintegrations. The number of "qualitative" changes observed for each labeling interval compared to the previous interval is plotted against time after fertilization and developmental stage.

coding for proteins whose synthesis increases upon hatching are newly synthesized (see below).

The mass distribution of hundreds of proteins can be analyzed by staining 2-D gels with silver [Oakley et al., 1980]. The patterns of proteins of eggs and plutei are remarkably similar, though a few polypeptides increase greatly in mass and others, probably including yolk proteins, decline during development [Bédard and Brandhorst, 1983]. Many proteins present throughout embryonic development are never detectably synthesized in eggs or embryos [Kuhn and Wilt, 1981; Bédard and Brandhorst, 1983]. Comparison of labeling and staining patterns indicates that individual proteins are metabolized in a variety of ways and have a range of stabilities [Bédard and Brandhorst, 1983]. The general constancy of mass of most polypeptides during embryonic development emphasizes that the sea urchin egg arrives well prepared for embryonic development.

Actin mRNA accumulates extensively in mass during sea urchin embryonic development and synthesis of actin increases concomitantly [Merlino et al., 1981; Durica and Crain, 1982]. The mass of actin proteins, however, remains essentially constant during embryonic development, even though the actin variants are among the major translation products [Bédard and Brandhorst, 1983]. By contrast, the synthesis of the small acidic ectodermal proteins and the mass of their respective mRNAs increases over 100-fold during development and this is reflected in a large increase in the mass of these proteins [Bruskin et al., 1982]. Thus there are major differences in the metabolism of various proteins, and the turnover of individual proteins probably changes during development. Posttranslational events may be of profound importance in establishing and maintaining appropriate levels of active gene products during embryonic development.

TISSUE SPECIFICITY OF EMBRYONIC PROTEINS

Sea urchin gastrulae can be dissociated and separated into ectodermal and endo/mesodermal fractions [McClay and Chambers, 1978; Bruskin et al., 1982]. Several newly synthesized proteins have been identified which are highly enriched or restricted to one of these fractions in either *S. purpuratus* or *Lytechinus pictus* embryos [Bruskin et al., 1982; our unpublished observations]. The synthesis of most of these tissue-specific proteins increases during development to the extent that they are among the major translation products in maturing embryos. Most ectodermal proteins are actively synthesized before gastrulation while the synthesis of most endo/mesodermal proteins begins after gastrulation. The major ectodermal proteins of *S. purpuratus* are not very similar in size and isofocusing points to those of *L. pictus*, though they tend to occur in clusters and are acidic. The major endo/mesodermal proteins are more similar in the two species.

Cloned cDNAs coding for a family of major ectodermal proteins of *S. purpuratus* have been isolated and characterized [Bruskin et al., 1981, 1982; see also Klein et al., this volume]. Some mRNAs coding for these proteins are detectable in eggs, suggesting that they might be segregated into presumptive ectodermal cells during cleavage. We have selected several cloned cDNAs from an *L. pictus* library which code for mRNAs enriched in ectoderm or endo/mesoderm, but have not yet carried out extensive analyses of these clones. Mesenchymal fractions can be isolated prior to gastrulation and have unique patterns of protein synthesis [Harkey and Whiteley, 1980; see also Harkey, this volume]. In all of these investigations it is clear that the vast majority of proteins do not show any significant enrichment in any of the three primary germ layers, just as most proteins do not undergo any significant changes in mass or relative rates of synthesis during embryonic development. Among the most extensively changing proteins are those enriched in one of these primary tissue layers.

VALIDITY OF ELECTROPHORETIC ANALYSES OF GENE EXPRESSION

Sensitivity

The proteins analyzed in our hands by 2-D electrophoresis are limited to those having isoelectric points (about 5–7) and molecular weights (about 10–200 $\times 10^3$ daltons) included within the gel system used. While this includes a large majority of newly synthesized proteins in sea urchin embryos [Brandhorst, 1976], some prominent proteins are excluded, notably the histones and other basic proteins. The number of newly synthesized, radioactive proteins detected by autoradiography increases with increasing exposure until major synthetic products begin to obscure adjacent spots. Normally about 1,000 polypeptides are resolved for a wide variety of organisms and cell types. This, however, is considerably less than the number of mRNAs usually detected in polysomes. For instance, in sea urchin embryos there are at least 10,000 actively translated mRNA molecules [Galau et al., 1974; Shepherd and Nemer, 1980; Duncan and Humphreys, 1981; Tufaro and Brandhorst, 1982]. Most of the mRNAs are rare, present in only about one copy per cell at pluteus stage (though they may be segregated into a subpopulation of the blastomeres), and it is commonly argued that translation products of only the more prevalent mRNAs are detected on 2-D gels [Brandhorst, 1976; Lasky et al., 1980]. This may be an oversimplification. Some major in vivo translation products, such as β-actin and some ectodermal proteins, are coded for by mRNAs which are present in less than 100 copies per cell in plutei [Xin et al., 1982; Brandhorst, unpublished observations]. These major proteins are more than 100-fold more intensely labeled than hundreds of minor polypeptides detected by autoradiography of 2-D gels. If some of the mRNAs coding for these minor spots are translated with efficiencies comparable to the translation of the mRNAs coding for these major spots, then they must be rare mRNAs. It is not yet clear why more polypeptides are not detected on 2-D gels. Evidence is accumulating that mRNAs are not all translated with similar efficiencies in sea urchin embryos [Infante and Heilman, 1981; Bruskin et al., 1982].

Very few polymorphic differences in patterns of protein synthesis are observed when comparing embryos from different sea urchin females of the same population. Nevertheless, we normally carry out comparisons on embryos of the same cross. Protein polymorphism can be a serious problem in other invertebrates [Brandhorst and Newrock, 1981].

Comparison With Results of Other Methods of Analysis

Populations of mRNA have been compared during sea urchin development by nucleic acid hybridization [Galau et al., 1976; Hough-Evans et al., 1977; Shepherd and Nemer, 1980]. Particularly informative have been comparisons using fractionated single-copy DNA probes hybridized to excess mRNA. These have shown that essentially all mRNAs detectable in polysomes during development are represented in maternal RNA of the egg, but there is a general decline in the sequence complexity of polysomal RNA during development, though there are stage-specific transcripts [Galau et al., 1976; Hough-Evans et al., 1977]. These comparisons are sensitive only to changes in at least hundreds of rare transcripts, while 2-D electrophoretic analyses are sensitive to changes in individual members of the population of prevalent translation products.

Another type of developmental analysis of populations of prevalent mRNAs is accomplished by hybridization of radioactive cDNA probes, complementary to these mRNAs, to arrays of cloned complementary DNAs. Such analyses have confirmed that most prevalent polyadenylated cytoplasmic RNAs are present in similar abundance in eggs and plutei but tend to decline in abundance at intervening stages [Lasky et al., 1980; Flytzanis et al., 1982]. Thus all types of analyses of RNA populations indicate that the majority of different mRNAs remain similar in relative prevalence at the beginning and end of embryonic development. A minority of mRNAs, as well as their translation products, change extensively during embryonic development. Cloned recombinant DNAs coding for several of these transcripts have been isolated [Bruskin et al., 1981; Flytzanis et al., 1982]. We have been identifying their corresponding polypeptides by translation of hybrid selected RNA [Bruskin et al., 1982; our unpublished observations].

Proteins whose synthesis and mass remain rather constant during development and which are not tissue specific in embryos might be ubiquitous, possibly performing essential, or "housekeeping," functions. However, the vast majority of both rare and prevalent mRNAs of embryos are not detectable in adult tissues [Galau et al., 1976; Xin et al., 1982]. The 2-D electrophoretic patterns of proteins synthesized by embryos and adult tissues are also quite distinct (our unpublished observations). Therefore most proteins detected on gels may not be housekeeping proteins.

LIMITED EXPRESSION OF THE PATERNAL GENOME IN INTERSPECIES HYBRID EMBRYOS

Evidence for the Persistence of Maternal RNA

While it is certain that maternal mRNA is eventually replaced in polysomes by mRNA synthesized in the embryo, the timing of this transition has been uncertain. The synthesis of proteins coded by the paternal genome in interspecies hybrid embryos can be used as an indication of the activity of the embryonic genome. We have analyzed protein synthesis in three hybrid crosses: the reciprocal crosses of *S. purpuratus* and *S. drobachiensis*, and the eggs of *S. purpuratus* fertilized with the sperm of *L. pictus* [Tufaro and Brandhorst, 1982]. All three crosses give rise to healthy plutei resembling larvae of the maternal species. As shown in Table I, a variety of polypeptides synthesized in embryos of the paternal species have distinct positions on 2-D

TABLE I. Detection of Synthesis of Paternal Proteins in Hybrid Embryos[a]

Hybrid cross[b]			Stage of development[d]					
			Cleavage		Hatching		Pluteus	
Egg	Sperm	No. of common proteins[c]	Obs.[e]	Pot.[f]	Obs.[e]	Pot.[f]	Obs.[e]	Pot.[f]
Sd	Sp	500	0	60	0	60	3	70
Sp	Sd	500	0	80	5	90	5	90
Sp	Lp	275	0	120	1	120	2	130

[a]Modified from Tufaro and Brandhorst [1982].
[b]Species used were *S. purpuratus* (Sp), *S. droebachiensis* (Sd), and *L. pictus* (Lp).
[c]This represents the number of newly synthesized proteins common to both species and detected on two-dimensional (2-D) gels. It was determined by overlapping autoradiographs of proteins of the two species compared and by confirming the identity of these spots on a gel containing a mixture of the proteins of both species. Numbers are approximate and are limited by the resolution of the gel having the poorest resolution. Only proteins which could be consistently and unambiguously analyzed were counted.
[d]Stage of embryonic development analyzed.
[e]The number of spots distinct to the paternal species observed by 2-D electrophoretic separations of proteins synthesized by hybrid embryos of that stage.
[f]The number of spots observed in embryos of the paternal species of that stage that are distinct electrophoretically from the maternal species.

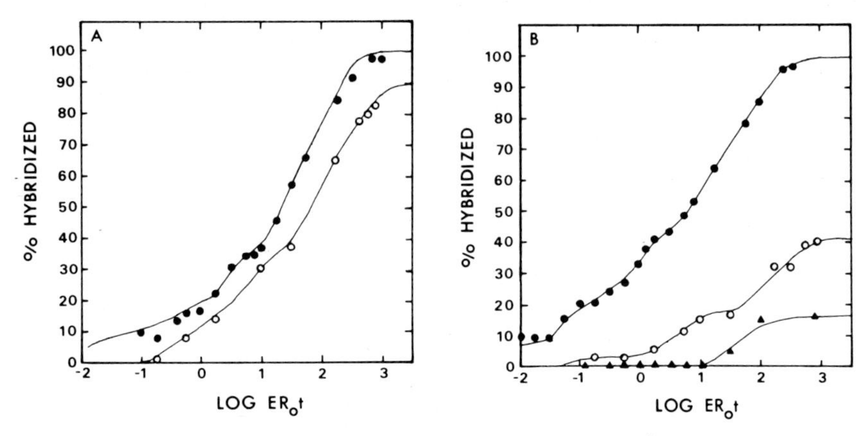

Fig. 4. Characterization of RNA populations of hybrid embryos. ^{3}H-cDNA was reverse transcribed from polysome-enriched cytoplasmic polyadenylated RNA purified from *S. purpuratus* and *L. pictus* gastrulae. The kinetics of hybridization of these cDNA probes with excess homologous, heterologous, or hybrid gastrula polysome-enriched, polyadenylated RNA were determined using an S1 nuclease assay. A. *S. purpuratus* cDNA reacted with excess *S. purpuratus* RNA (●) or *S. purpuratus* × *L. pictus* hybrid RNA (○). B. *L. pictus* cDNA reacted with excess *L. pictus* RNA (●), *S. purpuratus* × *L. pictus* hybrid RNA (○), or *S. purpuratus* RNA (▲). The final reactivity of each probe when driven with excess template RNA was 86% for *L. pictus* cDNA and 74% for *S. purpuratus* cDNA. All hybridization plots shown were standardized to 100% probe reactivity. Reproduced from Tufaro and Brandhorst [1982].

gels. The synthesis of none of these paternal proteins could be detected in rapidly cleaving embryos, consistent with the predominance of maternal mRNA in early embryos. Surprisingly, as shown in Table I, very few paternal polypeptides are detectably synthesized at any stage, including pluteus, in all three crosses. The failure of much of the paternal genome to be expressed normally is not due to a selective loss of paternal genes. The kinetics and extent of hybridization of a complementary DNA probe transcribed from paternal (*L. pictus*) mRNA indicates that paternal DNA coding for mRNA normally expressed in embryos of the paternal species is retained throughout embryonic development of the hybrid embryos. The kinetics of hybridization of this probe to excess polyadenylated mRNA indicates that most transcripts normally present in paternal embryos are greatly reduced in prevalence or absent, though many paternal transcripts can be detected as rare mRNAs (see Fig. 4). Paternal mRNA coding for histone H1 can be detected shortly after fertilization [Maxson and Egrie, 1980]; thus the paternal genome can be transcribed, at least partially, even in very early embryos.

These observations can be interpreted in either of two ways. Some paternal genes normally expressed during embryogenesis may not be expressed in the presence of cytoplasm of the maternal species (or in the absence of paternal cytoplasm). For example, species-specific cytoplasmic factors might be required for the synthesis of processing of mRNAs specified by these paternal genes. Alternatively, these paternal proteins may normally be translated from stored maternal mRNAs persistent through embryonic development.

Investigations of the kinetics of accumulation of newly synthesized mRNA in polysomes indicate that by gastrula stage it constitutes 3%–4% of the mass of RNA in polysomes [Brandhorst and Humphreys, 1972; Galau et al., 1977]. Since this is similar to estimates of the total mass of mRNA in polysomes for a variety of cells it was concluded that much or all of the RNA translated in the gastrula is transcribed from the embryonic genome. Recently the kinetics of accumulation of several individual newly synthesized mRNAs have been determined using cloned recombinant DNA probes [Cabrera et al., 1982]. Rare mRNAs have a half-life of several hours, comparable to that determined for total mRNA in polysomes [Brandhorst and Humphreys, 1972; Galau et al., 1977], while prevalent mRNAs are considerably more stable. Newly synthesized RNA corresponding to mRNA prevalent in embryos constitutes only a small fraction of RNA labeled with radioactive precursors. Moreover, the mass of some prevalent prism-stage mRNAs cannot be accounted for by the accumulation of newly synthesized RNA. Thus rare mRNAs in polysomes are mostly or entirely of embryonic origin by gastrula stage, while at least some prevalent maternal RNAs persist in the cytoplasm throughout most or all of the period of embryonic development. It is to be expected then, that rare mRNAs of paternal origin will be well represented in hybrid embryos, but that prevalent mRNAs which are normally of oogenetic (maternal) origin will be reduced or absent.

The mRNAs coding for ectodermal proteins accumulate extensively during embryonic development [Bruskin et al., 1981, 1982]. It is not surprising then, that the synthesis of the paternal forms of these proteins is detected in hybrid embryos [Tufaro and Brandhorst, 1982]. We have recently selected a set of recombinant DNA clones coding for mRNAs prevalent in *L. pictus* gastrulae but not in gastrulae of *S. purpuratus* × *L. pictus* hybrids. We plan to use them as probes to distinguish between the alternate explanations for the limited expression of the paternal genome in hybrid embryos.

Translational Regulation and the Structure of Maternal RNA

Some paternal specific proteins which fail to appear in interspecies hybrid embryos normally appear only at later stages of embryonic development. If these are indeed translated from maternal mRNAs they must be unavailable for translation in early embryos. One approach to the identification of translationally regulated mRNAs is to translate purified RNA from cytoplasmic fractions in a cell-free system and compare the translation products to proteins synthesized in vivo (or translated from polysomal RNA) at various stages using 2-D electrophoresis. Few, if any, mRNAs translatable in wheat germ or rabbit reticulocyte lysate cell-free systems but not in vivo have been identified in early sea urchin embryos though the efficiencies of translation of some mRNAs appear to change during development [Infante and Heilman, 1980; our unpublished observations]. It is possible though, that some maternal mRNAs are not in a form translatable either in cell-free systems or in vivo. Indeed, much of the mass of stored maternal RNA has a peculiar primary structure, consisting of colinear interspersed single-copy and repetitive DNA transcripts [Costantini et al., 1980]. Moreover, many maternal mRNA molecules contain multiple tracts of oligo(A); these molecules persist at least through early embryonic development but are not actively translated [Duncan and Humphreys, 1981b]. Thus the structure of much of the maternal mRNA resembles that of unprocessed nuclear RNA, though it resides in the cytoplasm. These maternal RNAs containing repeated sequences are not very efficiently translated in cell-free systems (D. Anderson and E.H. Davidson, unpublished observations) and quite possibly are not translatable in embryos either. They might be processed to a translatable form during development [Thomas et al., 1981], accounting for a translational level of regulation.

There is substantial evidence that selective translational regulation operates on expression of specific genes in eggs and embryos of other organisms (e.g., see chapter by Ruderman et al., this volume). Enucleate polar lobes of the mud snail *Ilyanassa* undergo developmental changes in patterns of protein synthesis which are a subset of those occurring in the nucleated control embryos [Brandhorst and Newrock, 1981; Collier and McCarthy, 1981]. Preliminary evidence suggests that these posttranscriptionally regulated changes may be mediated by a change in the translatability of stored maternal RNA (Brandhorst and Newrock, unpublished observations).

CYTOPLASMIC LOCALIZATION AND DETERMINATION IN SEA URCHINS

Determination of Micromeres

The fourth cleavage in sea urchins gives rise to three types of blastomeres distinguishable by size and position; macromeres, mesomeres, and micromeres. The micromeres normally give rise to primary mesenchyme cells which secrete skeletal spicules. Isolated micromeres can be cultured and eventually secrete skeletal elements as well [Okazaki, 1975]. The micromeres require no further interactions with other blastomeres to carry out their pathway of differentiation and are thus determined to become primary mesenchyme. The pattern of protein synthesis in micromeres is not distinguishable by 2-D electrophoresis from that of macromeres and mesomeres [Tufaro and Brandhorst, 1979]. The silver-stained patterns of 16-cell-stage blastomeres are not distinct either (our unpublished observations). Thus we have found no evidence for the segregation of either mRNAs or proteins during early sea urchin cleavage. On the other hand, Rodgers and Gross [1978] and Ernst et al. [1980] have reported that some rare transcripts of 16-cell embryos are absent in micromeres. While these transcripts do not appear to be in polysomes of 16-cell embryos [Ernst et al., 1980], it is possible that they are translated in more-advanced-stage embryos or that they serve some sort of regulatory role.

Maternal RNA, Cellular Commitment, and Cytoplasmic Localization

Since the establishment of the stored-maternal-mRNA hypothesis for sea urchin eggs and its extension to a wide variety of other organisms, it has been frequently proposed that cytoplasmic localization and determinative phenomena in eggs and embryos may be related to the localization of maternal RNA or factors regulating the translation thereof. Confirmatory evidence for such proposals has remained elusive. If commitment of a blastomere to a particular developmental fate is the result of, or accompanied by, establishment of a uniquely limited population of mRNA, a concurrent change in the pattern of protein synthesis might occur in that blastomere.

Protein synthesis has been examined by 2-D electrophoresis in several instances in which the developmental potential of a blastomere or embryo has been restricted. These instances include sea urchin micromeres [Tufaro and Brandhorst, 1979], vegetalized sea urchin embryos [Hutchins and Brandhorst, 1979], and delobed *Ilyanassa* embryos [Brandhorst and Newrock, 1981; Collier and McCarthy, 1981]. In these cases no definitive alterations in patterns of protein synthesis could be detected by 2-D electrophoretic separations of newly synthesized proteins at the time of commitment, though patterns change during the expression of that commitment. It is, of course, possible that the synthesis of proteins not detectable in these analyses is required for, or is the result of, these commitment phenomena.

The evidence that some prevalent maternal mRNAs are persistent and possibly translatable only in advanced embryos suggests that these mRNAs might be segregated during cleavage and that their translation could be required for the expression of the state of determination of the blastomeres containing them. Indeed, we have found that mRNA sequences which are enriched in ectoderm of gastrulae are present in eggs, albeit in considerably smaller quantities [Bruskin et al., 1982]. We anticipate that definition of the structural and metabolic properties, as well as the localization of persistent, maternal RNAs in embryos may provide important clues to the molecular nature of cytoplasmic localization and determination phenomena. Our detailed analyses of the properties of about 1,000 polypeptides synthesized in sea urchin eggs should be valuable in identifying cloned sea urchin genes useful for these investigations.

ACKNOWLEDGMENTS

Research reported from the authors' laboratory was supported in part by grants and scholarships from the National Science and Engineering Research Council of Canada, le Ministère de l'Education du Québec, and McGill University. Some of the research reported was performed in the embryology course at the Marine Biological Laboratory, which is supported by a grant from the NIH.

REFERENCES

Bédard P-A, Brandhorst BP (1983): Patterns of protein synthesis and metabolism during sea urchin embryogenesis. Dev Biol 96.

Brachet J, Ficq A, Tencer R (1963): Amino acid incorporation into proteins of nucleate and anucleate fragments of sea urchin eggs: Effect of parthenogenetic activation. Exp Cell Res 32:168–171.

Brandhorst BP (1976): Two dimensional gel patterns of protein synthesis before and after fertilization of sea urchin eggs. Dev Biol 52:310–317.

Brandhorst BP (1980): Simultaneous synthesis, translation, and storage of mRNA including histone mRNA in sea urchin eggs. Dev Biol 79:139–148.

Brandhorst BP, Humphreys T (1972): Stabilities of nuclear and messenger RNA molecules in sea urchin embryos. J Cell Biol 53:474–482.

Brandhorst BP, Newrock KM (1981): Post-transcriptional regulation of protein synthesis in *Ilyanassa* embryos and isolated polar lobes. Dev. Biol 83:250–254.

Brandhorst BP, Verma DPS, Fromson D (1979): Polyadenylated and nonpolyadenylated messenger RNA fractions from sea urchin embryos code for the same abundant proteins. Dev Biol 71:128–141.

Braude P, Pelham H, Flach T, Lobatto R (1979): Post-transcriptional control in the early mouse embryo. Nature 282:102–105.

Bruskin AM, Tyner AL, Wells ED, Showman RM, Klein WH (1981): Accumulation in embryogenesis of five mRNAs enriched in the ectoderm of the sea urchin pluteus. Dev Biol 87:308–318.

Bruskin AM, Bédard P-A, Tyner AL, Showman RM, Brandhorst BP, Klein WH (1982): A family of proteins accumulating in ectoderm of sea urchin embryos specified by two related cDNA clones. Dev Biol 91:317–324.

Cabrera CV, Ellison JW, Moore JG, Britten RJ, Davidson EH (1982): Regulation of cytoplasmic mRNA prevalence in sea urchin embryos: Rates of appearance and turnover for specific sequences. (submitted).

Childs G, Maxson R, Kedes LH (1979): Histone gene expression during sea urchin embryogenesis: Isolation and characterization of early and late messenger RNAs of *Strongylocentrotus purpuratus* by gene-specific hybridization and template activity. Dev Biol 73:153–173.

Collier JR, McCarthy ME (1981): Regulation of polypeptide synthesis during early embryogenesis of *Ilyanassa obsoleta*. Differentiation 19:31–46.

Costantini FD, Britten RJ, Davidson EH (1980): Message sequences and short repetitive sequences are interspersed in sea urchin egg poly(A)$^+$ RNAs. Nature 287:111–117.

Davidson EH (1976): "Gene Activity in Early Development, 2nd Ed." New York: Academic Press.
Denny PC, Tyler A (1964): Activation of protein biosynthesis in nonnucleate fragments of sea urchin eggs. Biochem Biophys Res Commun 14:245–249.
Duncan R, Humphreys T (1981a): Most sea urchin maternal mRNA sequences in every abundance class appear in both polyadenylated and nonpolyadenylated molecules. Dev Biol 88:201–210.
Duncan R, Humphreys T (1981b): Multiple oligo(A) tracts with inactive sea urchin maternal mRNA sequences. Dev Biol 88:211–219.
Durica DS, Crain WR (1982): Analysis of actin synthesis in early sea urchin development. Dev Biol 92:418–427.
Epel D (1967): Protein synthesis in sea urchin eggs: A late response to fertilization. Proc Natl Acad Sci USA 57:899–907.
Ernst SG, Hough-Evans BR, Britten RJ, Davidson EH (1980): Limited complexity of the RNA in micromeres of 16-cell sea urchin embryos. Dev Biol 79:119–127.
Flytzanis CN, Brandhorst BP, Britten RJ, Davidson EH (1982): Developmental patterns of cytoplasmic transcript prevalence in sea urchin embryos. Dev Biol 91:27–35.
Fry B, Gross PR (1970): Patterns and rates of protein synthesis. II. The calculation of absolute rates. Dev Biol 21:125–146.
Galau GA, Britten RJ, Davidson EH (1974): A measurement of the sequence complexity of polysomal messenger RNA in sea urchin embryos. Cell 2:9–20.
Galau GA, Klein WH, Davis MM, Wold BJ, Britten RJ, Davidson EH (1976): Structural gene sets active in embryos and adult tissues of the sea urchin. Cell 7:487–505.
Galau GA, Lipson EA, Britten RJ, Davidson EH (1977): Synthesis and turnover of polysome mRNA's in sea urchin embryos. Cell 19:415–432.
Goustin AS, Wilt FH (1981): Protein synthesis, polyribosomes, and peptide elongation in early development of *Strongylocentrotus purpuratus*. Dev Biol 82:32–40.
Greenhouse GA, Hynes RO, Gross PR (1971): Sea urchin embryos are permeable to actinomycin. Science 171:686–689.
Gross KW, Jacobs-Lorena M, Baglioni C, Gross PR (1973): Cell-free translation of maternal messenger RNA from sea urchin eggs. Proc Natl Acad Sci USA 70:2614–2618.
Gross PR, Cousineau GH (1963): Effects of actinomycin-D on macromolecular synthesis and early development of sea urchin eggs. Biochem Biophys Res Commun 10:321–326.
Gross PR, Malkin LI, Moyer WA (1964): Templates for the first proteins of embryonic development. Proc Natl Acad Sci USA 51:407–414.
Grunstein M (1978): Hatching in the sea urchin *Lythechinus pictus* is accompanied by a shift in histone H4 gene activity. Proc Natl Acad Sci USA 75:4135–4139.
Harkey MA, Whiteley AH (1980): Isolation, culture, and differentiation of echinoid primary mesencyme cells. W Rouxs Arch 189:111–122.
Hieter PA, Hendricks MB, Hemminki K, Weinberg ES (1979): Histone gene switch in the sea urchin embryo. Identification of late embryonic histone mesenger ribonucleic acids and the control of their synthesis. Biochemistry 18:2707–2716.
Hough-Evans BR, Wold BJ, Ernst SG, Britten RJ, Davidson EH (1977): Appearance and persistence of maternal RNA sequences in sea urchin development. Dev Biol 60:258–277.
Houk MS, Epel D (1974): Protein synthesis during hormonally induced meiotic maturation and fertilization of starfish oocytes. Dev Biol 40:298–309.
Hultin T (1950): The protein metabolism of sea urchin eggs during early development studied by means of ^{15}N-labelled ammonia. Exp Cell Res 1:599–602.
Humphreys T (1969): Efficiency of translation of messenger RNA before and after fertilization of sea urchin eggs. Dev Biol 20:435–458.
Humphreys T (1971): Measurements of messenger RNA entering polysomes upon fertilization of sea urchin eggs. Dev Biol 26:201–208.

Hutchins R, Brandhorst BP (1979): Commitment to vegetalized development in sea urchin embryos: Failure to detect changes in patterns of protein synthesis. W Rouxs Arch 186:95–102.

Infante AA, Heilmann LJ (1981): Distribution of messenger ribonucleic acid in polysomes and nonpolysomal particles of sea urchin embryos: Translational control of actin synthesis. Biochemistry 20:1–8.

Jefferey WE (1977): Polyadenyaltion of maternal and newly synthesized RNA during starfish oocyte maturation. Dev Biol 57:98–108.

Kanatani H, Shirai H (1967): In vitro production of meiosis inducing substance by nerve extract in ovary of starfish. Nature 216:284–286.

Kuhn O, Wilt FH (1981): Chromatin proteins of sea urchin embryos. Dev Biol 85:416–423.

Lasky LA, Lev Z, Xin J-H, Britten RJ, Davidson EH (1980): Messenger RNA prevalence in sea urchin embryos measured with cloned cDNAs. Proc Natl Acad Sci USA 77:5317–5321.

Maxson RE, Egrie JC (1980): Expression of maternal and paternal histone genes during early cleavage stages of echinoderm hybrid *Strongylocentrotus purpuratus* × *Lytechinus pictus*. Dev Biol 74:335–342.

McClay DR, Chambers AF (1978): Identification of four classes of cell surface antigens appearing at gastrulation in sea urchin embryos. Dev Biol 63:179–186.

Merlino GT, Water RD, Moore GP, Kleinsmith LJ (1981): Change in expression of the actin gene family during sea urchin development. Dev Biol 85:505–508.

Nakano E, Monroy A (1958): Incorporation of ^{35}S-methionine in the cell fractions of sea urchin eggs and embryos. Exp Cell Res 14:236–244.

Newrock KM, Cohen LH, Hendricks MB, Donelly RJ, Weinberg ES (1978): Stage-specific mRNAs coding for subtypes of H2A and H2B histones in the sea urchin embryo. Cell 14:327–336.

Oakley BR, Kirsch DR, Morris RN (1980): A simplified ultrasensitive stain for detecting proteins in polyacrylamide gels. Anal Biochem 105:361–363.

Okazaki K (1975): Spicule formation by isolated micromeres of the sea urchin embryo. Am Zool 15:567–581.

O'Farrell PH (1975): High resolution two-dimensional electrophoresis of proteins. J Biol Chem 250:4007–4021.

Raff RA (1980): Masked messenger RNA and the regulation of protein synthesis in eggs and embryos. In Prescott DM, Goldstein L (eds): "Cell Biology: A Comprehensive Treatise." New York: Academic Press, Vol 4, pp 107–136.

Raff RA, Showman RM (1983): Maternal messenger RNA: Quantitative, qualitative, and spatial control of its expression in embryos. In Metz CB, Monroy A (eds): "The Biology of Fertilization." New York: Academic Press (in press).

Raff RA, Colot HV, Selvig SE, Gross PR (1972): Oogenetic origin of messenger RNA for embryonic synthesis of microtubule proteins. Nature 235:211–214.

Rodgers WH, Gross PR (1978): Inhomogenous distribution of egg RNA sequences in the early embryo. Cell 14:279–288.

Rosenthal ET, Hunt T, Ruderman JV (1980) Selective translation of mRNA controls the pattern of protein synthesis during early development of the surf clam, *Spisula solidissima*. Cell 20:487–494.

Rosenthal ET, Brandhorst BP, Ruderman JV (1982): Translationally mediated changes in patterns of protein synthesis during maturation of starfish oocytes. Dev Biol 91:215–220.

Ruderman JV, Gross PR (1974): Histones and histone synthesis in sea urchin development. Dev Biol 36:286–298.

Schuetz AW, Biggers JD (1967): Regulation of germinal vesicle breakdown in starfish oocytes. Exp Cell Res 46:624–628.

Shepherd GW, Nemer M (1980): Developmental shifts in frequency distribution of polysomal mRNA and their posttranscriptional regulation in the sea urchin embryo. Proc Natl Acad Sci USA 77:4653–4656.

Skoultchi A, Gross PR (1973): Maternal histone messenger RNA: Detection by molecular hybridization. Proc Natl Acad Sci USA 70:2840–2844.

Terman SA (1970): Relative effect of transcription-level and translation-level control of protein synthesis during early development of the sea urchin. Proc Natl Acad Sci USA 65:985–992.

Terman SA, Gross PR (1965): Translation level control of protein synthesis during early development. Biochem Biophys Res Commun 21:595–600.

Thomas TL, Posakony WJ, Anderson DM, Britten RJ, Davidson EH (1981): Molecular structure of maternal RNA. Chromosoma 84:319–326.

Tufaro F, Brandhorst BP (1979): Similarity of proteins synthesized by isolated blastomeres of early sea urchin embryos. Dev Biol 72:390–397.

Tufaro F, Brandhorst BP (1982): Restricted expression of paternal genes in sea urchin interspecies hybrids. Dev Biol 92:209–220.

Wells DE, Showman RM, Klein WH, Raff RA (1981): Delayed recruitment of maternal mRNA in sea urchin embryos. Nature 292:477–478.

Xin J-H, Brandhorst BP, Britten RJ, Davidson EH (1982): Cloned embryo mRNAs not detectably expressed in adult sea urchin coelomocytes. Dev Biol 89:527–531.

Time, Space, and Pattern in Embryonic Development, pages 49–63

Spatial and Temporal Aspects of Gene Expression During *Spisula* Embryogenesis

Joan V. Ruderman, Terese R. Tansey, Eric T. Rosenthal, Tim Hunt, and Clarissa M. Cheney

Departments of Cell and Developmental Biology and Anatomy, Harvard Medical School, Boston, Massachusetts 02115 (J.V.R., T.R.T., E.T.R.), Department of Biochemistry, University of Cambridge, Cambridge CB2 1QW, England (T.H.), and Department of Biology, Johns Hopkins University, Baltimore, Maryland 21218 (C.M.C.)

Embryos of the bivalve mollusc *Spisula* (the common surf clam) are very suitable for investigating some aspects of the molecular mechanisms of early development. Two to 10 million oocytes (1–5 ml of packed cells) can be easily obtained from a single female and fertilized en masse in sterile seawater to give a large culture of embryos synchronously proceeding through development. Embryogenesis is rapid and essentially all of the embryos in a culture develop normally. The embryos have little endogenous nuclease or protease activity, they contain very low amounts of yolk, and they readily take up radioactive precursors. These features make *Spisula* amenable to many general kinds of smash-and-spin experiments designed to get at how cells switch on and off different sets of gene products during early development. In addition, *Spisula* embryos can be used to investigate some of the mechanisms by which embryonic cells embark upon different developmental pathways and express divergent patterns of gene expression. In this chapter, we shall present some experiments that examine the patterns of gene expression in the intact embryo and in the two embryonic lineages that arise at first cleavage.

Fertilization of the mature *Spisula* oocyte triggers breakdown of the large tetraploid germinal vesicle and the first and second meiotic reduction divisions. The first cleavage occurs 70 minutes after fertilization. Subsequent asynchronous cleavages occur rapidly without any significant increase in the overall mass of the embryo. About six hours after fertilization the embryo gastrulates, produces cilia, and becomes motile. Several complex morphological changes occur over the next 12 hours, resulting in the formation of the veliger larva. During this 24-hour period there are continuous changes in the pattern of protein synthesis. The earliest changes are due to stage-specific translation of different subsets of maternal mRNAs. As development proceeds and the maternal mRNA pool is depleted, newly transcribed mRNAs make increasing contributions to the pattern of protein synthesis [Rosenthal et al., 1980; Tansey and Ruderman, 1983].

As in many mosaic embryos, first cleavage is unequal (Fig. 1a) and results in the formation of two unequally sized blastomeres whose cell fates are quite different and are already specified at this early time. Classical lineage studies of closely related organisms show that the small AB cell gives rise to the highly ciliated ectoderm and specialized muscle cells of the larva whereas the CD cell gives rise to the endoderm, the adult muscle, and the shell gland [Meisenheimer, 1901; Wilson, 1925]. However, such studies of prospective cell fate say nothing about potential cell fate, nor do they give any information about when and how such differences arise. These questions can be approached by examining the developmental potentials of blastomeres that have been separated and cultured in isolation. A few years ago, we worked out a batch method for dissociating two-cell-stage embryos and isolating small quantities of AB and CD cell populations. The major difficulty is to remove the chorion of the embryo so that the two blastomeres can be separated without harming their ability to undergo further development. This chorion consists of a dense fibrillar extracellular matrix into which microvilli extend and branch [Rebhun, 1962; Longo and Anderson, 1970]. Brief exposure of intact embryos to an alkaline hypotonic solution releases some of the chorion material and the blastomeres begin to round up [Cheney et al., 1978] (Fig. 1a–c). The blastomeres can then be dissociated by gently pipetting the embryo suspension up and down a few times. The dissociated cells are next separated by size, by centrifugation through sucrose or bovine serum albumin (BSA) gradients (Fig. 1d,e). Different batches of embryos vary widely in their response to this protocol (most batches either fail to dissociate or they lyse during the pipetting step) but on occasion it is possible to obtain small preparations of purified cell types. When the separated blastomeres are raised in isolation, they develop into partial embryos with very different morphological features (Fig. 2). After 18 hours of development, the AB progeny are hollow spheres of highly ciliated ectodermlike cells; these go on to form several spindle-shaped contractile cells that span the interior of the sphere. The ciliary beat pattern of the AB half embryos is quite like that of the normal larva. The CD progeny look very different: They form a mass of yolky cells, some of which produce stumpy, quivering cilia. Some CD

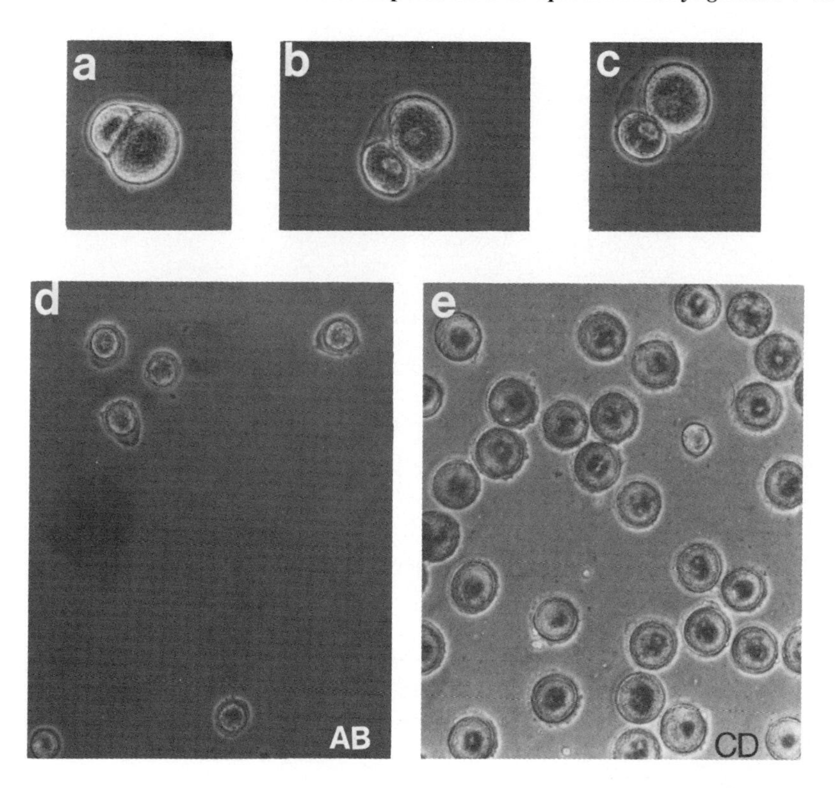

Fig. 1. The two-cell embryo and removal of the chorion. a. The two-cell embryo. b. Shortly after the embryo is placed in dechorionating solution the chorion begins to swell. c. When most of the chorion has dissolved the embryo can be dissociated by brief up-and-down pipetting. Batches of isolated AB (d) and CD (e) blastomeres.

progeny go on to produce a small mass of birefringent material suggestive of shell formation, but this is variable and has not been properly characterized. These results indicate that the fates of the AB and CD blastomeres are already different by the two-cell stage and that the progeny of these cells go on to produce several of the features that they would normally contribute to in the intact embryo.

What is segregated at this first cleavage and how do these two cell lineages compare in their subsequent patterns of gene expression? We isolated RNA from AB and CD cell preparations and examined them in vitro translation products encoded by these mRNAs [Cheney and Ruderman, 1978]. The concentrations of most mRNAs were equally represented in both cells but a few mRNAs appeared to be specific to either one cell or the other. This

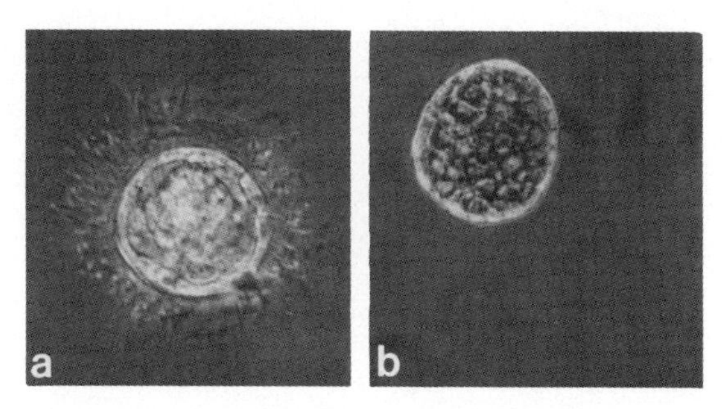

Fig. 2. Twenty-four hour-old progeny of AB (a) and CD (b) blastomeres.

interesting result suggests that differential localization or partitioning of maternal mRNA occurs and that it could be involved in specifying different programs of cell development, but further work will be required to substantiate this idea for *Spisula*. The evidence for spatial localization of maternal mRNA sequences is very compelling in certain other organisms, most notably the ascidians [Whittaker, 1977; Jeffery, 1982 and this volume].

In a second experiment, we compared the patterns of gene expression in the AB and CD half-embryo progeny at later stages with the pattern expressed by the whole embryo. Figure 3 shows the changes in protein synthesis that occur during the development of the intact embryo. Fertilization sets off a rapid switch in the pattern of protein synthesis. This change is regulated entirely at the translational level and is discussed below. Subsequent development is accompanied by continuous alterations in the pattern of protein synthesis. A major change in protein synthesis occurs at six hours after fertilization (the time of gastrulation) and is due to changes in mRNA levels that are the result of new transcription. For example, tubulin mRNA levels rise severalfold and tubulin synthesis increases markedly [Tansey and Ruderman, 1983]. As the embryo proceeds through the trochophore larva stage, the synthesis of other sets of proteins begins. At 12 hours, the embryo starts to synthesize and accumulate a 90,000-dalton doublet. Synthesis of this doublet continues for the next 24 hours. Since the AB- and CD-derived cells contribute to different tissues in the whole embryo and form several different structures in the half-embryo progeny, it seemed quite likely that the two lineages would synthesize rather different sets of proteins. However, when the protein synthetic patterns of the AB and CD half-embryo progeny were examined, this was not the result obtained. Small batches of AB and CD blastomeres were isolated, cultured for 18 hours, then labeled with ^{35}S-methionine, and the radioactive proteins were compared. Figure 4 shows that, although the two kinds of half embryos do indeed undergo changes in

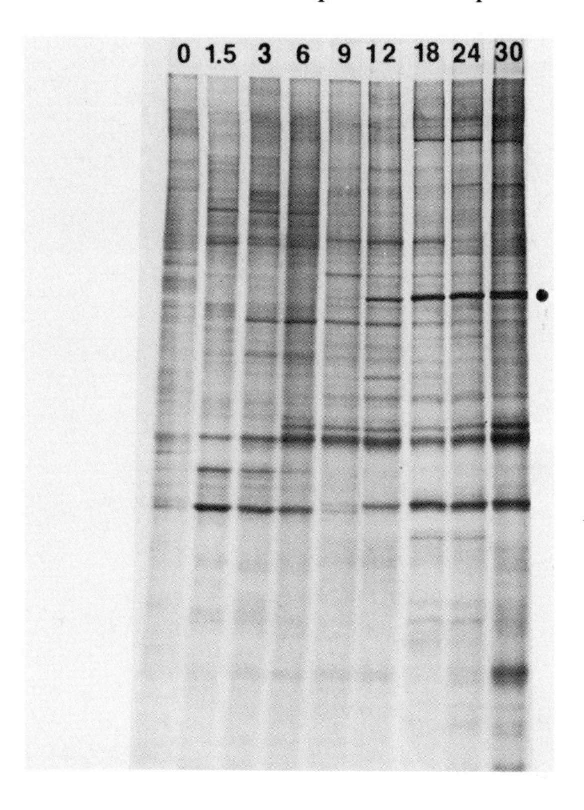

Fig. 3. Changes in the patterns of protein synthesis during early development of the intact embryo. Oocytes (0) and embryos (1 to 30 hours old) were labeled with ^{35}S-methionine for 60 minutes, and the labeled proteins were electrophoresed on 10% polyacrylamide-sodium dodecyl sulfate (SDS) gels and visualized by autoradiography. The position of the 90,000-dalton doublet is marked by a dot.

protein synthesis, each switches over to making the *same* sets of abundant proteins. One interpretation of this result is that the different features of the AB and CD lineages are due to differences in minor proteins that are not detectable in these experiments or to differences in the arrangement or disposition of the same basic groups of proteins. A second interesting finding is that when the AB and CD progeny develop in isolation, neither half embryo turns on the synthesis of the 90,000-dalton doublet that is characteristic of the veliger larva. One possible explanation for such a result is that this doublet is produced in response to some sort of inductive interaction between certain tissues of the two lineages, or in response to some other

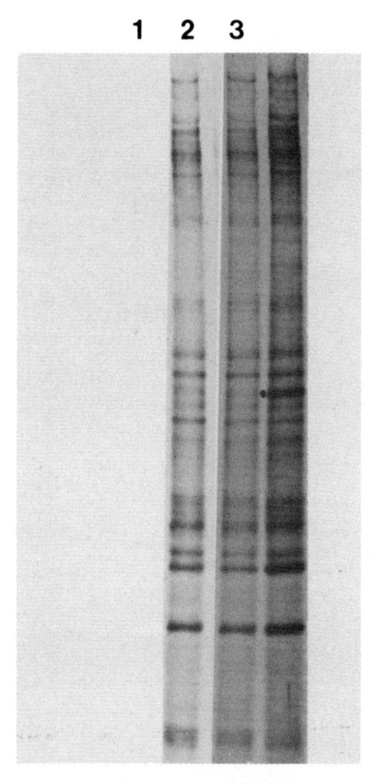

Fig. 4. The patterns of protein synthesis of AB and CD progeny. AB and CD blastomeres were isolated, cultured separately for 18 hours, and then labeled with ^{35}S-methionine for one hour. The labeled proteins were electrophoresed and visualized as in Figure 3. Proteins synthesized by: (1) AB progeny, (2) CD progeny, (3) intact embryos.

interaction that is disrupted when the lineages are not allowed to develop in their proper spatial orientations. Clearly some aspect of new transcription, mRNA processing, or protein stability is strongly influenced when normal cell-cell positioning is altered.

Spisula embryos provide an excellent opportunity to study another aspect of differential gene expression in very early development, namely, the changes in protein synthesis that occur as an immediate response to fertilization. The mature oocytes of most organisms, including *Spisula*, contain all the components required for protein synthesis, including ribosomes, amino acids, tRNA, protein synthetic enzymes, and mRNA, yet they generally show very low rates of protein synthesis. Fertilization in some species, or the resumption of meiosis in others, sets off a cascade of metabolic changes that usually include an increase in the *rate* of protein synthesis, a change in the *pattern* of protein synthesis, or, in some cases, both. These changes are due entirely to the recruitment of preexisting stores of maternal mRNA. They are independent of new transcription. The processes controlling these translational changes are not well understood. Most investigators have focused on the mechanisms involved in the increase in the overall rate of protein synthesis. In sea urchins, where this has been best studied, there is some very good evidence for the association of maternal mRNAs with masking proteins. Several experiments suggest that these proteins mask the translational activity of most maternal mRNAs in the egg and that fertilization leads to some change in this mRNA-protein (mRNP) complex that results in the derepression of mRNA template activity [Gross et al., 1973; Jenkins et al., 1978; Ilan and Ilan, 1978; Young and Raff, 1978]. For example, Kaumeyer et al. [1978] found that when mRNP complexes prepared from unfertilized eggs were added to a wheat germ cell-free protein-synthesizing system, they were not translated very well, whereas the phenol-extracted RNAs isolated from those RNP preparations were very active in directing protein synthesis. In contrast, RNPs prepared from embryos and their phenol-extracted RNAs were equally translatable in the cell-free system. Agreement on this point is not, however, complete. Moon et al. [1982] carried out very similar experiments and they found that egg RNP preparations exhibited just as much template activity in vitro as the phenol-extracted RNAs.

Physical sequestration of some mRNAs may also play a role in setting the rate and pattern of protein synthesis. A considerable portion of maternal histone mRNA appears to be localized within the sea urchin egg nucleus and remains associated with that structure until just before first cleavage, when it is released into the cytoplasm [Venezsky et al., 1981; Showman et al, 1982; Angerer and Angerer, this volume]. This cytological phenomenon fits well with the observation of Wells et al. [1981] that the entry of maternal histone mRNA into polysomes is delayed until 90 minutes after fertilization whereas the bulk of the newly recruited maternal mRNA enters the polysomes within five minutes of fertilization.

Certain aspects of the changes in protein synthesis that occur in *Spisula* oocytes after fertilization make this organism an especially favorable one for investigating the molecular mechanisms of stage-specific mRNA utilization. In *Spisula*, fertilization causes a two to fourfold rise in the rate of protein synthesis and a dramatic switch in the kinds of proteins being made. Both of these changes occur within ten minutes of fertilization, are independent of new transcription, and are mediated entirely at the translational level. For example, the proteins marked A, B, and C in Figure 5 are made at very low levels (if at all) in the oocyte, whereas they are prominently synthesized in the ten-minute embryo. Proteins X, Y, and Z are made by the oocyte, but synthesis of these proteins is undetectable after fertilization. These changes in the kinds of proteins made occur in the absence of any changes in the

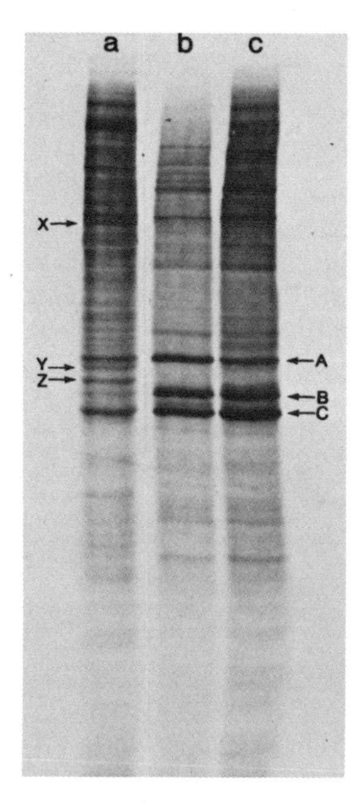

Fig. 5. Autoradiograms of proteins synthesized in vivo by oocytes and embryos. a. Oocytes labeled with ^{35}S-methionine for 20 minutes. b. Embryos labeled for 20 minutes after germinal vesicle breakdown. c. Embryos labeled for 20 minutes after first cleavage. From Rosenthal et al. [1980]. Reproduced with permission of MIT Press, Cambridge.

levels of their respective mRNAs. The mRNAs present in oocytes and zygotes were assayed by translating total cell phenol-extracted RNA in an mRNA-dependent reticulocyte lysate [Pelham and Jackson, 1976]. Figure 6 compares the in vitro translation products encoded by total RNA isolated from oocytes or embryos at 15 minutes (after germinal vesicle breakdown), 40 minutes (polar body formation), and 70 minutes (first cleavage) after fertilization. The RNAs extracted from oocytes and early embryos direct the synthesis of the same sets of proteins in vitro. This result demonstrates that the postfertilization switch in the pattern of protein synthesis cannot be due to either the transcription of new mRNAs or the selective degradation of preexisting mRNAs. Posttranslational modifications are also ruled out by the observation that the proteins synthesized in reticulocyte lysates in response to mRNA from either oocytes or embryos include essentially all of the proteins synthesized in vivo at both stages.

Since oocytes contain abundant amounts of the mRNAs for proteins A, B, and C, but do not make these proteins prior to fertilization, it appears that these mRNAs are translationally inactive in oocytes and become available for translation after fertilization. To test the idea that different developmental stages load different subsets of maternal mRNAs onto polysomes, 12,000g supernatants from oocyte or embryo homogenates were fractionated on sucrose gradients to separate polysomal from postribosomal structures. RNA was then purified from these gradient fractions and translated in vitro. Typical results, shown in Figure 7, demonstrate that mRNAs for proteins A, B, and C are not associated with polysomes in the oocytes, whereas they are almost completely recruited onto polysomes within ten minutes of fertilization. Other exposures, as well as slightly different kinds of gradient fraction-

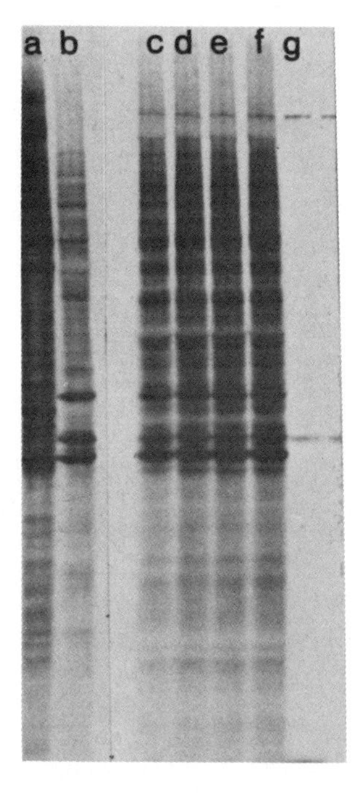

Fig. 6. Comparison of ^{35}S-methionine-labeled proteins synthesized in vivo by oocytes and embryos with those programmed in vitro by oocyte and embryo RNA. a. Oocyte proteins synthesized in vivo. b Two-cell embryo proteins synthesized in vivo. Lanes c–f, in vitro translation products encoded by: c. oocyte RNA; d. embryo RNA, 15 minutes postfertilization; e. embryo RNA, 50 minutes postfertilization; f. Top-cell embryo RNA; g. endogenous incorporation (no RNA added). From Rosenthal et al. [1980]. Reproduced with permission of MIT Press, Cambridge.

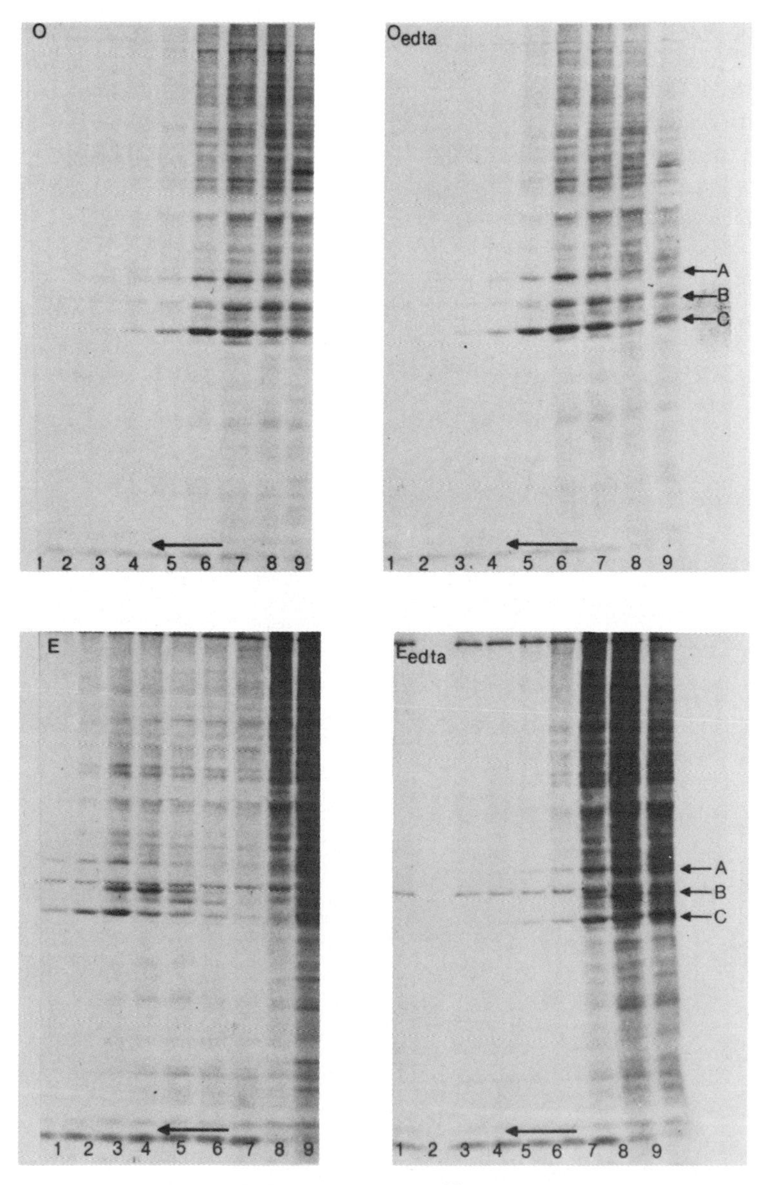

Fig. 7. In vitro translation products, labeled with ^{35}S-methionine and directed by RNA extracted from sucrose gradient fractions of oocyte and embryo 12K supernatants. Translation products directed by RNAs isolated from sucrose gradient fractionations of 12K supernatants from oocytes (O) and embryos (E), and ededtic acid (EDTA)-treated 12K supernatants from oocytes (O_{EDTA}) and embryos (E_{EDTA}). Arrow indicates direction of sedimentation; 60S subunits are found in fractions 7 and 8; 40S subunits are found in fractions 8 and 9.

ation schemes, show that other mRNAs that are engaged on polysomes in the oocyte are released from polysomes right after fertiliztion [Rosenthal et al., 1980, 1982]. Thus, it is quite clear the oocytes and embryos contain the same pool of mRNA sequences, but that oocytes utilize one subset of those mRNA sequences and embryos use another different (although overlapping) subset.

How is this stage-specific utilization of mRNA achieved? How does fertilization cause one group of mRNAs to be released from the polysomes and another group to be activated for translation? Descriptive studies and mechanistic investigation of translational control using in vitro translation assays are limited by two considerations. First, when the proteins encoded by two or more mRNAs comigrate, it is impossible to follow the individual behaviors of their respective mRNAs. Second, since we do not understand the molecular mechanisms responsible for stage-specific changes in mRNA translation, the in vitro translation assay may yield misleading results. As one approach to getting around these problems, we have constructed and cloned complementary DNA (cDNA) copies of some of these translationally regulated mRNAs. These cDNA probes can be used to monitor mRNA levels and polysome/nonpolysome localizations much more quantitatively and unambiguously than in vitro protein synthesis assays. They can also be used to study changes in mRNA structure, to isolate individual mRNPs, and to look for changes in mRNA-protein associations that could be functionally important in these translational switches. Two cDNA libraries were constructed by inserting double-stranded cDNA copies of poly(A)$^+$RNA into the Pst site of pBR322. One cDNA library contains copies of two-cell poly(A)$^+$RNA sequences, and the other contains 18-hour larval poly(A)$^+$RNA sequences. Among the clones used in the following experiments are those carrying sequences for protein A (clone 1T55), protein C (clone 1T43), and α-tubulin (clone 3V4). These clones were identified by their ability to selectively hybridize mRNAs that were found to program the synthesis of proteins A, C, and α-tubulin, respectively, in vitro. Two other clones, 1T9 and 6T21, were also used in this study. These two clones showed hybridization to polysomal mRNAs but did not yield detectable amounts of hybrid-selected translation products.

These clones were first used to test the previous conclusions that the levels of these mRNAs do not change at fertilization. Equal amounts of total RNA from oocytes and embryos were electrophoresed on an agarose gel, blotted to nitrocellulose, and hybridized to ^{32}P-labeled cloned cDNAs. The intensities of hybridization to oocyte vs. embryo RNA show directly and unambiguously that the amounts of these mRNAs do not change at fertilization (Fig. 8). Next, the predicted stage-specific polysomal localizations of these mRNAs were tested. RNA from polysome gradient fractions similar to those used in Figure 8 were electrophoresed on agarose gels, blotted, and hybridized with

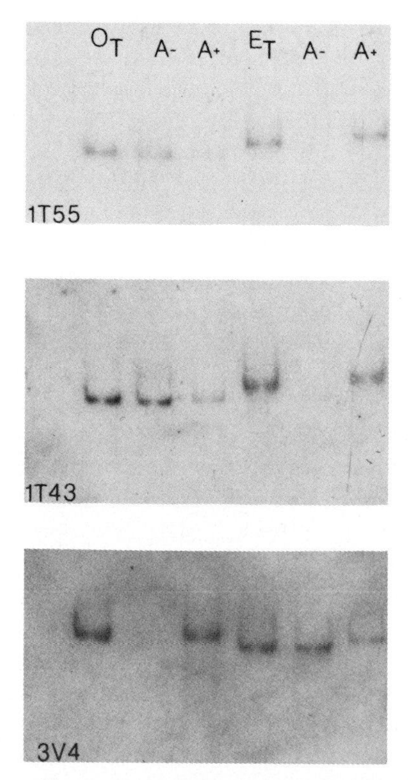

Fig. 8. Hybridization of ^{32}P-cloned DNA probes 1T55 (protein A), 1T43 (protein C) and 3V4 (α-tubulin) to total RNA, poly(A)$^{-}$RNA, and poly(A)$^{+}$RNA from oocytes and 30-minute postfertilization embryos. RNAs were electrophoresed in an agarose slab gel, blotted to nitrocellulose with ^{32}P-labeled cloned DNAs.

labeled cloned probes. The results showed quite clearly that mRNAs for proteins A and C are not present on polysomes in the oocyte and that they are completely loaded onto polysomes right after fertilization. RNAs complementary to clones 6T21 and 1T9 showed a different pattern of translational utilization: Both RNAs were predominantly nonpolysomal in oocytes but only some of these sequence moved onto polysomes after fertilization. In contrast, α-tubulin mRNA is found on polysomes in the embryo but is released from polysomes right after fertilization [Rosenthal et al., 1982]. Thus fertilization of *Spisula* oocytes causes at least three kinds of translational changes. Some mRNAs, such as those encoding proteins A and C, are not associated with polysomes in the oocyte and become completely recruited onto polysomes within 10 minutes of fertilization. Others, such as 6T21 and 1T9 mRNA, are only partially recruited onto polysomes after fertilization. Finally, others, like α-tubulin mRNA, are actively translated in oocytes and are released from polysomes right after fertilization.

What molecular features of these mRNAs change at fertilization? The recent work of Costantini et al. [1980] has revived the idea that many maternal RNA sequences reside in RNA molecules that have many features of nuclear precursor RNAs. These workers found that about 70% of the poly(A)$^+$RNA complexity in the mature sea urchin egg is found in very large molecules that are capable of annealing to form vast branching networks, but that such RNA molecules are not present in preparations of polysomal RNAs from later stages. This kind of explanation does not, however, appear to hold for the mRNAs we have been studying in *Spisula* oocytes. When the sizes of the pre- and postfertilization versions of these mRNAs are examined by hybridization of cloned probes to gel blots, the sizes of the mRNAs isolated from embryos are, if anything, larger than those isolated from oocytes! Thus, in *Spisula*, translational activation of these mRNAs does not depend on processing of larger, translationally inert precursors. We now know that the slight increase in the sizes of the mRNAs that are translationally activated after fertilization is due to the addition of poly(A) tails and that the polyadenylation status of these mRNAs is correlated with their translatability in vivo. Figure 8 shows that when mRNAs A and C are inactive (in the oocyte) they lack a poly(A) tail whereas they gain a poly(A) tail just at the time when they are recruited onto polysomes after fertilization. In contrast, the α-tubulin mRNA in the oocyte is active and is poly(A)$^+$ whereas this particular mRNA loses its poly(A) tail right after fertilization, just at the time when it loses its translational activity. Several other mRNAs also show this correspondence between possession of a poly(A) tail and translational activity in vivo [Rosenthal et al., 1982]. Finally, preliminary characterization of the sequences complementary to clones constructed using *oocyte* poly(A)$^+$RNA shows that many oocyte poly(A)$^+$RNAs become deadenylated after fertilization. This translational status of the poly(A)$^+$ and poly(A)$^-$ versions of these mRNAs should be known soon. Whether such correlations are functional or merely reflect some other aspect of mRNA metabolism remains to be tested directly. One telling test would be to fertilize oocytes in the presence of 3′-deoxyadenosine (cordycepin) in concentrations sufficient to block polyadenylation and ask if the switch in the pattern of protein synthesis is affected. Unfortunately, *Spisula* oocytes are not very permeable to cordycepin and we have been unable to use this drug to interfere with the postfertilization addition of poly(A) to maternal mRNAs.

Finally, the possibility should be discussed that the translation of specific mRNAs is modulated by masking proteins, an idea first proposed by Spirin and extended by many others [see Raff, 1980, for a recent review]. Earlier experiments from this lab attempted to reproduce the stage-specific translation of particular mRNAs in a *Spisula* cell-free system. Crude 12,000g supernatants were prepared from *Spisula* oocytes or embryos in a buffer that was developed by Winkler and Steinhardt [1981] for a sea urchin egg cell-free system. These 12,000g supernatants do not initiate protein synthesis very efficiently. However, when the crude supernatants are supplemented with mRNA-dependent reticulocyte lysate, this mixed cell-free system now initiates and synthesizes protein rather well [Rosenthal et al., 1980]. Furthermore, the translation products encoded by the oocyte homogenate resemble those made by the oocyte in vivo, whereas the embryo homogenate programs the synthesis of just the embryo-specific proteins. Thus, the initiations that occur when the mRNA-dependent reticulocyte lysate is presented with crude, unextracted homogenate are "correct," whereas the phenol-extracted mRNAs are translated indiscriminantly. Clearly, some translational information is present and preserved in the crude homogenates and is lost during phenol extraction. Very little is known about the nature of this information. When oocyte and embryo homogenates are mixed in varying proportions prior to adding them to the reticulocyte lysate, the pattern of proteins synthesized simply resembles the sum of the input oocyte and embryo patterns. This result suggests that there is no excess of either positive or negative diffusible regulators. The idea that individual mRNAs are complexed with regulatory, masking proteins is attractive and is obviously one of the interpretations of this experiment, but is by no means the only possibility. With cloned cDNAs complementary to some of the translationally regulated mRNAs now in hand, it should be possible to isolate specific, individual mRNPs free from cosedimenting ribosomes and other RNP fractions. This approach offers several opportunities for assessing the various contributions of mRNA structure, mRNA-associated proteins, ribosomes, and other components of the translational machinery to the regulation of mRNA activity.

ACKNOWLEDGMENTS

This work was supported by NSF grant PCM 7923046 and NIH grant K04HD00349 to J.V.R.

REFERENCES

Cheney CM, Ruderman JV (1978): Segregation of maternal mRNA at first cleavage in embryos of the mollusc *Spisula solidissima*. J Cell Biol 79:349a.

Constantini FD, Britten RJ, Davidson EH (1980): Message sequences and short repetitive sequences are interspersed in sea urchin egg poly(A)$^+$ RNAs. Nature 287:111–117.

Gross KW, Jacobs-Lorena M, Baglioni C, Gross PR (1973): Cell-free translation of maternal mRNA from sea urchin eggs. Proc Natl Acad Sci USA 70:2614–2618.

Ilan J, Ilan J (1978): Translation of maternal messenger ribonucleoprotein particles from sea urchin in a cell-free system from unfertilized eggs and product analysis. Dev Biol 66:375–385.
Jeffery WR (1982): Messenger RNA in the cytoskeletal framework: Analysis by in situ hybridization. J Cell Biol 95:1–21.
Jenkins NA, Kaumeyer JF, Young EM, Raff RA (1978): The template activity of messenger ribonucleoprotein particles isolated from sea urchin eggs. Dev Biol 63:279–298.
Kaumeyer JF, Jenkins NA, Raff RA (1978): Messenger ribonucleoprotein particles in unfertilized sea urchin eggs. Dev Biol 63:266–278.
Longo I, Anderson E (1970): An ultrastructural analysis of fertilization in the surf clam, *Spisula solidissima*. J Ultrastruct Res 33:495–527.
Meisenheimer J (1901): Entwicklungsgeschichte von Dreissensia polymorpha Pall. Wiss Zool 69:1–135.
Moon RT, Danilchik MV, Hille MB (1982): An assessment of the masked message hypothesis: Sea urchin messenger ribonucleoprotein complexes are efficient templates for in vitro protein synthesis. Dev Biol 93:389–403.
Pelham HRB, Jackson R (1976): An efficient mRNA-dependent translation system from reticulocyte lysates. Eur J Biochem 67:247–256.
Raff RA (1980): Masked messenger RNA and the regulation of protein synthesis in eggs and embryos. In Prescott DM, Goldstein L (eds): "Cell Biology, A Comprehensive Treatise." New York: Academic Press, p 107–136.
Rebhun LI (1962): Electron microscope studies on the vitelline membrane of the surf clam, *Spisula solidissima.* J Ultrastruct Res 6:107–124.
Rosenthal ET, Hunt T, Ruderman JV (1980): Selective translation of mRNA controls the pattern of protein synthesis during early development of the surf clam *Spisula solidissima*. Cell 20:487–496.
Rosenthal ET, Tansey TR, Ruderman JV (1983): Sequence-specific adenylations and deadenylations accompany changes in the translation of maternal mRNA after fertilization of *Spisula* occytes. J Mol Biol (in press).
Showman RM, Wells DE, Anstrom J, Hursh DA, Raff RA (1982): Message-specific sequestration of maternal histone mRNA in the sea urchin egg. Proc Natl Acad Sci USA 79:5944–5948.
Tansey TR, Ruderman JV (1983): Changes in abundant mRNAs during *Spisula* embryogenesis. Dev Biol (in press).
Venezky DL, Angerer LM, Angerer RC (1981): Accumulation of histone repeat transcripts in the sea urchin egg pronucleus. Cell 24:385–398.
Wells DE, Showman RM, Klein WH, Raff RA (1981): Delayed recruitment of maternal mRNA in sea urchin embryos. Nature 292:477–479.
Whittaker JR (1977): Segregation during cleavage of a factor determining endodermal alkaline phosphatase development in ascidian embryos. J Exp Zool 202:139–154.
Wilson EB (1925): "The Cell in Development and Heredity." New York: Macmillan and Co.
Winkler MM, Steinhardt RA (1981): Activation of protein synthesis in a sea urchin cell-free system. Dev Biol 84:432–439.
Young EM, Raff RA (1978): Messenger ribonucleoprotein particles in developing sea urchin embryos. Dev Biol 72:24–40.

Time, Space, and Pattern in Embryonic Development, pages 65–86

Localization and Temporal Control of Expression of Maternal Histone mRNA in Sea Urchin Embryos

Rudolf A. Raff

Program in Molecular, Cellular, and Developmental Biology, Indiana University, Bloomington, Indiana 47405 and Marine Biological Laboratory, Woods Hole, Massachusetts 02543

The existence in eggs of information, stored outside of the nuclear genome, that is required to direct the events and determinative steps of early development is well established. Two major traditions in embryology have contributed to our understanding of such stored information. The first, and older tradition, has been the study of the roles of "localized determinants" in a wide variety of embryos. These studies have been less concerned with the chemical nature of determinants than with their localization, times of function, and effects on development. The experimental approaches have been various, but most of our current understanding has resulted from experiments in which defined portions of eggs or blastomeres of embryos are removed, and the morphological and differentiative abilities of isolated portions of embryos subjected to these deletions are analyzed. A related experimental design has been to displace determinants to alter their locations or movements. A number of contributions in this volume exemplify the crucial information gained from these approaches.

The second tradition in the study of the stored information of eggs has focused directly on a known carrier of genetic information, messenger RNA (mRNA). The existence of a "maternal" program for early development of sea urchin embryos became apparent through studies of development of cross-species hybrids in the early decades of this century. Hybrids of many crosses between sea urchin species eliminate paternal chromosomes, but some are stable [e.g., Tennent, 1922]. Development of such hybrids follows the maternal species pattern until the mesenchyme blastula stage: Subsequently, paternal or hybrid characteristics appear. In the example of the cross between *Cidaris* (♀) × *Lytechinus* (♂), the maternal pattern governs the timing of mesenchyme formation and invagination of the archenteron. However, the paternal pattern is expressed in the site of origin of the primary mesenchyme cells.

It may be of some interest to note, in an age so steeped in the central role of DNA in development, that the reason many of the early experiments on cross-species hybrids were done was because an entirely different point of view was current during the first decades of the 20th century. A statement made by Conklin [1915] serves to make the point:

> At the time of fertilization the hereditary potencies of the two germ-cells are not equal, all the early stages of development, including the polarity, symmetry, type of cleavage, and the pattern, or relative positions and proportions of future organs, being foreshadowed in the cytoplasm of the egg-cell, while only the differentiations of later development are influenced by the sperm. In short, the egg cytoplasm fixes the general type of development and the sperm and egg nuclei supply only the details.

Tennent himself [1922], on the basis of his experiments, arrived at a conclusion more closely approximating that held today. "The thing inherited by offspring from parent is the capacity for development. What that development will be depends on the interactions between nucleus and cytoplasm and on adjustment to environment."

It was not until the early 1960s that evidence for the possible chemical nature of the maternal program became available. This evidence came from experiments in which eggs which had been physically enucleated and then artificially activated or, alternatively, embryos which had been cultured in the presence of the RNA synthesis inhibitor actinomycin D, were found to vigorously carry out protein synthesis. Although activated physically enucleated eggs do not exhibit any morphological development, sea urchin embryos functionally enucleated with actinomycin D continue to replicate DNA, cleave, and undergo sufficiently normal morphogenesis to produce a blastula and assemble cilia. There is thus no question but that maternal mRNAs are synthesized and stored during oogenesis and subsequently translated in early development. It has been a widely held hypothesis that maternal mRNAs provide the major part of the maternal program for early development and that they govern not only synthesis of various cellular proteins, but also govern early morphological events. This view has motivated much of the work done in my laboratory as well as that in many others. Yet, it is important to note that other modes of storage of information for a maternal developmental program exist, and that a role for maternal mRNA in controlling embryonic morphogenesis has not yet been demonstrated. There have been suggestive observations, perhaps the strongest of which have been made by Kalthoff and his co-workers on the anterior determinant of the dipteran insect *Smittia,* which provide most of our current evidence that RNAs can encode morphogenetic information. No analogous data exist for sea urchin embryos, although animal-vegetal polarity is established very early and is expressed in the fates and morphologies of the blastomeres of the 16-cell stage embryo.

If maternal mRNAs are to be convincingly implicated in the regulation or execution of morphogenesis, it must be shown that such RNA has a high information content. And, in fact, the maternal RNA of sea urchin eggs has a sufficient complexity to encode roughly 25,000 different protein species—assuming that all sequences represent mRNAs. About 75% of this sequence complexity is present in the polysomes of the 16-cell embryo [Galau et al., 1976; Hough-Evans et al., 1977].

The maternal mRNAs of sea urchin eggs may be envisioned as fulfilling three major roles: providing the zygote with a large amount of mRNA and providing localized, or temporally regulated, translation of specific sequences in the early embryo. The first of these functions is the best documented [Raff and Showman, 1983]. Although the quantitative supply of mRNA to support a large increase in protein synthesis, before enough nuclei have been generated to provide sufficient mRNA to support embryonic protein synthesis, is vital to the embryo, it appears to offer little prospect of providing any control over morphogenesis. Maternal mRNAs translated under spatial or temporal controls, however, would provide good candidates for regulatory functions.

Temporal regulation of translation of maternal mRNAs is not well understood, although several examples of such controls exist. Sequence-specific changes in translation have been found to occur shortly following maturation of oocytes in the clam *Spisula* [Rosenthal et al., 1980], the starfish *Asterias* [Rosenthal et al, 1982; Martindale and Brandhorst, 1982], the frog *Xenopus* [Ballantine et al., 1979], and mammals [van Blerkom, 1979]. Although sequence-specific translational controls commonly occur within minutes following maturation or fertilization, such controls also govern the timing of translation of maternal mRNAs in more advanced developmental stages several hours after fertilization. For example, Brandhorst and Newrock [1981] observed translation of maternal mRNAs in 21-hour-old embryos of the snail *Ilyanassa*. Similarly, Woodland and Ballantine [1980] and Flynn and Woodland [1980] have observed delayed translation of specific maternal mRNAs in *Xenopus* embryos.

Spatial localizations of maternal mRNAs have been sought in several organisms in which particularly early determinative events occur. Efforts to detect localized maternal mRNAs in sea urchin embryos have centered around the three dramatically distinct blastomere types produced at the fourth cleavage. Mesomeres, macromeres, and micromeres have different embryonic fates, and, because of their size differences, are readily isolated. However, studies of their maternal mRNAs have not revealed easily interpreted differences in mRNA distribution. Tufaro and Brandhorst [1979] were able to resolve about 1,000 protein species from 16-cell stage embryo blastomeres by use of two-dimensional gels. However, they detected no significant differences in patterns of synthesis of the proteins encoded by the relatively prevalent maternal mRNAs at this stage. Such a result might be interpreted to mean that prevalent sequences merely represent "housekeeping" proteins common to all cells. This kind of argument is probably risky because it ignores the evident existence of some complex regulatory patterns. For instance, isolated mesomeres, macromeres, and micromeres all synthesize tubulins, but only mesomeres and macromeres of *Arbacia* embryos form aggregates in vitro which produce a ciliated epithelium [Hynes et al., 1972]. Close examination of the expression of the tubulin multigene family in other organisms suggests the kinds of controls that may operate. In *Drosophila* the meiotic spindle apparatus of the male is constructed of a different β-subunit than that of the female. These two β-subunits are distinguishable on gels and so would be seen in a two-dimensional gel comparison of the two tissues. However, what would be missed in such an analysis is that the testis-specific β-subunit also plays a role in cytoplasmic microtubules and in the assembly of sperm tail axonemes [Kemphues et al., 1982]. Such a complex multifunctionality, if present in sea urchin embryo blastomeres, would allow different structures to be assembled even though structural protein synthesis pattern differences could not be detected. Nevertheless, the essential identity of pattern is discouraging to the hope of finding prevalent regulatory molecules. Prevalent maternal mRNAs do not appear to be localized in sea urchin embryos, and regional differences in prevalent mRNAs seem to arise late in development [Bruskin et al., 1981; Harkey and Whiteley, 1982] when most of the prevalent mRNAs are of embryonic origin [Hough-Evans et al., 1977].

Although the blastomeres of the 16-cell-stage sea urchin embryo provide little evidence for the segregation of prevalent maternal mRNAs, it does appear that there is a segregation of rare RNAs. Rodgers and Gross [1978] found that RNA extracted from micromeres contained a 20%–30% lower diversity of rare RNAs than the other blastomeres. The difference appears to be due to RNAs of maternal origin since blastomeres prepared from embryos cultured in the presence of actinomycin D exhibited the same nonequivalent distributions of rare RNAs. Ernst et al. [1980] observed a similar distribution of rare sequences. Hough Evans et al. [1977] had previously determined that, although the 16-cell stage embryo contains essentially the same diversity of rare sequence RNAs as the egg, only 75% of these sequences are found associated with polysomes. Ernst et al. [1980] found that both the intact 16-cell embryo and isolated micromeres have a polysomal mRNA complexity equal to about 75% of total embryo RNA complexity. Thus micromeres appear to translate the same fraction of maternal mRNAs as the mesomeres and macromeres, but they lack the additional RNA sequences found in the cytoplasm of the other blastomeres. Rare sequence RNAs clearly might serve as localized morphogenetic determinants, but there is no evidence that they in fact do so.

It should also be noted that if one could demonstrate a maternal mRNA localized in the cytoplasm, the specificity for such a localization must lie in the cytoskeleton as well as in the mRNA [Raff, 1977; Jeffery, 1983]. The establishment of a polarity in the cytoskeleton of the sea urchin embryo is distinctly seen in the orientation and locations of mitotic spindles which produce the characteristic pattern of blastomere formation at the fourth cleavage, and it is visible in the organization of the pigment granules of the egg [Schroeder, 1980]. The information required to establish oriented cytoskeletal systems is not yet known, but the fact that the orientation of the spindles of the early embryo of the snail *Lymnaea* is controlled by a maternal effect gene suggests that gene products produced during oogenesis can function in early embryonic morphogenesis.

In this chapter, I present the results of recent studies in my laboratory [Showman et al., 1982] that demonstrate a regulatory role for localization of certain maternal mRNAs in their expression in development. These studies center on the control of timing of translation and localization of maternal mRNAs for the major histone subtypes in early embryos of sea urchins. These mRNAs are sequestered into the female pronucleus of the unfertilized egg and only released for translation two hours after fertilization. The detailed mechanisms for this regulatory phenomenon are not presently understood, although the potential consequences for gene expression and for the proper remodeling of chromatin in the embryo are profound.

TIMING OF HISTONE SYNTHESIS

The histones constitute a multigene family with five classes: histones H1, H2A, H2B, H3, and H4. Each of these major classes contains variant genes which are expressed sequentially during sea urchin development [see recent discussions by Childs et al., 1979]. Thus, the cs (for cleavage stage) variants of H1, H2A, H2B, H3, and H4 histones are synthesized from maternal cs histone mRNAs in the unfertilized egg and from fertilization through the first few cleavages [Herlands et al., 1982]. The α-variants are synthesized from about the four-cell stage to blastula, when switches to synthesis of β-, γ-, and δ-variants occur.

The switching of histone synthesis from cs to α-subtypes is superimposed on a larger set of quantitative changes in translation of maternal mRNAs triggered by fertilization [reviewed by Raff and Showman, 1983]. Although unfertilized sea urchin eggs synthesize a diverse spectrum of proteins, their rate of protein synthesis is low, and only a small fraction of their available mRNAs and ribosomes are engaged in translation. Protein synthesis rises dramatically within a few minutes following fertilization. This acceleration is independent of mRNA synthesis by the zygote, but instead is dependent on the rate at which maternal mRNAs become available for translation, and the rate at which ribosomes transit mRNAs once engaged [Raff et al., 1981]. Messenger recruitment is qualitatively nonspecific: Brandhorst [1976] showed that the pattern of protein synthesis characteristic of unfertilized eggs is also typical of zygotes. Thus, the early and major quantitative rise in protein synthesis involves the indiscriminate mobilization of a large number of mRNA sequences.

Experiments in which the maternal mRNAs of sea urchin eggs have been translated in vitro indicated that as much as 4% of the total RNA of the egg is competent mRNA. Much of this RNA is capped and little or no processing seems to be necessary in the transition from egg to embryo, since mRNAs isolated from eggs when translated in vitro yield proteins apparently identical to those synthesized in vivo [reviewed by Raff and Showman, 1983]. Poly(A)$^+$ and poly(A)$^-$ editions of most nonhistone mRNA sequences are present and can be ultimately detected in the polysomes of embryos [Ruderman and Pardue, 1977; Brandhorst et al., 1979; Duncan and Humphreys, 1981]. Of course it should be noted that despite the existence of bona fide mRNAs in the cytoplasm of eggs, the possibility remains that not all stored RNAs are structurally equivalent to the mRNAs translated during development. Any structural differences might be of potential importance in the control of translatability, or at least provide clues to the events involved in storage of mRNAs during oogenesis.

Costantini et al. [1980] observed that although most of the poly(A)$^+$ RNA of the egg contains sequences derived from single-copy DNA, the presumptive single-copy coding regions are linked to segments of RNA derived from moderately repeated DNA. Egg poly(A)$^+$ RNAs with repeat transcript elements differ from gastrula mRNAs which are poor in these elements. Since both copies of both strands of repeats are present, egg poly(A)$^+$RNA can be annealed with itself to produce double-stranded hybrids. It is not clear whether such structures might exist in vivo in eggs, although Costantini et al. [1980] have suggested a number of control functions repeat transcripts attached to mRNAs might serve, including the suppression of mRNA translation before fertilization, sites for RNA processing events at fertilization, or signals for specific mRNA localization. An alternate possibilty is that these peculiar transcripts represent incomplete processing of mRNA precursors. The egg cytoplasm might in some respects be a museum containing the relics of oogenetic events.

The mRNAs for the α-subtype histones are also present as maternal mRNAs, and, in fact, constitute a considerable fraction (about 4%–8%) of the mass of maternal mRNAs stored in the egg [Davidson, 1976]. That the nontranslated, but competent histone mRNAs in sea urchin eggs are identical to histone mRNAs being translated after fertilization appears probable from the experiments of Lifton and Kedes [1976], who showed that histone mRNAs isolated from unfertilized eggs have electrophoretic mobilities identical to well-characterized and functioning embryonic histone mRNAs. These egg-derived histone mRNAs are translatable in vitro and produce several well-defined histones characteristic of early development [Childs et al., 1979]. However, the kinetics of recruitment of histone mRNAs of the α-subtypes into polysomes in early stages of sea urchin development are strikingly different from the kinetics of recruitment of the bulk of maternal mRNAs [Wells et al., 1981a,b]. This is illustrated in Figures 1 and 2. The early phase of recruitment of maternal mRNAs is traced by the kinetics of incorporation of ribosomes into polysomes. The increase in polysomal ribosomes (and mRNAs) begins at fertilization in a rapid and linear manner. By two hours after fertilization about 25% of the available ribosomes are found in polysomes. This level is maintained for several hours and suggests a steady state level of translationally active maternal mRNAs.

As noted above, the synthesis of the α-subtype histones does not begin at fertilization although the mRNAs for these proteins are present in the pool of maternal mRNAs and can be readily detected by hybridization or by in vitro translation techniques [Wells et al., 1981a; Winkler and Steinhardt, 1981]. In concordance with observed patterns of histone synthesis in embryos, titration of α-subtype maternal histone mRNA behavior with clones specific for H1, H2B, and H3 coding sequences reveals a very different recruitment behavior for these mRNAs than for the bulk of maternal mRNAs (Figs. 1, 2). Histone mRNAs can be detected in polysomes at only very low levels until 1.5 to two hours after fertilization. This time corresponds to about the period of the first cleavage at the temperature (13.5°C) used in these experiments. At that point a rapid recruitment of histone mRNAs begins such that by about 3.5 hours essentially 100% of detectable histone mRNAs have moved into polysomes. All of the mRNAs involved in this event are of maternal origin. No synthesis of histone mRNAs can be detected prior to about 4 to 5 hours of development [Wells et al., 1981a; and our unpublished data].

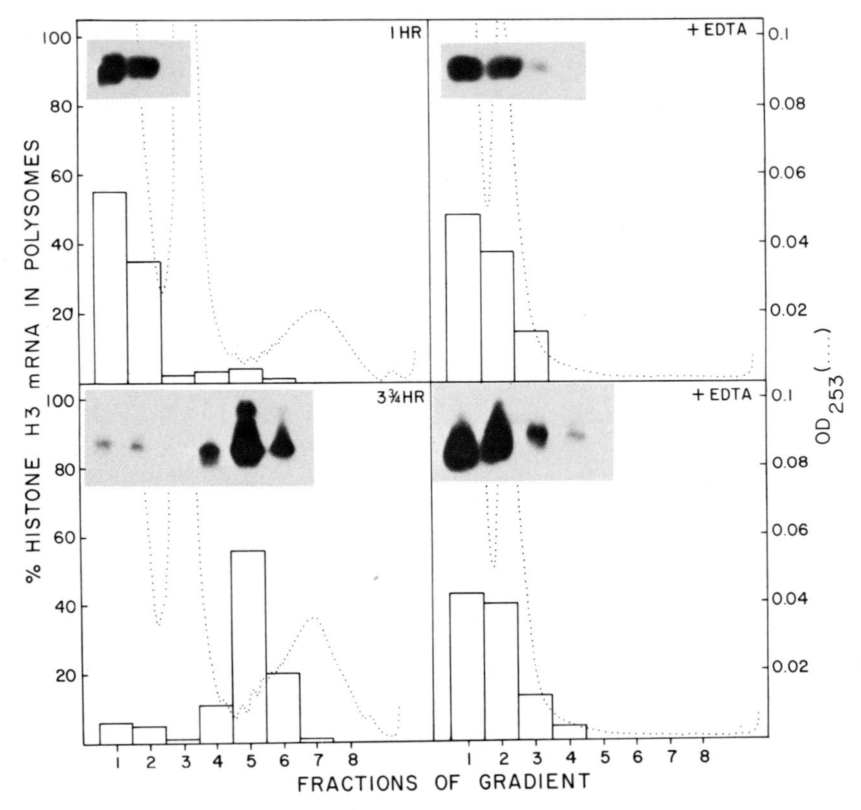

Fig. 1. Detection of maternal histone mRNA movement into polysomes of embryos of *Strongylocentrotus purpuratus*. Polysomes prepared as described by Wells et al. [1981a] were fractionated on sucrose gradients. RNA extracted from fractions was subjected to electrophoresis, blotted to a nitrocellulose filter, and hybridized to a nick-translated probe for α-histone H3 (provided by L. Kedes) as described by Wells et al. [1981a]. Autoradiographs of H3 hybridizations corresponding to the indicated sucrose gradient fractions are superimposed on the optical density traces of the gradients. Quantitations, the amount of probe hybridized to each fraction as a percentage of probe hybridized to all fractions of a gradient, are plotted as bars. Reproduced from Wells et al. [1981a].

These observations show that instead of exhibiting only a single event of early mobilization of maternal mRNAs at fertilization, sea urchin embryos recruit maternal mRNAs in two temporally distinct episodes. The early mobilization event includes thousands of mRNA sequences, each present in moderate to low prevalence; the second is limited to a few specific and highly prevalent sequences. Both events are set in motion by fertilization, but the timing and sequence-specific nature of the second event indicates that quite distinct mechanisms are involved in the two events. Four major issues can be

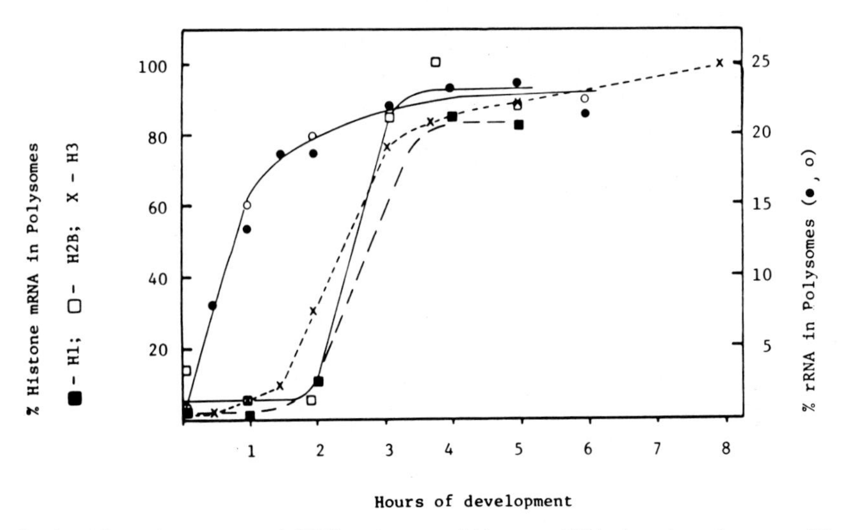

Fig. 2. Time of movement of rRNA and maternal histone mRNAs into the polysomes of *S. purpuratus* embryos. Polysome preparations and assays were performed as described by Wells et al. [1981a] and in Figure 1. Recruitment of rRNA (and thus ribosomes) was determined by optical density (-●-) and hybridization to a rDNA probe (provided by D. Stafford) (-○-). Histone mRNA recruitment was measured by hybridization to α-subtype gene probes (clones provided by L. Kedes).

envisaged in rationalizing the second event. These are (1) the spatial sequestration of α-subtype histone mRNAs as opposed to their presence as cytoplasmic mRNPs coexisting with other maternal mRNPs, but subject to a different translational control; (2) the recognition of specific maternal mRNA sequences by the cell; (3) the regulation of timing of expression; and (4) finally, the functional reason for a translational delay for the α-subtypes of histones. The questions raised by these issues are not yet fully answerable, but we are clearly able to establish the existence of a sequence-specific localization of mRNA and its functional role in control of timing of translation. We can at this point only speculate on the role of delay.

LOCALIZATION OF HISTONE mRNAs

The possibility that histone mRNAs might be physically sequestered was suggested by the in situ hybridization results of Venezky et al. [1981; see also Angerer and Angerer, this volume]. In situ hybridization of cloned histone probes to sections of sea urchin eggs reveals a high concentration of silver grains over the nucleus (Fig. 3). However, this does not necessarily mean that all or even a majority of histone mRNAs are present in the nucleus.

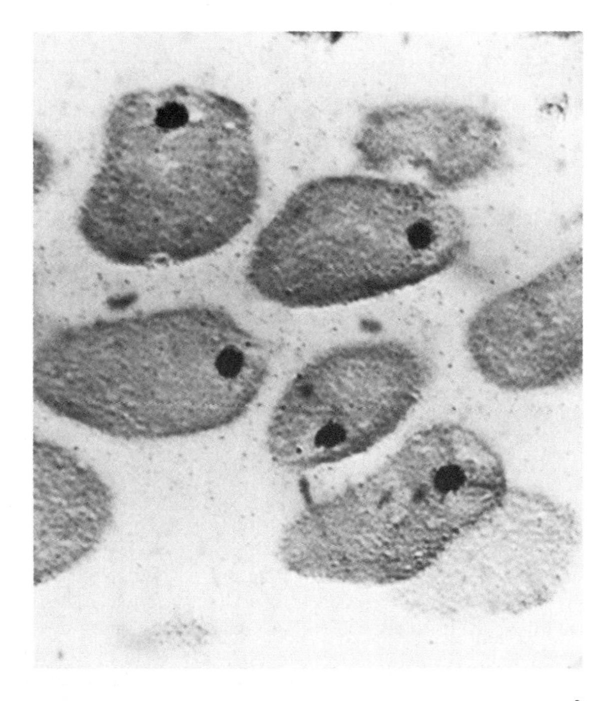

Fig. 3. In situ hybridization with a cloned α-subtype histone gene probe (^{3}H-labeled single-stranded RNA transcript copy of the entire five-gene cluster) to sections of *S. purpuratus* eggs in which the cytoplasmic components had been stratified by centrifugation. Note the dense concentrations of silver grains over nuclei and the lack of any concentrations of silver grains over the cytoplasm of the nuclear pole.

The nucleus occupies only a small volume of the egg, and when volume was taken into account, Venezky et al. [1981] calculated that although the density of grains was low in the cytoplasm as much as 88% of the α-subtype histone sequences were potentially cytoplasmic.

We [Showman et al., 1982] tested for a nuclear association for these mRNAs by physically splitting eggs into nucleate and enucleate fragments. When sea urchin eggs are subjected to a centrifugal force of 10,000–20,000g in a sucrose-seawater gradient their contents first stratify and then the eggs split, as shown in Figures 4 and 5. Nucleated fragments are separated readily from enucleated fragments in the gradients, and both types of fragments can be fertilized and will produce normal, albeit diminuitive embryos. Nucleated halves can be split again by centrifugation to produce nucleate and enucleate quarters. All of these fragments can be probed to identify the RNA species they contain. The results of some of these experiments are shown in Table I.

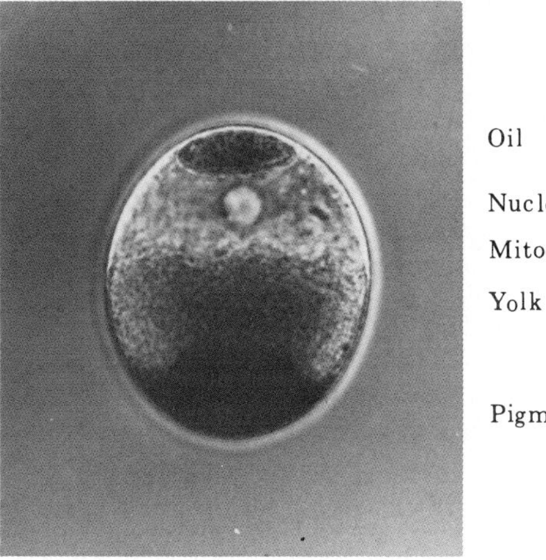

Fig. 4. Centrifugally stratified egg of *Arbacia punctulata.* The highest-density components, the pigment granules, are at the bottom, with progressively less dense components stratified above them. The nucleus is visible in a zone of clear cytoplasm.

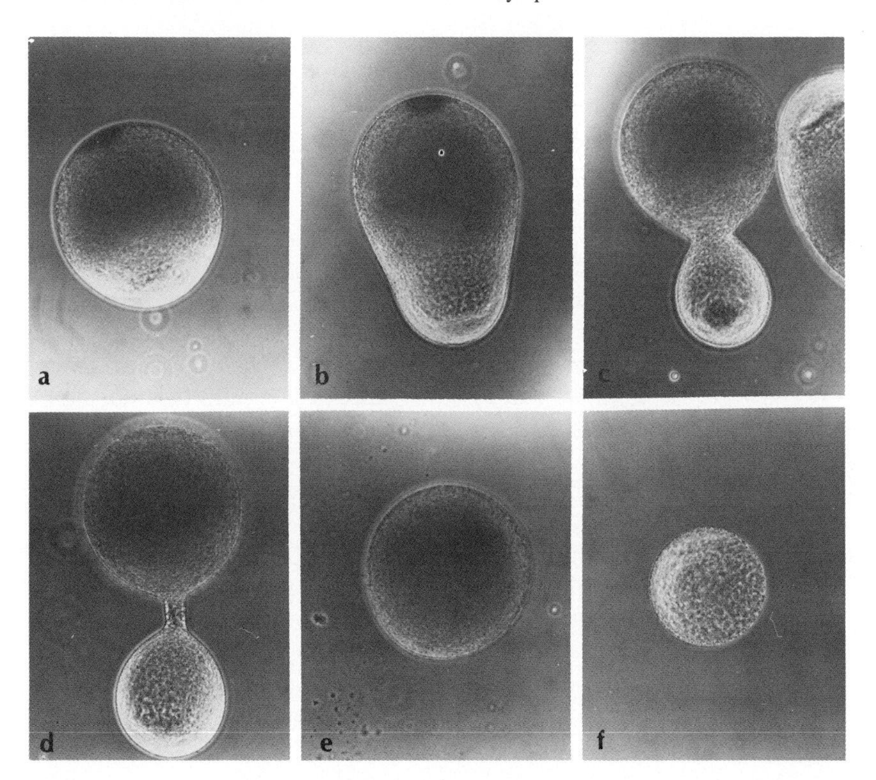

Fig. 5. Course of centrifugal splitting of the egg of *S. droebachiensis,* panels a–d. Panel e shows a nucleate fragment and panel f an enucleate fragment.

TABLE I. Distribution of RNAs in Centrifugally Split Egg Fragments[a]

	Whole egg	Anucleate half	Nucleate half	Anucleate quarter	Nucleate quarter
Relative volume	1.0	0.39	0.61	0.21	0.40
Total RNA	1.0	0.33	0.67	0.38	0.29
Poly(A)$^+$	1.0	0.31	0.69	0.39	0.30
Actin mRNA	1.0	0.33	0.67	(0.29)	(0.38)
α-tubulin mRNA	1.0	0.30	0.70	0.46	0.24
Histone mRNAs	1.0	0.02	0.98	<0.01	0.98
Mitochondrial RNA	1.0	0.77	0.23	0.23	0.005

[a]Egg and fragment volumes were calculated by using optimal measurements of the diameters of cells. Total RNA amounts were determined by absorbance at 260 nm. Poly(A)$^+$RNA values were obtained by hybridization with ^{3}H-polyuridylate. Histone mRNA probes used were H1, H2B, and H3 genes. Specific RNA measurements were done using ^{32}P nick-translated sequence-specific probes to measure the amount of each RNA present. All values are expressed as fraction of total per egg and are the average of two or more separate experiments. The values in parentheses were determined from a single experiment. Reproduced from Showman et al. [1982].

All numbers in the table are set relative to the corresponding value for the whole egg. Total RNA (representing primarily ribosomes) and poly(A)$^+$RNA (representing the bulk of maternal mRNAs in cytoplasmic mRNPs) distribute in approximate proportion to relative volumes. This is as expected since particles of the size of ribosomes and mRNPs would not be displaced by the centrifugal forces used. The somewhat lower than expected concentrations of these components in the nucleated quarters is apparently a result of exclusion of cytoplasm from the fragment by the large oil droplet that accumulates at the nuclear pole of the stratified egg. Two prevalent maternal mRNAs, those for actin and α-tubulin, confirm the behavior expected of mRNAs present as free cytoplasmic mRNPs. That RNAs attached to organelles are displaceable is indicated by the displacement of mitochondrial rRNA to the "heavy" enucleate half egg, in accord with the displacement of mitochondria.

The α-subtype maternal histone mRNAs behave totally unlike other maternal mRNAs in these experiments. These mRNAs are found exclusively in nucleated fragments. Such distribution implies either a nuclear localization or the presence of organelles that contain histone mRNAs and are displaced in a similar manner to nuclei by centrifugation. A third possibility is that histone mRNAs are associated with a particular region of the cortex of the egg. The animal-vegetal polarity of the sea urchin embryo is already established in the unfertilized egg [Horstadius, 1937; Schroeder, 1980]. However, reportedly both the location of the nucleus in the normal egg [Wilson, 1937] and the centrifugal/centripetal axis of centrifuged eggs [Morgan, 1927] seem to be random relative to the rigid cortex; it seems unlikely that the nucleus of centrifuged eggs would be consistently displaced to a hypothetical histone mRNA-bearing region of the cortex. The second and third possibilities have been rendered untenable by two other observations.

We have split eggs by an alternate method to centrifugation. Eggs contain histone mRNAs at a sufficient concentration that they can be detected in RNA extracted from as few as ten eggs. Thus, eggs were manually cut with a glass knife to produce nucleate and enucleate halves. Half eggs produced in this way differ significantly from centrifugally split eggs in that their contents have not been systematically reorganized by the splitting process. Since cutting is at random, a regular association of nuclei with any particular region of cortex is eliminated. Nucleate and enucleate fragments should have an equal probability of containing the hypothetical cortical region. In fact, only nucleated hand-split half eggs contain α-subtype histone mRNAs.

The possibility that these mRNAs are contained in organelles displaced into the nucleate half of the egg by centrifugation is also eliminated by this experiment, since no cytoplasmic stratification occurs in hand-split eggs. This is shown by the localization of mitochondrial rRNA in half eggs. Whereas in centrifugally split eggs most mitochondrial rRNA is displaced to the enucleate half, mitochondrial rRNA is equally distributed in both halves of hand-split eggs.

It would also be expected that if histone mRNAs were contained in displaceable cytoplasmic organelles, they should be observable by in situ hybridization to sections of centrifugally stratified eggs with a histone sequence probe. The results of such an experiment are presented in Figure 4, which shows no concentration of grains over the nuclear pole cytoplasm, but does show a strong hybridization to the nucleus. Since the isotope used to label the probe was ^3H, and since the diameter of the nucleus is 10 μm, a perinuclear association of histone mRNA with the nucleus should produce a ring of silver grains around the perimeter of the nucleus. The uniform density of grains over the section of the nucleus, combined with the other data discussed here, makes it most probable that all of the maternal α-subtype histone mRNAs of the sea urchin egg are located in the female pronucleus, and that the time of release of these mRNAs regulates the timing of their translation in the embryo.

It is of particular significance to note that the maternal mRNAs for cs-subtype histones are not sequestered into any displaceable structure, but are instead present in the cytoplasm like the nonhistone mRNAs. We have no cloned probe for the cs-subtype sequences, but we are able to detect them by gel electrophoresis of the histones synthesized by fertilized half eggs. The cs-subtype histones are synthesized by both nucleate and enucleate half eggs (Anstrom et al., unpublished data).

RELEASE OF HISTONE mRNAs FROM THE NUCLEUS

Recruitment of maternal α-subtype histone mRNAs into polysomes does not begin until about 30 minutes after nuclear breakdown preceding the first mitotic cleavage of the zygote. The extent of this process is illustrated by the rapid and nearly complete movement of histone mRNAs into polysomes within approximately one to 1.5 hours following the start of recruitment (Fig. 2). This observation correlates well with that of Venezky et al. [1981], who noted the virtual absence of nuclear silver grains in in situ examination of nuclei in two-cell embryos. The correlation between timing of recruitment and nuclear membrane breakdown is suggestive of a mechanistic connection.

We [Showman et al. 1982] have tested the requirement for nuclear membrane breakdown in making histone mRNAs available for translation by inhibiting nuclear membrane breakdown with dimethylaminopurine. A concentration of 0.4 mM dimethylaminopurine inhibits both nuclear membrane breakdown and cleavage. Inhibition is reversible since embryos treated with this compound will, upon removal of the inhibitor, resume cleavage and

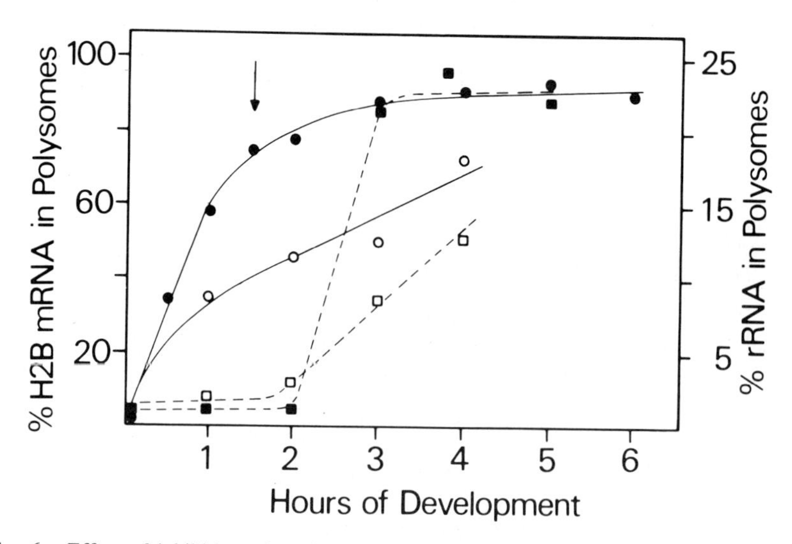

Fig. 6. Effect of inhibition of nuclear membrane breakdown by dimethylaminopurine on timing of recruitment of maternal mRNAs into polysomes. Untreated control embryos are indicated by solid symbols, treated by open symbols. Ribosomal RNA is indicated by circles, H2B histone mRNA by squares. The normal time of nuclear membrane breakdown is shown by the arrow. Reproduced from Showman et al. [1982].

ultimately produce ostensibly normal plutei. The effect of dimethylaminopurine on recruitment of α-subtype histone mRNAs is shown in Figure 6. Despite the complete inhibition of nuclear membrane breakdown, which normally occurs at 90 minutes (arrow in Fig. 6), histone mRNAs begin to be recruited at the normal time. Dimethylaminopurine does somewhat depress the rate at which histone mRNAs move into polysomes once the process has begun. However, this probably results from a nonspecific effect on maternal mRNA recruitment in general since the rate of movement of ribosomes into polysomes is also depressed. The release of maternal histone mRNAs from the nucleus may occur by a process akin to the normal transport of mRNAs through the nuclear membrane. The control of timing of release of α-subtype histone mRNAs from the nucleus has not yet been resolved: However, its connection to cytoplasmic changes induced by fertilization is suggested by preliminary experiments in which artificial activation induces an elevation of histone synthesis (unpublished data).

The specificity of mRNA release from the egg nucleus is a highly significant point. Ruderman and Schmidt [1981] have demonstrated that unfertilized sea urchin eggs synthesize histone mRNAs and that these mRNAs appear in the polysomes of unfertilized eggs. Herlands et al. [1982] have observed that although unfertilized eggs contain both cs- and α-subtype histone mRNAs, only those for cs-subtypes are translated in eggs. Thus, a consideration of these observations in conjunction with those on the localization of α-subtype histone mRNAs suggests that the nuclei of unfertilized sea urchin eggs may have the ability to selectively retain α-subtype histone mRNAs while simultaneously releasing those for cs-subtypes to the cytoplasm.

ISOLATION OF NUCLEI AND HISTONE mRNAs

The localization of histone mRNAs in the nucleus by methods such as centrifugal splitting and in situ hybridization is at odds with previously published data [e.g., Gross et al., 1973; Skoultchi and Gross, 1973] which consistently showed the bulk of the maternal histone mRNAs of the egg could be isolated in cytoplasmic fractions. Figure 1 presents data of this type in which nonpolysomal histone mRNAs appear to be present in cytoplasmic 25-30S mRNPs. We [Showman et al., 1982] attempted to isolate nuclei under conditions that would preserve nuclear structure and maintain intranuclear histone mRNAs. Eggs were accordingly disrupted by forcing them through a 10-μm Nitex mesh in an isotonic buffer. The disrupted eggs were fractionated by a series of centrifugation steps at increasing g forces. Pellets were collected at each centrifugation step. Nuclei were counted and RNA was extracted for determination of histone mRNA content. The results are shown in Figure 7. The quantitative recovery of nuclei in pellets was over 90%.

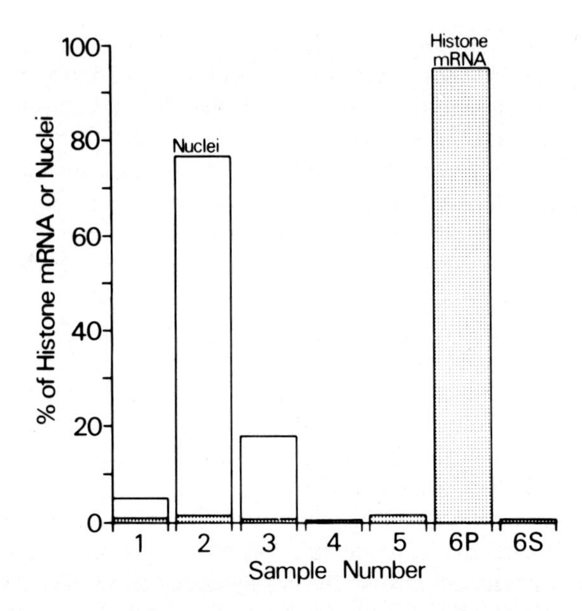

Fig. 7. Distribution of nuclei and maternal histone mRNAs in fractions of disrupted eggs. *S. purpuratus* eggs were lysed in an isotonic buffer by being forced through 10 μm Nitex cloth. The homogenate was centrifuged as indicated in a Beckman JA-17 rotor and the pellets and final supernatant examined visually for nuclei. Fractions were phenol extracted, the RNA separated on 0.8% agarose gels, blotted to nitrocellulose, and hybridized with a ^{32}P nick-translated probe specific for histone H2B. Data are expressed as percentage of total nuclei or histone mRNA in all fractions. Nuclei, open bars; histone mRNA, stippled bars. 1—200g pellet; 2—1,500g pellet; 3—5,000g pellet; 4—12,000g pellet; 5—20,000g pellet; 6P—100,000g pellet; 6S—100,000g supernatant.

These nuclei appeared round and intact by light microscopy, and they were a uniform 10 μm in diameter. Most (78%) of the nuclei were recovered in the 200g pellet, but only a negligible amount of histone mRNA was present in the pelleted fractions which contained the nuclei. Ninety-five percent of the α-subtype histone mRNAs was found in the 100,000g pelleted fraction, as was the maternal mRNA for α-tubulin, which is characteristically found in cytoplasmic mRNPs and exhibits a very different distribution from histone mRNAs in centrifugally split eggs (Table I). It would appear that although α-subtype histone mRNAs are stably sequestered into the nucleus in vivo, this association becomes extremely labile once the structure of the cell is disrupted, even in the presence of buffers that preserve ostensibly normal nuclei.

The release of sequestered α-subtype histone mRNAs in embryos occurs following nuclear breakdown associated with the first cleavage division. Release of these mRNAs may well be a simple spilling from the nucleus as it breaks down. However, the rapid release of histone mRNAs from isolated nuclei is in accord with the in vivo observations made with dimethylaminopurine. The ability of apparently intact nuclei to release α-subtype histone mRNAs may well represent an uncoupling of the mechanism(s) responsible for the normal retention of mRNAs by the egg nucleus. We are exploring other approaches for isolating nuclei that retain their histone mRNA pools.

HISTONE SWITCHING AND CHROMATIN REMODELING

Sea urchin embryos exhibit a series of switches in histone synthesis during early development. The sequence of variants expressed represents primarily the products of distinct genes for cs-, α-, and late subtypes of the major histone species [see discussion of Childs et al., 1979]. On the other hand, the condensed chromatin of sea urchin sperm nuclei contain histone variants distinct from those of the egg or embryo [Poccia et al., 1981]. By about 15 minutes after entry into the egg cytoplasm the sperm nucleus has largely decondensed, followed shortly thereafter by fusion of the pronuclei. DNA replication follows starting at about 30 minutes, and chromosome condensation in preparation for the first cleavage begins at about 90 minutes. Poccia et al. [1981] monitored the course of substitution of sperm histones by cs-subtype histones in eggs fertilized under conditions which resulted in polyspermy (approximately three to 20 male pronuclei per egg). This procedure facilitates the isolation of chromatin and renders the composition of the female pronucleus untroublesome. In such zygotes the male pronuclei undergo essentially the same course of morphological and biochemical transformations as in normally fertilized eggs (Fig. 8). The store of cs-subtype histones is large enough in the egg to accommodate the chromatin exchange of the supernumary pronuclei.

The earliest change observed by Poccia et al. [1981] was the extremely rapid ($<$ five minutes) exchange of sperm H1 by csH1 which occurred before decondensation. Savic et al. [1981] have observed that there is a rapid decline in intranucleosomal repeat length from the 245 nucleotide pair spacing of sperm chromatin to the 195–205 nucleotide pair spacing of early embryo chromatin within 90 minutes following fertilization (Fig. 8).

Changes in intranucleosomal spacing in embryo chromatin concomitant with the entry of late histone variants has also been documented [Arceci and Gross, 1980; Shaw et al., 1981]. However, unlike the chromatin modifications in the male pronucleus, these changes involve no displacement of earlier variants from chromatin—only addition of new variants with apparently different properties as DNA is replicated.

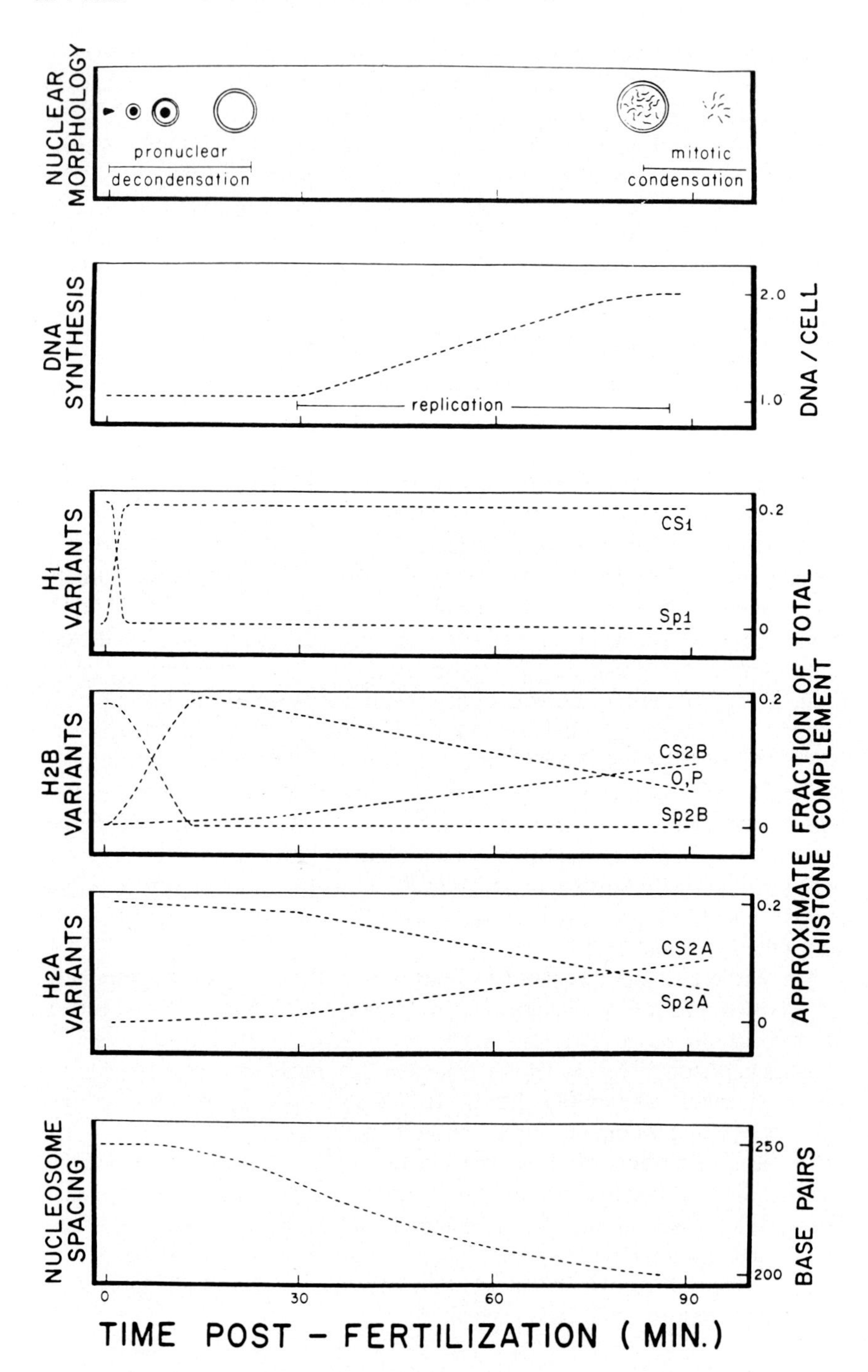
NUCLEAR MORPHOLOGY
pronuclear
decondensation
mitotic
condensation
DNA SYNTHESIS
replication
2.0
1.0
DNA / CELL
H1 VARIANTS
CS1
Sp1
0.2
0
H2B VARIANTS
CS2B
O,P
Sp2B
0.2
0
H2A VARIANTS
CS2A
Sp2A
0.2
0
APPROXIMATE FRACTION OF TOTAL HISTONE COMPLEMENT
NUCLEOSOME SPACING
250
200
BASE PAIRS
0
30
60
90
TIME POST - FERTILIZATION (MIN.)

Fig. 8.

POSSIBLE FUNCTIONS OF THE DELAY IN α-SUBTYPE TRANSLATION

The sequestration of α-subtype maternal histone mRNAs into the nucleus and the subsequent delay in their release and translation in the zygote can only be characterized as baroque. The evolution of this (thus far) unique mechanism implies a considerable functional constraint on the timing of synthesis of α-subtype proteins. The storage of roughly 10^6 copies of each α-subtype maternal mRNA and the extensive further production of α-subtype mRNAs by the embryo, beginning at about six hours following fertilization [Maxson and Wilt, 1982; Mauron et al., 1982], is consistent with a functional need in preblastula embryos for chromatin constructed using α-subtype histones. The nature of the requirement for stored α-subtype mRNAs is unclear since fertilized enucleate half eggs develop to produce normal, albeit small, plutei.

The replacement of sperm histones by cs-variants appears (although no direct experiments exist to support this idea) to be required for decondensation of the male pronucleus, its fusion with the female pronucleus, and entry of paternal chromosomes into the mitotic cycle. The question is then, why should cs-variants exist at all: Why could not sperm histones simply be replaced by α-variants? No experimental answer is currently available, but the lengths to which the egg goes to prevent premature synthesis of α-subtypes suggests that the presence of these histones may interfere with the correct remodeling of male pronuclear chromatin. The cs-subtypes would then serve in the transition to a chromatin structure compatible with entering mitosis. Presumably α-subtype histones play a subsequent role. We are testing the hypotheses raised in this chapter.

ACKNOWLEDGMENTS

The experiments from my laboratory described in this chapter were performed in collaboration with John Anstrom, Deborah Hursh, Richard Showman, and Dan Wells. Important technical assistance was provided by Carolyn Huffman. I also thank William Klein for his advice and collaboration in some aspects of this work, Robert and Lynne Angerer for open discussions and for their collaboration in in situ analysis of stratified eggs, and Richard Showman for his critical reading of the manuscript. Our work is supported by grant HD-06902 from the National Institutes of Health.

Fig. 8. Diagrammatic representation of pattern of nuclear events during the first mitotic cycle of *S. purpuratus* eggs of low polyspermy. Reproduced from Savic et al. [1982]. Figure courtesy of D. Poccia.

REFERENCES

Angerer RC, Angerer LM (1983): RNA localization in sea urchin embryos. In Jeffery WR, Raff RA (eds): "Time, Space, and Pattern in Embryonic Development." New York: Alan R. Liss, Inc. pp 101–129.

Arceci RJ, Gross PR (1980): Histone variants and chromatin structure during sea urchin development. Dev Biol 80:186–209.

Ballantine JEM, Woodland HR, Sturgess EA (1979): Changes in protein synthesis during the development of *Xenopus laevis*. J Embryol Exp Morphol 51:137–153.

Brandhorst BP (1976): Two dimensional gel patterns of protein synthesis before and after fertilization of sea urchin eggs. Dev Biol 52:310–317.

Brandhorst BP, Newrock KM (1981): Post-transcriptional regulation of protein synthesis in *Ilyanassa* embryos and isolated polar lobes. Dev Biol 83:250–254.

Brandhorst BP, Verma DPS, Fromson D (1979): Polyadenylated and nonpolyadenylated messenger RNA fractions from sea urchin embryos code for the same abundant proteins. Dev Biol 71:128–141.

Bruskin AM, Tyner AL, Wells DE, Showman RM, Klein WH (1981): Accumulation in embryogenesis of five mRNAs enriched in the ectoderm of the sea urchin pluteus. Dev Biol 87:308–318.

Childs G, Maxson R, Kedes LH (1979): Histone gene expression during sea urchin embryogenesis: Isolation and characterization of early and late messenger RNAs of *Strongylocentrotus purpuratus* by gene-specific hybridization and template activity. Dev Biol 73:153–173.

Conklin EG (1915): "Heredity and Environment in the Development of Men." Princeton University Press. Quoted by Tennent (1922).

Costantini FD, Britten RJ, Davidson EH (1980): Message sequences and short repetitive sequences are interspersed in sea urchin egg poly(A)$^+$RNAs. Nature 287:111–117.

Davidson EH (1976): "Gene Activity in Early Development." New York: Academic Press.

Duncan R, Humphreys T (1981): Most sea urchin maternal mRNA sequences in every abundance class appear in both polyadenylated and nonpolyadenylated molecules. Dev Biol 88:201–210.

Ernst SG, Hough-Evans BR, Britten RJ, Davidson EH (1980): Limited complexity of the RNA in micromeres of 16-cell sea urchin embryos. Dev Biol 79:119–127.

Flynn JM, Woodland HR (1980): The synthesis of histone H1 during early amphibian development. Dev Biol 75:222–230.

Galau GA, Klein WH, Davis MM, Wold BJ, Britten RJ, Davidson EH (1976): Structural gene sets active in embryos and adult tissues of the sea urchin. Cell 7:487–506.

Gross KW, Jacobs-Lorena M, Baglioni C, Gross PR (1973): Cell-free translation of maternal messenger RNA from sea urchin eggs. Proc Natl Acad Sci USA 70:2614–2618.

Harkey MA, Whiteley AH (1982): Cell-specific regulation of protein synthesis in the sea urchin gastrula: A two-dimensional electrophoretic study. Dev Biol 93:453–462.

Herlands L, Allfrey VG, Poccia D (1982): Translational regulation of histone synthesis in the sea urchin *Strongylocentrotus purpuratus*. J Cell Biol 94:219–223.

Horstadius S (1937): Investigations as to the localization of the micromere-, the skeleton-, and the endoderm-forming material in the unfertilized egg of *Arbacia punctulata*. Biol Bull 73:295–316.

Hough-Evans BR, Wold BJ, Ernst SG, Britten RJ, Davidson EH (1977): Appearance and persistence of maternal RNA sequences in sea urchin development. Dev Biol 60:258–277.

Hynes RO, Raff RA, Gross PR (1972): Properties of the three cell types in sixteen-cell sea urchin embryos. Aggregation and microtubule protein synthesis. Dev Biol 27:150–164.

Jeffery WR (1983): Messenger RNA localization and cytoskeletal domains in ascidian embryos. In Jeffrey WR, Raff RA (eds): "Time, Space, and Pattern in Embryonic Development." New York: Alan R. Liss, Inc., pp 241–259.
Kemphues KJ, Kaufman TC, Raff RA, Raff EC (1982): The testis-specific β-tubulin subunit in *Drosophila melanogaster* has multiple functions in spermatogenesis. Cell 31:655–670.
Lifton RP, Kedes LH (1976): Size and sequence homology of masked maternal and embryonic histone messenger RNAs. Dev Biol 48:47–55.
Martindale MQ, Brandhorst BP (1982): The role of the germinal vesicle in the 1-methyladenine-induced changes in protein synthesis in *Asterias* oocytes. Biol Bull 163:374.
Mauron A, Kedes LH, Hough-Evans B, Davidson EH (1982): Accumulation of individual histone mRNAs during embryogenesis of the sea urchin *Strongylocentrotus purpuratus.* Dev Biol 94:425–434.
Maxson Jr RE, Wilt FH (1982): Accumulation of the early histone messenger RNAs during the development of *S. purpuratus*. Dev Biol 94:435–440.
Morgan TH (1927): "Experimental Embryology." New York: Columbia University Press.
Poccia D, Salik J, Krystal G (1981): Transitions in histone variants of the male pronucleus following fertilization and evidence for a maternal store of cleavage-stage histones in the sea urchin egg. Dev Biol 82:287–296.
Raff RA (1977): The molecular determination of morphogenesis. Bioscience 27:394–401.
Raff RA, Showman RM (1983): Maternal messenger RNA: Quantitative, qualitative, and spatial control of its expression in embryos. In Metz CB, Monroy A (eds): "The Biology of Fertilization." New York: Academic Press (in press).
Raff RA, Brandis JW, Huffman CJ, Koch AL, Leister DE (1981): Protein synthesis as an early response to fertilization of the sea urchin egg: A model. Dev Biol 86:265–271.
Rodgers WA, Gross PR (1978): Inhomogenous distribution of egg RNA sequences in the early embryo. Cell 14:279–288.
Rosenthal ET, Hunt T, Ruderman JV (1980): Selective translation of mRNA controls the pattern of protein synthesis during early development of the surf clam, *Spisula solidissima*. Cell 20:487–494.
Rosenthal ET, Brandhorst BP, Ruderman JV (1982): Translationally mediated changes in patterns of protein synthesis during maturation of starfish oocytes. Dev Biol 91:200–215.
Ruderman JV, Pardue ML (1977): Cell-free translation analysis of messenger RNA in echinoderm and amphibian development. Dev Biol 60:48–68.
Ruderman JV, Schmidt MR (1981): RNA transcription and translation in sea urchin oocytes and eggs. Dev Biol 81:220–228.
Savic A, Richman P, Williamson P, Poccia D (1981): Alterations in chromatin structure during early sea urchin embryogeneis. Proc Natl Acad Sci USA 78:3706–3710.
Schroeder TE (1980): Expressions of the prefertilization polar axis in sea urchin eggs. Dev Biol 79:428–443.
Shaw BR, Cognetti G, Sholes WM, Richards RG (1981): Shift in nucleosome populations during embryogenesis: Microheterogeneity in nucleosomes during development of the sea urchin embryo. Biochemistry 20:4971–4978.
Showman RM, Wells DE, Anstrom J, Hursh DA, Raff RA (1982): Message-specific sequestration of maternal histone mRNA in the sea urchin egg. Proc Natl Acad Sci USA 79:5944–5947.
Skoultchi A, Gross PR (1973): Maternal histone messenger RNA: Detection by molecular hybridization. Proc Natl Acad Sci USA 70:2840–2844.
Tennent DH (1922): Studies on the hybridization of echinoids. Carnegie Inst Washington Publ 312:3–42.

Tufaro F, Brandhorst BP (1979): Similarity of proteins synthesized by isolated blastomeres of early sea urchin embryos. Dev Biol 72:390–397.

van Blerkom J (1979): Molecular differentiation of the rabbit ovum. III. Fertilization-autonomous polypeptide synthesis. Dev Biol 72:188–194.

Venezky DL, Angerer LM, Angerer RC (1981): Accumulation of histone repeat transcripts in the sea urchin egg pronucleus. Cell 24:385–391.

Wells DE, Showman RM, Klein WH, Raff RA (1981a): Delayed recruitment of maternal mRNA in sea urchin embryos. Nature 292:477–478.

Wells DE, Showman RM, Klein WH, Raff RA (1981b): Translational regulation in sea urchin embryos. Biol Bull 161:322.

Wilson EB (1937): "The Cell in Development and Heredity, 3rd Ed." New York: MacMillan.

Winkler MM, Steinhardt RA (1981): Activation of protein synthesis in a sea urchin cell-free system. Dev Biol 84:432–439.

Woodland HR, Ballantine JEM (1980): Paternal gene expression in developing hybrid embryos of *Xenopus laevis* and *Xenopus borealis.* J Embryol Exp Morphol 60:359–372.

Time, Space, and Pattern in Embryonic Development, pages 87–100

A Family of Genes Expressed in the Embryonic Ectoderm of Sea Urchins

William H. Klein, Clifford D. Carpenter, Angela L. Tyner, Lisa M. Spain, Elizabeth D. Eldon, and Arthur M. Bruskin

Program in Molecular, Cellular, and Developmental Biology, Department of Biology, Indiana University, Bloomington, Indiana 47405

Over the past decade the pattern and timing of gene expression during sea urchin embryogenesis have been the subject of many investigations [see Davidson et al., 1982, for review]. These studies have provided much information regarding the early molecular events of embryogenesis. Sea urchin eggs have large stores of maternal mRNAs which become available for translation following fertilization. At the time of hatching 20 hours later, embryonic transcription also contributes significantly to the total mRNA pool. Earlier work has shown that mRNA populations in sea urchin embryos can be divided into different abundance classes. The population of rare mRNAs changes substantially during the development of the embryo, the general pattern being a gradual decline in sequence complexity [Galau et al., 1976; Hough-Evans et al., 1977]. Populations of more prevalent mRNAs also change during sea urchin embryogenesis. The use of cloned DNA corresponding to a large number of unidentified mRNAs has confirmed that the prevalence of both rare and abundant mRNA sequences can change substantially during development [Lasky et al., 1980; Bruskin et al., 1981; Xin et al., 1982; Flytzanis et al., 1982].

In a recent comprehensive report, stage-specific patterns of protein synthesis were analyzed by two-dimensional gel electrophoresis [Bedard and Brandhorst, 1982; Brandhorst et al., this volume]. These experiments showed qualitative and quantitative changes in the synthesis of many individual proteins particularly during the interval from hatching blastula to midgastrula.

Although the developmental patterns of mRNA and protein populations are now known, nothing is known about the function of the proteins being synthesized nor what regulates their levels of expression in the embryo.

An inherent weakness in dealing with populations of mRNAs or proteins is the lack of detail which can be provided by investigating the expression of a single gene (or group of closely related genes). Therefore our strategy has been to investigate individual mRNAs that have potentially important functions in the developing embryo. Because the determination of tissue types takes place very early in sea urchin embryos, mRNAs that accumulate in localized regions of the embryo are strong candidates for playing important roles in development. Until now few specific sea urchin mRNAs have been investigated that are restricted to particular blastomeres or cell types. The recent development of methods for large-scale isolation of each of the primary germ layers of sea urchin embryos has provided a new approach to studies on differential gene expression [McClay and Chambers, 1978; Harkey and Whiteley, 1980]. Using this approach we have discovered a small family of mRNAs localized to dorsal ectodermal cells in the developing embryo [Bruskin et al., 1982; Angerer and Angerer, this volume]. These mRNAs (denoted Spec mRNAs for *Strongylocentrotus purpuratus* ectodermal mRNAs) code for a group of about ten low molecular weight acidic proteins. The mRNAs and their proteins are present at low or undetectable levels in the egg and early cleaving embryo and subsequently accumulate to be among the major gene products of the late-stage embryo. The function of the proteins is unknown, but their properties suggest they may have important developmental significance. In this chapter we report on some of the features of the Spec mRNAs and proteins and the genes which encode them.

SPEC1 AND SPEC2 mRNAs ARE TWO CLOSELY RELATED mRNAs PRESENT IN THE PLUTEUS ECTODERM

S. purpuratus pluteus larvae were fractionated into ectoderm and endoderm/mesoderm using the disaggregation procedures of McClay and Chambers (Fig. 1). Poly(A)$^+$RNA was isolated from total cellular RNA of either the ectoderm or endoderm/mesoderm fraction and ^{32}P cDNA was made from each. The cDNAs were used to probe a cDNA clone library. Two of the clones, pSpec1 and pSpec2, were found to hybridize to mRNAs greatly enriched in the pluteus ectoderm (see Fig. 7, lanes 1–4; Angerer and Angerer, this volume). The sequence of pSpec1 and pSpec2 show that the two clones share about 80% homology (see Fig. 8). This homology is clearly evident when gel blots of embryonic RNA are hybridized with pSpec1 or pSpec2 (Fig. 2). pSpec1 hybridizes strongly to a 1.5-kb RNA and weakly to a 2.2-kb and 3.2-kb RNA. pSpec2 hybridizes strongly to a 2.2-kb RNA and weakly

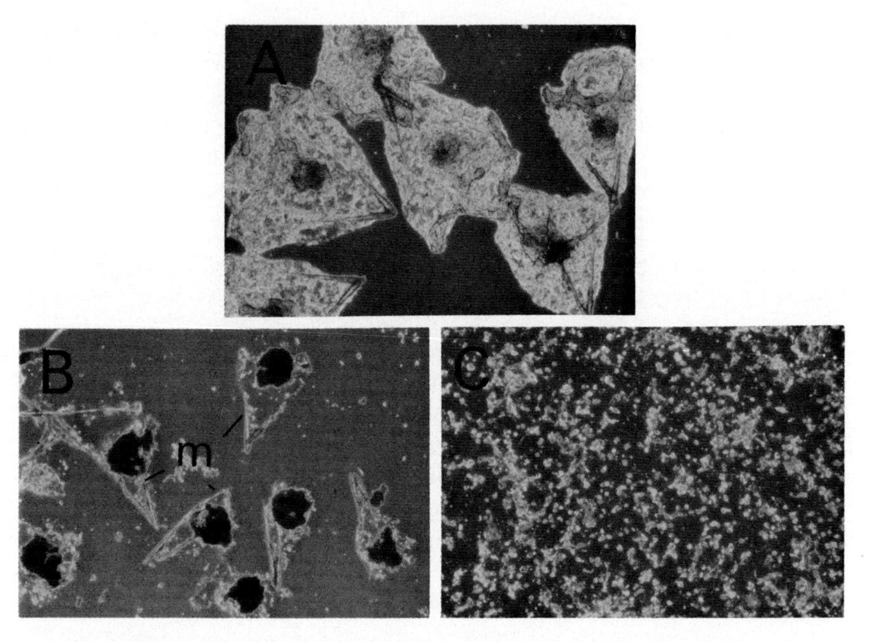

Fig. 1. Fractionation of pluteus larvae into ectoderm and endoderm/mesoderm. Embryos were disaggregated and fractionated through a 28-μm Nitex mesh. The resulting fractions were stained with leukocyte alkaline phosphatase (Sigma Chemical Co., histozyme kit 85L-2R) and photographed. A. Unfractionated plutei. B. Endoderm/mesoderm, lines point to mesenchyme cells (m). C. Ectoderm.

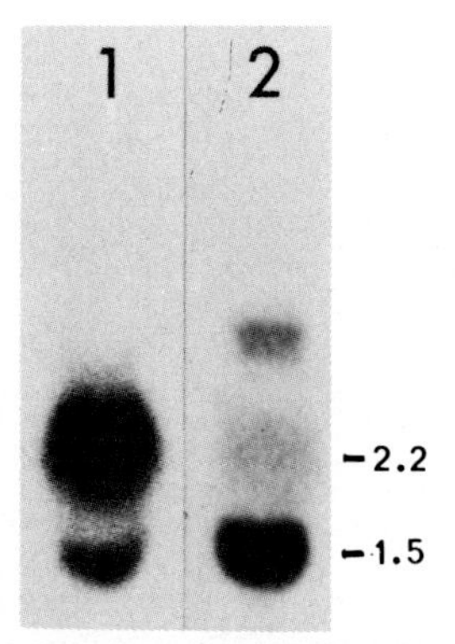

Fig. 2. Hybridization of probes pSpec1 and pSpec2 with gastrula RNA. Ten micrograms per lane of gastrula total cellular RNA was electrophoresed on a 1% agarose gel and the gel blotted to a nitrocellulose filter. The filter strips were hybridized with 1–2 × 10^7 cpm of [^{32}P]-labeled nick-translated plasmid DNA isolated from clone pSpec2, lane 1; clone pSpec1, lane 2. The autoradiogram was deliberately overexposed to visualize minor bands. The sizes of the transcripts, in kilobases, are listed on the right.

to a 1.5-kb and 3.2-kb RNA. pSpec1 and pSpec2 have been shown to be complementary to the 3′ terminal sequences of their respective RNAs (unpublished results). Inspection of the sequence of pSpec1 and pSpec2 reveals no open reading frames of any significant length extending toward the 5′ end of the mRNA. Thus pSpec1 is complementary to 0.5 kb of the 3′ untranslated region of a 1.5-kb ectoderm-specific mRNA and pSpec2 is complementary to 0.5 kb of the 3′ untranslated region of a 2.2-kb mRNA.

SPEC1 AND SPEC2 mRNAs ACCUMULATE IN MASS OVER 100-FOLD DURING EMBRYOGENESIS

Because Spec1 and Spec2 mRNAs are found highly localized to dorsal ectoderm cells in the pluteus (see Angerer and Angerer, this volume), we were interested in determining if they were present in developmental stages preceding the differentiation of dorsal ectoderm. The presence of the Spec1 and Spec2 mRNAs, monitored during embryogenesis by RNA gel blot analysis, is shown in Figure 3. Both mRNAs are present in low or undetectable levels in the unfertilized egg and early cleaving embryos. By 20 hours,

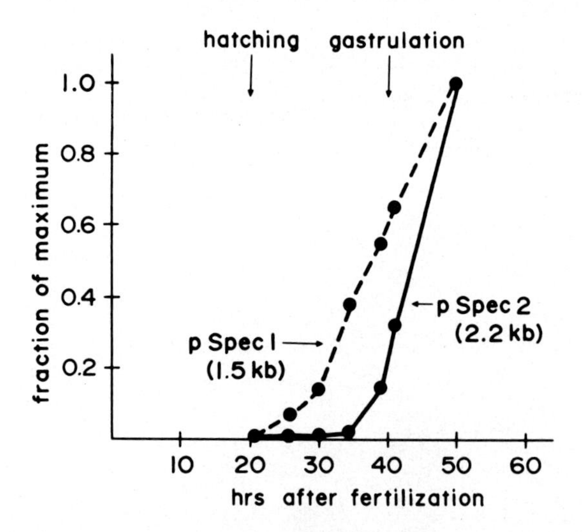

Fig. 3. Accumulation of pSpec1 and pSpec2 mRNAs during embryogenesis. Ten micrograms of total cellular RNA from the indicated stage or cell type was electrophoresed on 1% agarose formaldehyde gels and blotted to nitrocellulose filters. The filters were hybridized with 5×10^7 cpm of nick-translated pSpec1 and pSpec2 [^{32}P] DNA. Accumulations of 1.5- and 2.2-kb transcripts based on densitometric scans of RNA gel blots probed with pSpec1 or pSpec2 DNA from 4 to 50 hours after fertilization (points earlier than 20 hours were at background levels). The peaks from each scan were normalized to the most intense peak, which occurred at 50 hours.

hatching blastula stage, the Spec1 1.5-kb mRNA begins to accumulate. Its level increases over 100-fold from early cleavage to gastrula. The Spec2 2.2-kb mRNA also accumulates during embryogenesis starting about ten hours later than the Spec1 mRNA. Quantitation of the RNA gel blots with known standards suggests that the Spec1 mRNA is about 0.1% of the embryo mRNA at 50 hours and Spec2 mRNA is about one-tenth as prevalent. The dramatic rise in the levels of Spec1 and Spec2 mRNAs is localized to dorsal ectoderm and presumptive dorsal ectoderm. In this case the localization of two specific mRNAs arises from de novo synthesis in a defined cell type.

THE CELL-FREE TRANSLATION OF RNA HYBRID SELECTED BY pSPEC1 AND pSPEC2 YIELDS SEVERAL SMALL ACIDIC POLYPEPTIDES

The first step toward the elucidation of the function of these tissue-specific mRNAs is the characterization of the proteins. Spec1 and Spec2 mRNAs were hybrid selected and translated in vitro. The proteins were analyzed by two-dimensional electrophoresis, and results of a typical experiment are shown in Figure 4. RNA selected by either cloned DNA coded for approximately ten polypeptides which appeared on the acidic, low molecular weight region of the gel (pH 5.0–6.0, 14,000–17,000 daltons). Selection with either pSpec1 or pSpec2 resulted in essentially the same array of spots. Given the sequence homology between pSpec1 and pSpec2, we expected that they might select mRNAs that code for the same proteins; we were surprised that several polypeptides are among the translation products. The results suggest that pSpec1 and pSpec2 select a family of several related mRNAs.

THE ECTODERM-ENRICHED PROTEINS ARE SYNTHESIZED IN VIVO

Plutei were incubated for two hours with [^{35}S]-methionine and then separated into ectoderm and endoderm/mesoderm fractions. The newly synthesized proteins were analyzed by two-dimensional electrophoresis. Newly synthesized proteins enriched in the ectoderm fraction are pointed out by arrows in Figure 5A. The corresponding positions of these spots in the endoderm/mesoderm fraction are indicated by circles in Figure 5B. Some of the differences are very pronounced, others less obvious. Most of the very pronounced ectoderm-enriched proteins are small and acidic, having mobilities similar to the products of cell-free translation of Spec1 and Spec2 mRNAs.

The results presented in Figures 4 and 5 strongly suggest that some of the proteins whose synthesis is enriched in ectoderm in vivo are identical to the group of products of cell-free translation of hybrid-selected mRNAs corresponding to pSpec1 and pSpec2 DNAs. Coelectrophoresis experiments have

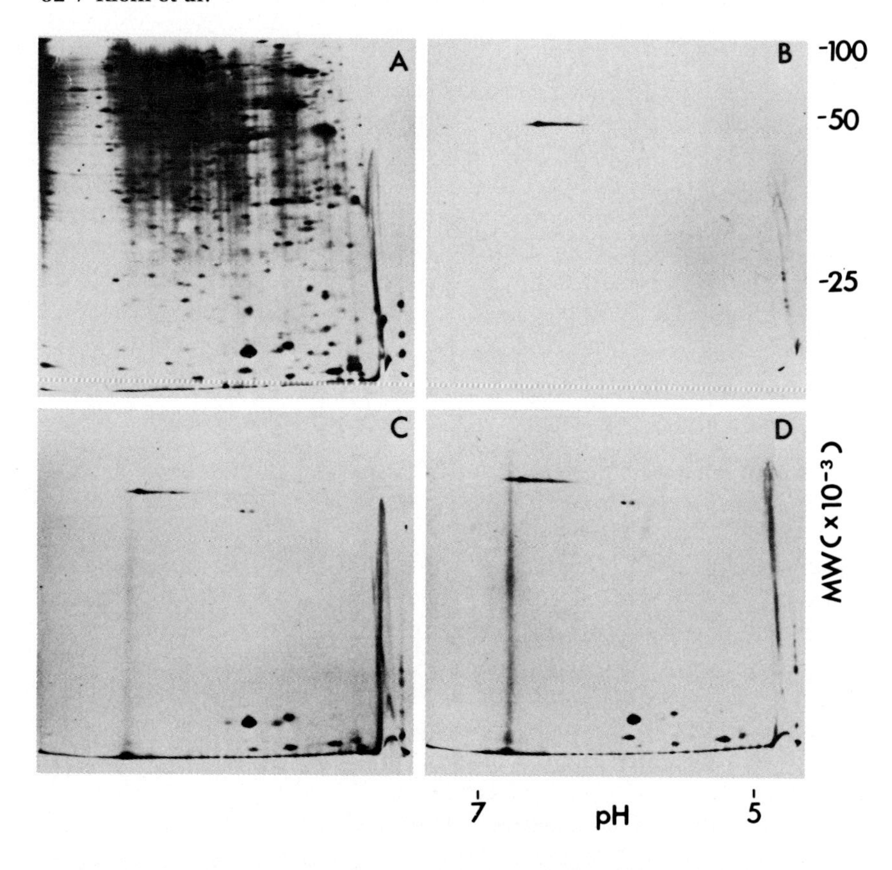

Fig. 4. Two-dimensional electrophoresis of products of translation of RNA selected by hybridization to pSpec1 and pSpec2 DNA. One hundred micrograms of total cellular RNA from gastrula-stage embryos was hybridized to 15 μg of filter-bound pSpec1 or pSpec2 DNA. The hybridized RNA was eluted and translated in a rabbit reticulocyte lysate cell-free system supplemented with [^{35}S]-methionine. The radiolabeled proteins were electrophoresed in two dimensions. The second dimension uses a 10% polyacrylamide gel. The gels were prepared for fluorography with Enhance (New England Nuclear) and exposed to X-Omat RP x-ray film for 2 weeks at −70°C. A. Translation of 10 μg of total cellular RNA from gastrula stage embryos. B. Translation with no exogenous RNA. C. Translation with gastrula RNA hybrid selected by pSpec1 DNA. D. Translation with gastrula RNA hybrid selected by pSpec2 DNA.

shown at least seven of the in vitro translation products comigrate with ectoderm-enriched proteins synthesized in vivo. From these data we conclude that pSpec1 and pSpec2 code for mRNAs which direct the synthesis of a family of similar but distinct ectoderm proteins.

In our initial attempts to purify some of the ectoderm proteins from pluteus-stage embryos, we lysed the embryos in nonionic detergents. Following centrifugation the most prominent ectoderm proteins are greatly enriched

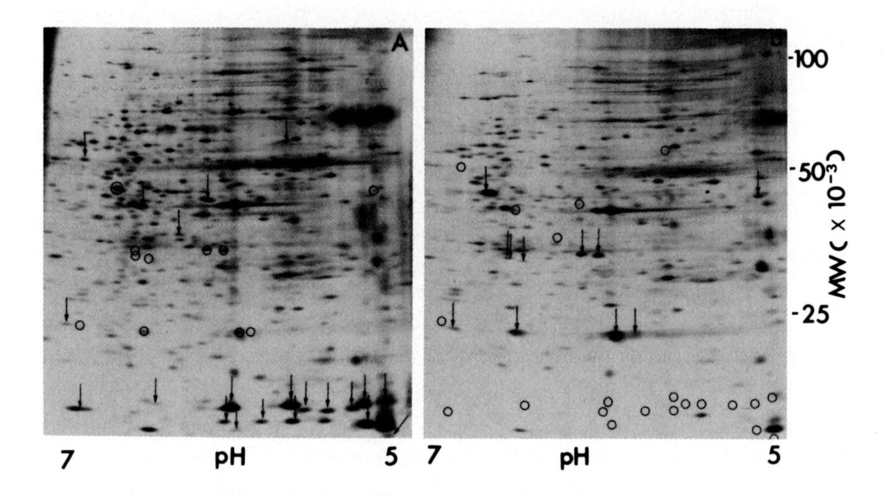

Fig. 5. Patterns of proteins synthesized in the ectoderm and endoderm/mesoderm fraction of plutei. Embryos were incubated with [^{35}S]-methionine and then separated into ectoderm and endoderm/mesoderm fractions. Proteins were separated by two-dimensional electrophoresis and detected by autoradiography. Second-dimension gels consisted of an exponential gradient of 10%–16% polyacrylamide. Arrows indicate spots which are considerably more intensely labeled for that tissue fraction, while circles indicate the corresponding spot or area of the gel for the other tissue fraction. Shown are ectoderm (A) and endoderm/mesoderm (B).

in a 5,000g pellet. This result is also obtained starting with isolated pluteus ectoderm cells. An intriguing possibility is that the proteins are either in the ectoderm nucleus or that they are part of some cytoskeletal structure.

SPEC1 AND SPEC2 mRNAs CONTAIN A REPETITIVE ELEMENT AT THEIR 3′ END

Given the relatively simple patterns we see in the Northern blots and the hybrid-selected in vitro translation experiments, we did not expect more than about ten genes in the ectoderm protein family. However, genomic Southern blots with pSpec1 or pSpec2 DNAs as probes yield very complex patterns, many bands, and a blurred background. This complexity is also seen when a Charon 4 λ sea urchin genomic DNA library is probed with either clone. A standard plaque hybridization screen of about 10,000 recombinants using nick-translated pSpec1 DNA as a probe shows that many more recombinants hybridize than are expected. Several hundred positives are seen in such a screen. Simple calculations from these data suggest that all or part of the pSpec1 and pSpec2 sequences are reiterated 2,000–3,000 times in the sea urchin genome.

The number of plaques scoring positive in such screens can be greatly reduced by increasing the wash criteria. In a typical experiment one or two positive clones per 10,000 are discernible. These experiments suggest that most of the λ clones have only weak homology with the pSpec1 probe.

To determine the relationship between the different λ-recombinants hybridizing with pSpec1, several different phages were isolated. DNA was purified from these phages and restriction maps were constructed. Figure 6 shows the restriction maps of three of these clones, λSpec1 isolated using the high-criteria wash and λrep1 and λrep2 isolated using the lower-criteria wash. Figure 6 also shows the region of homology with pSpec1 determined by Southern analysis: a 0.9-kb EcoR1-Sall fragment in λSpec1, a 0.9-kb EcoR1-Hind3 fragment in λrep1, and a 0.6-kb BamH1-EcoR1 fragment in λrep2. Besides these small regions of homology, the clones have no other homology as judged both by the restriction maps shown in Figure 6 and by cross-hybridization experiments. However, the three different λ-clones that have thus far been isolated by hybridizing with pSpec1 and washing at high criteria show extensive sequence homology (see below). Thus the λ-clones

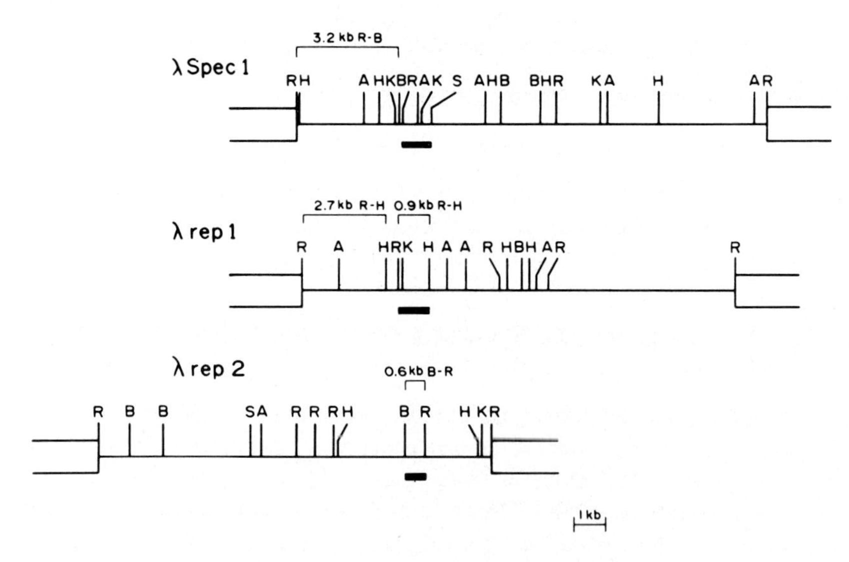

Fig. 6. Restriction maps of λSpec1, λrepl, and λrep2. Symbols for restriction sites are: R, EcoR1; H, Hind3; A, Xbal; K, Kpn1; B, BamH1; S, Sall. The EcoR1 sites bordering the inserted DNA derive from EcoR1 linkers ligated to partial HaeIII-digested sperm DNA used in construction of the library. The thick dark bars underneath each map represent the smallest restriction fragment tested which hybridized to nick-translated pSpec1 in a Southern analysis. The orientation of the gene of λSpec1 is 5′ to 3′ left to right on the figure. The restriction fragments shown in brackets have been isolated and subcloned into pBr322.

hybridizing with pSpec1 fall into two classes: those which appear at high frequency in the genome and are weakly homologous to pSpec1, and those which are far fewer in number, probably fewer than ten different genomic loci, and are closely related.

If any of these λ-clones contain genes for the 1.5-kb or 2.2-kb ectoderm mRNAs, they should hybridize with the RNAs in a Northern analysis (Fig. 7). As noted above, pSpec1 and pSpec2 hybridize to RNAs greatly enriched in the pluteus ectoderm (Fig. 7, lanes 1–4). λSpec1 hybridizes with a pattern very similar to that of the pSpec1 cDNA clone (lanes 5, 6). We have also hybridized with the 3.2-kb EcoR1-BamH1 subclone of λSpec1 (indicated in Fig. 6) and shown that this probe also hybridizes strongly with the 1.5-kb RNA. In addition, we have sequenced approximately 200 bp of λSpec1 corresponding to the pSpec1 homology and found it has the same sequence as pSpec1 (see Fig. 8). The simplest interpretation of these data is that λSpec1 contains the gene for the 1.5-kb mRNA (see below).

λrep1 gives a much different pattern than λSpec1 (Fig. 7, lanes 7, 8). λrep1 does not hybridize to any discrete RNAs but only to heterogeneous

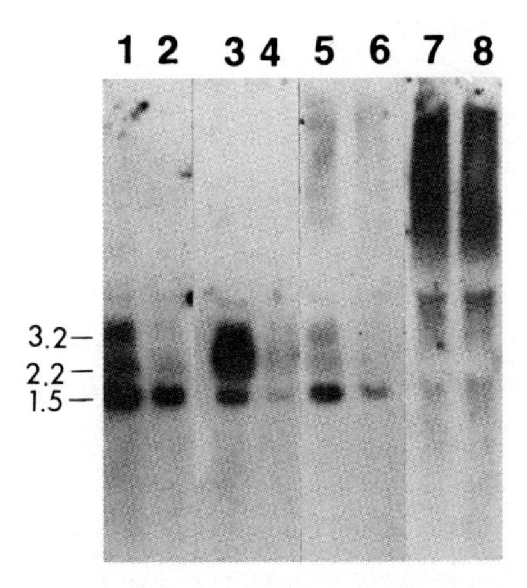

Fig. 7. RNA gel blots of pluteus ectoderm and endoderm/mesoderm RNA hybridized with pSpec1, pSpec2, λSpec1, and λrep1. Ten micrograms of total pluteus ectoderm RNA (lanes 1, 3, 5, 7) or endoderm/mesoderm RNA (lanes 2, 4, 6, 8) were electrophoresed on 1.0% agarose gels, blotted to nitrocellulose filters, and hybridized with 2×10^7 cpm of nick-translated [^{32}P] probe. Lanes 1 and 2, pSpec1; lanes 3 and 4, pSpec2; lanes 5 and 6, λSpec1; lanes 7 and 8, λrep1. The sizes of the transcripts in kilobases are indicated. The two interruptions seen in the pattern of hybridization with λrep1, lanes 7 and 8, correspond to the positions of the 28S and 18S ribosomal RNAs. The interruptions are due to the saturation of the nitrocellulose filters wih ribosomal RNA in these regions, thereby preventing normal hybridization. No such interruptions are apparent when poly(A)$^+$RNA is used for the blots rather than total RNA.

high molecular weight transcripts. We interpret this as hybridization to nuclear RNA. The hybridization of λrep2 with total RNA is identical to that of λrep1. Thus while λrep1 and λrep2 may be transcribed, they do not contain genes for the pSpec1 or pSpec2 mRNAs. These experiments suggest that λrep1 and λrep2 contain sequences that are only weakly homologous to the pSpec1 sequence.

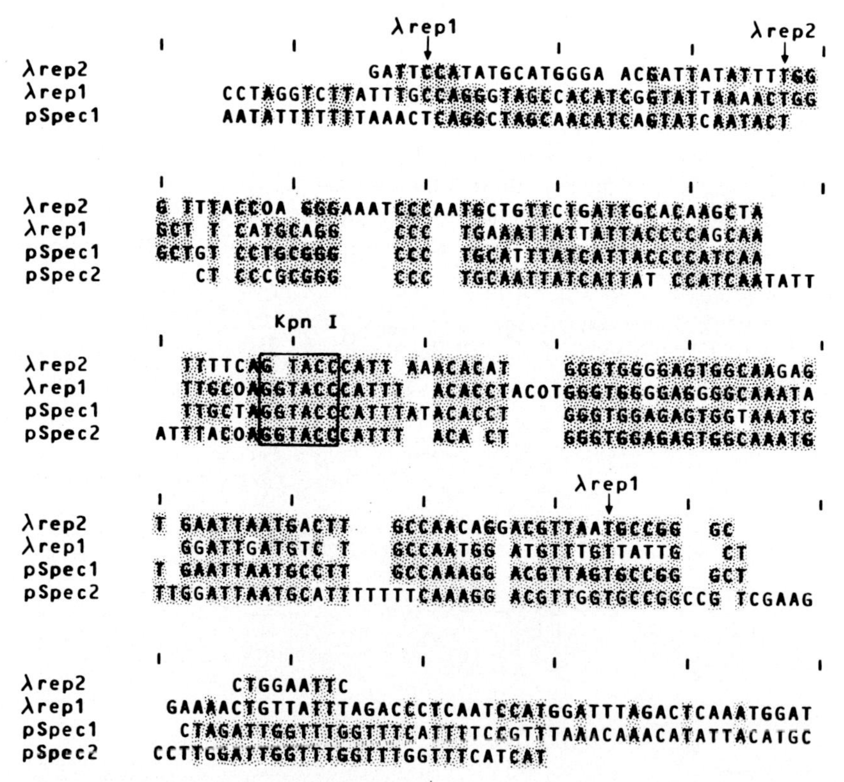

Fig. 8. Partial sequences of pSpec1, pSpec2, λrep1, and λrep2. pSpec1 has been completely sequenced (470bp) and pSpec2 has been partially sequenced; the regions homologous to λrep1 and λrep2 are shown. The subcloned 0.9-kb EcoR1-Hind3 fragment of λrep1 and the 0.6-kb BamH1-EcoR1 fragment of λrep2 were partially sequenced and their homologous regions are shown. The shaded areas are regions of homology. (O) denotes a gap in the sequencing ladder (this usually implies a methylated cytosine). The boxed region denotes a Kpn1 site found in pSpec and the 0.9-kb EcoR1-Hind3 subclone. The displayed strand is 5′ to 3′, left to right. λrep1 and pSpec1 (corresponding to λSpec1) are oriented opposite to the maps shown in Figure 6. The λrep2 sequence is oriented in the same way as the map in Figure 6. The vertical arrows mark the boundaries of homology for λrep1 and λrep2 with pSpec1. No boundary for λrep2 to the right of the Kpn1 site has been found.

The pSpec1 cDNA clone has been sequenced completely. Part of the sequence is shown in Figure 8. In order to ascertain the relationship between λrep1, λrep2, and pSpec1 and pSpec2, we have sequenced the appropriate regions of these phage DNAs as indicated in Figure 6. A summary of this sequence data is shown in Figure 8. When the sequences are compared, a short sequence homology of about 150 bp of both λrep1 and λrep2 with the pSpec1 and pSpec2 sequences can be clearly seen. This homology centers around a Kpn1 site present in pSpec1 and λrep1. We conclude from these experiments that the weak hybridization of λrep1 and λrep2 with pSpec1 and pSpec2 is due in both cases to a 150–200 bp sequence homology of about 80%.

We suggest that a repetitive sequence family, whose elements are approximately 150–200 bp in length, is present at 2,000–3,000 copies per haploid genome and that one member of this family is present in the 3′ untranslated region of the Spec1 mRNA and another is in the 3′ untranslated region of the Spec2 mRNA.

Although no direct evidence has been obtained as yet, we find no reason to believe this repetitive element has a function. Several mRNAs are known to have long 3′ untranslated tails, but the significance of these tails is obscure. We hypothesize that one repetitive element of the family discussed here is inserted into a region of the genome between the ancestral Spec amino acid coding sequence and its transcriptional termination site. Subsequent amplification of the Spec genes led to a family of genes coding for embryonic ectoderm proteins with a highly repetitive element toward the 3′ end of the genes. This simple model suggests an evolutionary origin for the repeat element in the mRNA without requiring any function for the repeat in transcription, processing, or translation.

THREE GENOMIC CLONES APPEAR TO CONTAIN SPEC1 GENES

Under the high-criteria wash conditions we were able to isolate three unique phages, designated λSpec1a, λSpec1b, and λSpec1c. Unlike the phages isolated with lower-criteria screens these phages give stong positive hybridization to the 1.5-kb pSpec1 mRNA. The phages show very close homology with each other within a 5.6-kb region as indicated in Figure 9, yet show significant differences beyond this region. The 5.6-kb homologous region contains the homology with pSpec1. The extensive homology on the 3′ side of the genes is striking, but its functional significance is unclear. Our tentative interpretation of these data is that λSpec1a, λSpec1b, and λSpec1c are genomic clones containing three members of the multigene family coding for ectoderm proteins.

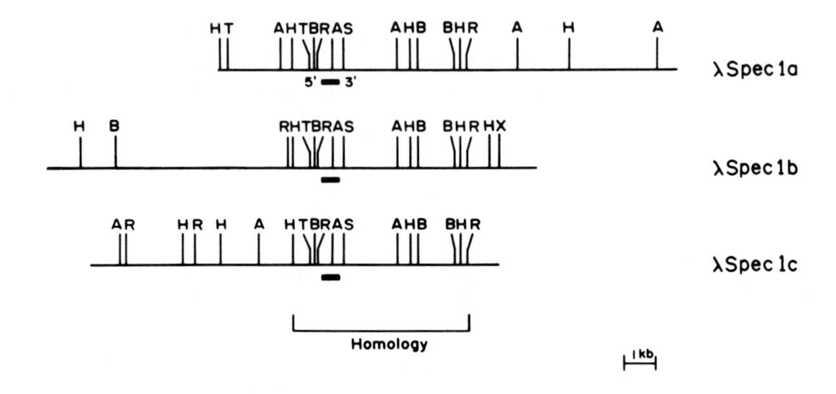

Fig. 9. Genomic sequences hybridizing with pSpec1. λSpec1a (designated λSpec1 in Fig. 6 and Fig. 10), λSpec1b, and λSpec1c were mapped with restriction enzymes by standard procedures. The bar underneath each map is the region complementary to pSpec1. The orientation of the pSpec1 mRNA is indicated under the map of λSpec1a. The bracket labeled "homology" is a 5.6-kb region that appears to be in common with all three phages. The maps diverge outside of this region. The restriction enzymes are: H, Hind 3; A, Xbal; B, BamH1; R, EcoR1; S, Sall; T, Sst1.

So far, we have investigated only one of these phages, λSpec1a, in more detail. Restriction fragments of λSpec1a mapping to the 5′ side of the pSpec1 homology hybridize strongly to the 1.5-kb Spec1 mRNA and also hybrid select the same set of ectoderm proteins. We have recently isolated a longer cDNA clone, pSpec1-1, which extends a few hundred base pairs farther toward the 5′ end of the mRNA than pSpec1. A restriction map of this clone along with a map of λSpec1a is presented in Figure 10. Sequencing these clones has shown that the gene for the Spec1 mRNA has a 2.6-kb intervening sequence beginning approximately 650 bp from the 3′ end. The exon 5′ to this invervening sequence is only 110 bp in length and is followed by a second intervening sequence of unknown length. We have not yet found genomic clones which are homologous to the remaining 200 bp of the pSpec1 1 cDNA clone. An interesting feature of this partial structure is that the two exons contained in λSpec1a have no amino acid coding sequence. Our sequencing data show pSpec1-1 is entirely a 3′ untranslated sequence. Thus the 3′ untranslated tail of the Spec1 mRNA appears to be spliced: a 650-nucleotide exon, containing the highly repeated element discussed above, is spliced to a 110-nucleotide exon containing noncoding sequence. Presumably these two 3′ exons are spliced to at least one more exon containing the amino acid coding region and the 5′ end of the mRNA. Because the length of the Spec1 mRNA is 1.5 kb and the molecular weight of the protein it codes for is at most 17,000 daltons we estimate that 5′ one-third of the mRNA will contain an amino acid coding sequence.

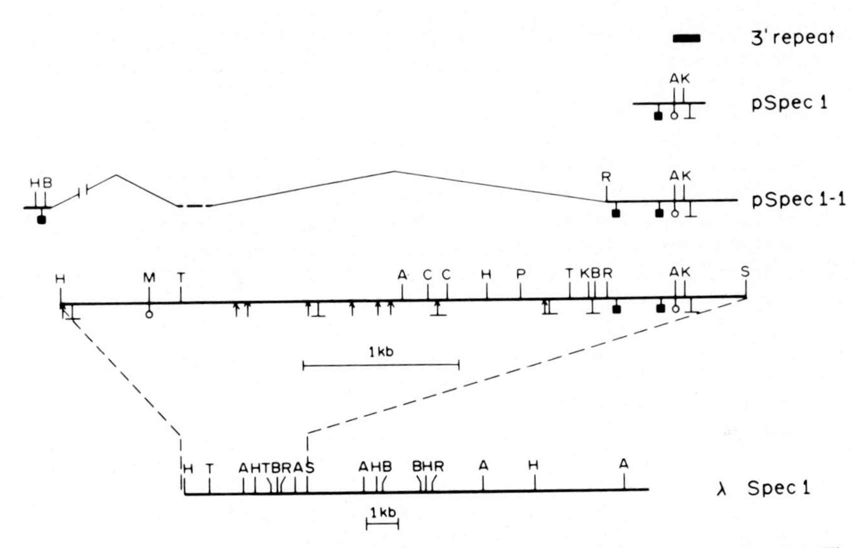

Fig. 10. Comparison of pSpec1, λSpec1 (designated λSpec1a in Fig.9), and pSpec1-1. The symbols for the restriction enzymes are as in Fig. 9. Additional enzymes are: M, Smal; C, Hind 2; P, Pst1; K, Kpn1; ⊥, HaeIII; ▲, Alu1; ♁, HpaII; ↑, Hinf. The direction of transcription on λSpec1 is 5′ to 3′ left to right. The pSpec1-1 cDNA clone has been mapped onto λSpec1 by Southern analysis with various restriction fragments and by DNA sequencing. The Spec1 gene appears to have at least three exons and two introns as shown in the figure. The region where the pSpec1-1 sequence diverges from λSpec1 is depicted by thin nonhorizontal lines. Inspection of the sequences shows good donor and acceptor concensus sequences bordering these regions. The 5′-most region of pSpec1-1 containing Hind 3 and Bam H1 sites is not present on λSpec1.

At this point we have not determined the complete structure of any of the genes for the ectoderm proteins. However, with the phage DNAs now isolated and partially characterized, we are in a good position to determine the structure of these ectoderm genes and to find other genes in the family.

ACKNOWLEDGMENTS

Research reported here is supported by grant PHS HD14182 (to W.H.K.) from the National Institutes of Health. C.D.C. and A.M.B. are supported by a predoctoral grant from the National Institutes of Health (GM7227).

REFERENCES

Bedard P-A, Brandhorst BP (1982): Patterns of protein synthesis and metabolism during sea urchin embryogenesis. Dev Biol (in press).

Bruskin AM, Bedard P-A, Tyner AL, Showman RM, Brandhorst BP, Klein WH (1982): A family of proteins accumulating in ectoderm of sea urchin embryos specified by two related cDNA clones. Dev Biol 91:317–324.

Bruskin AM, Tyner AL, Wells DE, Showman RM, Klein WH (1981): Accumulation in embryogenesis of five mRNAs enriched in the ectoderm of the sea urchin pluteus. Dev Biol 87:308–318.

Davidson EH, Hough-Evans BR, Britten RJ (1982): Molecular biology of the sea urchin embryo. Science 217:17–26.

Flytzanis CN, Brandhorst BP, Britten RJ, Davidson EH (1982): Developmental patterns of cytoplasmic transcript prevalence in sea urchin embryos. Dev Biol 91:27–35.

Galau GA, Klein WH, David MM, Wold BJ, Britten RJ, Davidson EH (1976): Structural gene sets active in embryos and adult tissues of the sea urchin. Cell 7:487–505.

Harkey MA, Whiteley AH (1980): Isolation, culture and differentiation of echinoid primary mesenchyme cells. W Rouxs Arch Dev Biol 189:111–122.

Hough-Evans BR, Wold BJ, Ernst SG, Britten RJ, Davidson EH (1977): Appearance and persistence of maternal RNA sequence in sea urchin embryos. Dev Biol 60:258–277.

Lasky LA, Lev Z, Xin J-H, Britten RJ, Davidson EH (1980): Messenger RNA prevalence in sea urchin embryos measured with cloned cDNAs. Proc Natl Acad Sci USA 77:5317–5321.

McClay DR, Chambers AF (1978): Identification of four classes of all surface antigens appearing at gastrulation in sea urchin embryos. Dev Biol 63:179–186.

Xin J-H, Brandhorst BP, Britten RJ, Davidson EH (1982): Cloned embryo mRNAs not detectably expressed in adult sea urchin coelomocytes. Dev Biol 89:527–531.

Time, Space, and Pattern in Embryonic Development, pages 101–129

RNA Localization in Sea Urchin Embryos

Robert C. Angerer and Lynne M. Angerer

Department of Biology, University of Rochester, Rochester, New York 14627

Classical Evidence for Localized Information in Sea Urchin Eggs

The embryos of echinoderms, and especially of sea urchins, have been favored systems for studying the localization of developmental information since the inception of experimental embryology and the first tests of developmental potency of individual blastomeres. Classical dye marking experiments established a fate map for intact embryos which relates regions of egg cytoplasm to major structures of differentiated pluteus larvae. There is clear evidence for an animal-vegetal axis of developmental potential in the unfertilized egg. Morphological indicators of this polarity are the extrusion of polar bodies and the presence of the jelly canal at the animal pole, and localized pigment "bands" in some unfertilized eggs of *Paracentrotus* [reviewed in Schroeder, 1980]. With these exceptions, the developmental polarity is not manifested by any corresponding ultrastructural polarity. The developmental axis is strikingly displayed in 16-cell embryos with the appearance of eight mesomeres in the animal half and four macromeres and four polar micromeres in the vegetal half. In several species, formation of micromeres is normally temporally and spatially coordinated with a cytoplasmic rearrangement visible as a withdrawal of cortical pigment granules, leaving a clear area of cytoplasm which appears to be quantitatively inherited by micromeres [Schroeder, 1980]. Appearance of this clear area, however, may be experimentally unlinked from cell division.

Demonstration of the animal-vegetal developmental axis and analysis of its properties have been pursued by various techniques of experimental embryology over a period of almost a century. Elegant microsurgical techniques have been employed to test the fate of isolated fractions of embryos and their interactions in various abnormal combinations. Although a wealth of interesting detail is lost in such a summary, several major conclusions of these studies are given here. The reader is referred to the review by Hörstadius [1973] for a more thorough discussion of this extensive literature.

Many studies which demonstrate an animal-vegetal developmental axis in early embryos have been carried out using 16-cell embryos where the distinct morphology of the three blastomere types provides easy identification of the axis, or at the 32- or 64-cell stage when derivatives of mesomeres and macromeres, respectively, separate into more animal and vegetal tiers of eight cells each. The development of animal and vegetal halves separated at these stages is quite variable, but characteristic: Animal halves contain only presumptive ectoderm and give rise to "blastulae" with enlarged tufts of cilia. They subsequently fail to gastrulate, induce formation of the stomodeum (mouth), or form a ciliated band. Vegetal halves lack about two thirds of the prospective ectoderm. They fail to form animal tufts of cilia and most typically produce an ovoid "larva" with a straight, rather than bent, tripartite gut. No mouth is induced and the oral field is poorly differentiated.

Experiments in which the cytoplasmic contents of unfertilized eggs are grossly rearranged by centrifugation imply that the axis is determined by relatively immobile (presumably cortical) components, since the location of micromeres at the vegetal pole is random with respect to the axis of stratification [see Harvey, 1956, for a review of centrifugation experiments]. Surgical separation of centrifugally elongated eggs into two halves yields results consistent with the idea that this axis is already established in unfertilized eggs and not alterable by centrifugation.

The development of embryo fragments cultured separately or in abnormal combinations demonstrates that they may give rise to, or participate in the formation of, structures outside their normal developmental pathway. In addition to regulation within parts, regulatory interactions may operate between different cell types. One of the most impressive of these is the "inductive" effect of a determined cell type, the micromeres, in producing relatively normal development when implanted on the tier of eight cells from the animal pole of 32-cell embryos—a combination which excludes all prospective endoderm. (In such combinations micromeres appear to pursue their normal destiny of becoming primary mesenchyme cells and do not contribute to formation of other parts of the embryo.) Such results, of course, are the experimental basis for classifying sea urchin embryos as "regulative," and for the distinction between "prospective significance" (normal fate of a cell lineage) and "potency" (developmental potential of such a lineage in abnormal circumstances created by clever experimental embryologists).

Much experimental data relevant to determination along the animal-vegetal axis has been interpreted in a manner consistent with the idea that morphogenesis in urchin embryos is regulated by a balance between reciprocal "animalizing" and "vegetalizing" gradients. Despite a number of studies which have searched for (and sometimes found) physiological or chemical inhomogeneities, none has been directly related to developmental processes.

Determination of the dorsal-ventral axis and bilateral symmetry are less easily demonstrated. This is, in part, because the absence of any morphological indicators of such polarity in eggs or early embryos makes experiments such as those described above more difficult. For example, prospective dorsal and ventral regions of 16-cell embryos cannot be identified and separated. Hörstadius [1973] argues that a dorsal-ventral axis is already established in eggs since halves of surgically bisected eggs allowed to develop in pairs show complementing patterns consistent with such an axis. In addition, eggs of some other echinoderms have morphological indications of the future dorsal-ventral axis. However, determination along this axis appears to be less rigid than that along the animal-vegetal axis of early embryos. For example, halves of 16-cell embryos divided along the animal-vegetal axis and stained to mark the cut side may give rise to two rather normal embryos, with development less complete (or retarded) on the side (left or right) of the missing blastomeres. Other pairs, presumably separated along a plane at right angles to the former, develop less perfectly, and the dorsal-ventral axis of one of the pair reverses. This and other experiments are logically consistent with Lindahl's proposal [1932] that there are two potential ventral "centers" in embryos, one of which is normally dominant. Again, such putative centers have no known biochemical identity.

In summary, evidence from experimental embryology defines phenomena for which we would like to find molecular explanations. The fate of individual blastomeres is rather precisely specified in normal embryos, and this appears to be at least initially dependent on differential distribution of maternal information. ("Information" is used here in the broadest sense short of "vital force.") At some point in development this must result in different patterns of gene expression in different cell lineages. In addition, the regulatory ability of blastomeres in isolation or collages and inductive interactions which occur in normal embryos would seem to require elaborate mechanisms whereby cells monitor their local environment and alter their genetic activity.

Expression of Genetic Information During Development

At present, the sea urchin embryo is probably the best-characterized embryo with respect to rates of synthesis of RNA and protein, and qualitative and quantitative analyses of these gene products during the course of development. There is as much molecular as morphological detail in this system.

Earlier results have been discussed by Davidson [1976], and the current status of many molecular aspects of sea urchin development has been summarized recently by Davidson et al. [1982]. Here we briefly review some of the major conclusions, surprises, and puzzlements.

Maternal RNA. The mature sea urchin egg contains 50–100 pg of maternal RNA. Maternal RNA designates transcripts synthesized during oogenesis, many of which are polyadenylated, excluding species such as ribosomal and transfer RNAs. About 70% of different maternal sequences are implicated as being associated with messenger RNA, since this fraction is found represented in polysomal RNA of 16-cell embryos [Ernst et al., 1980]. Other transcripts may exist which provide "information" to early embryos in a manner other than by coding for specific proteins. However, compared to mRNA of later stages, maternal RNA has several surprising and intriguing properties. At least 70% of the mass of poly(A)$^+$RNA contains transcripts of short repetitive sequences interspersed with single-copy sequence, reflecting the sequence organization of a major fraction of the urchin genome [Costantini et al., 1980]. As a class maternal RNA is longer than mature mRNA of later stages (about 3 kb vs. 2 kb) [Costantini et al., 1980] and transcripts which bear repeat sequences are even longer, up to 15 kb [Davidson et al., 1982]. Many of the latter contain two or more repetitive sequence elements [Costantini et al., 1980]. Most interestingly, maternal RNA contains high concentrations of transcripts of specific repeat-sequence families [Costantini et al., 1978]. Calculations based on the number of different maternal transcripts and repeats and their concentration in maternal RNA indicate that each of these abundant repeat transcripts is represented on a diverse set of maternal transcripts [Costantini et al., 1980]. Specific repetitive sequences must be concentrated in regions of the genome coding for maternal RNA since their high concentration in this RNA is not correlated with a higher genomic repetition frequency [Costantini et al., 1978] nor is it due to their inclusion on only a few very abundant maternal transcripts [Costantini et al., 1980]. Such abundant repetitive sequence transcripts are not found in mRNA populations of late stages of development (e.g., gastrula) which derive largely from postfertilization transcription [Davidson et al., 1982].

Thus, much of maternal RNA has a length and structure resembling that of unprocessed or partially processed nuclear transcripts, although its sequence complexity is severalfold lower than that of nuclear RNA at all later stages so far examined. It contains "extra" sequences, transcribed from both single-copy and repetitive DNA, whose arrangement with respect to protein coding sequences is not yet known. Since 5′ termini of nuclear transcripts are conserved [Nemer et al., 1980], major alternatives are that the extra repeats and single-copy sequences are contained either in introns or long 3′ untranslated tails, or both. Repeat transcripts per se are probably rarely, if ever, part of protein coding sequences since they contain frequent translation termination codons [Posakony et al., 1981]. Furthermore, it is not yet clear whether all such long transcripts are precursors to functional mRNAs and, if so, whether they must be processed to normal message length before assuming that function. It is interesting that actin mRNAs in unfertilized eggs appear only as fully processed messages although their nuclear transcripts appear to be spliced [Durica et al., 1980; Scheller et al., 1981].

The structure and sequence content of maternal RNA allow speculation on a number of interesting potential functions for the noncoding sequences [discussed in Davidson et al., 1982]. Among these are the control of maternal messenger utilization at the level of processing, use of repeat sequences as "address labels" for specific localization of individual mRNAs, or perhaps regulation in blastomere nuclei. It is apparent that detailed elucidation of the structure and function of maternal RNA is a major question in the study of early development.

Temporal and spatial patterns of gene expression. The messenger RNA complement of embryos has been analyzed both by nucleic acid hybridization assays (using both whole RNA populations and individual cloned sequences) and by two-dimensional (2-D) gel electrophoresis of proteins synthesized in vivo. Most of these studies, except as noted below, have examined RNAs or proteins isolated from whole embryos. Messenger RNAs of sea urchin embryos are typical of those of all eukaryotes in that different messages are present at different concentrations. In urchins the rare (or "complex") mRNA fraction includes the vast majority of different mRNAs, each present at about 1,000–3,000 copies per embryo, or about one to three copies per cell at gastrula-pluteus. In contrast to the many thousands of rare sequences, there are only several hundred prevalent mRNAs at most stages of development, present at about 100-fold higher concentrations [Shepherd and Nemer, 1980].

The sequence complexity of maternal RNA is about 3.7×10^7 nucleotides (NT), corresponding to expression during oogenesis of about 6% of the genomic complexity [Hough-Evans et al., 1977]. Because of the unusual structure and length of maternal RNA, it is difficult to relate its complexity to the number of functional mRNAs. An initial, surprising result was the observation that the total polysomal mRNA complexity decreases to 2.4×10^7 and 1.7×10^7 NT at blastula and gastrula, respectively, while the morphological complexity of the embryo increases. Furthermore, essentially all rare sequences present in the gastrula to pluteus period are also represented in maternal RNA [Galau et al., 1976]. Since mRNAs at later stages are shorter, it now seems likely that a major fraction of the decrease in complexity observed between the unfertilized egg and later stages is due to turnover and/or processing of long maternal RNAs, coupled with new synthesis and accumulation of the same message sequences on transcripts of typical mRNA length. Nonetheless, it is clear that differentiation is not accompanied by the appearance in whole embryos of a large set of new rare mRNAs. Despite the reduced complexity at later stages, there are still 12,000 and 8,000 different mRNAs in blastula and gastrula embryos, respectively—values which are severalfold higher than found in most adult urchin tissues [Galau et al., 1976]. Tests of sequence overlaps show that most of these mRNAs are absent from individual adult tissues.

Analysis of proteins translated from prevalent mRNAs [Brandhorst, 1976] shows that about 80% are synthesitec in whole embryos throughout development, although quantitative modulations occur. This agrees with measurements of the prevalence of individual RNAs using cDNA clones, which show that the concentration of most moderately abundant mRNAs varies less than tenfold throughout development [Flytzanis et al., 1982]. Some new abundant mRNAs begin to accumulate around blastula stage and comprise about 10%–20% of total moderately abundant sequences at later stages. Identified examples of individual mRNAs which exhibit this behavior are those coding for actins [Crain et al., 1981], ectoderm-specific proteins of as yet undefined functions [Bruskin et al., 1981], and the late histone variants [Childs et al., 1979; Hieter et al., 1979].

The major difficulty in interpreting measurements of whole embryo RNA concentration is, of course, that they provide no information on possible changes in prevalence in individual cells. Marked changes in distribution could occur in the absence of observable changes in whole embryo abundance. Conversely, the functional significance of observed changes in whole embryo concentration would be quite different if all cells of the embryo showed that change than if it were a direct reflection of an altered pattern of localization. At present, there are only two cell or tissue types which can be isolated from urchin embryos in reasonable purity, and in both cases only after their fate is determined. Each case has, however, led to interesting observations.

Micromeres, isolated by differential centrifugation [Hynes and Gross, 1970], continue to differentiate biochemically and deposit skeletal material [Okazaki, 1975; Mintz et al., 1981]. However, one important aspect of this process, pattern formation, has not been duplicated in isolated cells since it presumably requires interactions with cells in the wall of the gastrula. Surprisingly, both hybridization experiments [Ernst et al., 1980] and protein analysis [Tufaro and Brandhorst, 1979] show that micromeres have the same polysomal RNA sequences as the macromeres + mesomeres. About a 20% reduction in cytoplasmic RNA complexity is observed in micromeres compared to that of other blastomeres [Rodgers and Gross, 1978] but this is attributable to their lack of nonpolysomal maternal sequences present in the remainder of the embryo [Ernst et al., 1980]. It is not currently known whether these sequences are absent from the vegetal cytoplasm in unfertilized eggs, physically excluded at fourth cleavage, or very rapidly destroyed by processing and/or degradation. However, this phenomenon, like that of pigment granule exclusion, marks vegetal pole cytoplasm as unique in some manner.

The fractionation method of McClay and Marchase [1979] permits separation of ectoderm cells from endoderm-mesoderm at later developmental stages. Using this procedure, Bruskin et al. [1981] screened libraries of cDNA clones prepared from gastrula and pluteus RNAs and identified five different clones coding for mRNAs enriched to various extents in ectoderm. The detailed localization of one such sequence is described in the last section of this chapter.

From Molecules to Morphology

As we have outlined, detailed morphological and biochemical studies have described the development of sea urchin embryos. The problem is, of course, that the relationships between these two sets of observations are entirely unclear. There are several approaches which might relate the molecular biology with events of determination and differentiation. One approach is that followed by Bruskin et al. [1981], our laboratory, and others: to find mRNAs and proteins which are cell-type specific and elucidate their function. The converse approach is also being pursued: to analyze identified proteins with expected morphogenetic function (e.g., actins, cell surface components) in order to determine when and where they are required in embryos, and how their synthesis is regulated.

Our approach to this kind of information has been to develop and adapt techniques of in situ hybridization to detect RNA sequences in individual cells of embryos. We believe these techniques will allow us to begin to answer the following general kinds of questions: What fraction of embryonic mRNA sequences is cell-type specific at some stage of development? Are most mRNAs present in all cells throughout development, or might different sets appear in different cell lineages at different times? When do lineage-specific sequences appear? Are there certain critical stages (perhaps blastula) where many different cell types are differentiated at the mRNA level, although only determined at the morphological level? Are any RNAs (mRNAs, repetitive sequence transcripts, etc.) localized as early in development as the unfertilized egg? Does a given cell lineage begin to express its unique set of mRNAs simultaneously or sequentially? For multigene families such as those encoding actins, histones, and tubulins, to what extent are individual members expressed tissue specifically? The answer may help elucidate the function of such protein variants.

IN SITU HYBRIDIZATION TECHNIQUE

Detection of cellular RNAs by in situ hybridization imposes the somewhat conflicting requirements of attaining good fixation, which preserves cellular morphology and provides high retention of target RNAs throughout the entire protocol, and of maintaining accessibility of the RNAs to labeled probes without chemical damage which would interfere with either the extent or specificity of hybridization. In the past few years several investigators [Capco and Jeffery, 1978; Godard and Jones, 1978; Brahic and Haase, 1978; Angerer and Angerer, 1981] have developed useful protocols in a variety of biological systems. The techniques are somewhat diverse, and the technological conclusions sometimes conflicting. Here we present important features of the methodology we have developed for embryos of the sea urchin *Strongylocentrotus purpuratus*. It seems likely that most aspects of these techniques will apply to similar biological material, but they may require some significant alterations when used, for example, on such systems as the yolky eggs of amphibia. Our current methodology is summarized in Table I.

Fixation and Embedding

The most commonly used fixative for in situ hybridization has been Carnoy (ethanol:acetic acid, 3:1). Godard and Jones [1980] demonstrated superior retention of RNAs labeled in vivo using a cross-linking fixative, glutaraldehyde. We have reported similar observations for urchin eggs and embryos [Angerer and Angerer, 1981]. Thus, when embryos were continuously labeled with ^{3}H-uridine for four hours beginning at fertilization and carried through an in situ protocol, retention of cytoplasmic RNA was about 20-fold higher for glutaraldehyde- than for Carnoy-fixed embryos. Under these conditions much of the label is incorporated in tRNA [Gross et al., 1964] so this result implies that even very short RNAs may be retained in glutaraldehyde-fixed preparations. On the other hand, Brahic and Haase [1978] have reported efficiencies of detection of viral RNAs of 100% using Carnoy fixation. This apparent difference may be explained by better retention of very large viral RNA targets. An additional advantage of glutaraldehyde is that it affords better preservation of cellular morphology. We routinely employ fixation in 1% glutaraldehyde, which results in good hybridization efficiency (see below) and better morphological preservation than with lower concentrations. We have used glutaraldehyde-fixed material (stored at 4°C either in 70% ethanol or embedded in paraffin) over periods of several years without noticeable changes in hybridization signals. This allows experiments to be carried out when fresh material is not available and facilitates comparisons of hybridization patterns for different probes on the same set of embryos.

TABLE I. In Situ Hybridization Using RNA Probes[a]

Fixation and embedding
  a. 1 hr, 1% glutaraldehyde, 3% NaCl, 50 mM sodium phosphate, pH 7.4, 0°C
  b. Wash 2 times 30 minutes in buffer, 0°C
  c. Dehydrate
  d. Embed in paraffin
  e. Section 5 μm nominal thickness
Prehybridization treatments
  a. Deparaffinize and hydrate
  b. 1 μg/ml proteinase K in 0.1 M TRIS, 50 mM EDTA, pH 8.0, 30 min, 37°C
  c. 0.25% (v/v) acetic anhydride, 0.1 M triethanolamine, pH 8.0, 10 min, 25°C
Probe preparation
  a. In vitro RNA synthesis from recombinants in RVIIΔ7 vector, using Sp6 RNA polymerase
  b. DNAase to remove template
  c. Limited alkaline hydrolysis to average length of ~ 150 NT
Hybridization
  a. 50% formamide, 0.3M NaCl, 10mM TRIS, 1 mM EDTA, 0.02% each Ficoll, BSA, polyvinylpyrrolidone, 500 μg/ml tRNA, 10% dextran sulfate (500 μg/ml poly(A) added for inserts cloned by A-T homopolymer tailing)
  b. Saturation achieved by ~0.2–0.25 μg/ml per kb probe complexity, 16 hr
  c. Hybridize at (usually) Tm −25°C for RNA-RNA duplexes (normally 45–50°C)
Posthybridization washes
  a. Brief washes 4 × SSC
  b. 20 μg/ml RNAase A in 0.5 M NaCl, 10 mM TRIS, 1 mM EDTA, 30 min, 37°C
  c. 2 × SSC and 0.1 × SSC, each 30 min, 25°C
Autoradiography
  a. Dip in NTB-2 emulsion (diluted with equal volume 0.6 M ammonium acetate)
  b. Air dry
  c. Incubate in moist chamber 3 hr
  d. Air dry
  e. Expose at 4°C in vacuum dessicator

[a]EDTA, ethylenediamine tetraacetic acid; BSA, bovine serum albumin; Tm, temperature of the midpoint in the thermal denaturation transition; SSC, 0.15 M NaCL, 0.015 M trisodium citrate.

In all of our studies we have employed standard histological techniques of paraffin embedding and sectioning at 5 μm nominal thickness. We routinely stain only with eosin y to provide the best visibility of autoradiographic grains. It should be noted that staining protocols which require prolonged exposure to low pH (such as Delafield hematoxylin) may lead to artifactual loss of grains [Rogers, 1979; and our unpublished observations].

Prehybridization Treatments

Our initial hybridization of ^{3}H poly(U) to glutaraldehyde-fixed material resulted in low hybridization efficiencies [Angerer and Angerer, 1981]. When we employed prehybridization treatments with proteinase K as suggested by Brahic and Haase [1978] an eight to tenfold increase in specific hybridization (signal with poly(U) minus background binding of poly(C)) was achieved at an optimum concentration of 1–3 μg proteinase K/ml (30 minutes, 37°C). Under these conditions there is no detectable loss of in vivo-labeled RNA from sections. Concentrations of proteinase K above these produce noticeable deterioration of cellular morphology. After proteinase K digestion, sections are treated with acetic anhydride to decrease nonspecific binding of the probe [Hayashi et al., 1978].

Choice of Probe

A priori it was arguable what kind of probe would give the best signals. In hybridizations with non-strand-separated (which we term symmetric) DNA probes, self-reassociation of the probe in solution competes with hybridization in situ. Therefore, single-stranded (asymmetric) probes might hybridize more efficiently. On the other hand, symmetric probes might form multistranded hybrids in situ, thus amplifying the signal. Comparison of our early results using an asymmetric probe (poly(U)) and a symmetric probe (nick-translated DNA of the early histone repeat clone pCO2; [Overton and Weinberg, 1978]) to detect targets of known concentration implied approximately tenfold higher efficiency for the asymmetric probe. While the targets were not identical and hybridization conditions were somewhat different, it seemed likely that probe self-reassociation formed structures unable to penetrate to target RNAs and were therefore removed from the in situ reaction. In support of this were observations that the efficiency of in situ hybridization was higher with shorter probes [Angerer and Angerer, 1981].

We have recently compared directly the hybridization efficiency of asymmetric and symmetric RNA probes (Hughes et al., manuscript in preparation). For this analysis the histone early repeat unit from *S. purpuratus* was transferred from pCO2 to an RNA transcription vector, RVIIΔ7 (RVII), whose relevant properties are illustrated in Figure 1. RVII was constructed in the laboratory of Dr. T. Maniatis by excision of the smaller BamHI-EcoRI fragment from pBR322, and insertion of a promoter from the *Salmonella* phage Sp6 [Kassavetis et al., 1982] with an adjacent downstream multiple cloning site. The histone repeat fragment was inserted at the HindIII site, and clones containing the insert in both orientations relative to the promoter were identified. These are designated pCO2/R^+ (coding strand transcript) and pCO2/R^- (mRNA strand transcript). Each recombinant DNA was cleaved downstream from the insert with BglII, and asymmetric run-off transcripts were synthesized in an in vitro system by purified Sp6 RNA polymerase [Butler and Chamberlin, 1982]. The fragment size of the probe was adjusted to ~150 NT and hybridized in situ at increasing concentrations to sections of 12-hour embryos until saturation was achieved. A symmetric probe of similar length was prepared by reassociation of complementary RNA tran-

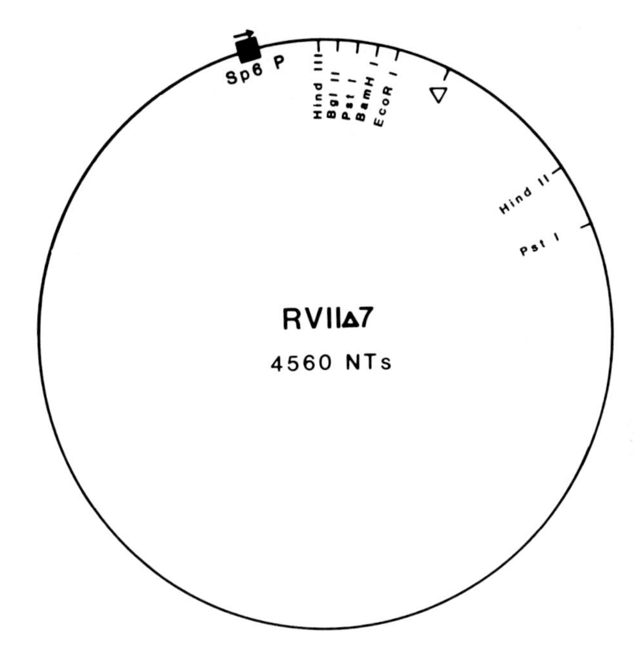

Fig. 1. Map of the RNA transcription vector RVIIΔ7. The position of the promoter for Sp6 RNA polymerase and the direction of transcription are indicated by the square and arrow, respectively.

scripts followed by RNAase digestion of unhybridized RNA. This symmetric probe was used to construct a similar saturation curve, using identical hybridization conditions. Representative examples are shown in Figure 2. At apparent saturation the specific hybridization was eightfold higher for the asymmetric probe.

This experiment demonstrates that probe self-reassociation decreases in situ hybridization signals significantly. Several investigators have reported that the rate of hybridization in situ is approximately tenfold lower than the solution rate of hybridization [Brahic and Haase, 1978; Szabo et al., 1977]. Thus, most probe fragments are in first-collision duplexes early in the in situ reaction. Although such duplexes contain single-stranded regions, it is likely that the aggregate fragment size prevents penetration to target sites. Our experiments with symmetric DNA probes (K. Hughes, unpublished observations) show that their hybridization in situ is reciprocally dependent on both probe concentration and hybridization time over a tenfold range in each parameter. It is therefore not possible to increase the relative rate of hybridization in situ by altering either of these variables. The difference in signals with symmetric and asymmetric probes might diminish with very short

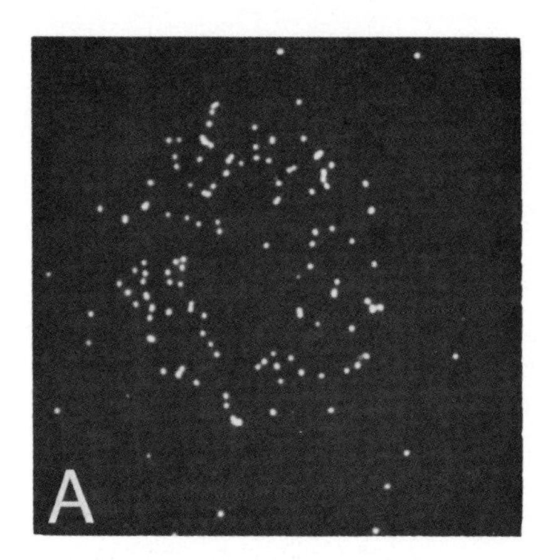

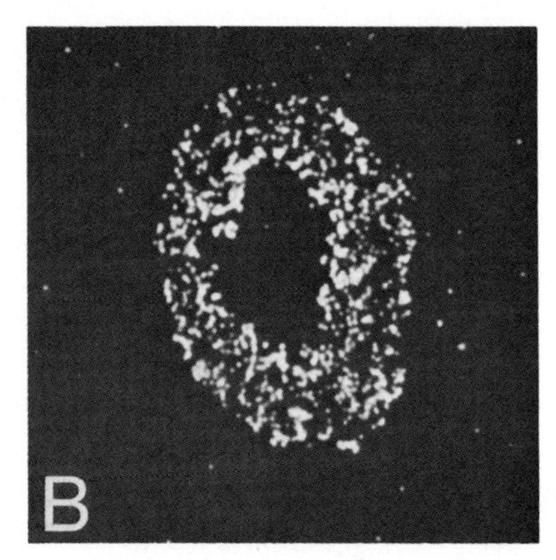

Fig. 2. Comparison of saturation levels of hybridization with symmetric and asymmetric RNA probes. A. Hybridization to 12-hour embryos of equimolar quantities of RNA transcribed from pCO2/R$^+$ and pCO2/R$^-$, corresponding to coding strand and mRNA transcripts, respectively. B. An identical hybridization with coding strand transcripts only. Both hybridizations were carried out under optimum conditions at saturating probe inputs. The signal is eightfold higher with the asymmetric RNA probe. Photographed under darkfield illumination.

probes, but those much shorter than 150 NT contain many fragments too short to hybridize, or which form duplexes of reduced stability. Finally, we note that while hybridizations with symmetric probes generate apparent saturation curves [Venezky et al., 1981], these terminate at values considerably below saturation of available RNA targets. Such saturation curves appear to merely reflect the rate of removal of probe from the reaction by self-reassociation.

At present, the major advantage of symmetric DNA probes is that they can be easily prepared by nick translation of existing clones. Our results suggest that individual RNA species comprising about 1% of embryo messenger RNA (1.5–3 $\times$ 10^{-4} of total RNA) can be detected using probes of maximum specific activity, and less abundant ones may be localized if restricted to a portion of the embryo. Since most moderately abundant mRNAs in urchin embryos have concentrations lower than this [Shepherd and Nemer, 1980; Flytzanis et al., 1982] we routinely use asymmetric probes and will largely confine further discussion to their use and advantages.

The Sp6 RNA polymerase is a very stable enzyme, and activities sufficient to synthesize milligram quantities of RNA in vitro can be easily prepared from a single liter of phage-infected cells [Butler and Chamberlin, 1982]. Depending on the specific template, its concentration, and the concentration of deoxyribonucleoside triphosphates (dNTPs), incorporations of 30–70% of input nucleotides are typical. We routinely synthesize run-off transcripts using restricted templates as described above. For both symmetric and asymmetric probes we have observed that nonspecific binding of heterologous probes is a linear function of probe input. Thus, the absence of contaminating vector sequence in run-off transcripts allows saturation at lower total probe input, with proportionately lower backgrounds. Since occasional weak termination signals occur in DNA inserts, RNA transcripts do not contain equimolar amounts of all sequences. For most relatively short inserts (300–1,000 NT) this is a minor effect, and up to 80% of the product is full length. Heterogeneity of sequence concentration in the probe is expected to introduce corresponding heterogeneity in hybridization kinetics but should not affect final saturation levels. For insert sequences which are asymmetrically expressed in vivo (e.g., mRNA coding sequences), probes representing the mRNA strand provide control sequences of identical GC content for determination of levels of nonspecific binding.

Hybridization Criterion and Kinetics

Since duplex RNAs are biological rarities, comparatively little is known about RNA-RNA hybridization. Available data indicate that in aqueous solutions RNA-RNA duplexes are 10–15°C more stable than DNA duplexes of similar base composition [Wetmur et al., 1981]. We employ formamide-containing hybridization mixtures to decrease the temperature required for hybridization specificity. It has been demonstrated that the stability of DNA-RNA hybrids relative to that of DNA duplexes increases as a function of formamide concentration [Casey and Davidson, 1977]. We compared the Tm reduction/% formamide in 0.3 M Na^+ for DNA and RNA duplexes, both representing the pCO2 histone sequence (Hughes et al., manuscript in preparation). DNA duplexes show a Tm reduction of 0.65°C/% formamide as previously reported [McConaughy et al., 1969]. RNA duplexes show a reduction of only 0.35°C/% formamide. Thus, in our standard hybridization condition of 50% formamide, 0.3 M NaCl, RNA-RNA duplexes have Tm's about 25°C higher than corresponding DNA duplexes.

Hybrids formed in situ to RNAs in 12-hour embryo sections with the pCO2/R^+ transcripts have Tm's slightly lower ($\sim$5°C) than hybrids formed in solution with RNA isolated from unfertilized eggs or 12 hour embryos (Hughes et al., manuscript in preparation). This suggests that the length of duplexes formed in situ may be somewhat less than 100 NT. Measurements of signals as a function of temperature indicate that the optimum temperature for in situ hybridization rate is approximately 25°C below the Tm measured for hybrids formed in situ (Hughes et al., manuscript in preparation). Although Brahic and Haase [1978] reported higher signals at Tm −40°C, their reference temperature was the solution Tm and no correction was made for the effect of duplex length. Saturation is achieved in 5 hours at a probe concentration of about 1.5 μg/ml. The probe input required to saturate is inversely proportional to probe complexity; we therefore anticipate that all probes will achieve saturation when inputs are 0.2–0.25 μg/ml per kb probe complexity and hybridization times are $\geqslant$ 5 hours. As has been reported in different systems [Wahl et al., 1979; Lederman et al., 1981], addition of dextran sulfate during hybridization markedly increases the hybridization rate both in solution and in situ.

Posthybridization Washes

The use of RNA probes offers the distinct advantage that RNAase A effectively removes unhybridized probe, resulting in very low backgrounds. No major difference in nonspecific binding is discernible using RNAase concentrations between 5 and 50 μg/ml. Under these conditions, RNAase A appears to digest most unhybridized tails of probe fragments so that only probe in actual RNA-RNA hybrid contributes to the signals [Angerer and Angerer, 1981]. RNAase does not appear to destroy significantly mismatched duplexes. Thus, when pCO2/R^+ was hybridized to RNAs in sections of early (12 hour) and late (pluteus) embryos, melts of these hybrids showed the same difference in Tm (11–13°C) as observed for melts of hybrids formed in solution (Hughes et al., manuscript in preparation). Significant destruction of mismatched hybrids would have reduced this difference.

Signal, Noise, and Specificity

We have estimated the efficiency of in situ hybridization from signals obtained for hybridization of pCO2/R^+ to RNAs in 12-hour embryo sections (Hughes et al., manuscript in preparation). The data are derived from several separate saturation hybridization experiments utilizing three different probe preparations with a tenfold range in specific activity. All hybridizations were carried out within about 7°C of the in situ optimum temperature (43–58°C) and exposure times varied from nine to 144 hours. The average saturation value is 3.3×10^{-2} grains/μ^2 section area/day/10^6 dpm per μg probe specific activity. Separate estimates show a range of two to threefold in saturation values similar to that observed for separate hybridizations with poly(U). We have previously discussed possible sources of this variation [Angerer and Angerer, 1981].

Twelve-hour embryos contain 12.7 pg of histone mRNA [Goustin, 1981] in a volume (after fixation and dehydration) of 5×10^4 μm^3. Therefore, each μm^2 of a 5-μm-thick section contains 1.3×10^{-3} pg of histone mRNA (since histone mRNA is essentially randomly distributed in 12-hour embryos).

Assuming the efficiency of autoradiography is 0.02 grain/disintegration for tritium randomly distributed in a 5-μm-thick section [discussed in Angerer and Angerer, 1981] the expected grain yield at saturation is 3.7×10^{-2} grains/μm^2/day/10^6 dpm per μg. This is close to the average observed value of 3.3×10^{-2}. Although corrections for autoradiographic efficiency are somewhat uncertain, this indicates that a relatively large fraction of target RNAs are hybridized at saturation.

In order to determine the potential level of sensitivity of such probes, we have estimated the level of nonspecific binding of heterologous RNAs using run-off transcripts synthesized from the RVII vector (Hughes et al., manuscript in preparation). Nonspecific backgrounds are a linear function of probe input and are slightly higher over sections than in surrounding emulsion. The average of 14 separate determinations with autoradiographic exposure times of nine to 15 hours was $2.4 \pm 0.7 \times 10^{-4}$ grains/μm^2/day/10^6 dpm per μg probe specific activity/μg per ml probe input. Less extensive data for longer exposure times (six days) indicate backgrounds severalfold lower, suggesting that a significant fraction of the background results from latent grains in the emulsion and that background grains do not increase strictly in proportion to exposure time.

The current sensitivity of the method is illustrated in the last section of this chapter where we demonstrate the pattern of localization of a set of mRNAs with a probe complementary to about 0.05% of total embryo mRNA nucleotides (6×10^7/embryo). The signal and noise levels indicate that patterns can be determined for most moderately prevalent mRNAs.

We have employed a number of standard controls to demonstrate the specificity of hybridization in situ. Backgrounds with heterologous probes such as poly(C) [Angerer and Angerer, 1981] or RNA transcribed from RVII (described above) are low. Backgrounds with homologous, but noncoding probes, such as RNA transcripts representing mRNA strands of the histone repeat or of the pSpec1 sequence (see below), are similar to those for heterologous probes. Signals with poly(U) and pCO2/R^+ have been shown to be RNAase sensitive and DNAase insensitive [Angerer and Angerer, 1981; Venezky et al., 1981; our unpublished observations].

In two cases we have demonstrated that in situ grain density is proportional to known content of target sequences. Data from one such case [Angerer and Angerer, 1981] is shown in Figure 3. Using poly(U) probes the same increase in poly(A) content between unfertilized eggs and four-cell embryos was obtained by in situ measurements of grain density and by solution titrations. These data also illustrate the fact that the four-cell/egg ratio is the same over a tenfold range in grain densities achieved, in part, by variations in probe input and length. Thus, comparisons of relative grain density over different sections within a slide are valid even when saturation is not achieved. We

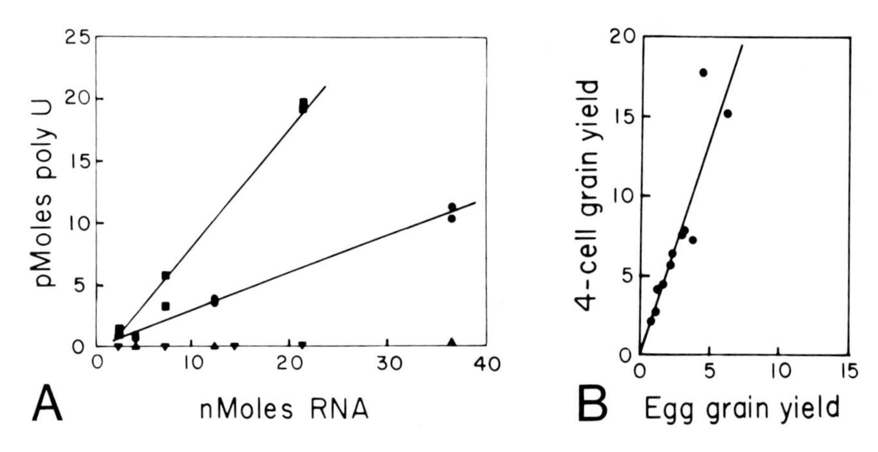

Fig. 3. Quantitation of poly(A) content of four-cell embryos vs. unfertilized eggs. A. Standard solution titration of poly(A) content of RNA from four-cell embryos (■) and unfertilized eggs (●) in the presence of excess labeled poly(U). The difference in slopes demonstrates that the poly(A) content of four-cell RNA is 2.8-fold higher. B. Determinations of grain densities over four-cell embryos and eggs after hybridization in situ with labeled poly(U), using embryos from the same culture as in A. The slope of the line demonstrates a 2.8-fold difference in poly(A) content. (Courtesy of Nucleic Acids Research.)

have also used the pCO2/R^+ probe to measure relative grain densities over sections of two-cell and 12-hour embryos (Hughes et al., manuscript in preparation). Although available estimates of the ratio of 12-hour/unfertilized egg histone mRNA are somewhat variable, the value is about tenfold [Davidson et al., 1982]. There is little or no increase in histone mRNA content between egg and two-cell embryo, so a similar ratio is expected for 12-hr/2-cell. The average of six measurements made by in situ hybridization is ninefold. (We measured the 12-hr/2-cell ratio rather than the 12-hr/egg ratio because of difficulties in quantitating egg concentrations introduced by the accumulation of these transcripts in egg pronuclei [Venezky et al., 1981]). Thus, in two cases in situ hybridization measures accurately three to tenfold differences in target concentration. We suspect that most significant biological localizations will be at least this great.

Additional evidence for specificity of hybridization comes from thermal denaturation studies with the early histone repeat (pCO2/R^+) probe. Such melts show transitions as sharp as observed for solution melts of RNA/RNA hybrids. As mentioned above, the same difference in Tm is observed for hybrids formed in situ to sections of early or late stages of development as is observed for comparable hybrids formed in solution.

DISTRIBUTION OF POLY(A)$^+$RNA IN EGGS AND EMBRYOS

Interpretation of the patterns of distribution of individual mRNAs requires knowledge of bulk mRNA distribution. For example, it is possible that some cells, especially at later stages of development, might have a lower mRNA content per se and that such differences would be reflected in lower concentrations of individual mRNA species in these cells. By "concentration" we mean here molecules per volume of cytoplasm, which corresponds to the parameter grains/μm^2. It is an interesting question whether cells regulate mRNA molecules per cell or per cytoplasmic volume. Naively, one might expect that mRNAs for proteins required in per cell amounts, such as histones, might be regulated differently from those for proteins required at specific cellular concentration, e.g., structural proteins or general metabolic enzymes.

We have used poly(U) to determine the distribution of poly(A) tracts, and presumably of poly(A)$^+$RNA at several stages of development. Analysis of poly(A) distribution in unfertilized eggs [Angerer and Angerer, 1981] required serial sections, since there are no morphological reference points. One such series is shown in Figure 4 and demonstrates that there is no significant localization of bulk poly(A)$^+$RNA. In particular, there is no evidence for a general gradient of mRNA concentration along the animal-vegetal or any other axis, and no discernible difference in cortical vs. central cytoplasm. Thus, at a resolution of a few microns, bulk mRNA appears to be rather uniformly distributed throughout egg cytoplasm.

A similar analysis for 16-cell embryos showed no major differences in poly(A) content in different regions of the embryo. However, comparisions of grain densities over micromere cytoplasm to those for adjacent macromeres revealed a 20% lower poly(A) concentration in micromeres. While this difference could merely reflect a shorter average poly(A) tract in micromeres, it seems more likely that the content of poly(A)$^+$RNA is lower in these cells. It is interesting to compare this result with the exclusion of specific maternal sequences and pigment granules from micromeres.

Between midcleavage and late gastrula, poly(A)$^+$RNA appears to be rather uniformly distributed throughout the embryo (see Fig. 8). The distribution becomes nonuniform in early plutei (Fig. 8). We observe consistently higher concentrations of poly(A) in the small, densely packed cells of the growing arms and gut. Labeling is lowest over the large cells of dorsal ectoderm, and an intermediate level is observed in ventral ectoderm. Thus a general inverse relationship exists between cell size and poly(A) concentration in different cells of plutei.

We have observed one unexpected localization of specific mRNAs in unfertilized eggs. Hybridization of nick-translated DNA of the histone repeat

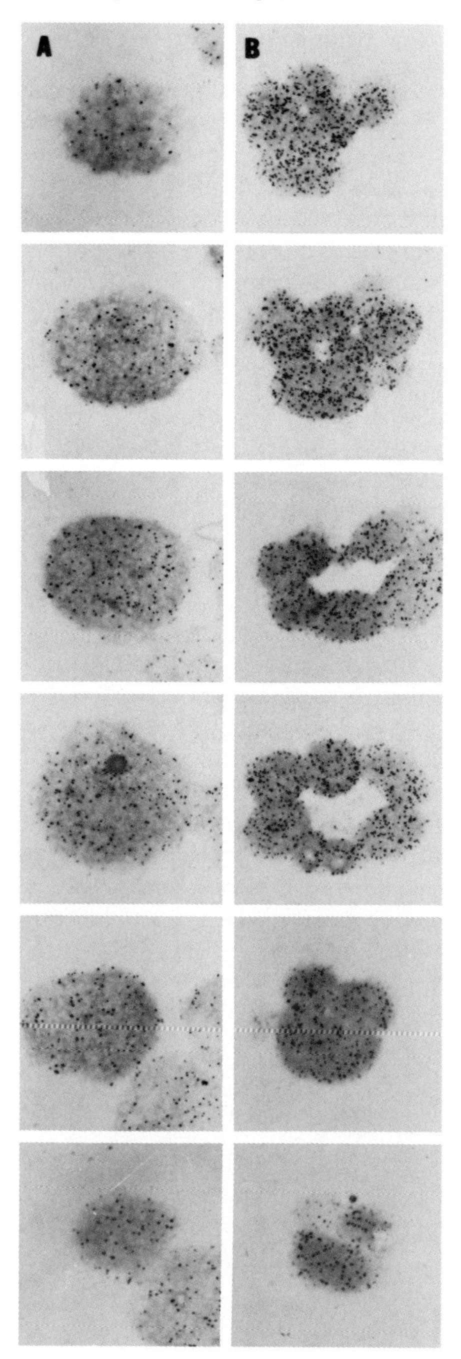

Fig. 4. Poly(A) distribution in unfertilized eggs and 16-cell embryos. Labeled poly(U) was hybridized in situ to serial sections of unfertilized eggs (A) and 16-cell embryos (B). Approximately alternate sections of each series are shown. (Courtesy of Nucleic Acids Research.)

(pCO2) demonstrated a nuclear concentration of complementary RNAs at least 50-fold higher than in the cytoplasm [Venezky et al., 1981]. Recent studies in our laboratory (DeLeon et al., manuscript in preparation) indicate that nuclear transcripts include histone mRNA, but not spacer sequences. Original estimates based on rather low cytoplasmic signals and large corrections for relative volumes of nucleus and cytoplasm indicated that 12% or more of total histone mRNA of eggs is contained in nuclei. Recent estimates with the more sensitive RNA probes (DeLeon et al., manuscript in preparation) suggest that the fraction of maternal histone mRNA in the pronucleus may be considerably higher. This high nuclear concentration persists until nuclear membrane breakdown at first cleavage [Venezky et al., 1981], about which time histone mRNAs are observed to load rapidly on polysomes [Wells et al., 1981]. The functional relationship between these two events, if any, is unknown.

The accumulation of histone transcripts in pronuclei does not appear to be a general property of maternal RNAs. The hybridization pattern with the histone RNA probe is compared to that observed with poly(U) in Figure 5. Hybridization of poly(U) to pronuclei is very low, and we have presented arguments [Angerer and Angerer, 1981] that the level of polyadenylation of RNA in pronuclei is much lower than at later stages of development. It is interesting that a somewhat reciprocal pattern is evident during oogenesis. Figure 5 also shows a comparison of hybridization of poly(U) and pCO2/R$^+$ to oocytes. At this stage the concentration of poly(A) in the germinal vesicle is higher than that in the surrounding cytoplasm, whereas the content of histone mRNA in the germinal vesicle (and in the whole oocyte) is rather low.

IN SITU LOCALIZATION OF ECTODERM-SPECIFIC mRNAs

We have collaborated with Dr. William Klein and co-workers to determine the detailed localization of an ectoderm-specific mRNA set designated Spec1 [Lynn et al., 1983]. The properties of this set of genes and their expression are discussed in detail by Klein et al., this volume. Here we summarize a few relevant features of these genes and their mRNAs, as revealed by data accumulated by these workers [Bruskin et al., 1981; Bruskin et al., 1982].

The set of five pSpec clones was identified by screening libraries of cDNA clones with radioactively labeled cDNAs complementary to RNAs isolated from tissue fractions of plutei enriched for ectoderm vs. endoderm-mesoderm. Two of these clones, pSpec1 and pSpec2, are closely related, but not identical sequences. When used in hybrid selection experiments, each of these clones selects the same set of mRNAs which, on in vitro translation, yields about ten different small acidic polypeptides. Cell fractionation studies

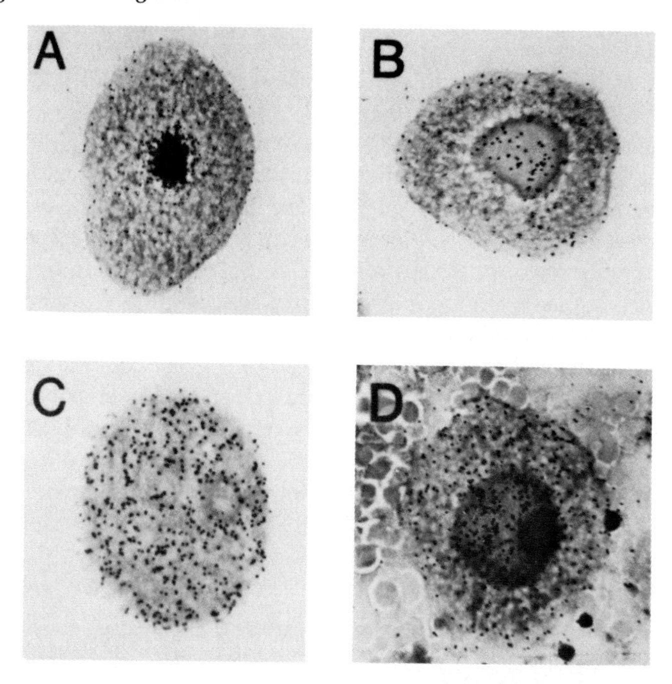

Fig. 5. Histone mRNA and poly(A) distributions in eggs and oocytes. Labeled RNA transcripts of the histone repeat (pCO2/R^+) (A and B) and labeled poly(U) (C and D) were hybridized in situ to sections of unfertilized eggs (A and C) or vitellogenic oocytes (B and D). (Photographed under phase contrast.)

show that these proteins are contained in structures which sediment at low g force, possibly nuclei (Klein, personal communication). Hybridization of pSpec1 and pSpec2 probes to Northern blots of embryo RNAs reveal two prominent bands at 1.5 and 2.2 kb which, of course, may represent more than two mRNA species. The pSpec1 probe reacts strongly with the 1.5-kb band and weakly with the 2.2-kb band, while the pSpec2 probe shows a reciprocal pattern. Sequencing studies and comparisons with genomic sequences show that the 500-NT pSpec1 cDNA clone consists entirely of 3′ untranslated sequences of the mRNA.

We transferred the pSpec1 sequence in both orientations to the HindIII site of RVII. Hybridizations of in vitro transcripts to pluteus polysomal RNA identified the orientation producing a coding strand (designated pSpec1/R^+) or a mRNA strand (pSpec1/R^-) transcript. pSpec1/R^+ transcripts of specific activity 1.8×10^8 dpm/μg were hybridized in situ at probe inputs expected to yield saturation at Tm $-25°$C. Hybridizations to early (68 hour) plutei confirm the identification of Spec1 as an ectoderm-specific sequence and demonstrate that these mRNAs are highly restricted to a subset of ectoderm cells.

Two partial serial sections of plutei from such hybridizations are shown in Figure 6A and B. The series in A consists of sections cut approximately parallel to the plane of bilateral symmetry of the pluteus. The series in B is cut approximately perpendicular to that in A, and is close to parallel to the anterior side of the embryo. (We define the anterior-posterior axis as perpendicular to the dorsal-ventral (i.e., mouth to apex) axis and lying in the plane of bilateral symmetry, with anterior on the side of the oral lobe and posterior on the side of the anus; this axis is slightly oblique to the original animal-vegetal axis of the embryo at earlier stages.) The entire cone-shaped epithelium of dorsal ectoderm cells is heavily labeled. This is shown by sections A2 and 3, which approximately bisect the embryo and show labeling on both anterior and posterior sides, and by sections such as B3 and 4 which show that both right and left sides of dorsal ectoderm are labeled. Section B4 passes through part of this sheet of cells on the anterior surface, all of which appear to be labeled. Labeling over all other regions of the pluteus is distinctly lower. This includes both endodermal and mesodermal derivatives, i.e., the entire gut and the coelomic rudiments on either side of the anterior region of the gut. (Individual cells involved in skeleton production are difficult to identify in these sections, but see below.) Perhaps most interestingly, the entire ventral face of the pluteus, which is also derived from ectoderm, shows very low labeling. This includes the sheet of cuboidal cells around the mouth, as well as ridges of small irregularly packed cells which form anal and oral arms.

Bruskin et al. [1980] have shown that Spec1 mRNAs are present at very low concentrations in unfertilized eggs and early embryos, and these mRNAs increase more than 100-fold in concentration beginning at late blastula. In agreement with this, we have not detected any signals over sections of unfertilized eggs with the pSpec1/R$^+$ probe. At early gastrula, these mRNAs are already strikingly localized. Two partial series of sections of gastrulae are shown in Figure 7. The area of labeled cells forms a continuous region, i.e., a single line can be drawn on the gastrula surface which separates labeled and unlabeled regions. Although we have not examined stages between gastrula and pluteus, we interpret the labeled regions as the future dorsal ectoderm. The pattern of labeling is bilaterally symmetric. The ectoderm of a gastrula is crudely divisible into dorsal (labeled) and ventral (unlabeled) hemispheres, but the pattern shows several deviations from this simple geometry. There is an area of cells at the animal pole (i.e., around the apical tuft or acron) which is not labeled. At the vegetal pole unlabeled cells completely surround the blastopore. Viewed from the side, the area of labeled cells projects laterally from the dorsal side past the plane through the

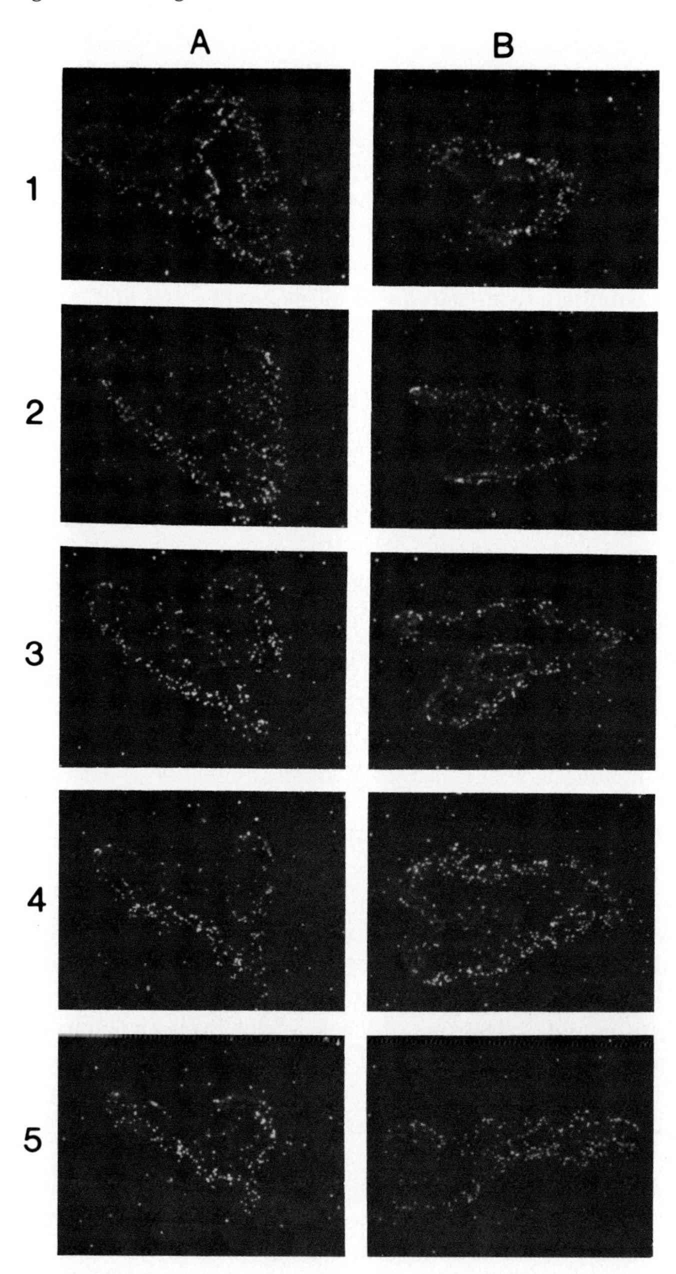

Fig. 6. Distribution of Spec1 mRNAs in pluteus larvae. Labeled RNA transcribed from pSpec1/R$^+$ was hybridized in situ to serial sections of 68-hour pluteus larvae. Approximately alternate sections from two series (A and B) are shown. The plane of the sections and labeling patterns are described in the text. (Photographed under darkfield illumination.)

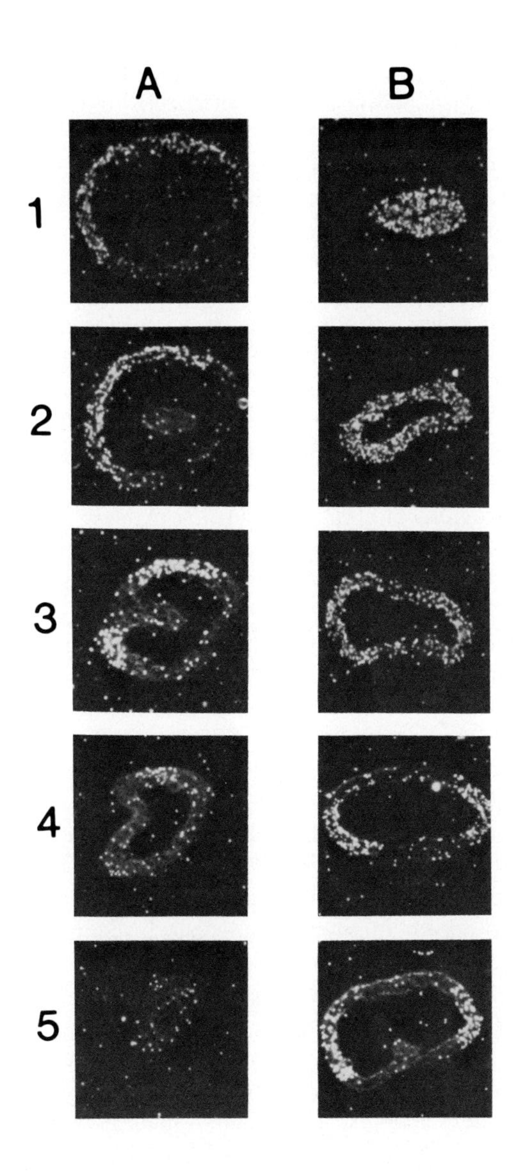

Fig. 7. Distribution of Spec1 mRNAs in gastrulae. Labeled RNA transcribed from pSpec1/ R^+ was hybridized in situ to serial sections of 40-hour gastrulae. Approximately alternate sections from two series (A and B) are shown. The plane of the sections and labeling patterns are described in the text. These sections were contained on the same slides as the pluteus sections shown in Figure 6, and thus grain densities are directly comparable. Photographed under darkfield illumination. (From Lynn et al., 1983)

animal-vegetal axis dividing the embryo into dorsal and ventral halves, forming blunt projections whose width at their tip is about one fourth the circumference of the embryo. In Figure 7A sections are cut approximately parallel to the animal-vegetal axis and perpendicular to the dorsal-ventral axis. Section A5 consists entirely of unlabeled cells on the ventral side. As sections move deeper into the embryo (A4 and 3) the lateral projections of labeled cells appear on either side of the archenteron. As sections move past the archenteron (A2), labeling becomes continuous through the vegetal pole, and gradually extends farther toward the animal pole (A1). Figure 7B shows sections of a different embryo cut in a similar orientation, approaching from the dorsal side. The first sections (B1 and 2) show continuous labeling around the circumference. In section B4 the plane of the section passes through the lateral projections and there are unlabeled cells at both the animal and vegetal poles. Note that this occurs before the level of the archenteron is reached (in the following section). This type of pattern demonstrates the existence of the lateral projections as well as unlabeled areas at the animal and vegetal poles.

We see no evidence for significant heterogeneity of Spec1 mRNA concentration in different regions of dorsal ectoderm at pluteus. Indeed, all heavily labeled cells have quite similar morphology: they are large, lightly staining epithelial cells with a large cytoplasmic/nuclear volume ratio. The only exceptions are the apical cells which are smaller, rounder, and more closely packed. No other cells in the pluteus have this morphology, and borders between this cell type and structurally distinct cells on the ventral face appear to be congruent with the borders between labeled and unlabeled regions of the embryo. It therefore seems likely that Spec1 mRNAs are highly concentrated in a single morphological cell type at pluteus. There is no such obvious correlation between morphology and localization in the gastrulae we examined. It appears that molecular differentiation of these cells as defined by expression of Spec1 mRNAs precedes overt morphological differentiation. In some gastrula sections we have seen occasional regions contiguous with labeled areas which appear to have intermediate levels of labeling, and in some cases two adjacent sections show the same such pattern. This raises the possibility that the concentration of these RNAs may be established in different regions of the embryo at slightly different times.

The distribution of Spec1 mRNAs at gastrula and pluteus would be much less interesting if they merely reflected corresponding differences in bulk mRNA content of these regions. The results obtained with poly(U) imply that poly(A)$^{+}$mRNA is homogeneously distributed in the gastrula (Fig. 8). Furthermore, at pluteus poly(A) is *least* concentrated in dorsal ectoderm cells. An early histone probe (pCO2/R^{+}) hybridized at permissive criterion also shows much higher grain densities over gut and growing arms (K. Hughes, unpublished observations). These observations also show that the Spec1 pattern is not merely the result of an artifact of differential accessibility to probes of different cell types.

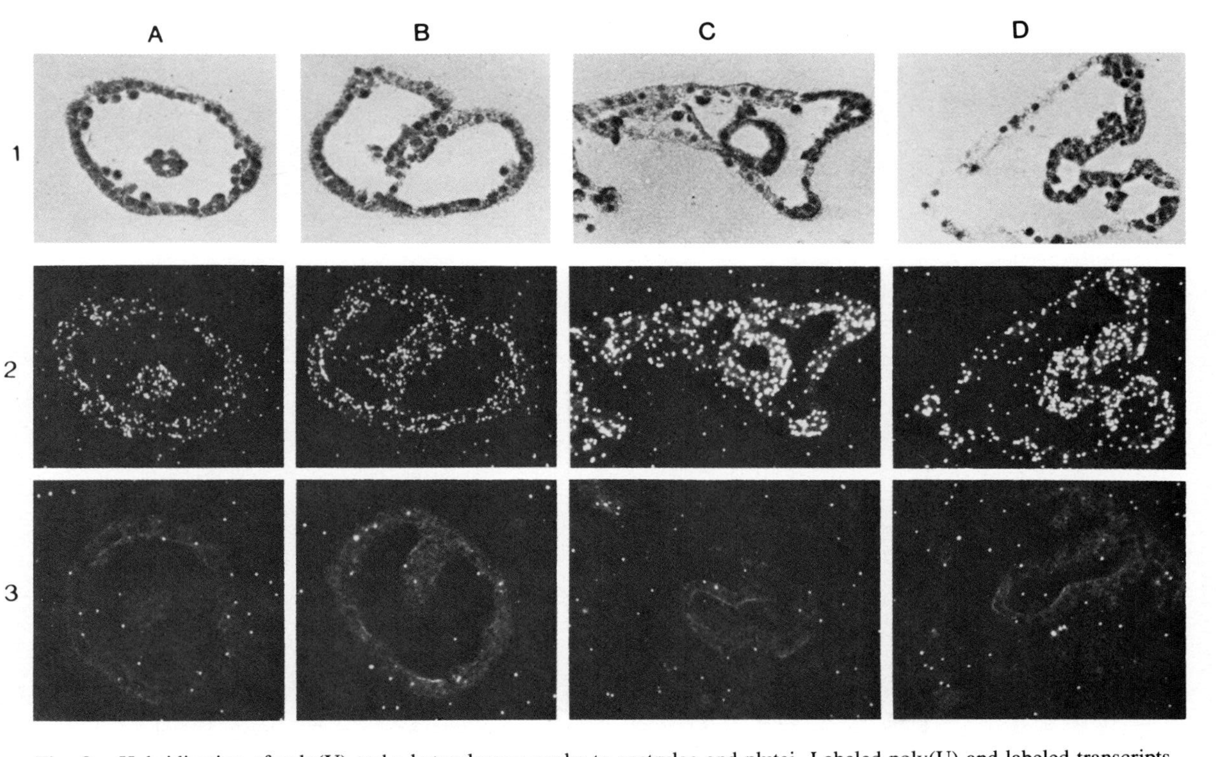

Fig. 8. Hybridization of poly(U) and a heterologous probe to gastrulae and plutei. Labeled poly(U) and labeled transcripts from the histone repeat clone pCO2/R$^-$ (representing the nonhybridizing mRNA strand) were hybridized under identical conditions to sections of 40-hour gastrulae (columns A and B) or 68-hour plutei (columns C and D). Rows 1 and 2: hybridization with labeled poly(U). These are pairs of photographs of the same sections under phase-contrast (row 1) and darkfield (row 2) illumination. Row 3: nonspecific background with the pCO2/R$^-$ probe. Photographed under darkfield illumination. (From Lynn et al., 1983)

Grain densities over unlabeled regions of both gastrulae and plutei are indistinguishable from randomly distributed backgrounds observed with the control probe (pSpec1/R^-) [Lynn et al., 1983]. Therefore, it seems likely that the concentration of these mRNAs (i.e., fraction of mRNA) is much greater than threefold higher in dorsal ectoderm cells than it is in other cell types. Estimates of the number of dorsal ectoderm cells and of the concentration of Spec1 mRNAs per embryo [Lynn et al., 1983] indicate that there are about 400–500 of these mRNAs per cell at pluteus.

The distribution of Spec1 mRNAs at gastrula and pluteus appears to be consistent with the fate map for sea urchin embryos. The unlabeled cells at the animal pole of gastrulae are destined to form the oral lobe of the pluteus, which is also unlabeled. As expected, the archenteron is unlabeled. Primary mesenchyme cells at gastrula are clearly visible in some sections and unlabeled, and therefore we expect that spicule-forming cells at pluteus will also be unlabeled. While a much greater fraction of the surface area of ectoderm is labeled at pluteus than at gastrula, this is due to expansion of dorsal ectoderm cells during morphogenesis. Preliminary estimates indicate that the number of labeled cells does not increase appreciably between gastrula and pluteus [Lynn et al., 1983].

We conclude that Spec1 mRNAs are most likely specific for a single cell type, are markers at earlier stages for cells determined to become dorsal ectoderm, and are probably expressed very early in the differentiation of these cells. The curious structure of the genes and mRNAs of this family, the suggestive properties of the proteins for which they code, and the early expression of the mRNAs during differentiation of a uniform cell type in a restricted region of the embryo will make this an interesting system for exploring the relationship between differential gene expression and the events of determination and/or differentiation.

ACKNOWLEDGMENTS

Work in the authors' laboratory was supported by the National Institute of General Medical Sciences and the United States Public Health Service. It is a pleasure to acknowledge the collaboration of members of our laboratory: Kathleen Hughes, David Lynn, and Donna DeLeon.

REFERENCES

Ada GL, Humphrey JH, Askonas BA, McDevitt HO, Nossal GV (1966): Correlation of grain counts with radioactivity (^{125}I and tritium) in autoradiography. Exp Cell Res 41:557–572.

Angerer LM, Angerer RC (1981): Detection of poly A^+ RNA in sea urchin eggs and embryos by quantitative in situ hybridization. Nucleic Acids Res 9:2819–2840.

Brahic M, Haase AT (1978): Detection of viral sequences of low reiteration frequency by in situ hybridization. Proc Natl Acad Sci USA 75:6125–6129.
Brandhorst BP (1976): Two-dimensional gel patterns of protein synthesis before and after fertilization of sea urchin eggs. Dev Biol 52:310–317.
Bruskin A, Tyner AL, Wells DE, Showman RM, Klein WH (1981): Developmental regulation of six mRNAs enriched in ectoderm of sea urchin embryos. Dev Biol 87:308–318.
Bruskin AM, Bedard P-A, Tyner AL, Showman RM, Brandhorst BP, Klein WH (1982): A family of proteins accumulating in ectoderm of sea urchin embryos specified by two related cDNA clones. Dev Biol 91:317–324.
Butler ET, Chamberlin MJ (1982): Bacteriophage Sp6-specific RNA polymerase. I. Isolation and characterization of the enzyme. J Biol Chem 257:5772–5778.
Capco DG, Jeffery WR (1978): Differential distribution of poly A-containing RNA in the embryonic cells of Oncopeltus fasciatus. Dev Biol 67:137–151.
Casey J, Davidson N (1977): Rates of formation and thermal stability of RNA:DNA and DNA:DNA duplexes at high concentrations of formamide. Nucleic Acids Res 4:1539–1552.
Childs G, Maxson R, Kedes LH (1979): Histone gene expression during sea urchin embryogenesis: Isolation and characterization of early and late messenger RNAs of Strongyolocentrotus purpuratus by gene-specific hybridization and template activity. Dev Biol 73:153–173.
Costantini FD, Scheller RH, Britten RJ, Davidson EH (1978): Repetitive sequence transcripts in the mature sea urchin oocyte. Cell 15:173–187.
Costantini FD, Britten RJ, Davidson EH (1980): Message sequences and short repetitive sequences are interspersed in sea urchin egg poly $(A)^{+}$RNAs. Nature 287:111–117.
Crain Jr WR, Durica DS, Van Doren K (1981): Actin gene expression in developing sea urchin embryos. Mol Cell Biol 1:711–720.
Davidson EH (1976): "Gene Activity in Early Development." New York: Academic Press.
Davidson EH, Hough-Evans BR, Britten RJ (1982): Molecular biology of the sea urchin embryo. Science 217:17–26.
Durica DS, Schloss JA, Crain, Jr WR (1980): Organization of active gene sequences in the sea urchin: Molecular cloning of an intron-containing DNA sequence coding for a cytoplasmic actin. Proc Natl Acad Sci USA 77:5683–5687.
Ernst SG, Hough-Evans BR, Britten RJ, Davidson EH (1980): Limited complexity of the RNA in micromeres of 16-cell sea urchin embryos. Dev Biol 79:119–127.
Flytzanis CN, Brandhorst BP, Britten RJ, Davidson EH (1982): Developmental patterns of cytoplasmic transcript prevalence in sea urchin embryos. Dev Biol 91:27–35.
Galau GA, Klein WH, Davis MM, Wold BJ, Britten RJ, Davidson EH (1976): Structural gene sets active in embryos and adult tissues of the sea urchin. Cell 7:487–505.
Gall JG, Pardue ML (1971): Nucleic acid hybridization in cytological preparations. Methods Enzymol 38:470–480.
Godard C, Jones JW (1979): Detection of AKR MuLV-specific RNA in AKR mouse cells by in situ hybridization. Nucleic Acids Res 6:2849–2861.
Goustin AS (1981): Two temporal phases for the control of histone gene activity in cleaving sea urchin embryos (S purpuratus). Dev Biol 87:163–175.
Gross PR, Malkin LI, Moyer WA (1964): Templates for the first proteins of embryonic development. Proc Natl Acad Sci USA 51:407–414.
Harvey EB (1956): "The American Arbacia and Other Sea Urchins." Princeton, New Jersey: Princeton University Press.
Hayashi S, Gillam IC, Delaney AD, Tener GM (1978): Acetylation of chromosome squashes of Drosophila melanogaster decreases the background in autoradiographs with ^{125}I-labeled RNA. J Histochem Cytochem 36:677–679.
Hieter PA, Hendricks MB, Hemminki K, Weinberg ES (1979): Histone gene switch in the sea

urchin embryo. Identification of late embryonic histone messenger ribonucleic acids and the control of their synthesis. Biochem 18:2707–2716.
Hörstadius S (1973): "Experimental Embryology of Echinoderms, Chaps 6, 7." Oxford: Clarendon Press.
Hough-Evans BR, Wold BJ, Ernst SG, Britten RJ, Davidson EH (1977): Appearance and persistence of maternal RNA sequences in sea urchin development. Dev Biol 60:258–277.
Hynes RO, Gross PR (1970): A method for separating cells from early sea urchin embryos. Dev Biol 21:383–402.
Kassavetis GA, Butler IT, Roulland D, Chamberlin MJ (1982): Bacteriophage SP6-specific RNA polymerase II. Mapping of SP6 DNA and selective in vitro transcription. J Biol Chem 257:5779-5788.
Lederman L, Kawasaki ES, Szabo P (1981): The rate of nucleic acid annealing to cytological preparations is increased in the presence of dextran sulfate. Anal Biochem 117:158–163.
Lindahl PE (1932): Zur Experimentellen Analyse der determination der dorsoventralachse beim seeigelkeim I. Versuch mit gestrechten eiern. II. Versuche mit zentrifugierten eiern. W Roux Arch Entw Mech Org 127:300–338.
Lynn DA, Angerer LM, Bruskin AM, Klein WH, Angerer RC (1983): Localization of a family of mRNAs in a single cell type and its precursors in sea urchin embryos. Proc Natl Acad Sci USA (in press).
McClay DR, Marchase RB (1979): Separation of ectoderm and endoderm from sea urchin pluteus larvae and demonstration of germ layer-specific antigens. Dev Biol 71:289–296.
McConaughy BL, Laird CD, McCarthy BJ (1969): Nucleic acid reassociation in formamide. Biochem 8:3289–3295.
Mintz GR, DeFrancesco S, Lennarz WJ (1981): Spicule formation by cultured embryonic cells from the sea urchin. J Biol Chem 256:13105–13111.
Nemer M, Ginzberg I, Surrey S, Litwin S (1980): Rates of synthesis and turnover of 5′ cap structures of hnRNA and messenger RNA and their change during sea urchin development. Dev Genet 1:151–165.
Okazaki K (1975): Spicule formation by isolated micromeres of the sea urchin embryo. Am Zool 15:567–581.
Overton C, Weinberg ES (1978): Length and sequence heterogeneity of the histone gene repeat unit of the sea urchin, S. purpuratus. Cell 14:247–258.
Pelc SR, Welton MGE (1967): Quantitative evaluation of tritium in autoradiography and biochemistry. Nature 216:925–927.
Posakony JW, Scheller RH, Anderson DM, Britten RJ, Davidson EH (1981): Repetitive sequences of the sea urchin genome III. Nucleotide sequences of cloned repeat elements. J Mol Biol 149:41–67.
Rodgers WH, Gross PR (1978): Inhomogeneous distribution of egg RNA sequences in the early embryo. Cell 14:279–288.
Rogers AW (1979): "Techniques in Autoradiography." Elsevier/North Holland Biomedical Press, p 143.
Scheller RH, McAllister LB, Crain WR, Durica DS, Posakony JW, Thomas TL, Britten RJ, Davidson EH (1981): Organization and expression of multiple actin genes in the sea urchin. Mol Cell Biol 1:609–624.
Schroeder TE (1980): Expressions of the prefertilization polar axis in sea urchin eggs. Dev Biol 79:428–443.

Shepard GW, Nemer M (1980): Developmental shifts in frequency distribution of polysomal mRNA and their posttranscriptional regulation in the sea urchin embryo. Proc Natl Acad Sci USA 77:4653–4656.

Szabo P, Elder R, Steffensen DM, Uhlenbeck OC (1977): Quantitative in situ hybridization of ribosomal RNA species to polytene chromosomes of Drosophila melanogaster. J Mol Biol 115:539–561.

Tufaro F, Brandhorst BP (1979): Similarity of proteins synthesized by isolated blastomeres of early sea urchin embryos. Dev Biol 72:390–397.

Venezky DL, Angerer LM, Angerer RC (1981): Accumulation of histone repeat transcripts in the sea urchin egg pronucleus. Cell 24:385–391.

Wahl GM, Stern M, Stark GR (1979): Efficient transfer of large DNA fragments from agarose gels to diazobenzyloxymethyl-paper and rapid hybridization by using dextran sulfate. Proc Natl Acad Sci USA 76:3683–3687.

Wells DE, Showman RM, Klein WH, Raff RA (1981): Delayed recruitment of maternal mRNA in sea urchin embryos. Nature 292:477–478.

Wetmur J, Ruyechan WT, Donthart RJ (1981): Denaturation and renaturation of Penicillium chrysogenum mycophage double-stranded ribonucleic acid in tetraalkylammonium salt solutions. Biochemistry 20:2999–3002.

Time, Space, and Pattern in Embryonic Development, pages 131–155

Determination and Differentiation of Micromeres in the Sea Urchin Embryo

Michael Alan Harkey

Departments of Zoology and Microbiology and Friday Harbor Laboratories, University of Washington, Seattle, Washington 98195

One of the most dramatic and well-documented events in the development of the echinoid embryo is the differentiation of the pluteus skeleton. A small group of cells, known as the primary mesenchyme, secretes and shapes the structure, which in turn supports the elaborate form of the pluteus larva. Since Boveri [1901], it has been known that the primary mesenchyme is derived from four micromeres formed at the 16-cell stage.

The development of micromeres into skeletogenic cells is easily followed by direct observation of living embryos. This development involves dramatic changes in morphology, physiology, and behavior. The micromeres are the first cells to exhibit a highly restricted developmental potential and the first cells to become morphologically specialized during the development of the embryo. In addition to their role in skeletogenesis, the micromeres and their derivatives play an important inductive role in early development.

The segregation of the micromeres at the fourth cleavage is a highly determinative event. If these cells are transplanted to other parts of the embryo they retain both their inductive properties and their presumptive fate [for review see Hörstadius, 1975]. Furthermore, when micromeres are cultured in complete isolation from other cell types, they still differentiate spicules with the same timing as in whole embryos [Okazaki, 1975a].

The development of sea urchin micromeres is well characterized. Yet, in spite of almost a century of experimentation and speculation, we have almost no understanding of the mechanism of determination of these cells. In this chapter I will first describe the known details of micromere development at both the morphological and molecular levels. I will then explore the major facts and theories that exist regarding their early determination.

DEVELOPMENT OF MICROMERES

The Morphological Events

Cleavage. The micromeres are formed by an unequal cleavage at the vegetal pole of the embryo at the fourth cleavage cycle. Typically, the cleavage period of development consists of a rapid series of ten cleavage cycles. The three cleavages preceding micromere formation are synchronous, equal, and radial. Thus the eight-cell stage consists of two tiers of four cells, one facing the animal pole and the other facing the vegetal pole. During the fourth cleavage, the four animal cells divide equally and radially, forming a tier of eight cells called mesomeres. The vegetal cells divide unequally and obliquely, forming a cluster of four small cells at the vegetal pole, called micromeres, and a tier of four large cells called macromeres (Fig. 1A). This unequal cleavage is preceded by, and perhaps caused by (1) migration of the nuclei to the vegetal regions of the four vegetal cells [Dan, 1979] and (2) formation of an oblique and asymmetric mitotic apparatus in each of the vegetal cells [Dan and Nakajima, 1956].

The lineage of the micromeres through subsequent development has been carefully worked out with cinematography by Endo [1966]. During the fifth cleavage cycle, the micromeres again divide unequally and obliquely, forming a set of four small micromeres and a set of four large micromeres (Fig. 1B). The small micromeres divide once more, producing a total of eight equal cells situated at the vegetal pole. The large micromeres typically divide three times, forming a ring of 32 cells around the small micromere derivatives. According to Endo, only the large micromere derivatives will differentiate into skeletogenic cells. The daughters of the small micromeres will be carried on the tip of the archenteron during gastrulaton and will eventually form part of the pharynx. The mesomeres and macromeres continue to divide at each cleavage cycle. Thus, by the blastula stage the embryo contains about 800 cells, only 32 of which will develop into primary mesenchyme cells. Some species appear to undergo an additional round of cleavage, producing 60–70 primary mesenchyme cells [Gustafson and Wolpert, 1961; Harkey and Whiteley, 1980; unpublished observations].

Fig. 1. Morphological events in micromere development in *Strongylocentrotus drobachiensis*. A. Sixteen-cell stage embryo showing micromeres (arrow) at the vegetal pole. B. Thirty-two-cell stage showing small (s) and large (l) micromeres. C. Blastula prior to hatching. D. Ingression of primary mesenchyme (pm) at early mesenchyme blastula stage. E. Migration phase of primary mesenchyme cells. F. Early gastrula with mesenchymal ring. G. Higher magnification of mesenchymal ring showing multiple pseudopodia of interacting primary mesenchyme cells. H. Skeleton formation in pluteus larva. The ectoderm was removed from the larva by brief exposure to 0.03% sodium dodecyl sulfate (SDS) in seawater in order to more easily observe the skeleton and associated primary mesenchyme cells [Harkey and Whiteley, 1982c].

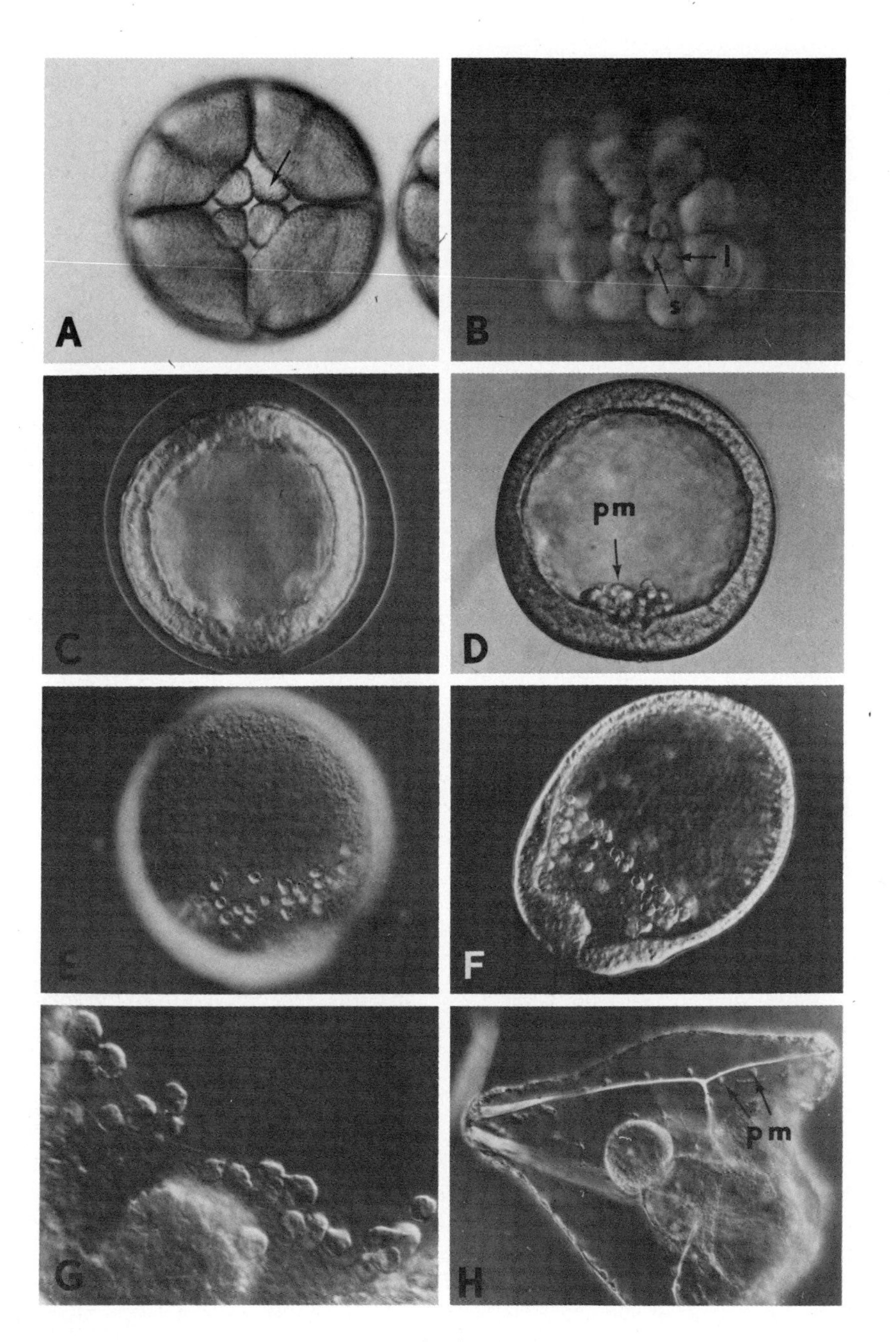

Fig. 1.

In addition to their own developmental fate, the micromeres have been implicated as pacemakers for mitotic activity during the cleavage stage [Parisi et al., 1978, 1979] and as inducers of endoderm development [for review, see Hörstadius, 1975].

Blastula formation. By the ninth or tenth cleavage the cells of the embryo assemble as a spherical epithelium (Fig. 1C). All of the cells, including the derivatives of the micromeres, assume a columnar or conical shape and exhibit a characteristic arrangement of organelles [Wolpert and Mercer, 1963; Katow and Solursh, 1980]. The cells become tightly associated by septate desmosomes and basal interdigitations [Balinski, 1959; Wolpert and Mercer, 1963; Endo, 1966; Gibbins et al., 1969; Katow and Solursh, 1980]. At this time the epithelium becomes an osmotic barrier between the blastocoel and the external environment [Moore, 1940; Dan, 1952].

The basal or blastocoel surfaces of the cells are smooth (i.e., lacking microvilli) and are in contact with a basal lamina [Wolpert and Mercer, 1963; Okazaki and Niijima, 1964; Endo, 1966; Gibbins et al., 1969]. The apical surfaces, which contact the hyaline layer, exhibit numerous microvilli [Gibbins et al., 1969; Katow and Solursh, 1980]. Each cell bears a single cilium.

Whether or not the micromere-derived cells are ciliated at the blastula stage is a point of controversy. Cilia have been reported on these cells in *Arbacia punctulata* [Gibbins et al., 1969], *Lytechinus pictus* [Katow and Solursh, 1980], and *Temnopleurus toreumaticus* [Masuda, 1979]. However, they never appear in *Mespilia globulus* [Endo, 1966] and *Hemicentrotus pulcherrimus* [Masuda, 1979]. Thus, the presence or absence of cilia on micromere derivatives may be species specific.

Ingression of the primary mesenchyme. The embryo typically hatches shortly after the tenth cleavage. At this time, it is spherical and the animal-vegetal (A-V) axis is identifiable only by a tuft of long cilia (the apical tuft) at the animal pole. However, the embryo soon elongates along the A-V axis. The vegetal region of the epithelium becomes thickened and flattened, thus reestablishing an easily definable axis. The eight cells derived from the small micromeres form the central region of the vegetal plate, while the descendants of the large micromeres encircle them [Katow and Solursh, 1980].

The ingression of the primary mesenchyme into the blastocoel has been described in detail by several investigators [Gustafson and Kinnander, 1956; Gibbins et al., 1969; Tilney and Gibbins, 1969; Davidson, 1974; Katow and Solursh, 1980]. Just prior to ingression the presumptive primary mesenchyme cells undergo lively pulsations, repeatedly extending and retracting lobes and pseudopodia from their basal surfaces. During this period, these cells lose their septate desmosomes and cilia. Their apical ends become narrower and their basal ends bulge into the blastocoel giving the cells a bottle shape.

Ultimately they penetrate the basal lamina, release their contacts with the hyaline layer and with neighboring cells, and slip into the blastocoel. Once inside, they round up and accumulate as a loose pile of cells on the floor of the blastocoel (Fig. 1D). At this time, they are termed primary mesenchyme cells. Only the ring of large micromere descendants moves into the blastocoel. The eight descendants of the small micromeres remain in the epithelium [Katow and Solursh, 1980].

Migration and ring formation. The migratory behavior of echinoid primary mesenchyme cells has been studied extensively [for reviews, see Gustafson and Wolpert, 1963; Gustafson, 1975; Okazaki, 1960, 1975a,b]. Shortly after they enter the blastocoel, these cells begin to move along the surface of the epithelium (the blastocoel wall]. This activity is mediated by continuous extension and retraction of multiple filopodia that seem to explore the blastocoel wall. When a filopod fails to adhere to the wall, it collapses back into the cell. When it does attach, its retraction pulls the cell toward the new anchor point. These filopodia contain numerous microtubules aligned parallel to their long axis [Gibbins et al., 1969]. Agents that disrupt microtubules also block the formation of filopodia and the migration of primary mesenchyme cells [Tilney and Gibbins, 1969].

Initially the primary mesenchyme cells appear to migrate at random. They quickly become scattered over the vegetal portion of the blastocoel wall (Fig. 1E). They do not associate with the roof or animal third of the wall. Within one to several hours, depending upon the temperature and the species, these cells become organized into a subequatorial ring, the mesenchyme ring (Fig. 1F,G). Most of the cells in the ring become concentrated in two ventrolateral clusters, the mesenchymal aggregates. An additional chain of cells eventually extends from each aggregate toward the animal pole. Thus, each mesenchymal aggregate takes the shape of a triangle with chains of cells extending from each corner. This pattern is important in skeletogenesis. The two spicules that comprise the major parts of the pluteus skeleton originate in the mesenchymal aggregates and grow by extension along the cellular chains.

Skeletogenesis. The formation of skeletal spicules (Fig 1H) occurs within a syncytium [Gibbins et al., 1969; Millonig, 1970]. This syncytium is formed by, and remains continuous with, the primary mesenchyme cells (Fig. 2A). As these cells accumulate in their characteristic annular pattern, they interact extensively with each other by filopodial exploration [Okazaki, 1960, 1965; Wolpert and Gustafson, 1961; Gustafson and Wolpert, 1961, 1963] (Fig. 1G). Within the mesenchymal aggregates a syncytial mass is formed by fusion of the filopodia from the cells in the aggregates. The detailed observations of Okazaki [1960, 1965] showed that the syncytium is produced and enlarged by the repeated extension and fusion of filopodia from the participating cells. Each filopod contains a "lump" of refractile material that becomes

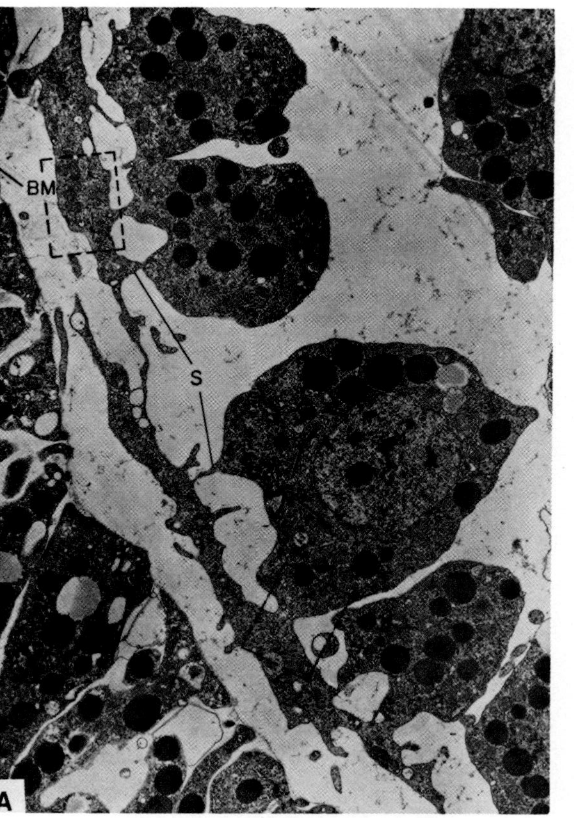

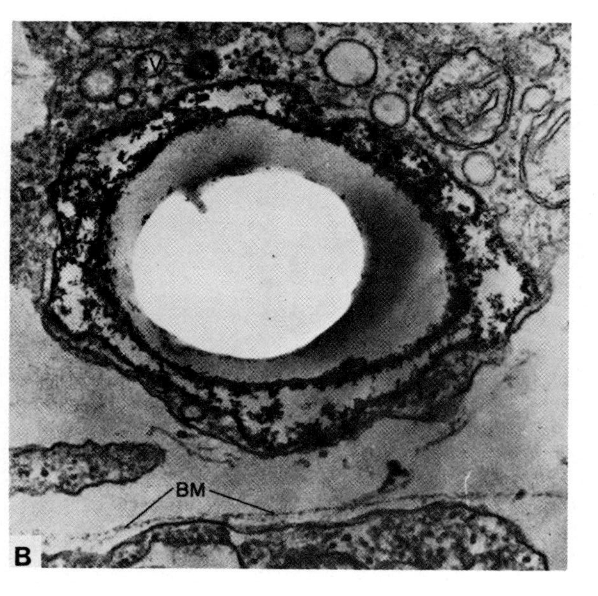

Fig. 2. The syncytial nature of the primary mesenchyme. A. Low-magnification electron micrograph showing the arrangement of the primary mesenchyme cells shortly after cable formation in the embryo of *Arbacia punctulata*. The cable runs from the upper left to the lower right corners of the picture. Several mesenchyme cell bodies, seen to the right of the cable in the blastocoel space, are attached to the cable by cytoplasmic stalks (S). The cable maintains a position near the ectoderm (to the left) by short cytoplasmic processes. B. transverse section through a cable in the process of spicule growth. The skeleton, represented by the hole in the center of the micrograph, is contained within a membrane-limited vacuole in the cytoplasm of the cable. The ectoderm and associated basement membrane are visible at the bottom of the micrograph. ×60,000. From Gibbins et al [1969].

incorporated into the growing mass. This process of fusion continues out from the mesenchymal aggregates into the chains of cells. In this way the syncytium is elaborated into cables. The entire structure is flattened like a thin envelope against the blastocoel wall, and it is firmly anchored to the wall by numerous filopodia.

The spicule rudiments appear as birefringent granules of crystalline calcium/magnesium carbonate in the central mass of the scynctium. These rudiments develop into planar triradiate structures which become oriented with the rays pointing toward the cables leading away from the central mass. Ultimately, spicule growth continues into the cables and appears to be guided by them.

At the electron microscope level it is apparent that spicule formation and growth occurs within a membrane-bound vesicle inside the syncytium [Gibbins et al., 1969; Millonig, 1970] (Fig. 2B). The primary mesenchyme cells remain connected to the syncytium by numerous thin cytoplasmic stalks. There are some indications that the growing spicule vesicle is supplied by fusion with smaller vesicles that are produced in the primary mesenchyme cell bodies and are transported through the cytoplasmic bridges [Gibbins et al., 1969; Uemura, unpublished data cited by Okazaki and Inoué, 1976].

Growth of the skeleton involves the continuous extension of the cables which, in turn, requires that the primary mesenchyme cells move with the growing tips of the spicules. According to Okazaki [1965], these cells frequently disengage from the syncytium and rejoin it at other points. Thus the major accumulations of primary mesenchyme cells in pluteus larvae are at the tips of the spicules. However, the entire skeleton remains inside the spicule vesicle which, in turn, remains inside the syncytium. Thus, the skeleton continues to thicken as well as lengthen during larval growth.

Specialized Molecular Properties of the Primary Mesenchyme

The micromere/primary mesenchyme cell line becomes visibly specialized at the time of ingression, when the cells abandon their epithelial morphology and environment, and acquire the ability to migrate. They also begin to exhibit specialized molecular properties at this time.

Acid mucopolysaccharides and migration. Just prior to ingression acid mucopolysaccharides accumulate on the basal surface of the presumptive primary mesenchyme cells. This has been repeatedly demonstrated by various histochemical methods [Motomura, 1960; Immers, 1961; Sugiyama, 1972]. Labeling with $^{35}SO_4^{=}$ indicates that this material is highly sulfated and that only the presumptive primary mesenchyme cells produce it [Sugiyama, 1972). Incorporation of $^{35}SO_4^{=}$ is first detectable at the late blastula stage, just prior to ingression [Immers, 1961; Karp and Solursh, 1974]. It rises to a sharp peak during ingression, falls off, then rises slowly through subsequent development [Karp and Solursh, 1974]. Presumptive primary mesenchyme cells undergo a cell-specific rise in surface negative charge at the time of ingression [Sano, 1977], and this result is consistent with the apparent secretion of sulfated mucopolysaccharides.

These sulfated molecules are thought to be important in the migration of primary mesenchyme cells. When embryos are grown in sulfate-free medium, the primary mesenchyme cells enter the blastocoel but fail to migrate [see Sugiyama, 1972; Karp and Solursh, 1974]. Sulfated glycoproteins may also be involved in migration. Karp and Solursh tentatively identified such proteins among the $^{35}SO_4^{=}$-labeled material of mesenchyme blastulae. In addition, tunicamycin, which inhibits glycosylation of proteins, also blocks migration of the primary mesenchyme [Schneider et al., 1978].

The spicules. The major differentiated product of the primary mesenchyme is the skeleton. This structure is composed of individual spicules (two in the early pluteus) made of magnesium and calcium carbonates in the proportions of 1:20 [Okazaki, 1970].

The uptake and deposition of $^{45}Ca^{++}$ by echinoid embryos has been demonstrated [Nakano et al., 1963; Bevelander and Nakahara, 1960]. Calcium uptake rises rapidly from the time of spicule initiation and almost all of it is incorporated into the spicules. The size, shape, and rate of growth of spicules are sensitive to the concentrations of Ca^{++}, Mg^{++}, and H^{+} in the seawater [Okazaki, 1956, 1961; Bevelander and Nakahara, 1960].

Each spicule behaves optically as if it were carved from a single calcite crystal. However, scanning electron microscopy fails to reveal crystal faces on the surfaces of the spicules, even after fracturing or acid etching [Okazaki and Inoué, 1976]. Thus it appears that each spicule is composed of exceedingly small crystals arranged such that their crystal axes are in perfect alignment throughout the entire structure.

The spicule matrix. The alignment of microcrystals in the spicules suggests that they are embedded in some kind of organic matrix. The presence of an organic material associated with the spicules is easily demonstrated. If spicules are isolated, cleaned in NaOH or hypochlorite, and then demineralized by the addition of acid or edetic acid (EDTA), a strand of material is left that stains with Nile blue sulfate [Okazaki, 1960] and with toluidine blue [Uemura, cited by Okazaki and Inoué, 1976]. Ellis and Winter [1967] demonstrated the incorporation of labeled amino acids into the spicules of plutei. Since electron microscopic observations of pluteus skeletons [Gibbins et al., 1969; Millonig, 1970] have not revealed an internal matrix, the material in question may be in the form of a sheath around the spicule. In this discussion I will refer to it as organic matrix.

According to Uemura [cited by Okazaki and Inoué, 1976], the organic matrix accounts for about 1% of the total weight of the spicule. He has partially characterized the matrix as 36% protein, 2.3% hexose, 2.4% amino sugars, and 3.6% reducing sugars. An analysis by Pucci-Minafra et al. [1980] indicates that for every gram of protein in the matrix there exist 0.9 grams hexuronic acid, 26 mg hexosamine, and 0.57 mg hydroxyproline. This analysis, in addition to an amino acid analysis of the matrix proteins, suggested to the authors that the matrix contains a small amount of collagen and a large amount of glycoprotein.

Considerable evidence exists for the presence and importance of collagen in the spicule matrix. (1) Inhibitors of collagen processing prevent spicule formation and $^{45}Ca^{++}$ incorporation by embryos [Davidson, 1974] and by cultured micromeres [Mintz and Lennarz, 1982]. (2) Selective inhibition of collagen synthesis by low doses of actinomycin D inhibits spicule formation and $^{45}Ca^{++}$ incorporation by embryos [Peltz and Giudice, 1967; Gould and Benson, 1978]. (3) ^{3}H-proline is converted to hydroxyproline and incorporated into spicules [Pucci-Minafra et al., 1972; Golob et al., 1974; Ellis and Winter, 1967]. (4) An hydroxyproline-containing material with the solubility characteristics of collagen can be extracted from spicules [Pucci-Minafra et al., 1972; Golob et al., 1974; Pucci-Minafra et al., 1975]. (5) This material can be induced to form typical collagen fibers in vitro [Pucci-Minafra and Minafra, 1980]. These authors were unable to demonstrate the natural occurrence of collagen fibers in the matrix. The amino acid and sugar composition of the matrix suggested to them that less than 5% is collagen and that a large proportion is glycoprotein.

The importance of glycoproteins in skeletogenesis is further suggested by the effects of tunicamycin. Inhibition of protein glycosylation by this drug stops incorporation $^{45}Ca^{++}$ into spicules both in vivo [Schneider et al., 1978] and in differentiated cultures of isolated micromeres [Mintz and Lennarz, 1982].

Other proteins localized in the primary mesenchyme. Cytochrome C reductase, demonstrated by histochemical staining at the mesenchyme blastula or pluteus stages, exhibits high activity exclusively in the primary mesenchyme [Czihak, 1962].

Alkaline phosphatase activity is found in the gut and in the primary mesenchyme of gastrula and older embryos [Evola-Maltese, 1957: Hsiao and and Fujii, 1963]. This activity is low prior to gastrulation but rises sharply during and after gastrulation [Pfol and Giudice, 1967; Gustafson and Hasselberg, 1951; Hsiao and Fujii, 1963]. At the time of ingression a new isozyme appears in the embryo [Pfol, 1965]. The cellular origin of the isozyme is not known.

Acetylcholinesterase activity is localized predominantly in the primary mesenchyme and secondarily in the gut of prism and pluteus larvae [Ozaki, 1974]. This activity is low until the end of gastrulation, at which time it rises sharply.

Primary mesenchyme-specific antigens have been demonstrated using monoclonal antibodies (McClay et al., this volume). These antigens become detectable suddenly and specifically on the surface of primary mesenchyme cells immediately after ingression and retain this association throughout skeletogenesis.

Specialization of Protein Synthesis

In collaboration with Dr. Arthur Whiteley I have examined several aspects of the program of protein synthesis that is played out during the development of micromeres. Recently developed procedures have made it possible to examine these cells in isolated culture at almost any stage of development. Micromeres can be isolated at the 16-cell stage by virtue of their small size, and thus, relatively slow sedimentation rate [Spiegel and Tyler, 1966]. Under the proper conditions [Okazaki, 1975a], these cells can develop in culture according to their normal in vivo program. In addition, we have established procedures for the isolation of primary mesenchyme cells from early gastrulae and for the enrichment of these cells from postgastrula stages [Harkey and Whiteley, 1980]. These procedures exploit the fact that primary mesenchyme cells are biologically segregated from most of the other cells in the embryo by a basal lamina. These cells also develop normally in culture. We have examined the populations of proteins synthesized by the micromere/primary mesenchyme cell line throughout its development using these isolation and culture methods in conjunction with two-dimensional polyacrylamide gel electrophoresis and fluorogaphy of ^{3}H-valine-labeled proteins [Harkey and Whiteley, 1982a,b, 1983].

The electrophoretic pattern of newly synthesized proteins changes dramatically during the development of cultured micromeres (compare Figs. 3A, 3B). This change involves increases and decreases in relative intensity as well as de novo appearances and disappearances of spots. Of 161 analyzed spots, more than 50% exhibit such changes and one third of these changes are qualitative.

A distinctive feature of the fully differentiated pattern (Fig. 3B) is the concentration of label among a small number of very intense spots. Most of these spots are either absent or of minor importance at the 16-cell stage (Fig. 3A). Thus, the majority of protein synthetic machinery in differentiated micromeres is engaged in the synthesis of stage-specific products. An identical pattern is produced when the cells are isolated at a later stage in development using a different isolation procedure (Fig. 3C). Thus, we believe that this pattern represents the normal differentiated pattern of protein synthesis for primary mesenchyme cells rather than an artifact of isolation.

At the 16-cell stage, the pattern of protein synthesis in micromeres is industinguishable from that of the other blastomeres [Tufaro and Brandhorst,

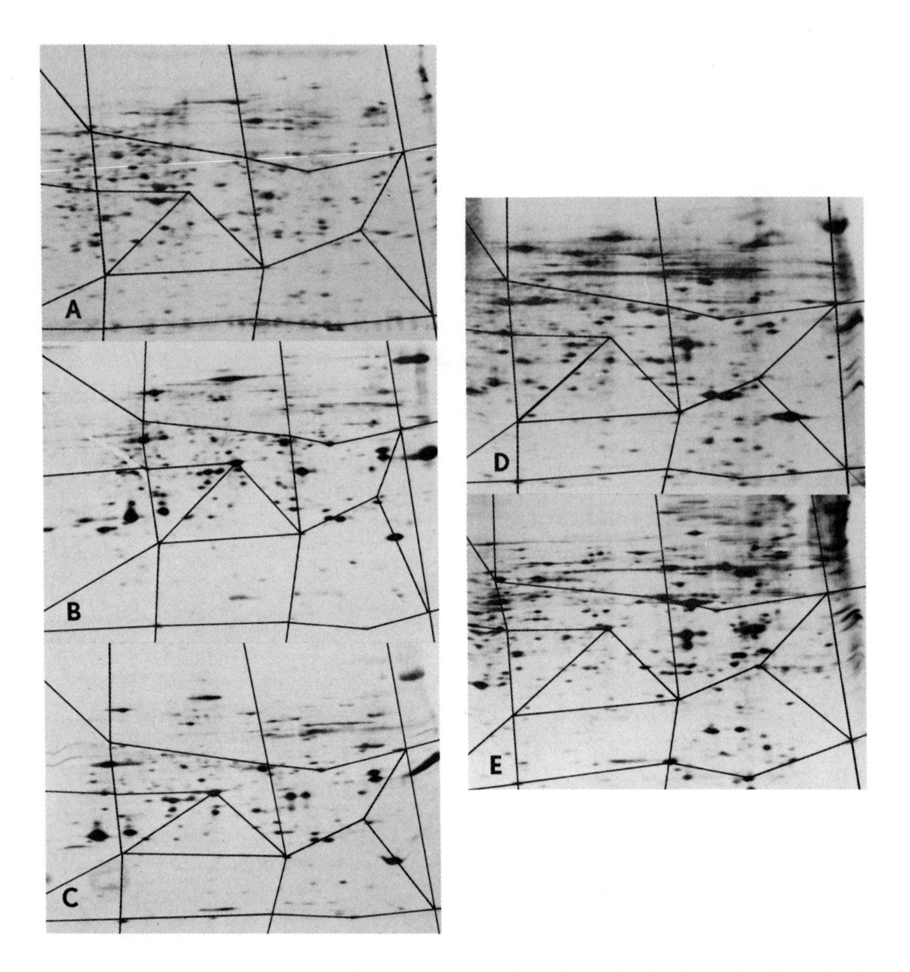

Fig. 3. Two-dimensional electrophoretic patterns of newly synthesized proteins from developing micromeres of *Strongylocentrotus purpuratus*. Each fluorograph represents the total cellular proteins from a given cell preparation labeled during a four-to-five hour pulse with ^{3}H-valine, and electrophoretically separated according to O'Farrell [1975]. The pH range for the first dimension was 4.5–6.5. the second dimension was in a 11%–14% exponential polyacrylamide gradient. A. Micromeres labeled immediately after isolation at the 16-cell stage. B. Isolated micromeres labeled after 90 hours of culture at 11°C. The cultured cells were producing spicules and control embryos had reached the pluteus stage. C. Primary mesenchyme cells isolated from early gastrulae and labeled after 40 hours of culture at 11°C. The cultured cells were producing spicules and control embryos were plutei. D. Primary mesenchyme cells labeled immediately after isolation from early gastrulae. E. Epithelial cells (ectoderm plus endoderm) from early gastrulae. These cells were isolated immediately after labeling of the whole embyros.

1979; Harkey and Whiteley, 1983]. However, the micromere pattern becomes increasingly specialized during development. In a comparison of the proteins synthesized by primary mesenchyme cells)Fig. 3D) and epithelial cells (Fig. 3E) isolated from early gastrulae [Harkey and Whiteley, 1983] 28% of the detectable proteins exhibit either qualitative or quantitative cell specificity with respect to labeling intensities. This value rises to about 50% by the pluteus stage [Harkey and Whiteley, 1983]. Thus, as it turns out, the heavily labeled stage-specific proteins that characterize the differentiated micromere pattern of protein synthesis are also largely specific to the micromere/primary mesenchyme cell line.

We have examined the timing of the changes in protein synthesis during micromere development by following the labeling histories of 161 individual proteins: Eight separate stages were examined [Harkey and Whiteley, 1983]. Almost all of the observed changes are initiated in the short period between hatching and the start of gastrulation. As shown in Figure 4, the frequency with which changes are initiated shows a dramatic peak during this period, corresponding to the time of ingression of primary mesenchyme cells in the intact embryo.

Some changes occur even earlier. Several of the major proteins that characterize the cell-specific pattern of protein synthesis of fully differentiated micromeres appear prior to hatching. Since these proteins do not appear in other cells of the embryo, this result indicates that the program of protein synthesis in the developing micromere/primary mesenchyme cell line begins to specialize even before the cells are morphologically distinguishable.

DETERMINATION OF MICROMERES

The Mechanism of Determination

It is clear from the development of micromeres in ectopic implants [Hörstadius, 1935] and in isolation [Okazaki, 1975a] that these cells are determined at the 16-cell stage to follow a narrow developmental pathway. This pathway is uniquc to these cells when observed at the morphological, enzymatic, and protein synthetic levels.

Two basic theories have evolved regarding the mechanism of this determination. The first theory states that the size of the cells is important. The importance of size is suggested by the strict conservation of the unequal fourth cleavage among almost all the echinoids and by the insensitivity of that unequal cleavage to drastic manipulations of the three previous cleavages [Rustad, 1960; Dan and Ikeda, 1971; Dan, 1972]. According to Tanaka [1976], the fourth cleavage can be modified to occur symmetrically, resulting in 16 equal cells, if the embryo is treated with certain surfactants at the 16-cell stage. The resultant micromereless embryos do not develop primary mesenchyme cells or spicules.

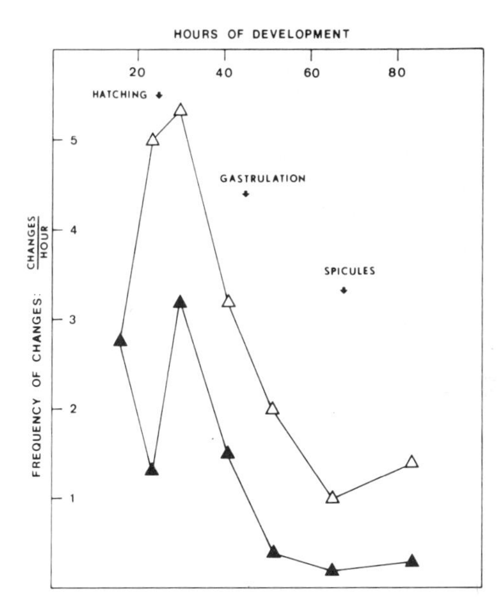

Fig. 4. The frequency of changes in the synthesis of proteins during the development of cultured micromeres from *S. purpuratus*. Newly synthesized proteins were labeled and analysed at eight stages of development. Following the behaviors of 161 individual proteins, we recorded at each interval both the total number of proteins that changed in relative labeling intensity, and the number that changed for the first time during that interval. The former category contained all of the members of the latter plus those changes which continued progressively over several intervals of development. These data are presented as frequencies of changes by dividing the numbers of changes in each interval by the elapsed time of that interval. ▲, initiations of changes; △, total changes. From Harkey and Whiteley [1983].

It has been suggested that the high volume ratio in micromeres of nuclear or perinuclear material to cytoplasm may be important in determination (Dan, personal communication). Extending this idea Senger et al. [1978] proposed that an especially high ratio of newly synthesized RNA to maternal RNA might develop in micromeres soon after the fourth cleavage. Even assuming (1) a homogeneous distribution of maternal RNA to the 16-cell stage blastomeres and (2) equivalent rates and types of RNA synthesis in each of the blastomere nuclei, the ratio of maternal and newly synthesized RNA would differ according to the volume of cytoplasm in each cell type. These differences could lead to quantitative differences in the synthesis of specific proteins among the cell types. The authors provided evidence consistent with this theory regarding the synthesis of H1 histones.

Two recent experiments indicate that the size of micromeres may not be critical in the determination of their developmental fate. First, in at least one species, *Dendraster excentricus*, inhibition of the unequal fourth cleavage by the method of Tanaka does not prevent the appearance of primary mesenchyme cells or the differentiation of spicules [Langelan and Whiteley, 1980]. Second, unusually large micromeres can be produced by inducing their precocious appearance at the eight-cell stage [Dan, 1972]. These large micromeres differentiate in isolated culture in the same manner as normal micromeres [Kitajima and Okazaki, 1980; Takahashi and Okazaki, 1979].

The second theory of determination states that (1) certain materials are inhomogeneously distributed in the egg and early embryo, particularly with respect to the animal-vegetal axis, and (2) micromeres receive a unique assortment of these materials, an assortment that specifies their developmental fate. Judging from the specialization of protein synthesis in developing micromeres, these segregated materials would have to include either (1) messenger RNAs or (2) factors that influence selective translation of mRNAs or (3) factors that influence the concentration of specific mRNA sequences by selective synthesis, degradation, or processing.

The sea urchin egg is endowed with an animal-vegetal polarity. During oogenesis, the vegetal pole is established as the site of attachment of the oocyte to the ovary wall, while the animal pole is the site of polar body formation [Jenkinson, 1911; Lindahl, 1932]. A gradient of developmental potential coincides with the animal-vegetal axis of eggs and early embryos [Hörstadius, 1935; see Hörstadius, 1975, for review]. As would be expected from the fate map of the egg (Fig. 5), egg fragments taken from the animal region develop exaggerated ectodermal structures and reduced or absent endoderm. Vegetal fragments tend to develop exaggerated endodermal structures and reduced ectoderm.

However, this gradient is very labile: no particular region of the egg can be said to have a fixed potential. A classic series of experiments by Hörstadius [1935] (Fig. 6) revealed that while an entire pluteus larva can develop from small fragments of the early embryo, normal development requires that a balance of animal and vegetal materials be present. This evidence, plus the fact that a large number of unrelated chemicals can shift whole embryos toward more animal or more vegetal types of development [see Lallier, 1975 for review] have led to the double gradient theory of Runnström [1928; see also Runnström, 1975, for review]. He proposed that the development of any given region of the egg is directed by its relative exposure to two factors, vegetal and animal, which exist in two opposing gradients within the egg. The micromeres contain the highest ratio of vegetal to animal factors in this scheme.

This gradient system is not disturbed by centrifugal stratification of cytoplasmic inclusions within the egg [Lyon 1906a,b; Morgan and Lyon, 1907; Morgan and Spooner, 1909; Hörstadius, 1953]. The micromeres always form

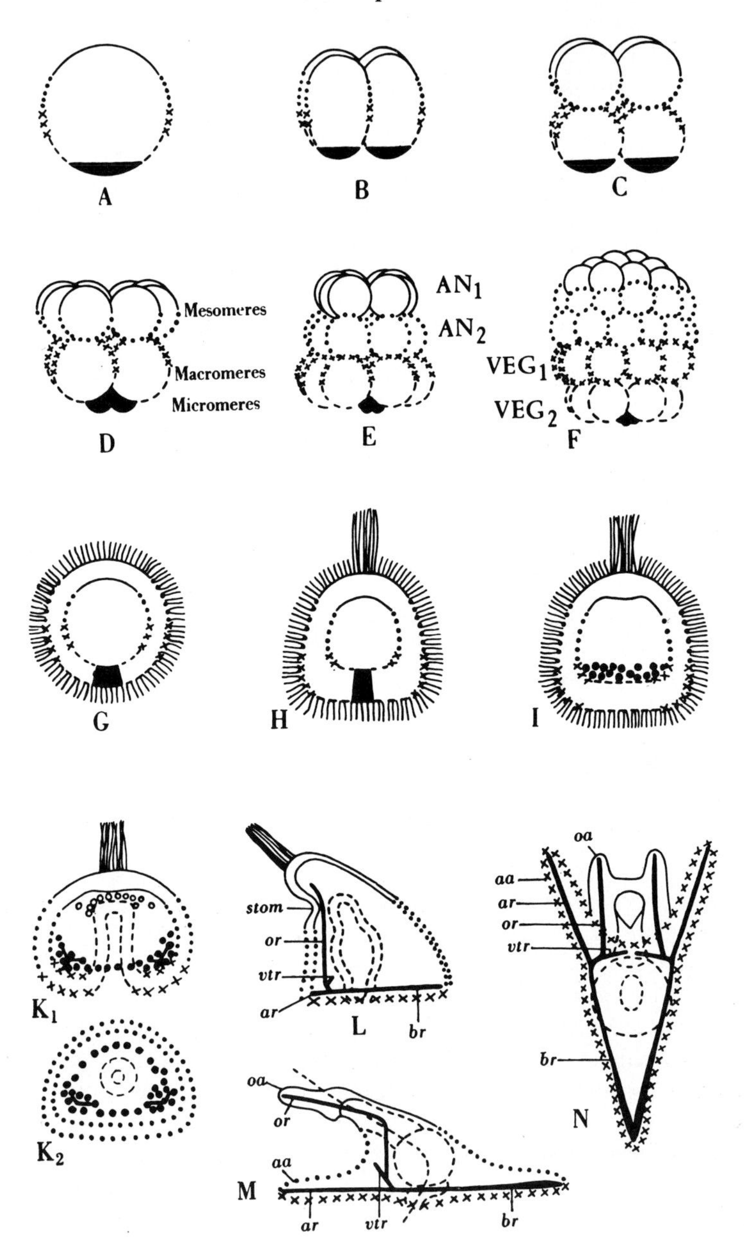

Fig. 5. Developmental fate of various regions of the sea urchin egg. Regions are defined in this drawing on the basis of their positions in the cell layers of a 64-cell embryo: animal 1 layer, continuous lines; animal 2, dotted; vegetal 1 crosses; vegetal 2, broken lines; micromeres, filled spaces. A. Uncleaved egg. B. Four-cell stage. C. Eight-cell stage. D. Sixteen-cell stage. E. Thirty-two-cell stage. F. Sixty-four-cell stage. G. Spherical blastula. H. Elongate blastula. I. Mesenchyme blastula. K. Gastrula. L. Prism. M,N. Pluteus. In all figures except K_2 and N, the orientation is animal pole up. From Hörstadius [1939].

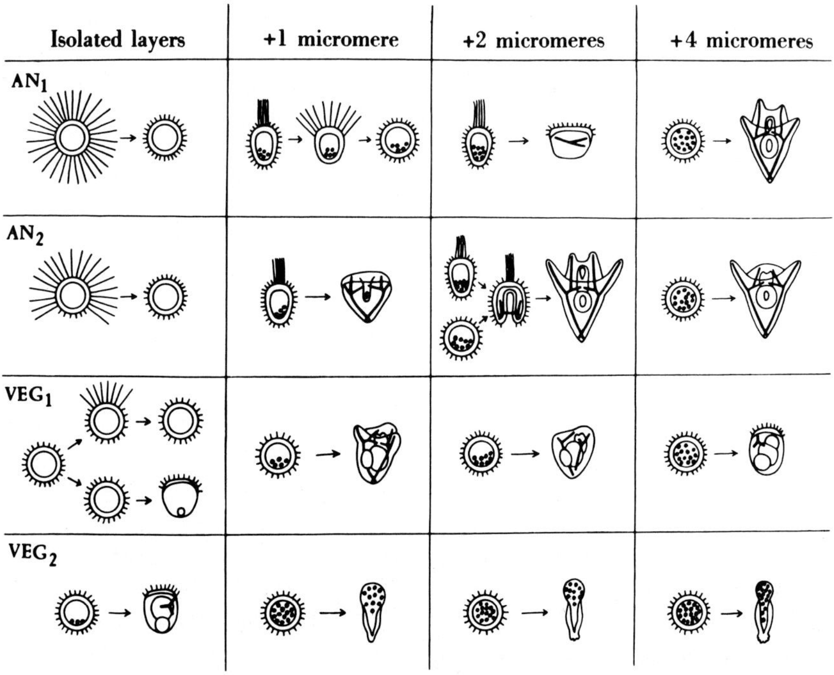

Fig. 6. Diagram of the development of the various layers (AN_1, AN_2, VEG_1, and VEG_2) from 64-cell embryos. Each layer is allowed to develop either in isolation (left column) or with one-to-four implanted micromeres. From Hörstadius [1935].

at the vegetal pole and a normal pluteus develops regardless of the axis of stratification. Thus, if specific determinants from the egg are segregated into or away from the micromeres, they are not easily centrifugable cytoplasmic inclusions. They may be either (1) associated with the cortex, plasma membrane, or extracellular coats of the egg, or (2) arranged in the cytoplasm in some noncentrifugable form, perhaps as a dynamic gradient of soluble material or in association with a hypothesized cytoskeletal network that is stable to centrifugation.

The cortex and surface of micromeres are specialized in several ways. (1) The micromere cortex is lacking in pigment vesicles [Harvey, 1956; Dan, 1954]. Those species that do not have cortical pigment do have homologous vesicles and those vesicles show the same cortical distribution [for review, see Showman, 1979]. In the fertilized egg these vesicles are uniformly distributed in the cortex. However, they leave the vegetal region prior to micromere formation. (2) Microvilli are much less abundant on micromeres than on mesomeres and macromeres [Schroeder, 1982]. (3) Concanavalin A (Con A) binding sites on the surfaces of micromeres behave differently than those on mesomeres and macromeres. In the presence of Con A, micromeres agglutinate more readily [Roberson and Oppenheimer, 1975] and their Con A binding sites exhibit a higher tendency to cap [Neri et al., 1975]. (4) In reaggregation experiments micromeres sort out from the other cells, indicating the presence of cell-specific surface receptors [Spiegel and Spiegel, 1978]. These unique properties suggest that the cortex and surface layers of the vegetal pole may play a role in determination of the micromeres.

Several studies have compared the blastomeres of 16-cell embryos at the molecular level, particularly with regard to the distribution and synthesis of macromolecules. The development of methods for mass separation of the three size classes of blastomeres [Spiegel and Tyler, 1966; Hynes and Gross, 1970; Mizuno et al., 1974; Chamberlain, 1977] has made this type of comparison feasible.

In many respects micromeres have proved to be remarkably similar to the other blastomeres. (1) The three cell types are essentially identical with respect to the rates of synthesis of DNA [Hynes and Gross, 1970], RNA [Hynes and Gross, 1970; Hynes et al., 1972a; Spiegel and Rubinstein, 1972], protein in general [Spiegel and Tyler, 1966; Hynes and Gross, 1970; Spiegel and Rubinstein, 1972], and tubulin specifically [Hynes et al., 1972b]. (2) Two-dimensional electrophoretic analysis of over 900 individual proteins indicates that the major accumulated proteins of 16-cell embryos are uniformly distributed among the blastomeres [Tufaro and Brandhorst, 1979]. (3) Similar electrophoretic analysis of newly synthesized proteins shows no qualitative differences among the cell types [Tufaro and Brandhorst, 1979; Harkey and Whiteley, 1983] with the possible exception of a single micromere-specific protein [Chamberlain, 1977]. (4) RNA transcribed from reiterated DNA during oogenesis and during development up to the 16-cell stage is uniformly distributed among the blastomeres at the fourth cleavage [Mizuno et al., 1974].

The 16-cell stage blastomeres do exhibit some important differences at the molecular level. (1) The three cell types show quantitative differences in the synthesis of lysine-rich proteins [Senger and Gross, 1978; Senger et al., 1978]. In particular, the relative rates of synthesis of two H1 histones vary systematically with cell size. (2) A portion of the RNA transcribed in 16-cell embryos from reiterated DNA is synthesized only in the micromeres [Mizuno et al., 1974]. These micromere-specific sequences are not detected in embryos before the 16-cell stage. (3) About one quarter of the RNA sequences derived from single-copy DNA and found in the cytoplasm of eggs and 16-cell embryos are missing from the cytoplasm of micromeres [Rodgers and Gross, 1978; Ernst et al., 1980]. This result is not altered when development takes place in the presence of actinomycin D. The implication is that a large number of maternal RNA sequences are selectively excluded from the micromeres during the fourth cleavage. (4) A large fraction of the so-called "complex" nuclear RNA in mesomeres and macromeres is not detectable in micromeres [Ernst et al., 1980].

It is not technically feasible to analyze specific portions of the animal-vegetal axis of sea-urchin embryos prior to the 16-cell stage using standard biochemical methods. Thus we can not examine directly when RNAs or other molecules may become segregated to or from the presumptive micromere cytoplasm. However, recent findings of animal-vegetal gradients of specific proteins [Moen and Namenwirth, 1977] and RNAs [Carpenter and Klein, 1982] in frog eggs, and of localized mRNAs in ascidian eggs (Jeffery, this volume) lend plausibility to the existence of such localizations in sea urchin eggs and early embryos.

The determination of micromeres might be viewed, not as an acquisition or segregation of skeletogenic potential, but rather as a selective loss of all other potentials. If the micromeres are removed from a 16-cell embryo, the remaining cells can develop into a normal pluteus larva with primary mesenchyme cells and spicules [Hörstadius, 1935; R. Langelan, unpublished results]. With the use of "vegetalizing" agents such as LiCl one can induce the differentiation of spicules by cells that would ordinarily become ectoderm [for review, see Hörstadius, 1975]. Thus, rather than possessing some unique determinant of skeletogenic potential, micromeres may be specialized by the lack of determinants for other potentials. It is interesting in this light that most of the evidence for segregation of materials at the fourth cleavages is negative with respect to the micromeres: Microvilli, pigment vesicles, and maternal RNAs are specifically excluded from these cells.

The Extent of Determination

As a final point, it is of interest to consider the *extent* to which micromeres are determined. In its strictest sense, the term "determined" indicates that the development of a cell or tissue is totally independent of the influences of other components of the embryo. This concept may not be entirely applicable to any developing system. It certainly does not describe the "determination" of micromeres. For these cells can be influenced by other cells to delay, modify, or abort their normal developmental program.

A delay in skeletogenesis occurs during the reaggregation of cells from dissociated gastrulae [Harkey and Whiteley, 1980]. Primary mesenchyme cells differentiate spicules 48 hour later in aggregates of total gastrula cells than they do in isolation or in undissociated embryos. Thus, the presence of reaggregating endoderm and ectoderm cells appears to temporarily block or reset the micromere program of development.

Both the annular pattern of primary mesenchyme cells within the gastrula and the shape of the spicules are influenced by the embryonic ectoderm. This has been inferred from observations and manipulations of whole embyros by many investigators [for reviews see Gustafson and Wolpert, 1963; Okazaki, 1975a]. With regard to ring formation, several authors have observed a subequatorial annular pattern of thickened ectoderm in the blastula which matches the pattern assumed later by the primary mesenchyme [Motomura, 1960; Okazaki, et al., 1962; Gustafson and Wolpert, 1961]. These authors suggested that the thickened ectoderm, which reflects a lengthened apical-basal axis of the individual cells in that region, may result from increased mutual adhesiveness among these cells. If this heightened adhesiveness also effects epithelial-mesenchymal interactions, the primary mesenchyme cells would be expected to accumulate along the ring of thickened epithelium. Such epithelial-mesenchymal interactions would secondarily govern the shape of the spicules since spicule growth follows the pattern of the cells. The ectodermal template theory is further supported by the fact that experimental displacement of the ectodermal thickening causes an analogous shift in the position of the primary mesenchyme ring [Czihak, 1962; Gustafson and Wolpert, 1961].

Observations of micromeres and primary mesenchyme cells in isolated culture [Harkey and Whiteley, 1980] have shown that (1) these cells do not accumulate as rings in the absence of other cell types, and (2) the spicules they produce differ in a characteristic manner from those made within embryos (Fig. 7). Furthermore, recombination of primary mesenchyme cells with epithelial cells (ectoderm and endoderm) can lead to ring formation and to the production of more normal spicules. Thus, morphogenesis of the primary mesenchyme is clearly influenced by extrinsic forces.

The complete absence of spicule formation by micromere derivatives has been noted in at least two kinds of experiments. First, isolated micromeres do not produce spicules in culture unless horse serum is added [Okazaki, 1975a; Harkey and Whiteley, 1980]. The horse serum may substitute for some essential component of the blastocoele matrix within which skeletogenesis normally occurs. Supporting this idea, primary mesenchyme cells can differentiate spicules in culture in the absence of horse serum or other cells if they are first isolated by a procedure that preserves the surrounding ectodermal basal lamina [Harkey and Whiteley, 1980]. This structure, which forms a closed bag around the cells, probably retains some of the components of the blastocoel matrix as well. Thus, it would appear that the ectoderm confines and perhaps produces a permissive environment for skeletogenesis. Second, extreme animalization or vegetalization of embryos, either by chemical agents or by dissection, can block skeletogenesis [see Hörstadius, 1975]. For example, the transplantation of isolated micromeres into the Veg 2 layer

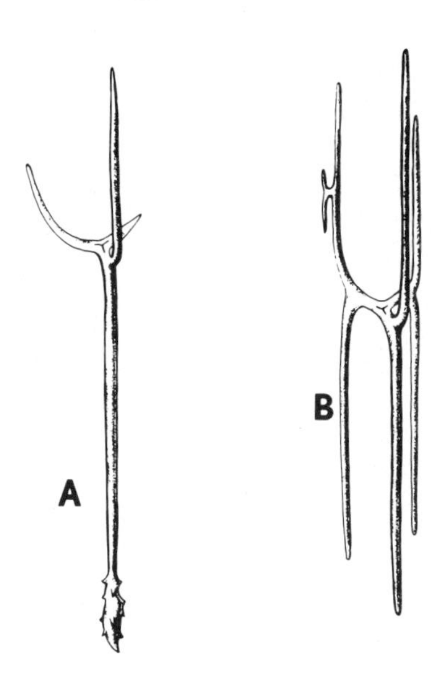

Fig. 7. Comparison of typical spicules grown in normal embryos (A) and in cultures of primary mesenchyme cells isolated from early gastrulae of *Strongylocentrotus purpuratus* (B). From Harkey and Whiteley [1980].

of a 64-cell stage embryo (see Fig. 5) results in the development of an exogastrula without spicules [Hörstadius, 1935] (Fig. 6).

In conclusion, although the formation of micromeres is a highly determinative event, this event does not dictate every detail of micromere development. Skeletogenesis is controlled by a combination of the unique properties given to these cells at the fourth cleavage, and interactions with other cells in the embryo.

ACKNOWLEDGMENTS

This work was supported in part by grant No. HL-10312 to Arthur L. Whiteley and training grant No. HD-00266 from the National Institute of Child Health and Human Development.

REFERENCES

Balinksy BI (1959): An electron microscopic investigation of the mechanisms of adhesion of the cells in sea urchin blastula and gastrula. Exp Cell Res 16:429–433.

Bevelander G, Nakahara H (1960): Development of the skeleton of the sand dollar (*Echinarachnius parma*). In: "Calcification in Biological Systems," Washington: Amer Assoc Adv Sci, pp 41–56.

Boveri T (1901): Die Polaritat von Oocyte, E: und Larve des *Strongylocentrotus lividus*. Zool Jahrb Abt Anat Ontol 14:630–653.

Carpenter CD, Klein WH (1982): A gradient of poly $(A)^+$ RNA sequences in *Xenopus laevis* eggs and embryos. Dev Biol 91:43–49.

Chamberlain JP (1977): Protein synthesis by separated blastomeres of sixteen-cell sea urchin embryos. J Cell Biol [Abstr] 75:33a.

Czihak G (1962): Entwicklungsphysiologische Untersuchungen an Echiniden. (Topochemie der Blastula und Gastrula, Entwicklung der Bilateral-und der Coelomdivertikel). W Rouxs Arch Entwicklungsmech 154:29–211.

Dan K (1952): Cyto-embyrological studies of sea urchins. II. Blastula stage. Biol Bull 102:74–89.

Dan K (1954): The cortical movement in *Arbacia punctulata* eggs through cleavage cycles. Embryologia 2:115–122.

Dan K (1972): Modified cleavage pattern after suppression of one mitotic division. Exp Cell Res 72:69–73.

Dan K (1979): Studies on unequal cleavage in sea urchins. I. Migration of the nuclei to the vegetal pole. Dev Growth Differ 21:527–535.

Dan K, Ikeda M (1971): On the system controlling the time of micromere formation in sea urchin embryos. Dev Growth Differ 13:285–301.

Dan K, Nakajima T (1956): On the morphology of the mitotic apparatus isolated from Echinoderm eggs. Embryologia 3:187–200.

Davidson JM (1974): "On the Role of the Epithelial Basal Lamina in Echinoid Morphogenesis." PhD thesis: Department of Biological Sciences, Stanford University.

Driesch M (1896): Die taktische Reizbarkeit der Mesenchymzellen von *Echinus microtuberculatus*. W Rouxs Arch Entwicklungsmech 3:362–380.

Driesch H (1898): Uber rein-mutterliche Charaktere an Bastardlarven von Echiniden. Arch. Entwicklungsmech Org 7:65.

Ellis CH Jr, Winter RJ (1967): Protein synthesis and skeletal spicule formation in the sea urchin larva. Am Zool [Abstract] 7:750.

Endo Y (1966): Fertilization, cleavage and early stages in development. In Isemura et al (eds): "Contemporary Biology. Vol 4. Development and Differentiation." Tokyo: Iwanami Shoten, pp 1–61.

Ernst SG, Hough-Evans BR, Britten RJ, Davidson EH (1980): Limited complexity of the RNA in micromeres of sixteen-cell sea urchin embryos. Dev Biol 79–119–127.

Evola-Maltese C (1957): Histochemical localization of alkaline phosphatase in the sea urchin embryo. Acta Embryol Morphol Exp 1:99.

Gibbins JR, Tilney LG, Porter KR (1969): Microtubules in the formation and development of the primary mesenchyme in *Arbacia punctulata*. I. The distribution of microtubules. J Cell Biol 41:201, 227.

Golob R, Chetsanga CJ, Doty P (1974): The onset of collagen synthesis in sea urchin embryos. Biochim Biophys Acta 349:135–141.

Gould D, Benson SC (1978): Selective inhibition of collagen synthesis in sea urchin embryos by a low concentration of actinomycin D. Exp Cell Res 112:73–78.

Gustafson T (1975): Cellular behavior and cytochemistry in early stages of development. In G Czihak (ed): "The Sea Urchin Embryo." Berlin: Springer-Verlag, pp 175–232.

Gustafson T, Hasselberg I (1951): Studies on enzymes in the developing sea urchin egg. Exp Cell Res 2:642–672.

Gustafson T, Kinnander H (1956): Gastrulation in the sea urchin larva studied by aid of time-lapse cinematography. Exp Cell Res 10:733–756.

Gustafson T, Wolpert L (1961): Studies on the cellular basis of morphogenesis in the sea urchin embryo. Directed movements of primary mesenchyme cells in normal and vegetalized larvae. Exp Cell Res 24:64–79.

Gustafson T, Wolpert L (1963): The cellular basis of morphogenesis and sea urchin development. Int Rev Cytol 15:139, 214.

Harkey MA, Whiteley AH (1980): Isolation, culture, and differentiation of echinoid primary mesenchyme cells. W Rouxs Arch 189:111–122.

Harkey MA, Whiteley AH (1982a): The translational program during the differentiation of isolated primary mesenchyme cells. Cell Differ 11:325–329.

Harkey MA, Whiteley AH (1982b): Cell-specific regulation of protein synthesis in the sea urchin gastrula: A two-dimensional electrophoretic analysis. Dev Biol 93:453–462.

Harkey MA, Whiteley AH (1983): The program of protein synthesis during the development of the micromere-primary mesenchyme cell line in the sea urchin embryo (submitted).

Harvey EB (1956): "The American Arbacia and Other Sea Urchins." Princeton University Press, Princeton.

Herbst C (1896): Experimentelle Untersuchungen Uber den Einfluss der veranderten chemischen Zusammensetzung des umgebenden Mediums auf die Entwicklung der Tiere. III. Uber das Ineinandergreifen von normaler Gastrulation und Lithiumentwicklung. W Rouxs Arch Entwicklungsmech 2:455–516.

Hörstadius S (1935): Uber die Determination im Verlaufe der Eiachse beir Seeigeln. Pubbl Staz Zool Napoli 14:251–479.

Hörstadius S (1939): The mechanisms of sea urchin development, studied by operative methods. Biol Rev 14:132–179.

Hörstadius S (1953): The effect of lithium ions on centrifuged eggs of *Paracentrotus lividus*. Pubbl Staz Zool Napoli 24:45–60.

Hörstadius S (1975): Isolation and transplantation experiments. In Czihak D (ed): "The Sea Urchin Embryo." Berlin: Springer-Verlag, pp 364–406.

Hsiao SC, Fujii WK (1963): Early ontogenic changes in the concentration of alkaline phosphatase in Hawaiian sea urchins. Exp Cell Res 32:217–231.

Hynes RO, Gross PR (1970): A method for separating cells from early sea urchin embryos. Dev Biol 21:383–402.

Hynes RO, Greenhouse GA, Minkoff R, Gross PR (1972a): Properties of the three cell types in sixteen-cell sea urchin embryos: RNA synthesis. Dev Biol 27:457–478.

Hynes RO, Raff RA, Gross PR (1972b): Properties of the three cell types in sixteen-cell sea urchin embryos: Aggregation and microtubule protein synthesis. Dev Biol 27:150–164.

Immers J (1961): Comparative study of the localization of incorporated ^{14}C-labeled amino acids and ^{35}SO in the sea urchin ovary, egg, and embryo. Exp Cell Res 24:356, 378.

Immers J, Runnström J (1965): Further studies on the effects of deprivation of sulfate on the early development of the sea urchin *Paracentrotus lividus*. J Embryol Exp Morphol 14:289–305.

Jenkinson JW (1911): On the origin of the polar and bilateral structure of the egg of the sea urchin. Arch Entw. Mech. 32:699–716.

Karp GC, Solursh M (1974): Acid mucopolysaccharide metabolism, the cell surface, and primary mesenchyme cell activity in the sea urchin embryo. Dev Biol 41:110–123.

Katow H, Solursh M (1980): Ultrastructure of primary mesenchyme cell ingression in the sea urchin, *Lytechnicus pictus*. J Exp Zool 213:231–246.

Kitajima T, Okazaki K (1980): Spicule formation in vitro by the descendants of precocious micromere formed at the 8-cell stage of sea urchin embryo. Dev Growth Differ 22:265–279.

Lallier R (1975): Animalization and vegetalization. In G. Czihak (ed): "The Sea Urchin Embryo." Berlin: Springer-Verlag, pp. 473–509.

Langelan RE, Whiteley AH (1980): The importance of cell size in echinoid determination. Am Zool [abstract] 20:65.

Lindahl PE (1932): Zur Kenntnis des Ovarialeies bei dem Seeigel. W Rouxs Arch Entwicklungsmech. Organismen 126:373–390.

Lyon EP (1906a): Some results of centrifugalizing the eggs of *Arbacia*. Am J Physiol 15:21.

Lyon EP (1906b): Results of centrifugalizing eggs. Arch Entwisklungsmech 23:151.

Masuda M (1979): Species specific pattern of ciliogenesis in developing sea urchin embryos. Dev Growth Differ 21:545–552.

Millonig G (1970): A study on the formation of the sea urchin spicule. J Submicro Cytol 2:157–165.

Mintz GR, Lennarz WJ (1982): Spicule formation by cultured embryonic cells from the sea urchin. Cell Differ 11:331–333.

Mizuno S, Lee YR, Whiteley AH, Whiteley HR (1974): Cellular distribution of RNA populations in 16-cell stage embryos of the sand dollar, *Dendraster excentricus*. Dev Biol 37:18–27.

Moen TL, Namenwirth M (1977): The distribution of soluble proteins along the animal-vegetal axis of the frog egg. Dev Biol 58:1–10.

Moore AR (1940): Osmotic and structural properties of the blastular wall in *Dendraster excentricus*. J Exp Zool 84:73–79.

Morgan TH, Lyon EP (1907): The relation of the substances of the egg, separated by a strong centrifugal force, to the location of the embryo. Arch Entwicklungsmech 24:147.

Morgan TE, Spooner GB (1909): The polarity of the centrifuged egg. Arch Entwicklungsmech 28:104.

Motomura J (1960): Secretion of mucosubstance in the gastrula of the sea urchin embryo. Bull Biol Stn Asamuchi 10:165–169.

Nakano E, Okazaki K, Iwamatsu T (1963): Accumulation of radioactive calcium in larvae of the sea urchin *Pseudocentrotus depressus*. Biol Bull 125:125–133.

Neri A, Roberson M, Connolly DT, Oppenheimer SB (1975): Quantitative evaluation of concanavalin A receptor site distributions on the surfaces of specific populations of embryonic cells. Nature 258:342–344.

O'Farrell PH (1975): High resolution two-dimensional electrophoresis of proteins. J Biol Chem 250:4007–4021.

Okazaki K (1956): Skeleton formation of sea urchin larvae. I. Effect of Ca concentration of the medium. Biol Bull III:320–333.

Okazaki K (1960): Skeleton formation of sea urchin larvae. II. Organic matrix of the spicule. Embryologia 5:283–320.

Okazaki K (1961): Skeleton formation of sea urchin larvae. III. Similarity of effect of low calcium and high magnesium on spicule formation. Biol Bull 120:177–182.

Okazaki K (1965): Skeleton formation of sea urchin larvae. V. Continuous observation of the process of matrix formation. Exp Cell Res 40:585–596.

Okazaki K (1970): Growth of the spicules in sea urchin larvae. Collagen Symp 8:113–129. (In Japanese).

Okazaki K (1971a): In vitro culture of the micromeres and primary mesenchyme cells isolated from sea urchin embryos and larvae. In: Jpn Soc Dev Biol (eds): "Cells in Early Stages of Development." Tokyo:Iwanami Shoten. (In Japanese).

Okazaki K (1971b): Spicule formation in sea urchin larvae; observations in vivo and in vitro. Symp Cell Biol 22:163–171. (In Japanese).

Okazaki K (1975a): Spicule formation by isolated micromeres of the sea urchin embryo. Am Zool 15:567–581.

Okazaki K (1975b): Normal development to metamorphosis. In Czihak G: "The Sea Urchin Embryo." Berlin: Springer-Verlag, pp 175–232.

Okazaki K, Fukushi T, Dan K (1962): Cyto-embryological studies of sea urchins. IV. Correlation between the shape of the ectodermal cells and the arrangement of the primary mesenchyme cells in sea urchin larvae. Acta Embryol Morphol Exp 5:17–31.

Okazaki K, Inoué S (1976): Crystal property of the larval sea urchin spicule. Dev Growth and Differ 18:413–434.

Okazaki K, Niijima L (1964): Basement membrane in sea urchin larvae. Embyrologia 8:89–100.

Ozaki H (1974): Localization and multiple forms of acetylcholinesterase in sea urchin embryos. Dev Growth and Differ 16:267–279.

Parisi E, Filosa S, De Petrocellis B, Monroy A (1978): The pattern of cell division in the early development of the sea urchin, *Paracentrotus lividus*. Dev Biol 65:38–49.

Parisi E, Filosa S, Monroy A (1979): Actinomycin D- disruption of the mitotic gradient in the cleavage stages of the sea urchin embryo. Dev Biol 72:167–174.

Peltz R, Guidice G (1967): The control of skeleton differentiation in sea urchin embryos. A molecular approach. Biol Bull [abstract] 133:479.

Pfohl RJ (1965): Changes in alkaline phosphatase during the early development of the sea urchin, *Arbacia punctulata*. Exp Cell Res 39:496–503.

Pfohl RJ, Giudice G (1967): The role of cell interaction in the control of enzyme activity during embryogenesis. Biochim Biophys Acta 142:263–266.

Pucci-Minafra I, Casano C, La Rosa C (1972): Collagen synthesis and spicule formation in sea urchin embryos. Cell Differentiation 1:157–165.

Pucci-Minafra I, Fanara M, Minafra S (1980): Chemical and physical changes in the organic matrix of mineralized tissues from embryo and adult *Paracentrotus lividus*. J Submicrosc Cytol 12:267–273.

Pucci-Minafra I, Minafra S, Gianguzza F, Casano C (1975): Amino acid composition of collagen extracted from the spicules of sea urchin embryos (*Paracentrotus lividus*). Boll Zool 42:201–204.

Roberson M, Oppenheimer SB (1975): Quantitative agglutination of specific populations of sea urchin embryo cells with concanavalin A. Exp Cell Res 91:263–268.

Rodgers WH, Gross PR (1978): Inhomogeneous distribution of egg RNA sequences in the early embyro. Cell 14:279–288.

Rustad RC (1960): Dissociation of the mitotic time-schedule from the micromere "clock" with x-rays. Acta Embryol Morphol Exp 3:155–158.

Runnström J (1928): Plasmabau und Determination bei dem E; von *Paracentrotus lividus* LK. W Rouxs Arch Entwicklungsmech. Organismen 113:556–581.

Runnström J (1975): Integrating factors. In Czihak GH (ed): "The Sea Urchin Embryo." Berlin: Springer-Verlag, pp 646–670.

Sano K (1977): Changes in cell surface charges during differentiation of isolated micromeres and mesomeres from sea urchin embryos. Dev Biol 60:404–415.

Schneider EG, Nguyen HT, Lennarz WJ (1978): The effect of tunicamycin, an inhibitor of protein glycosylation, on embryonic development in the sea urchin. J Biol Chem 253:2348–2355.

Schroeder TE (1982): In Poste G, Nicolson GL (eds): "Cytoskeletal Elements and Plasma Membrane Organization. Cell Surface Reviews." Amsterdam: Elsevier/North Holland, Vol 7, pp 167–214.

Senger DR, Arceci RJ, Gross PR (1978): Histones of sea urchin embryos. Transients in transcription, translation, and the composition of chromatin. Dev Biol 65:416–425.

Senger DR, Gross PR (1978): Macromolecule synthesis and determination in sea urchin blastomers at the sixteen-cell stage. Dev Biol 65:404–415.

Showman RM (1979): "The Nature of the Echinoid Hatching Enzyme and the Origin of its Messenger RNA." PhD Thesis, Department of Zoology, University of Washington.

Spiegel M, Rubinstein NA (1972): Synthesis of RNA by dissociated cells of the sea urchin embyro. Exp Cell Res 70:423–439.

Spiegel M, Spiegel E (1978): Sorting out of sea urchin embyronic cells according to cell type. Exp Cell Res 117:269–171.

Spiegel M, Tyler A (1966): Protein synthesis in micromeres of the sea urchin egg. Science 151:1233–1234.

Sugiyama K (1972): Occurrence of mucopolysaccharides in the early development of the sea urchin embryo and its role in gastrulation. Dev Growth Differ 14:63–73.

Takahashi MM, Okazaki K (1979): Total cell number and number of primary mesenchyme cells in whole, ½ and ¼ larvae of *Clypeaster japonicus*. Dev Growth Differ 21:553–566.

Tanaka Y (1976): Effects of the surfactants on the cleavage and further development of the sea urchin embryos. I. The inhibition of micromere formation at the fourth cleavage. Dev Growth Differ 18:113–122.

Tilney LG, Gibbins JR (1969): Microtubules in the formation and development of the primary mesenchyme in *Arbacia punctulata*. J Cell Biol 41:227–250.

Tufaro F, Brandhorst BP (1979): Similarity of proteins synthesized by isolated blastomeres of early sea urchin embryos. Dev Biol 72:390–397.

Wolpert L, Gustafson T (1961): Studies on the cellular basis of morphogenesis of the sea urchin embryo. Development of the skeletal pattern. Exp Cell Res 25:311–325.

Wolpert L, Mercer EH (1963): An electron microscope study of the development of the blastula of the sea urchin embryo and its radial polarity. Exp Cell Res 30:280–300.

Time, Space, and Pattern in Embryonic Development, pages 157–169

Patterns of Antigenic Expression in Early Sea Urchin Development

David R. McClay, Gail W. Cannon, Gary M. Wessel, Rachel D. Fink, and Richard B. Marchase

Departments of Zoology (D.R.M., G.W.C., R.D.F.) and Anatomy (G.M.W., R.B.M.), Duke University, Durham, North Carolina 27706

Morphogenetic movements at gastrulation have been modeled by Gustafson and Wolpert [1963, 1967] to include a series of mechanical movements accompanied by changes in cell contacts. The latter were thought to be molecular-based alterations in cell-cell or cell-extracellular matrix affinities. Thus, the model tacitly assumes that molecular changes in the cell surface somehow direct the fundamental changes in cell affinities which affect cell shape change and subsequent tissue movement. The experimental support for this notion is largely indirect and is based on the capacity of cells in aggregates to sort out into layers appropriate to normal germ layer associations [Townes and Holtfreter, 1955; Guidice, 1962]. Information on the molecular detail of cell recognition remains fragmentary in most systems although several laboratories have made progress by describing suspected recognition components [Marchase et al., 1976].

In the sea urchin embryo, developmental changes in the cell surface have been examined by several investigators. Studies by Lennarz and colleagues [Schneider et al., 1978] using inhibition by tunicamycin demonstrated a potential role for N-glycosidically linked cell surface glycoproteins in gastrulation. Extracellular matrix components known to be cellular substrates in other systems have been described either in the intercellular space [Spiegel et al., 1980] or in the basal lamina that lines the blastocoele [Crise-Benson and Benson, 1979; Wessel et al., 1982]. When embryos are grown in sulfate-free seawater or in inhibitors of collagen synthesis, they fail to gastrulate [Wessel et al., 1982]. This suggests a role for extracellular matrix components in gastrulation. Primary mesenchyme cells have been shown to lose an affinity for the extraembryonic hyaline layer at the beginning of their delamination movements [McClay and Fink, 1982]. Cell aggregation of dissociated sea urchin embryonic cells has been shown to be affected by a butanol extract of cell membranes [Noll et al., 1979]. The cells treated with low concentrations of butanol lack the ability to reaggregate while addition of the extract restores adhesive capacity. An antibody to the extract has a selective effect on aggregation. Recognition specificity was shown to change at gastrulation by the expression of new antigens on the cell surface [McClay et al, 1977; McClay and Chambers, 1978]. In addition, changes in lectin binding [Oppenheimer, 1977; Spiegel and Burger, 1982] support the notion that cell surface components change during development. While not always bearing directly on specific cell surface molecules, these studies illustrate that significant changes in the cell surface occur at gastrulation, coincident with the dramatic morphogenetic rearrangements which characterize the gastrula stage. Since systematic studies on protein changes in membranes of the sea urchin have not been reported, the complexity and temporal diversity of the cell surface is not known.

Previous studies on proteins during development have focused on changes in whole embryo homogenates as seen with the protein separation techniques. The highest resolution yet achieved by two-dimensional (2-D) gel electrophoresis [Brandhorst, 1976], has shown that of about 10^3 proteins resolved, there is very little change in the spectrum synthesized up to the blastula stage and only about 15%–20% change (mostly quantitative) thereafter [Brandhorst, 1976].

The protein studies and similar studies on RNA have not localized molecules within the embryo so that information on spatial distribution is lacking. Such information would be of value for morphogenetic considerations. Therefore, spatial and temporal localization of proteins can provide an important developmental perspective which complements the studies on protein synthesis.

This chapter describes the patterns of expression of a panel of cell surface and extracellular matrix components during development in the sea urchin embryo using the monoclonal antibody technique [Kohler and Milstein, 1975]. The purpose of this study is to demonstrate specific antigen distribution at a level of gene expression which cannot be seen in biochemical studies on RNA or protein diversity during development. From these studies, it appears that maternal templates play a dominant role in early development, a finding that supports the recent studies on the molecular biology of sea urchin development [Davidson et al., 1982]. A level of order and complexity is demonstrated which suggests that gradients and pattern cues are going to be important to the understanding of gene expression in the three-dimensional (3-D) microenvironment of the embryo.

Before describing the antigens and their patterns of expression, it is important to understand the strengths and the limitations of monoclonal antibodies. In this technology a lymphocyte is fused with a myeloma cell to form a hybridoma. The lymphocyte contributes the ability to produce a specific antibody, and the myeloma heritage enables the fused hybrid to grow in culture. The investigator can isolate clones of hybridomas that produce antibodies against single determinants. The ability to harvest antibodies in culture also enables one to isolate large quantities of antibodies which, by comparison, might be present in diminishingly small amounts in polyclonal sera. A limitation of monoclonal technology that must be appreciated is that it still relies on the ability of the immune system (in this case, mouse) to respond to the antigen. Thus, rare antigens in a mixture of immunizing material and antigens that are poor immunogens are less likely to provide hybridoma clones than antigens in abundance or antigens with good immunogenicity. For this reason, a panel of monoclonals may overrepresent some antigens and may underrepresent others.

The monoclonal antibodies used in this study were raised against cell membranes and extracellular matrix from gastrula-stage sea urchin embryos. The antibodies were screened for germ layer specificity first by enzyme-linked immunosobent assays (ELISA), using membranes prepared from isolated germ layers [McClay and Marchase, 1979]. Those with a germ layer preference were cloned and examined by immunofluorescence on formalin-fixed, paraffin-sectioned material. Most of the monoclonal antibodies retained activity on the fixed material, and it was by far, the most satisfactory approach for gaining a 3-D impression of antigen localization. Of the antigens examined, several may be represented on more than a single molecule. For instance, an antibody to a common oligosaccaride chain may be shared by several different glycoproteins.

The antigens localized in this study have been subdivided into two groups: those that are present in the egg, and those that are expressed de novo during development.

Antigens Expressed in the Egg

The antigens in this class are present in the egg and therefore presumed to be expressed by maternal templates. Most of these antigens later become distributed to a localized area of the embryo. It has not yet been determined whether (1) the original antigens are compartmentalized, (2) localization requires expression of maternal mRNA, or (3) localization requires expression of embryonic mRNA. Nevertheless, the patterns illustrated below show that localization occurs so that an allocation mechanism must exist at some level. As a class, these antigens are localized to a germ layer or an area of the embryo on a time table that appears to be independent of other antigens. Several examples are given below to illustrate the behavior of this class.

Antigens that ultimately become localized to the endoderm are shown in Figure 1. All of them are present in the egg and remain intracellular through the blastula stage. Some become confined to the endoderm during gastrulation or in the early pluteus stage (Fig. 1c); others are localized much later in the pluteus stage (Fig. 1d).

Antigens in the egg that become confined to ectoderm can be distinguished first by their movement to the cell surface as part of the fertilization response (Fig. 2). During early developmental stages these antigens often are at least partially retained intracellularly (Fig. 2b) though not always (Fig. 2c). The

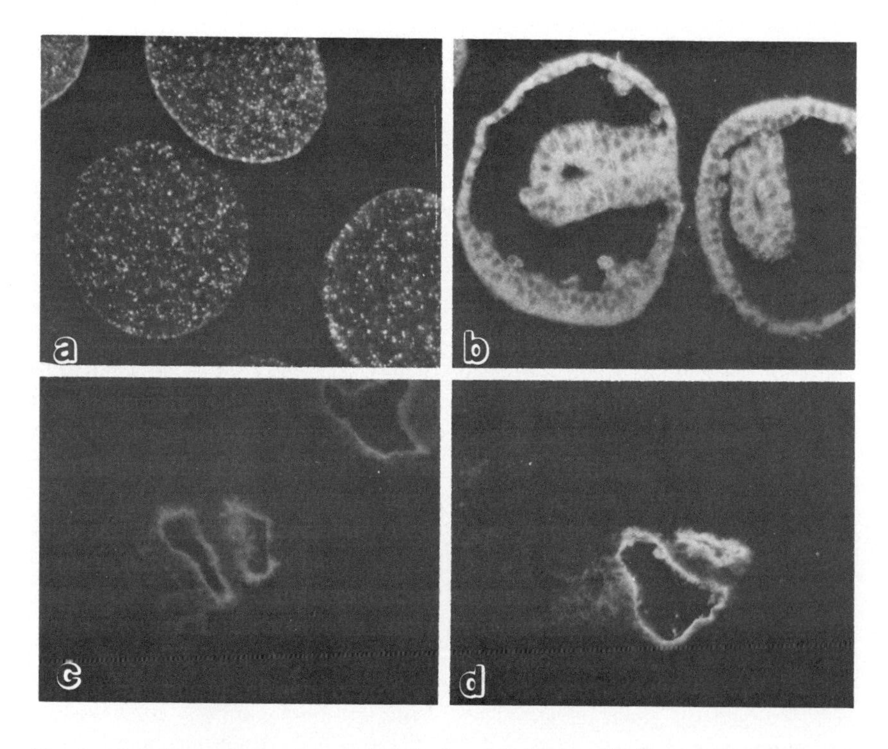

Fig. 1. Endoderm antigens that are initially found in the egg. (a) These antigens appear to be contained in discrete particulate depositions in unfertilized eggs. The antigens in this class that were observed to be eventually localized to the endoderm remained intracellular in all embryonic cells through the gastrula stage as in (b). The antigens then were segregated to the midgut in the prism stage as in (c) or to the foregut and midgut as in (d). Other endoderm antigens segregated to the hindgut or to the entire gut at times that were independent of the segregation of other endoderm antigens. All figures approximately ×250.

presumptive ectodermal antigens continue to be found along the surface of the archenteron during gastrulation. They then are lost from the endodermal surface at different times and to different extents depending on the antigen (Fig. 2d).

Some antigens present in the egg become localized to the basal lamina or to the matrix of the blastocoele. One is released from cells at the blastula stage (Fig. 3a) and others are released later (Fig. 3). Although the pattern of deposition is somewhat variable (Fig. 3), each antigen becomes localized to the wall of the blastocoele. It should be noted that the same cells release the basal lamina material to the basal surface and other antigens to the apical surface of the embryo. This polarized expression of surface- or matrix-

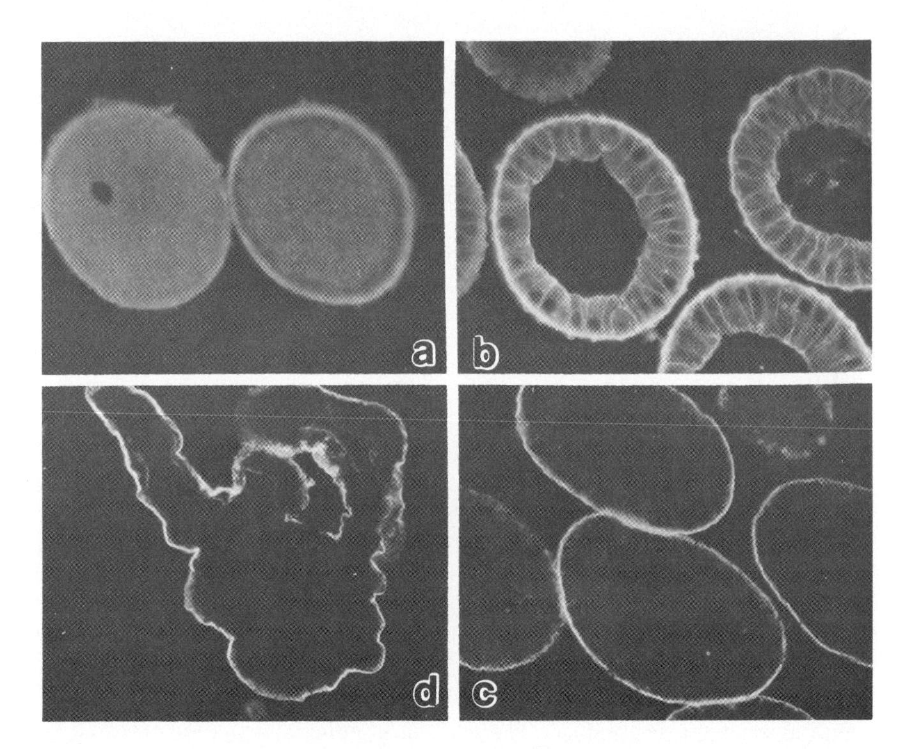

Fig. 2. Antigens of the ectoderm that are localized in the egg. The antigens observed that fall into this class are distributed evenly as particulate spots in the egg (a). At fertilization some of the antigen becomes concentrated at the surface of the zygote leaving an area of apparent antigen depletion beneath. Some of the antigens that will ultimately become confined to the ectoderm retain the antigens inside cells as well as on the surface (b). Other antigens are localized on the embryo surface almost exclusively (c). The antigen shown in (c) becomes confined to the ectoderm surface and to the surface of the foregut (d).

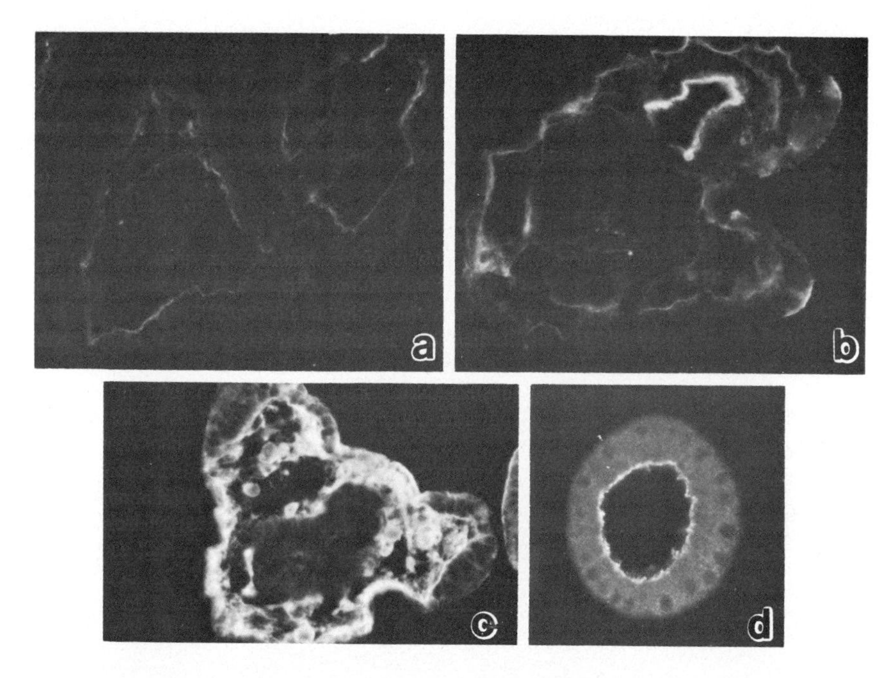

Fig. 3. Antigens of the basal lamina. Four different antigens are shown. (a). Only the basal lamina is stained. This antigen was not observed within cells at any stage. In (b) the basal lamina is stained but the antigen is also found at high concentrations in the foregut. (c) is an antigen that is secreted by primary mesenchyme cells and becomes localized to a matrix throughout the blastocoele. (d) is an antigen that is in the egg and is released basally beginning in the blastula stage.

related antigens is a dominant theme for most of the antigens observed, and it will be of importance to learn how intracellular trafficking moves proteins in such a polarized fashion. Figure 4 shows embryos with apical intracellular particulate arrays of antigens yet the antigens in Figure 4a and b are deposited on the basal surface while the antigen in Figure 4c is deposited on the apical surface of the embryo.

Of the antigens localized in the egg several, including hyalin (Fig. 5), are found in the cortical granules. The cortical granule contents are released at fertilization and hyalin becomes an extraembryonic matrix (Fig. 5). The presumptive ectoderm, endoderm, and mesoderm antigens described above with monoclonal antibodies have not yet been localized ultrastructurally. Thus, it is not known how they are distributed—only that the eggs have a speckled appearance suggesting some kind of intracellular localization.

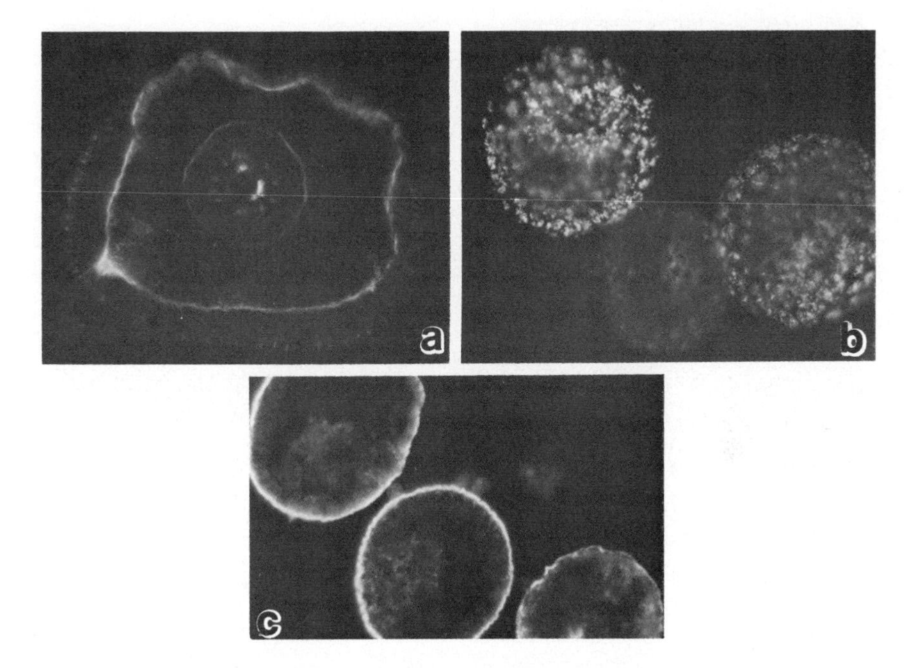

Fig. 4. Examples of intracellular particulate depositions of antigen. (a). At the gastrula stage each cell contains one apical area of localized antigen. The antigen is released into the blastocoele at the basal end of the cells. The particulate material for different antigens sometimes appears as a single deposit (as in a) though for other antigens it appears as multiple deposits in the apical region of the cells (b). (c). Another antigen that is secreted apically also has antigen localized in the apical end of the cells (c).

Antigens Expressed de Novo

The earliest antigen observed to be expressed de novo is a mesodermal antigen (Fig. 6) that is expressed on primary mesenchyme cells at the mesenchyme blastula stage. The antigen does not appear until cells are fully delaminated from the cell layer of the surrounding blastula. Cells fixed in the process of delamination do not have the antigen on their surface while cells in the same section that have already completed their transit into the blastocoele express the antigen. The antigen is also expressed on micromeres in cultures at a time corresponding to the mesoderm antigen expression in vivo. As can be seen in Figure 6f, the antigen covers a network of cell extensions and covers the surface or envelope of the growing spicule.

The next antigen observed is an endodermal antigen which appears on the

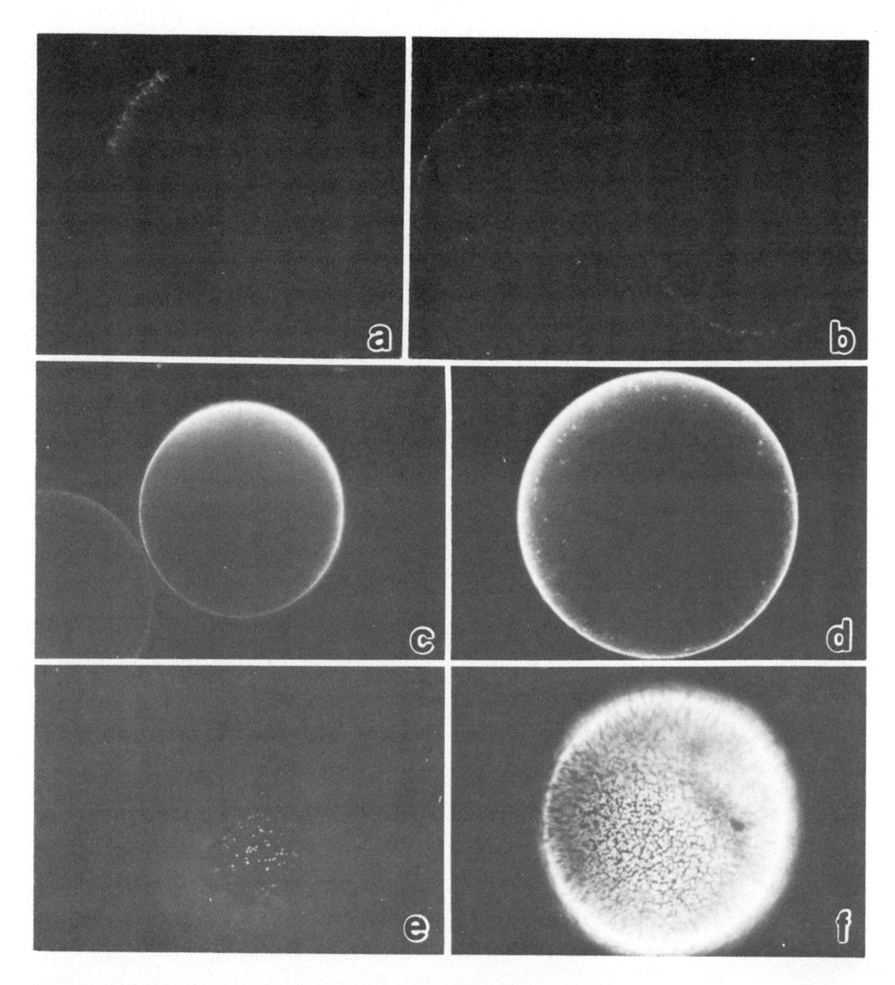

Fig. 5. Hyalin release. Hyalin is one of the proteins localized to cortical granules prior to fertilization. It is released to the embryo surface in a wave beginning at the point of sperm entry (a–d). In a surface view of the eggs the release can be seen as spots the size of cortical granules (e). The hyalin spreads to form a complete layer by about two minutes postfertilization (f).

vegetal plate just prior to invagination of the archenteron. The antigen is localized on the apical surface of endodermal cells (Fig. 7) and is expressed on the posterior two thirds of the gut. This restricted pattern is revealed early in gastrulation long before there is an apparent morphological restriction in the gut. The antigen persists only in part of the archenteron and is not present on most of the anterior third of the gut, nor is it present on secondary mesenchyme (Fig. 7).

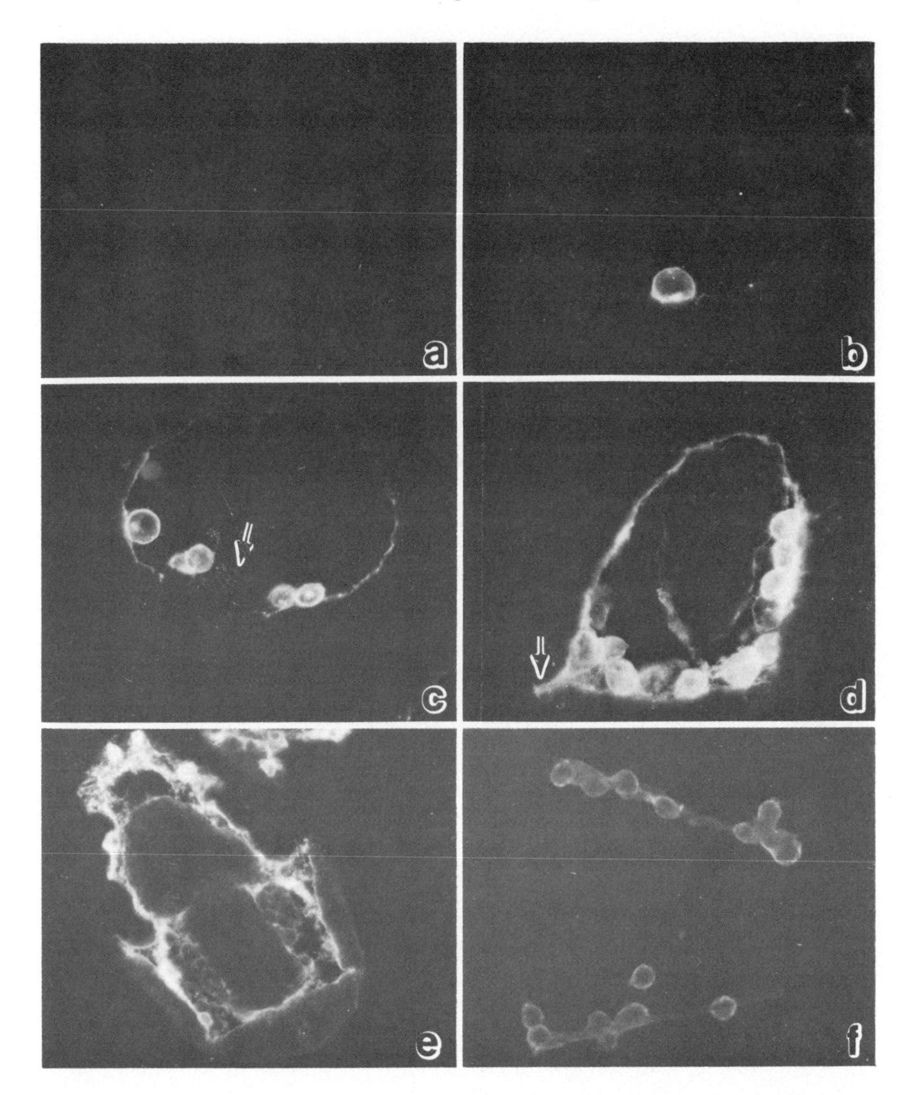

Fig. 6. Mesodermal antigen. This antigen is one of a family that is localized to the primary mesenchyme cells. It does not stain eggs (a) and first appears as primary mesenchyme cells delaminate from the monolayer of the blastula wall (b). The antigen is not present on cells fixed in the process of delamination (arrow in c) but is present on cells that have already delaminated. The cells in (c) also have small intracellular spots of stain. The antigen is released from the cells to be deposited on the blastocoele surface, first at the vegetal pole (b), then progressing toward the animal pole (c). At two points on the prism stage the antigen defines a mesenchymal extension into the overlying ectoderm (arrow in d). This location lies beneath the area where the triradiate spicule begins its growth. Eventually the antigen forms part of the matrix that fills the blastocoelic area of the pluteus larva (e). In cultures of micromeres the antigen appears at the same time as it appears in vivo. It covers the surface of the primary mesenchyme cells and the spicule envelope (f).

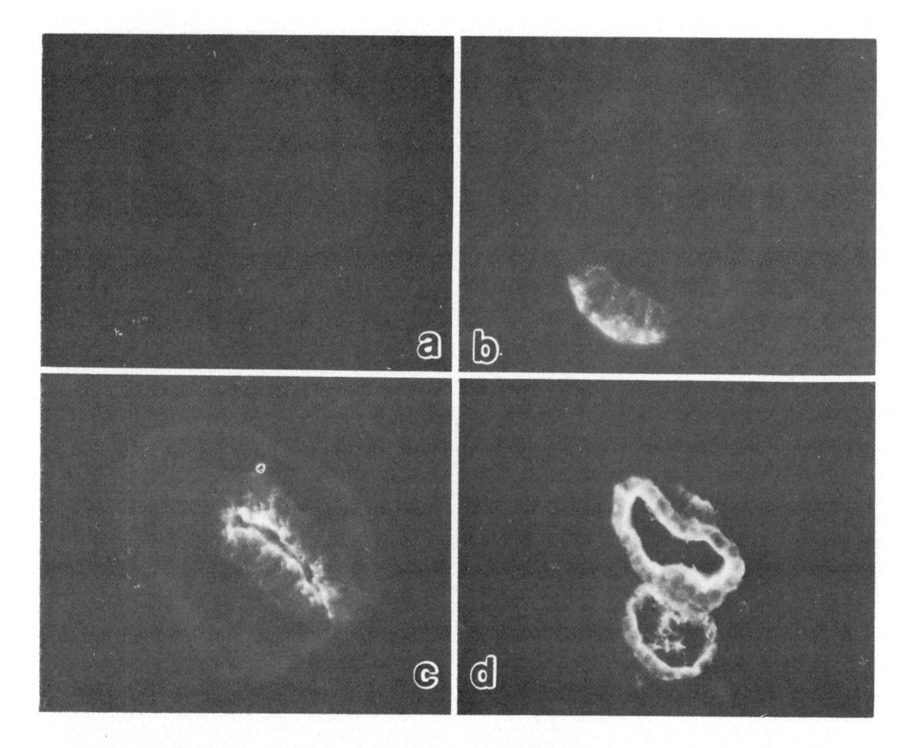

Fig. 7. Endodermal antigen. This antigen is not present during early development (a) and first appears at the vegetal plate during the mesenchyme blastula stage (b). The antigen is confined to the posterior two thirds of the archenteron (c) where it appears confined to the apical surface of the endoderm cells. The antigen does not stain the anterior one third of the archenteron nor does it stain secondary mesenchyme cells. At the pluteus stage the antigen is confined to the hindgut, midgut, and to a few cells in the floor of the foregut (d).

Protein expression in development need not be confined to germ layer. Two examples are given in Figures 8 and 9. One of the antigens observed stains only the area surrounding the stomodeal opening (Fig. 8). The antigen seems to appear shortly after induction of the stomodeal opening and it is found only in the area of the stomodeal opening and in the foregut. The cells containing the antigen are of ectodermal *and* endodermal origin. A second antigen of this type is confined to the vegetal half of the embryos (Fig. 9). This antigen appears at gastrulation and is distributed in a gradient (Fig. 9a). The antigen persists on the posterior hindgut and on the ventral ectoderm in the pluteus larva.

The de novo expression of proteins was first observed at the mesenchyme blastula stage. This does not mean that the proteins are necessarily products

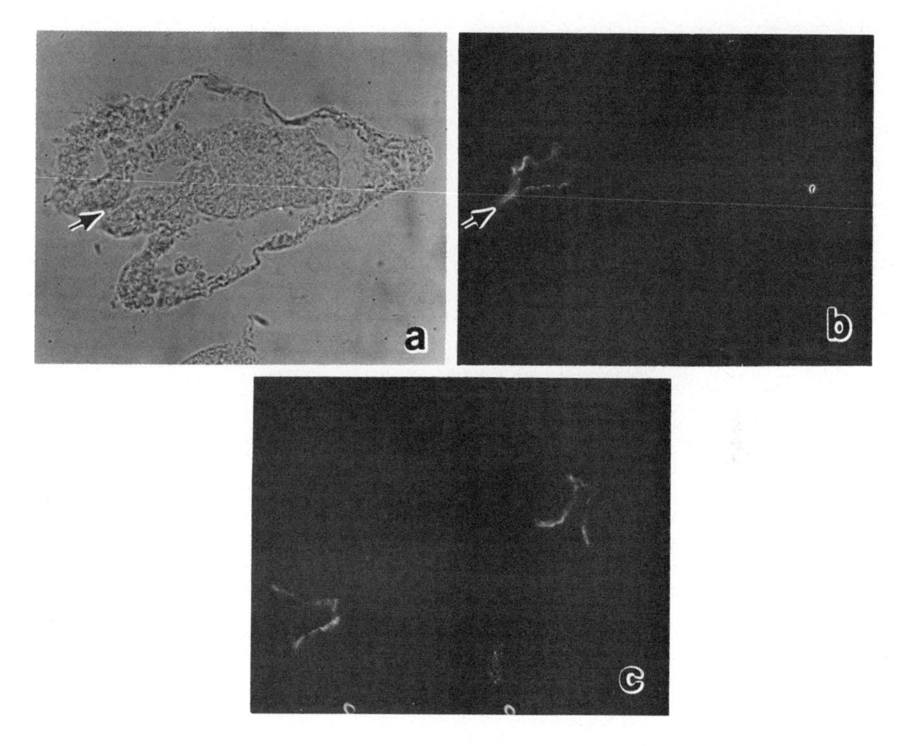

Fig. 8. Stomodeal antigen. This antigen appears following induction of the stomodeum (arrow) and is confined to the ectoderm and endoderm immediately surrounding the stomodeal opening. (b) is a fluorescent section of the embryo shown in brightfield in (a). (c) is a cross section through the stomodeum of two embryos.

of embryonic mRNAs however. Indeed, Davidson et al. [1982] have recently shown that some maternal mRNAs are not translated until late in development. Thus "de novo" refers only to the appearance of a protein that cannot be detected in the egg.

To summarize, the kinds of localization seen in the examples given above include antigens that were present in the egg and gradually became compartmentalized. There was a group of antigens that, although presumably synthesized by the same cell, became localized either apically on the extraembryonic layers, or basally on the matrix lining the blastocoele. Antigens were observed to be germ layer specific and appear de novo at the exact time of germ layer delineation at gastrulation. Another antigen defined a gradient over one entire axis of the embryo. And finally, a group of antigens were expressed only in specific locations on the embryo.

The patterns observed in this study may represent only a small subset of

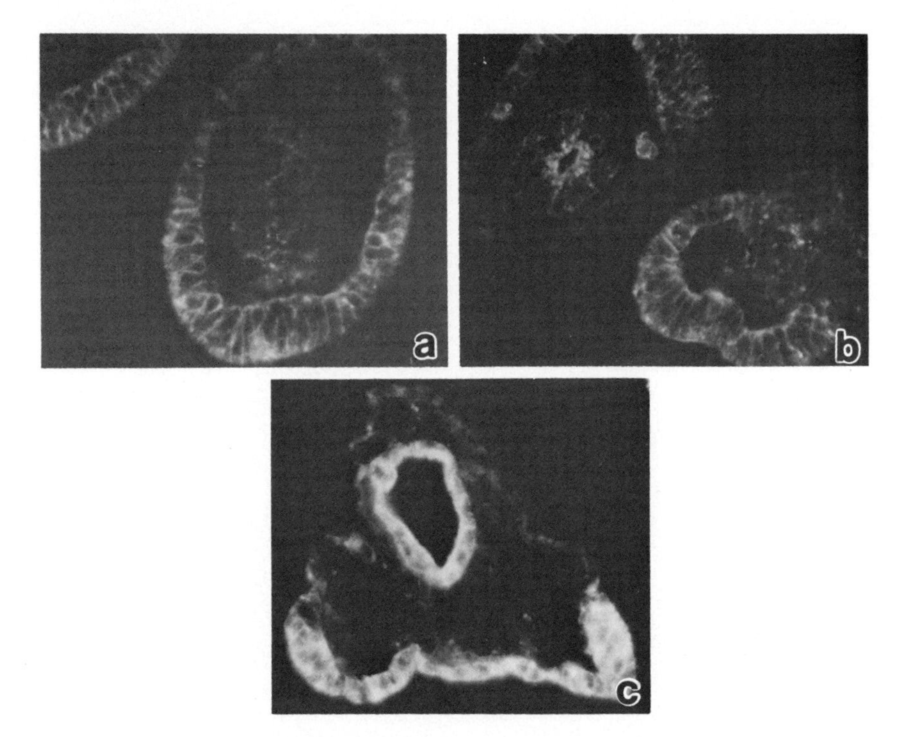

Fig. 9. Vegetal antigen. This antigen appears during the gastrula stage (a) where it is found in an increasing gradient from the animal to the vegetal pole. The gradient persists through the prism stage (b) and then becomes localized (c) to ectoderm cells of the ventral plane surrounding the anus and the hindgut.

the variety expressed by the embryo but the expression of these antigens seems to support the existence of an extensive repertoire of potential morphogenetic function. Some of the antigens observed, though highly localized spatially and temporally, appear conserved. The immunizing membranes were derived from *Arbacia punctulata* and of those antigens tested, most were found to be similarly present on *Lytechinus variegatus* and *Tripneustes esculentus*.

REFERENCES

Brandhorst BP (1976): Two dimensional gel patterns of protein synthesis before and after fertilization of sea urchin eggs. Dev Biol 52:310–317.

Crise-Benson N, Benson SC (1979): Ultrastructure of collagen in sea urchin embryos. W Roux Arch Dev Biol 186:65–70.

Davidson EH, Hough-Evans BR, Britten RJ (1982): Molecular biology of the sea urchin embryo. Science 217:17–26.

Guidice G (1962): Restitution of whole larvae from disaggregated cells of sea urchin embryos. Dev Biol 5:402–411.

Gustafson T (1963): Cellular mechanisms in the morphogenesis of the sea urchin embryo. Cell contacts within the ectoderm and between mesenchyme and ectoderm cells. Exp Cell Res 32:570–589.

Gustafson T, Wolpert L (1967): Cellular movement and contact in sea urchin morphogenesis. Biol Rev 42:442–498.

Kohler G, Milstein C (1975): Continuous cultures of fused cells secreting antibody of predefined specificity. Nature 256:495–497.

Marchase RB, Vosbeck K, Roth S (1976): Intercellular adhesive specificity. Biochem Biophys Acta 457:385–416.

McClay DR, Chambers AF (1978): Identification of four classes of cell surface antigens appearing at gastrulation in sea urchin embryos. Dev Biol 63:179–186.

McClay DR, Chambers AF, Warren RH (1977): Specificity of cell-cell interactions in sea urchin embryos. Dev Biol 56:343–355.

McClay DR, Fink RD (1982): Sea urchin hyalin: Appearance and function in development. Dev Biol 92:285–293.

McClay DR, Marchase RB (1979): Separation of ectoderm and endoderm from sea urchin pluteus larvae and demonstration of germ layer specific antigens. Dev Biol 71:289–296.

Noll H, Matranga V, Cascino D, Vittorelli L (1979): Reconstitution of membranes and embryonic development in dissociated blastula cells of the sea urchin by reinsertion of aggregation-promoting membrane proteins extracted with butanol. Proc Natl Acad Sci USA 76:288–292.

Oppenheimer SB (1977): Interactions of lectins with embryonic cell surfaces. In Moscona AA, Monroy A (eds) "Current Topics of Developmental Biology." New York: Academic Press, Vol II, pp 1–11.

Schneider EG, Nguyen HT, Lennarz WJ (1978): The effect of tunicamycin, an inhibitor of protein glycosylation, on embryonic development in the sea urchin. J Biol Chem 253:2348–55.

Spiegel M, Burger M (1982): Cell adhesion during gastrulation. A new approach. Exp Cell Res 139:377–382.

Spiegel E, Burger M, Spiegel M (1980): Fibronectin in the developing sea urchin embryo. J Cell Biol 87:309–313.

Townes PL, Holtheter J (1955): Directed movements and selective adhesion of embryonic amphibian cells. J Exp Zool 128:53–120.

Wessel GM, Marchase RB, McClay DR (1982): Spatial and temporal localization of extracellular matrix components in the sea urchin embryo. J Cell Biol 95:134a.

Time, Space, and Pattern in Embryonic Development, pages 171–196

The Role of Egg Organization in the Generation of Cleavage Patterns

Gary Freeman

Department of Zoology, University of Texas at Austin, Austin, Texas 78712

The eggs of many kinds of animals generate a stereotypic cleavage pattern during early embryogenesis. In many animals there is a link between the cleavage pattern and the axial properties of the developing organism, and in certain cases, the distribution of determinants which bias different pathways of cell differentiation [Wilson, 1925; Freeman, 1979].

There are four elements that make up a given cleavage pattern. These elements are (1) the plane of a given cleavage with respect to some reference point such as the site of polar body formation or the plane of a previous cleavage, (2) the relative sizes of the daughter blastomeres that form as a consequence of a given cleavage, (3) the time interval that occurs between cleavages, and (4) the number of cleavages which occur in a given cell lineage. The placement of a plane of cleavage is the most important element in defining a cleavage pattern. Cleavage is normally coupled with mitosis. The plane of a given cleavage is always at right angles to the axis defined by the spindle of the mitotic apparatus; this relationship reflects the fact that the interaction of the mitotic apparatus with the cell surface defines the site of contractile ring formation which mediates cleavage [Rappaport, 1974]. This means that the different elements that comprise cleavage patterns can be referred to the factors that orient and position the nucleus and mitotic apparatus, determine the size of the mitotic apparatus and its components, and control when a given cell will have a mitotic apparatus.

Different kinds of cleavage patterns vary considerably in complexity. One of the simplest patterns is seen during embryogenesis in several species of hydrozoans. In these eggs the first cleavage is invariably initiated at the site of polar body formation (Fig. 1) [Freeman, 1980]. Cleavage is unipolar; the

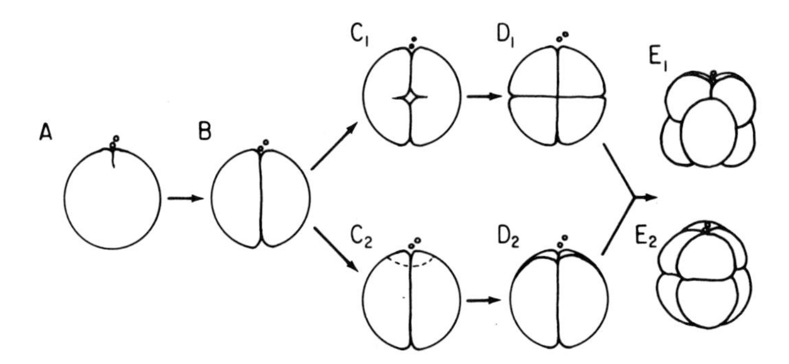

Fig. 1. A typical hydrozoan cleavage pattern. A. Unipolar cleavage beginning at the site of polar body formation. B. Two-cell stage. C_1. Unipolar second cleavage beginning at the equator. C_2. Unipolar second cleavage beginning at the site of polar body formation. Both C_1 and C_2 can occur in eggs from a single individual. D_1. The four-cell stage that forms after a C_1 cleavage. D_2. The four-cell stage that forms after a C_2 cleavage. E_1. Eight-cell stage in which the blastomeres have a pseudospiral distribution. E_2. Eight-cell stage in which the blastomeres have a radial distribution. The eight-cell embryos E_1 and E_2 can form from both D_1 and D_2 embryos. All embryos are viewed from the side.

furrow moves from its site of origin to the opposite side of the egg. The site of origin of the first cleavage specifies the oral pole of the oral-aboral axis in these embryos [Freeman, 1980]. After the first cleavage there is no set orientation for subsequent cleavage furrows, even though certain cleavage planes are more probable than others. Only one cleavage pattern element, the placement of the first cleavage furrow, has been used in this case and it is only used once.

The spiralians exhibit a more complex cleavage pattern (Fig. 2). Typically the planes of the first two cleavages run along the animal-vegetal axis of the egg dividing it into four quadrants. These four blastomeres take on a modified tetrahedral arrangement so that in most cases two opposite blastomeres

Fig. 2. Normal spiral cleavage patterns and the cleavage patterns after blastomere isolation. The spiral cleavage pattern can be either dextral (A) or sinistral (B); these cleavage patterns are mirror images of each other. A_1, B_1. Two-cell stage (polar view). A_2, B_2. Four-cell stage (side view). $A_{2'}$, $B_{2'}$. Four-cell stage (polar view). A_3, B_3. Eight-cell stage; in A_3 the first quartet is being given off in a clockwise direction, in B_3 it is being given off in a counterclockwise direction (side view). $A_{3'}$, $B_{3'}$. Eight-cell stage (polar view). A_4, B_4. Sixteen-cell stage; in A_4 the blastomeres have divided in a counterclockwise direction, in B_4 they have divided in a clockwise direction. C. The cleavage of isolated blastomeres from a dextral embryo (all views from the side). C_2. Operation to produce isolated blastomeres at the four-cell stage. C_3. Micromere formation as a consequence of the next cleavage. $C_{3'}$. Macromere formation as a consequence of the next cleavage. The blastomeres isolated at the four-cell stage (C_2) either divide to produce the blastomere configuration seen in C_3 or the configuration seen in $C_{3'}$. $C_{4'}$. The blastomere configuration generated by $C_{3'}$ after the next cleavage; this cleavage is clockwise.

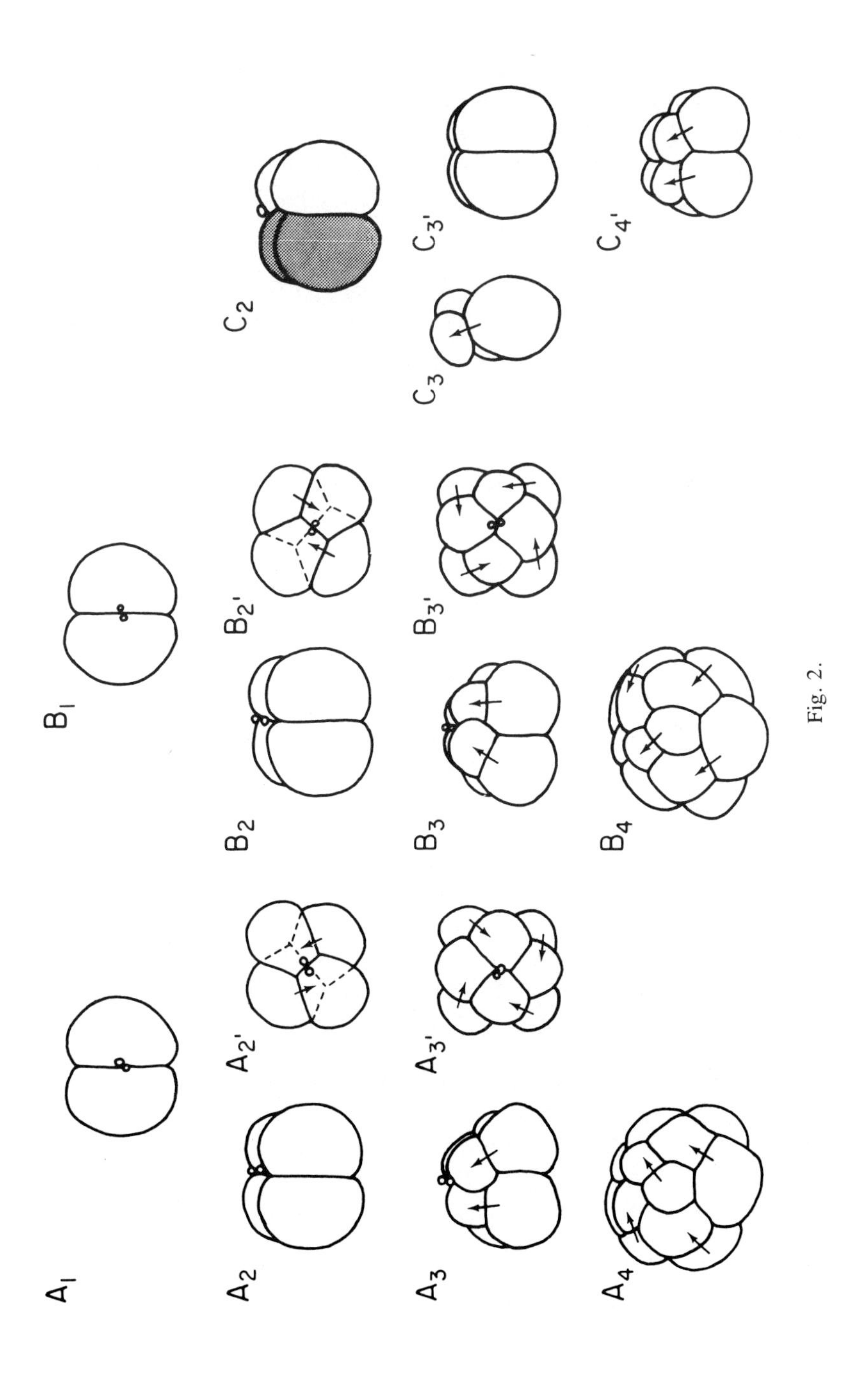

Fig. 2.

(typically A and C) meet along the animal cross furrow while the other two (B and D) meet along the vegetal cross furrow. During successive cleavages blastomeres are given off along the animal-vegetal axis in quartets. These quartets do not lie exactly above the basal blastomeres in each quadrant but are displaced obliquely toward one side or the other so as to alternate with them in a regular order. The first quartet is typically rotated in a clockwise direction when one views the egg by looking down on the animal pole while the second quartet is rotated in the opposite direction. This cleavage pattern reflects the fact that the mitotic apparatus takes up an oblique orientation with reference to the equator of the egg or its animal-vegetal axis during each cleavage and the fact that its orientation shifts 90° at each cleavage. There are several modifications of this basic cleavage pattern. The spiral cleavage pattern is asymmetrical. Figure 2A shows the dextral form of this pattern in which the first quartet is given off in a clockwise direction while 2B shows the sinistral form of this pattern in which the first quartet is given off in a counterclockwise direction. In some genera of gastropod molluscs the cleavage pattern is dextral (*Ilyanassa*), in other genera it is sinistral (*Physa*), while in some species there are dextral and sinistral individuals (*Lymnaea peregra*). There is a one to one relationship between the cleavage pattern in these forms and the spiral asymmetry of the visceral mass and the shell of the adult [Crampton, 1894]. Another modification of the spiral cleavage pattern involves cleavage by duets (acoel turbellarians) and monets (cirripedes). In these animals blastomeres form along the animal-vegetal axis of the embryo one or two cleavages earlier than they do in forms with four quartets [Costello and Henley, 1976]. There are also modifications of this cleavage pattern involving the relative sizes of the daughter blastomeres that form, and the timing of cleavage. In some annelids and molluscs the macromeres which make up the four-cell-stage embryos are equal (*Lymnaea*) while in other species the first two cleavages occur in such a way that the D quadrant is larger than the other quadrants (*Nereis*). The cleavage rhythm of the cells derived from the D quadrant is frequently different from the cells of the other three quadrants [see Clement, 1952, for a well-documented example involving *Ilyanassa*]. Typically spiral cleavage utilizes every cleavage pattern element, and each of these elements is used during several cleavages.

When the pioneering cell lineage work was being done on the embryos of invertebrates, there was a great deal of interest in the cellular mechanisms which specified the function of a given cleavage pattern element [see chapter 13 in Wilson, 1925, for a review of this earlier literature]. During the past 50 years there have been a number of contributions to this problem; however, the frequency of these contributions could be best described as sporadic. This review will summarize what we know about the mechanisms that underlie one cleavage pattern element—the orientation of the plane of a given cleavage.

LOCAL FACTORS WHICH POSITION NUCLEI

Presumably polar body formation and some cleavages occur where they do because of local egg properties which position nuclei at that site. In many oocytes the germinal vesicle has a more or less central position and there is some kind of external mark on the egg surface such as a pigment spot or bleb. When polar body formation takes place it does not occur at some random point on the surface, but at a definite site with reference to the external marker (e.g., formation of the polar bodies at the center of the pigment hemisphere of the amphibian egg or at the side of the egg opposite the vegetal peduncle in nemertines). In some oocytes the behavior of the germinal vesicle and/or meiotic apparatus can be followed during oocyte maturation in living material. The oocyte of the sea cucumber, *Holothuria leucospilota*, is translucent and has an external marker in the form of a small surface bleb, the micropyle. Polar body formation always takes place at the micropyle. Maturation can be induced by treating these eggs with dithiothreitol. A few minutes after this treatment the germinal vesicle moves to the micropyle where it breaks down. The movement of the germinal vesicle does not constitute a random walk which is interrupted by attachment at the micropyle site; it is a straight movement, as if the germinal vesicle were being guided to the micropyle [Maruyama, 1981].

Another example of this phenomenon is micromere formation during the fourth cleavage of sea urchin and sand dollar embryogenesis. At the eight-cell stage there are four animal and four vegetal blastomeres. During the fourth cleavage the animal cells divide equally and horizontally while the vegetal blastomeres divide unequally and vertically into macromeres and micromeres (Fig. 3A). Observations on suitable living embryos during the interphase period prior to the fourth cleavage indicate that the nuclei of the four vegetal blastomeres migrate directly to the vegetal end of each cell where the micromeres will be given off; there is no movement of the nuclei in the four animal blastomeres [Dan, 1979]. This translocation is colcemid sensitive, but not cytochalasin sensitive, suggesting that microtubules play a role in the process [Lutz and Inoué, 1982].

These two examples involving observations on living material are supported by a large number of cytological studies on fixed and sectioned material involving other kind of eggs [e.g., Coe, 1899; Conklin, 1902; Lillie, 1906; Longo, 1973].

In most cases it is not clear how these special regions of the egg which position nuclei arise and how they function. In starfish, oocyte maturation is mediated by 1-methyladenine. In *Asterina pectinifera* the germinal vesicle has an eccentric position in the oocyte. Marking experiments have established that the presumptive site of polar body formation in this species is the region of oocyte surface nearest the germinal vesicle. Shirai and Kanatani [1980] have demonstrated that when 1-methyladenine is locally applied to the oocyte

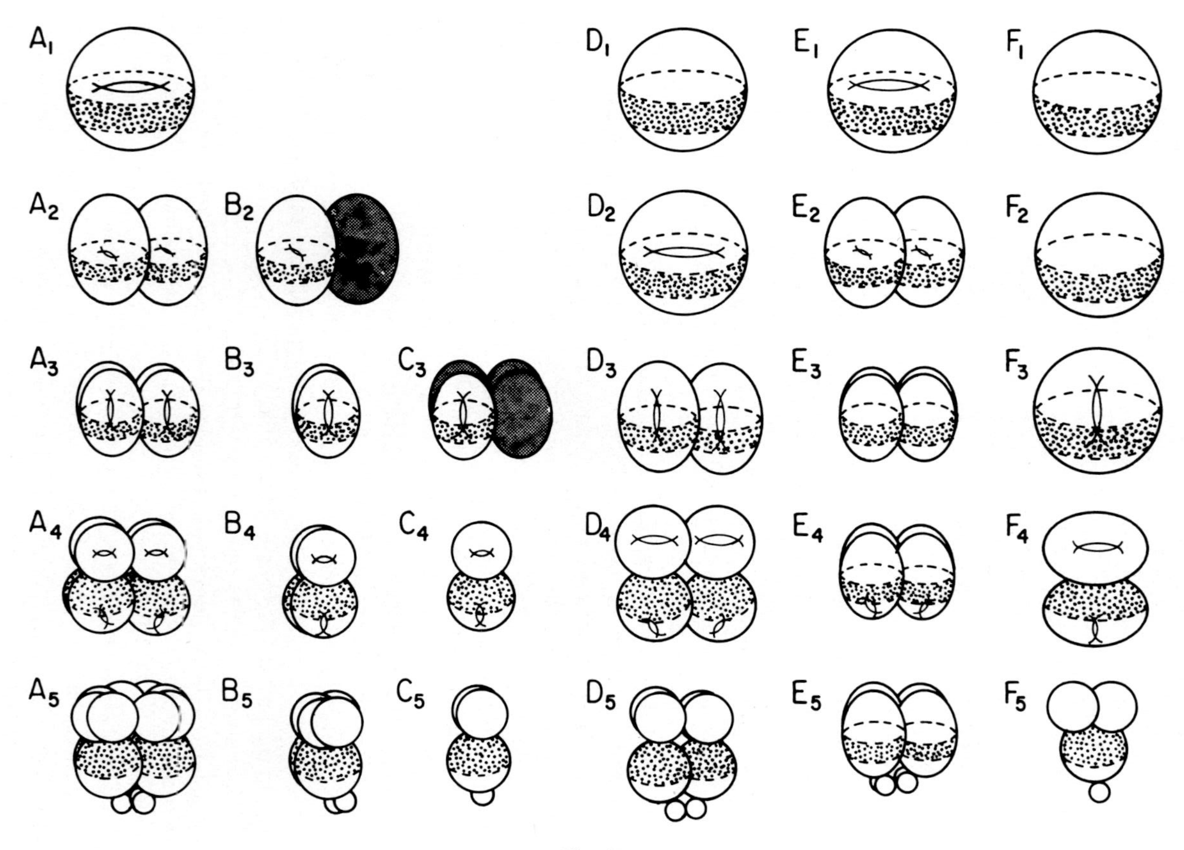
A1
B2
C3
D1
E1
F1
A2
D2
E2
F2
A3
B3
D3
E3
F3
A4
B4
C4
D4
E4
F4
A5
B5
C5
D5
E5
F5

Fig. 3.

at sites where the polar bodies would not normally form, polar body formation frequently takes place at the site of application. One interpretation of this experiment is that 1-methyladenine, by virtue of its action on the cell surface, sets up a site which attracts the meiotic apparatus. Under normal circumstances the meiotic apparatus is attracted here because of a special property of the site which makes lower concentrations of 1-methyladenine more effective.

In sea urchins the site of micromere formation is specified prior to fertilization. This point has been established by cutting unfertilized sea urchin eggs into animal and vegetal halves and fertilizing both halves. While both halves cleave, only the vegetal half forms micromeres [Hörstadius, 1937a]. Another component of micromere formation concerns the timing of this event. Normally micromere formation takes place at the fourth cleavage. Various experiments in which cleavage has been reversibly inhibited and/or the plane of a given cleavage has been changed have shown that the amount of time elapsed after fertilization is the decisive factor in determining when micromere formation occurs rather than the prior cleavage history of the embryo [Morgan and Spooner, 1909; Dan and Ikeda, 1971]. There also appears to be a change in the state of the region, which attracts nuclei, that takes place between the second and third cleavage. When embryos are treated with a low concentration of sodium lauryl sulfate during this period, micromere formation is suppressed [Tanaka, 1976]. Treatment with this surfactant at other time periods prior to the fourth cleavage has no effect on micromere formation.

FACTORS WHICH ORIENT THE FIRST MITOTIC SPINDLE

The position of the nucleus does not explain the positioning of the plane of a given cleavage. In those cases in which nuclei take up an eccentric position because of a local factor, the spindle could form parallel to the arc of the cell surface; this would lead to unipolar cleavage (Fig. 1). The

Fig. 3. The normal cleavage pattern of the sea urchin embryo (A), the cleavage pattern of isolated blastomeres (B, C), and the cleavage pattern of eggs in which cytokinesis has been reversibly inhibited (D–F) (all views from the side). A_{1-5}. Normal cleavage-uncleaved egg to 16-cell stage. Note the formation of the micromeres at the 16-cell stage. B_{2-5}. Isolation of one blastomere at the two-cell stage and the subsequent cleavages of the blastomere. C_{3-5}. Isolation of one blastomere at the four-cell stage and the subsequent cleavages of that blastomere. D_{1-5}. The reversible inhibition of the first cleavage and the subsequent cleavages of this embryo. E_{1-5}. The reversible inhibition of the second cleavage and the subsequent cleavages of this embryo. F_{1-5}. The reversible inhibition of the first and second cleavages and the subsequent cleavages of this embryo. In these figures the subequatorial pigment band and the position of the spindle is indicated. Adapted from Hörstadius [1973].

probability of this kind of configuration is much higher than the perpendicular arrangement of the spindle with reference to the cell surface that occurs during micromere or polar body formation. The perpendicular arrangement of the spindle reflects the fact that it attaches to the cell surface at one pole [see Conklin, 1917; and Shimizu, 1981, for direct evidence that one pole of the spindle is attached to the cell surface during certain stages of the meiotic reduction divisions]. The attachment of one pole of the spindle to the cell surface makes that aster smaller and changes its shape, causing the interaction between the poles, which defines the site of cytokinesis, to take place much closer to the pole that is attached to the egg surface.

In many kinds of eggs, such as molluscan eggs with equal cleavage (Fig. 2) or sea urchin egg (Fig. 3), the spindle of the mitotic apparatus for the first cleavage is invariably aligned perpendicular to the animal-vegetal axis of the egg and the plane of the first cleavage passes through the animal-vegetal axis (the position of the axis being defined by the polar bodies in the case of the mollusc or the micropyle in the jelly coat in the case of the sea urchin). The basis for this unique spindle orientation is not clear. However, it is relatively easy to change the orientation of this spindle by stratifying the contents of the egg or by changing the shape of the egg. If the contents of the egg of the sea urchin, *Arbacia*, are stratified using low centrifugal forces (ca. 3,000g) the egg remains round. By examining the position of the micropyle in the jelly coat with reference to the axis of stratification for a population of these eggs one can establish that the eggs have oriented at random in the centrifuge. When the eggs cleave, the plane of the first cleavage usually coincides with the axis of stratification, indicating that the spindle has oriented along the discontinuity created by the stratification of the egg contents [Morgan and Spooner, 1909] (Fig. 4A). If *Arbacia* eggs are centrifuged using higher forces (ca. 10,000g) the contents of the eggs stratify and the eggs elongate.

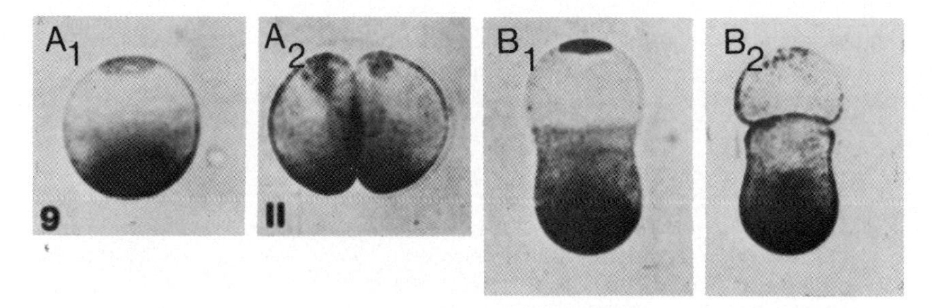

Fig. 4. Cleavage in centrifuged eggs of the sea urchin *Arbacia*. A_1. Stratification of the contents of an egg centrifuged at 3,000g. A_2. Two-cell stage of centrifuged egg in which the furrow coincides with the axis of stratification. B_1. Stratification and elongation of egg centrifuged at 10,000g. B_2. Two-cell stage of centrifuged egg in which the furrow is perpendicular to the axis of stratification. From plate VIII in Harvey [1956].

If these eggs undergo their first cleavage while still elongated the plane of the cleavage is perpendicular to the axis of stratification. In this case the dumbbell-like shape of the egg has constrained the spindle so that it orients along the long axis of the egg [Harvey, 1956] (Fig. 4B).

While it is relatively easy to alter the orientation of the mitotic spindle for the first cleavage, the fact remains that in an undisturbed egg the spindle is oriented in a fixed plane. One would like to identify the developmental factors or events that have led to this stereotypy. The only developmental event that has received serious consideration in this context is polar body formation. One way in which the role of polar body formation in setting up the orientation of the first cleavage furrow has been studied is by shifting the site of the maturation divisions. Guerrier [1968] did a set of experiments on the egg of the gastropod mollusc, *Limax*, in which the first maturation division was allowed to occur under normal conditions and then the embryo was compressed in such a way that the second polar body was frequently given off roughly 90° from the first polar body. In these eggs the plane of the first cleavage occurred along the axis that was specified by the site where the second polar body was given off (Fig. 5A). However, not all eggs behave the way the *Limax* egg does. In another gastropod, *Crepidula*, Conklin [1917] did a comparable experiment in which eggs were centrifuged at about 600g after the first polar body was given off. This treatment stratifies the contents of the egg; if the egg is oriented in the right way and centrifuged at the appropriate time the second polar body can form some distance from the first polar body. In this case the plane of the first cleavage occurred along the axis specified by the site where the first polar body was given off (Fig. 5B). In

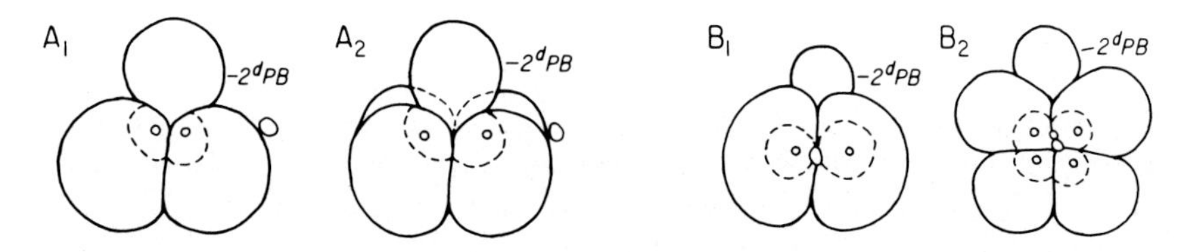

Fig. 5. Cleavage in *Limax* (A) and *Crepidula* (B) where the first and second polar bodies have formed at different sites. In both of these species a giant second polar body (2^{d}PB) has formed. A_1. Two-cell stage in which the furrow goes through the site where the second polar body was given off. A_2. Four-cell stage in which the axis of both furrows goes through the site where the second polar body was given off. Both embryos are viewed from the side. The clear cytoplasm containing the nuclei is adjacent to the second polar body. B_1. Two-cell stage in which the furrow goes through the first and second polar bodies. B_2. Four-cell stage in which the axis of both furrows goes through the site where the first polar body was given off. Both embryos viewed looking down on the first polar body. The clear cytoplasm containing the nuclei is adjacent to the first polar body. A is adapted from Guerrier [1968]; B is adapted from Conklin [1917].

both of these eggs a process of visible cytoplasmic segregation takes place during the maturation divisions and early cleavages. In *Limax* a change in the site of second polar body formation changes the axis along which segregation occurs; in *Crepidula* a change in the site of second polar body formation does not have this effect.

Another way to approach this problem is by preventing polar body formation. In the nemertine *Cerebratulus*, the unfertilized egg frequently has a transitory peduncle which marks the vegetal pole; after the egg has been fertilized the polar bodies are given off at the side of the egg opposite the peduncle. It is possible to mark an egg at the site of the peduncle with a vital dye and cut the egg equatorially into an animal and a vegetal half. When the vegetal half is fertilized it does not produce polar bodies, but it cleaves. The plane of the first cleavage is invariably parallel to the animal-vegetal axis as defined by the stained region [Freeman, 1978, and unpublished work]. This suggests that a global egg property is orienting the spindle.

FACTORS ORIENTING CLEAVAGE PLANES FOLLOWING FIRST CLEAVAGE

There are several factors which play a role in generating the pattern of cleavages that follow the initial cleavage of embryogenesis. The importance of a given factor probably depends upon the kind of embryo being studied and even the cleavage stage being studied.

One experiment which has been used to define these factors involves the isolation of a blastomere from an embryo at an early stage of development. The subsequent cleavages of this isolated blastomere are then followed. If the blastomere cleaves in the same way it would have if it had been in the embryo it is possible to argue that the positioning of the mitotic spindle is an autonomous property of the cell and that the daughter cells will also cleave in a plane which is an autonomous property of each of these cells and so on. The other possibility is that the cleavage pattern of the isolated blastomere will be altered in some way. One can infer in this case that normally some extrinsic factor, such as a chemical signal from a neighboring blastomere, or a physical factor, such as a blastomere shape change due to the way a given blastomere packs with respect to its neighbors, plays a role in defining the cleavage pattern. While this kind of experiment sounds straightforward in practice, it is difficult to do in a rigorous manner. The blastomeres used for these experiments must either contain appropriate natural markers or markers must be added to the blastomere so that the position of the next cleavage can be assayed. Once a cleavage has occurred the possibility that the cleavage plane will influence the next plane of cleavage has to be controlled by isolating both daughter blastomeres. The time when a blastomere is isolated after it has been generated also has to be controlled because there may be a critical time period during the cell cycle when the spindle site is set up.

Figure 3(B and C) outlines blastomere isolation experiments involving the blastomeres of the two- and four-cell stage embryo of the sea urchin *Paracentrotus lividis* [Hörstadius, 1973]. Frequently eggs of this species have a subequatorial pigment band which can be used as a natural marker. The results of the experiments show that the isolated blastomeres continue to cleave just as they would have in a normal embryo. Most of the blastomere isolation experiments in which the cleavage pattern of the isolated blastomeres has been followed have been interpreted in this way [e.g., Conklin, 1905a, ascidians; Costello, 1945, annelids; Freeman, 1976, ctenophores].

Not all isolated blastomeres show this kind of cleavage pattern. One well-documented case involves the behavior of isolated blastomeres from pulmonate gastropods [Meshcheryakov, 1976; Morrill et al., 1973]. When two blastomeres are isolated at the four-cell stage in *Lymnaea*, at the next cleavage micromeres and macromeres can be produced as though the blastomeres are part of the whole embryo, or the cleavage can be meridional and equal, forming four blastomeres. In the latter case, the next cleavage of these embryos produces four micromeres and macromeres (Fig. 2C). A discussion of the mechanisms which underlie the behavior of the isolated blastomeres in these different kinds of embryos follows.

Extrinsic Factors

Meshcheryakov [1976] has argued that the two types of cleavage that occur following blastomere isolation in *Lymnaea* reflect the topography of intercellular contacts between blastomeres at the time the spindle forms. The kind of cleavage that occurs in this experiment depends upon when the blastomeres are isolated. If they are isolated prior to the formation of the mitotic apparatus they cleave equally and meridionally. If they are isolated at metaphase they cleave unequally. During the two- and four-cell stages the nuclei are found near the animal pole; their position reflects the segregation of clear cytoplasm from yolky cytoplasm which occurs at this pole of the embryo. When spindles form during the two-cell stage they lie beneath the animal surface parallel to the zone where the two blastomeres are in contact. After the second cleavage the four blastomeres pack in such a way that they are elongated with respect to their polar axis (Fig. 2A,B). When spindles form they parallel the polar axis; because of the eccentric position of the spindle at the animal pole the cleavage is unequal. At both the two- and four-cell stages the spindles orient in the way they do because spindle growth is least constrained in that axis by the shape of the blastomere. When two blastomeres are isolated from a four-cell-stage embryo prior to spindle formation the blastomeres change their shape, becoming like the blastomeres of a two-cell-stage embryo. When spindles form in these blastomeres they orient in an axis appropriate for the blastomere's shape. When two blastomeres are isolated at metaphase there is so little time until cytokinesis begins that cleavage occurs as it would have in the whole embryo.

The role of blastomere shape in specifying spindle orientation has been examined in another way by altering the degree of contact between two blastomeres [Meshcheryakov, 1978a]. In pulmonate gastropods a contact zone develops between blastomeres during the period between cleavages. Figure 6A shows the contact zone of a normal two-cell embryo. At prophase preceding the second cleavage the axes of developing spindles form diverse angles with the contact zone. Between prophase and metaphase the spindle doubles in diameter and increases in length. At metaphase and anaphase virtually all of the spindles line up parallel to the contact zone. In Meshcheryakov's experiments the size of the contact zone between the blastomeres was decreased to different extents by lowering the level of calcium in the culture solution (Fig. 6B–D). The effect of this shape change on spindle orientation with reference to the contact zone is summarized in Figure 6E. As the size of the contact zone gets smaller the percentage of cases in which the spindle orients parallel to the contact zone decreases and the variability of spindle orientation increases.

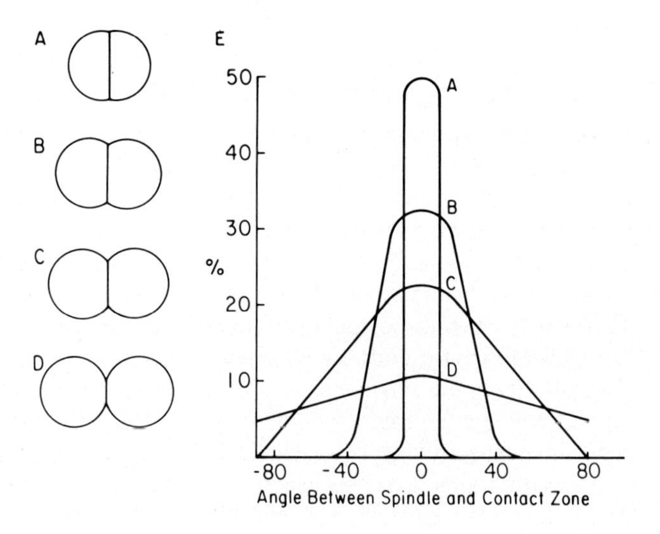

Fig. 6. Summary of experiment in which the contact zone between blastomeres was decreased. A–D. Two-cell-stage embryos showing the different degrees of contact zone reduction between blastomeres. E. Graph showing the angles between the metaphase and anaphase spindles in embryos with different degrees of contact zone reduction. The spindles in opposite blastomeres behave independently of each other. Adapted from Meshcheryakov [1978a].

In spirally cleaving embryos the way in which the blastomeres pack can be described in terms of the theory of surface tension [see Thompson, 1942, chapters 7 and 8]. This theory states that an assemblage of fluid films will be in equilibrium when the sum of the free surface areas of the film is minimal. There are only certain packing arrangements that fit these conditions. For example, at the animal pole of the four-cell-stage embryo, the cell membranes that delimit the blastomeres do not intersect at right angles but meet at two foci composed of three edges connected by a cross furrow (Fig. 2, $A_{2'}$ and $B_{2'}$). The configuration in which the cell membranes intersect at right angles does not exist because it reflects an unstable state, while both the dextral and sinistral four-cell stage packing arrangements shown in Figure 2 are stable because the free surface areas have been minimized. The packing arrangement of soap bubbles can only conform to the laws of surface tension. Robert [1902] made soap bubble models of the different cleavage stages of the mollusc *Trochus* in which the bubbles were the same relative size as the blastomeres from that stage (Fig. 7). By building up the models in tiers, he found that he could produce a packing arrangement of the bubbles that was identical to the packing arrangement of the blastomeres at each cleavage stage.

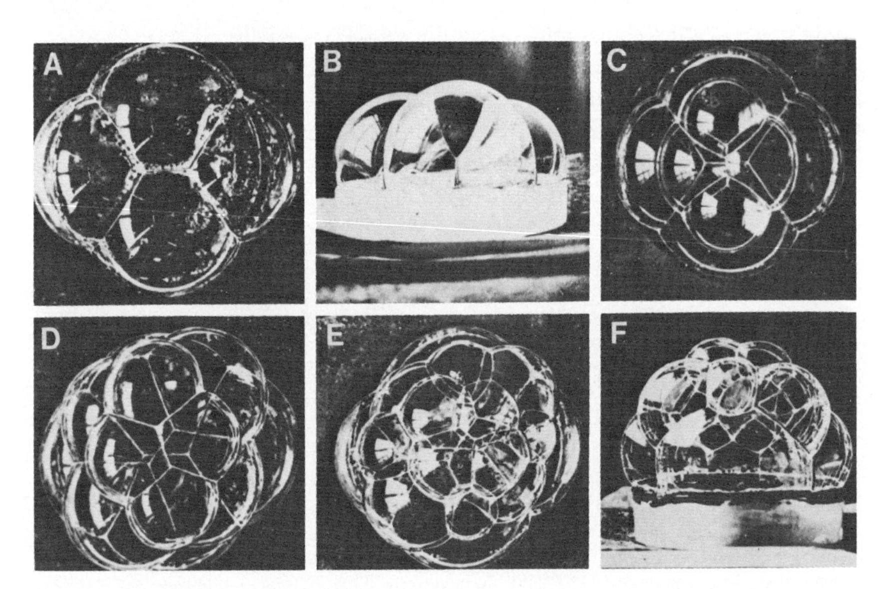

Fig. 7. Arrangement of bubbles which correspond to the first stages of segmentation in *Trochus*. A. Polar view of the four-cell stage; note the polar cross furrow. B. Side view of the four-cell stage. C. Polar view of the eight-cell stage. D. Polar view of the 12-cell stage prior to the cleavage of the first quartet. E. Polar view of the 16-cell stage. F. Side view of the 16-cell stage. From Robert [1902], Plate XII.

From the eight-cell stage on in embryos which show spiral cleavage, the packing arrangement of blastomeres around the animal-vegetal axis creates asymmetric blastomere shapes. Meshcheryakov [1978b] has argued that blastomere shape at these stages plays a role in establishing spindle orientation much as it does during earlier stages. At the eight-cell stage a given micromere of the first quartet lies not immediately over, but above and slightly to the side of its sister macromere (Fig. 2, A_3). In the example that follows we will only consider eggs with a dextral cleavage pattern. This displacement of the first quartet micromeres presumably reflects the asymmetrical rotation of these cells with reference to the macromeres during this cleavage (see next section). One consequence of this rotation is that the adjacent clockwise macromere has part of its external surface in contact with this micromere. This edge of the macromere will be called the cross-over edge. This macromere will also form a contact zone with its own sister micromere. This zone will be referred to as a related edge. By the end of the eight-cell stage the related edge of the macromere will be considerably longer than its cross-over edge (Fig. 8A). The nucleus of the macromere is in its most animal region near the external surface. As the mitotic apparatus forms for the next cleavage it lines up parallel to the related edge of the macromere. The mitotic apparatus behaves in a similar way in the micromeres. The spindle configuration generates the counterclockwise fourth cleavage (Fig. 2, A_4).

The role of blastomere shape in aligning the mitotic apparatus at this stage has been tested by changing the contact relationships between blastomeres just after the third cleavage by treating embryos with a low concentration of trypsin. This treatment drastically reduces the contacts between blastomeres and they become spherical (Fig. 8B). When these embryos are removed from

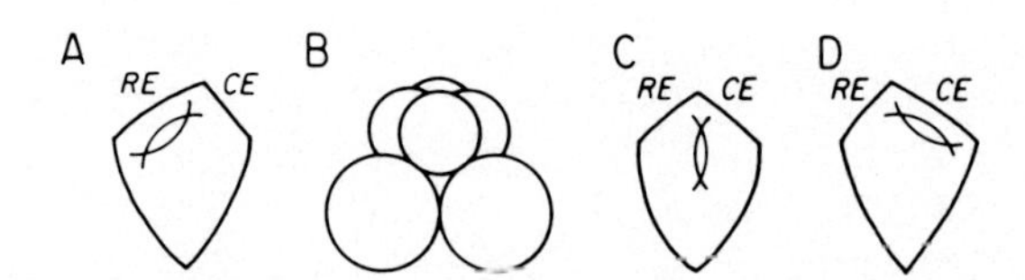

Fig. 8. Micromere shapes in normal and trypsinized eight-cell-stage embryos. All views are from the side. A. A typical macromere in a late eight-cell-stage embryo showing the counterclockwise orientation of the spindle. The crossover edge of the blastomere (CE) is much shorter than its related edge (RE). B. Trypsinized eight-cell-stage embryo; the contacts between the blastomeres are minimal (compared with Fig. 2, A_3, C, and D). Trypsinized eight-cell-stage embryos in which the contacts between blastomeres have reformed. In C, CE and RE are the same length. In many of these macromeres the spindle has a vertical orientation. In D, CE > RE and the spindle has a clockwise orientation. Adapted from Meshcheryakov [1978b].

TABLE I. Number of Macromeres With Different Orientations of the Fourth Cleavage Spindles as a Function of Blastomere Shape in *Lymnaea*

Length ratio of RE to CE	Spindle orientation[a]			
	Counterclockwise	Clockwise	Vertical	Horizontal
RE > CE	108			2
RE < CE		4		
RE = CE	47	9	7	19

[a]Spindle orientation measured at prophase through anaphase.

the trypsin solution and placed in culture medium with an excess of calcium the zone of contact between the blastomeres reforms. However, under these conditions the relative lengths of the cross-over edge and a related edge are frequently changed. In most cases they are nearly equal, while in a few cases the cross-over edge is longer than the related edge (Fig. 8C,D). Table I shows how the spindles in these blastomeres orient prior to the fourth cleavage. Spindle orientation is clearly influenced by blastomere shape.

While it has been clear for some time that blastomere shape can be used to control the positioning of a cleavage plane, there have been very few studies like those of Meshcheryakov in which this phenomenon has been examined in the context of normal embryogenesis in animals. More studies like this need to be done.

Intrinsic Factors

A number of intrinsic factors that do not reflect local conditions have been demonstrated or postulated that are thought to specify different elements of cleavage patterns. The first of these which will be discussed is the chirality factor in spiralians (Fig. 2A,B).

In pulmonate molluscs, Meshcheryakov [1978a,b] and Meshcheryakov and Beloussov [1975] have shown that during the early cleavages of the embryo an asymmetrical rotation of the blastomeres occurs. This rotation can be demonstrated by placing carbon particles on denuded eggs or blastomeres isolated at the two-, four-, or eight-cell stages and watching the way in which the particles change positions with reference to one another as the egg cleaves. Uncleaved eggs or blastomeres isolated from a given type of embryo, be it dextral or sinistral, always rotate in the same direction at each cleavage. However, the directions of rotation in dextral and sinistral eggs are mirror images of each other (Fig. 9). Meshcheryakov and Beloussov [1975] have argued that the direction of blastomere rotation reflects the way in which the contractile ring which mediates cytokinesis is organized. They have postulated a spiral contractile ring in which the pitch of the spiral is of opposite twist in dextral and sinistral eggs.

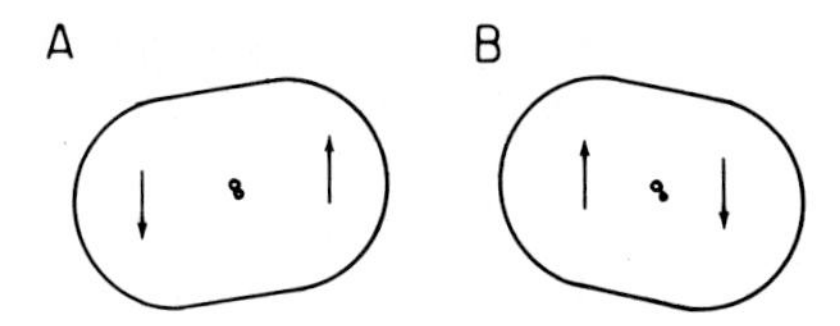

Fig. 9. The rotation of blastomeres during the first cleavage in dextral (A) and sinistral (B) eggs. The eggs are viewed from the animal pole; note the polar bodies. The arrows indicate the direction of blastomere rotation. Adapted from Meshcheryakov and Beloussov [1975].

In *Lymnaea peregra*, the directionality of blastomere rotation is under genetic control. Chirality is controlled by a single locus which shows a maternal mode of inheritance. This pattern of inheritance argues that gene activity at this locus during oogenesis specifies the pattern of cleavage during embryogenesis. The dextral alleles of the chirality gene behave as dominants with respect to the sinistral alleles. When cytoplasm from dextral eggs is injected into uncleaved sinistral eggs prior to second polar body formation it causes the recipient egg to cleave in a dextral manner [Freeman and Lundelius, 1982]. Cytoplasm from sinistral eggs has no effect on dextral eggs. This work suggests that there is some kind of developmental event that plays a role in determining the directionality of cleavage which occurs around the time of second polar body formation. In the presence of a dextral gene product cleavage occurs in a dextral manner; in the absence of this gene product it occurs in a sinistral manner.

Another intrinsic factor, the cycle of centriolar replication, has been postulated to play a role in the positioning of cleavage planes. Costello [1961] has argued that the orientation of the centrioles at any given division determines the position and the path of separation of the daughter centrioles with respect to each other. Since the centrioles will become the poles of the mitotic spindle, their positions will determine the position of the plane of cleavage. Costello has also argued that the right-angle orientation of the centrioles at the two poles of the spindle provides the basis for the alternating clockwise and counterclockwise cleavages in eggs which show spiral cleavage. One of the problems with this hypothesis is that almost nothing is known about the paths of centriole separation following centriole duplication. As a consequence, it is not clear that the pathways of centriolar movement bear any relation to the orientation of the spindle at the next cleavage. The best evidence that speaks to this hypothesis is based on studies that have examined the relationship between the pathway taken by the sperm aster and the orientation of the spindle for the first cleavage. In a variety of different eggs [e.g., anurans, Ancel and Vintemberger, 1948; ascidians, Conklin, 1905b; sea urchins, Schatten, 1981; and various spiralians, Guerrier, 1970b] the first cleavage spindle develops perpendicular to this path. This relationship is thought to reflect the paths taken by the centrioles derived from the sperm aster as they move away from each other to become the poles of the spindle (if that is in fact what happens).

While this hypothesis is consistent with many observations regarding the positioning of cleavage planes, there are certain experiments that are hard to rationalize in terms of it. The cleavage patterns of blastomere pairs isolated at the four-cell stage in *Lymnaea* have already been described (Fig. 2C). The centriolar hypothesis would predict the cleavage pattern in Figure 2, C_3; it would not predict the cleavage pattern in Figure 2, $C_{3'}$, and $C_{4'}$; instead one would expect the micromeres to be given off in a counterclockwise direction. Meshcheryakov and Varyasova [1979] have tested Costello's hypothesis by inhibiting either the first or the second or the third cleavage of *Lymnaea* embryos with cytochalasin B; this treatment creates binucleate cells. When these cells formed mitotic apparatuses the positions of the four poles were noted to see if they reflected the centriole movement rules. While pole configurations that correspond to Costello's hypothesis are found, they are not more prevalent than other classes of configurations that would not be predicted by the hypothesis.

The last intrinsic factor which will be discussed that has been implicated in the positioning of cleavage planes is referred to as the cleavage clock. There are a number of studies of this phenomenon in sea urchins [see Hörstadius, 1973, for a review]. A cleavage clock that controls spindle orientation and positioning has also been demonstrated in ctenophores [Freeman, 1976] and in some molluscan species [Conklin, 1912, 1917; Cather and Render, unpublished observations]. A set of cleavage clock experiments on sea urchin eggs are summarized in Figure 3D–F. In this set of experiments, either the first or the second or both the first and second cleavages have been reversibly inhibited. When the inhibitor is removed, the spindle(s) in the embryo orients in the same way it would orient in a control embryo of the same age that has not been inhibited (Fig. 3, compare A_2 with D_2, A_4 with E_4, and A_3 with F_3). In a variation on this experiment, embryos which have had cleavage reversibly inhibited were treated in such a way that cleavage was initiated while untreated control embryos were between cleavages. If the cleavage takes place at a transition where the spindles were in a horizontal plane for a prior cleavage of control embryos, and will be in a vertical position at the next cleavage (e.g., Fig. 3, A_2 to A_3), the spindles of the experimental embryos will frequently orient at an oblique angle with reference to the animal-vegetal axis. These experiments demonstrate that a clock which is activated at some point in development plays a role in orienting the spindles and that the orienting influence operates independently of mitosis and/or cleavage.

There are a number of treatments that reversibly inhibit cleavage in sea urchin embryos. Some of these treatments (e.g., 2,4-dinitrophenol) allow the cleavage clock to proceed, while other treatments (e.g., sodium azide) block both cleavage and the cleavage clock. In sea urchin embryos Sakai [1968] has characterized two protein fractions whose -SH contents change as a function of the cell cycle. One fraction consists of a KCl-soluble contractile protein which appears to be localized primarily in the cell cortex. The -SH content of the fraction is lowest at interphase and highest just before cleavage. The second protein precipitates from the water-soluble fraction of egg homogenate after the addition of calcium. Its -SH content is highest at interphase and lowest at cleavage. Sakai thinks that these two proteins are part of a thiol-disulfide exchange system. Dan and Ikeda [1971] have argued that the -SH exchange system is the cleavage clock. This argument is based on the demonstration that cleavage inhibitors that do not block the cleavage clock do not affect the -SH cycle while those inhibitors that block the cleavage clock also block the -SH cycle.

By analyzing the plane of the second cleavage in embryos in which the plane of the first cleavage is either appropriate or inappropriate with reference to the animal-vegetal axis, one can deduce that the cytoplasmic machinery that is run by the cleavage clock appears to orient spindles primarily with reference to the animal-vegetal axis and to a lesser extent with reference to the plane of the previous cleavage. In sea urchins the site where the micromeres form corresponds to the vegetal pole of the egg; the animal-vegetal axis of the embryo is set up prior to fertilization and is not changed by low-speed centrifugation. If eggs are centrifuged at low speed, the contents stratify and the first cleavage plane parallels the axis of stratification (Fig. 4A). Since sea urchin eggs orient at random in a centrifugal field, the axis of stratification can coincide with the animal-vegetal axis or it can occur at an angle with reference to the axis. Figure 10 presents the results of an experiment by Morgan and Spooner [1909] in which the cleavage patterns are diagrammed and the number of eggs showing a particular cleavage pattern is given for a set of embryos where the plane of the first cleavage was either parallel to or perpendicular to the animal-vegetal axis. When the first cleavage is parallel to the animal-vegetal axis (Fig. 10A,B), in the majority of cases the second cleavage is also parallel to the animal-vegetal axis and the axis of stratification. When the first cleavage is perpendicular to the animal-vegetal axis (Fig. 10C,D), the second cleavage is parallel to the animal-vegetal axis and perpendicular to the axis of stratification in the majority of cases. This comparison indicates that the animal-vegetal axis is orienting the second cleavage. It would be interesting to repeat these experiments of Morgan and Spooner's under conditions in which one or more of the early cleavages was reversibly inhibited.

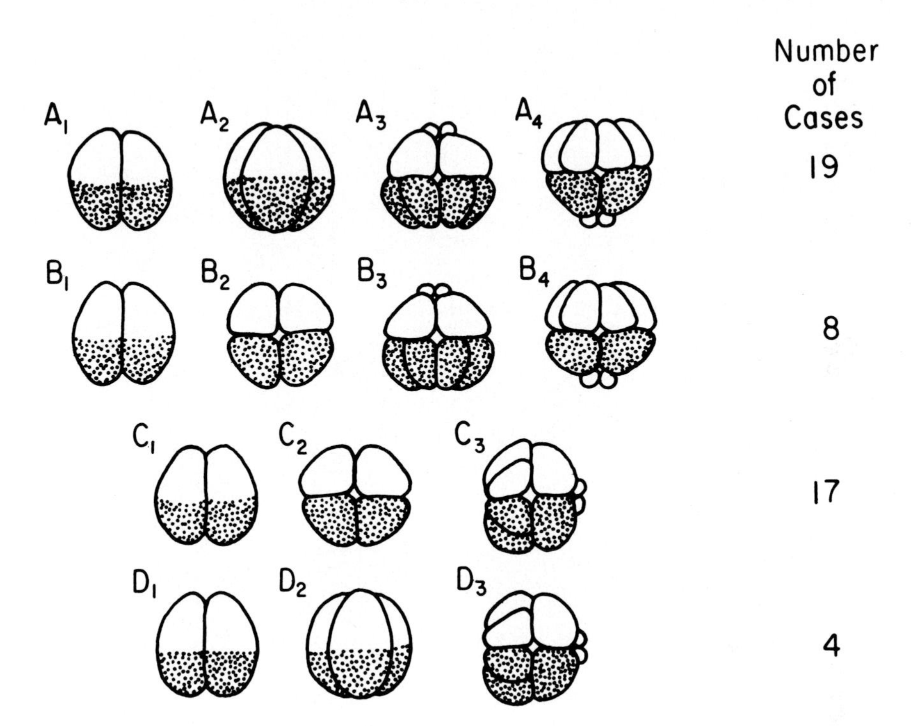

Fig. 10. Cleavage patterns of centrifuged sea urchin eggs. The stippling represents the pigment granules that have accumulated in the centrifugal half of the egg. All embryos are viewed from the side. The micromeres mark the vegetal pole. In A and B the plane of the first cleavage parallels the animal-vegetal axis. In A_2 the plane of the second cleavage parallels the animal-vegetal axis while in B_2 it is perpendicular to the animal-vegetal axis. In C and D the plane of the first cleavage is perpendicular to the animal-vegetal axis. In C_2 the plane of the second cleavage is parallel to the animal-vegetal axis, while in D_2 it is perpendicular to the animal-vegetal axis. Adapted from Morgan and Spooner [1909].

Many of the properties of the cleavage clock of sea urchins are found in ctenophore and molluscan embryos. However, there appear to be some differences with regard to the clock phenomena in these three groups of animals: (1) The cleavage clock is activated at fertilization in sea urchins, while it is not activated until first cleavage in ctenophores [Freeman, 1976]. (2) Treatment with 2,4-dinitrophenol will inhibit cleavage but allows the clock to run in both sea urchin and ctenophore embryos; however, this treatment inhibits both cleavage and the cleavage clock in molluscan embryos (Cather and Render, unpublished observations). This suggests that the cleavage clock may have a different metabolic basis in these two groups of animals. (3) In sea urchin embryos the timing of micromere formation is

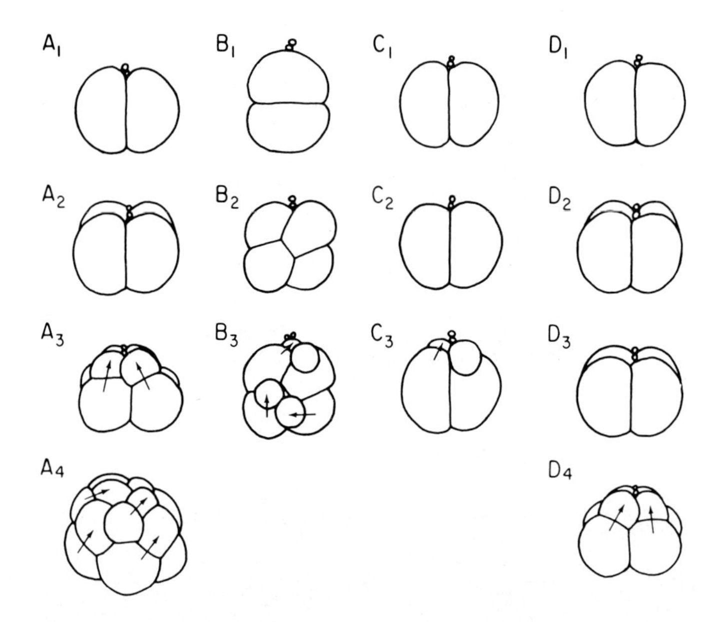

Fig. 11. Cleavage clock experiments on molluscs. All embryos are viewed from the side. A. Normal cleavage. B. The cleavage of an egg centrifuged after second polar body formation. The stratification of the egg contents has caused the first cleavage plane to form at right angles to the animal-vegetal axis as defined by the site of polar body formation. B_3 shows the micromeres; they were given off in a clockwise manner. Adapted from Conklin [1917]. C. The reversible inhibition of the first cleavage and the subsequent cleavage of this embryo; C_3 shows the micromeres; they were given off in a clockwise direction. Adapted from Conklin [1912]. D. The reversible inhibition of the second cleavage and the subsequent cleavage of this embryo; D_4 shows the micromeres; they have been given off in a counterclockwise direction. Based on unpublished results of Cather and Render; these investigators have used cytochalasin B and high pressure to reversibly inhibit cleavage.

specified by the cleavage clock but the site of micromere formation is specified by the vegetal region of the embryo. In molluscan embryos the timing of micromere formation is specified by the cleavage clock, but the site of micromere formation is not specified. Figure 11B illustrates an experiment which Conklin [1917] carried out on *Crepidula* embryos. In this experiment eggs were centrifuged after second polar body formation. As a consequence of centrifugation the first cleavage furrow sometimes develops perpendicular to the animal-vegetal axis. After centrifugation the stratified cytoplasm tends to become redistributed in such a way that the clear cytoplasm becomes localized at the animal pole region. In the vegetal half of the embryo the cleavage plane forms a barrier to this redistribution process so that the clear cytoplasm accumulates in the regions of the vegetal blastomeres

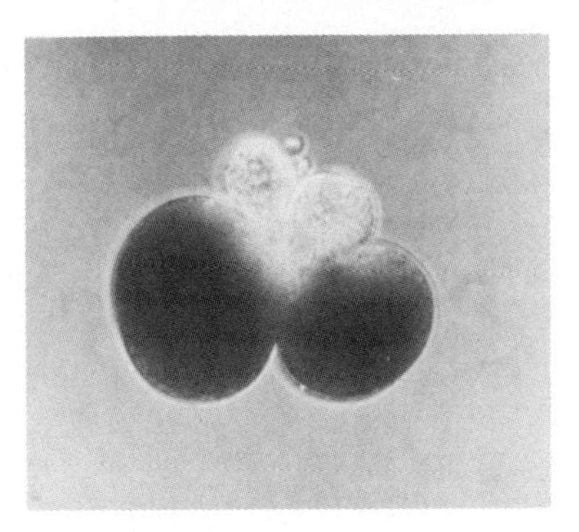

Fig. 12. *Ilyanassa* embryo in which the second cleavage has been reversibly inhibited followed by micromere formation at the next cleavage (courtesy of James Cather and JoAnn Render).

which are adjacent to the animal blastomere. At the third cleavage (Fig. 11, B_3) the two micromeres that form in the animal half of the embryo tend to be given off in the most animal regions of the blastomeres where they would normally form. In the vegetal half of the embryo two micromeres also form; here they form adjacent to the equator. This is not their normal site of formation. The time of micromere formation does not depend upon the previous cleavage history of the molluscan embryo. This can be demonstrated by reversibly inhibiting the second cleavage (Figs. 11C, 12). Under these circumstances embryos composed of two macromeres form micromeres at the next cleavage which is the time when micromere formation is normally initiated. This experiment has been done on *Crepidula* [Conklin, 1912] and on *Ilyanassa* (Cather and Render, unpublished observations). When the time for micromere formation arrives, presumably some kind of global change occurs in the embryo which affects spindle postion and/or size.

One special feature of the spiral cleavage pattern is the alternating clockwise and counterclockwise spindle orientation that is especially prominent during quartet formation. Cather and Render (unpublished observations) have recently examined the cleavage clock phenomenon in *Ilyanassa*. In one experiment that they did the third cleavage was reversibly inhibited (Fig. 11D). When cleavage began again, the first cleavage occurred in a counterclockwise manner; this is the same cleavage direction that untreated control embryos utilize when they cleave at this time to form the second quartet. This result does not fit the asymmetrical contractile ring hypothesis of Meshcheryakov and Beloussov. They would predict that the first group of micromeres to arise should show a clockwise mode of cleavage regardless of the cleavage cycle. The cleavage clock phenomenon is consistent with Costello's centriolar replication hypothesis; however, it doesn't speak to this set of ideas in a direct way.

SPINDLE PLACEMENT MECHANISMS AND CLEAVAGE PATTERNS

It is one thing to catalog the different mechanisms that might play a role in determining spindle placement; it is another thing to use these mechanisms to explain a given cleavage pattern. These mechanisms will now be utilized in an attempt to explain the cleavage patterns of a sea urchin embryo and a typical molluscan embryo.

In considering the cleavage pattern of a sea urchin embryo most of the major features of the pattern can be explained by invoking the factors discussed here. At the time the egg is fertilized its cleavage clock begins to function. This clock puts into motion the cytoplasmic machinery which is responsible for orienting the spindle for the first cleavage. The machinery operates by using the animal-vegetal axis of the embryo as a reference point even though a specific animal or vegetal region does not have to be present for it to work. It functions to orient the spindles at subsequent cleavages. After the third cleavage the clock begins to function in a new way by activating or unmasking a vegetal region which attracts nuclei to it causing micromere and macromere formation. These events would not occur if this vegetal region were not present.

If one attempts to explain the cleavage pattern in forms with spiral cleavage, the situation becomes much more complex. All of the investigators who have thought about the basis for cleavage patterns in these forms would agree that there is some kind of cytoplasmic organization related to the animal-vegetal axis of the egg that is responsible for defining the site of polar body formation and spindle orientation for the first cleavage. Meshcheryakov, on the basis of his studies on freshwater pulmonates, has argued that there are two additional factors that generate the other elements of the cleavage pattern: (1) an asymmetric blastomere rotation mechanism of a given direction and (2) the pattern of intercellular contacts between blastomeres which define blastomere shape and thereby constrain spindle orientation. The chirality of first quartet formation can be explained by asymmetric blastomere rotation. Clearly one cannot use this mechanism to explain second quartet formation. Meshcheryakov argues that once the first quartet has formed, the blastomere shapes that have been created constrain spindle orientation in such a way that in spite of an unchanged blastomere rotation the next cleavage takes place in the opposite direction. This hypothesis can be tested by allowing the first quartet to form and then removing these cells. The direction of second quartet formation can then be studied. Meshcheryakov would predict that in embryos in which the first quartet forms in a clockwise direction, the removal of the quartet should cause the second quartet to form in a clockwise direction rather than its normal counterclockwise direction.

Meshcheryakov [1976] reported that in *Lymnaea* and *Physa*, when two macromeres are isolated at the early eight-cell stage, these macromeres form micromeres in the same direction that they are given off during first quartet formation. Guerrier [1970a] isolated all four macromeres at the eight-cell stage in *Limax*. The majority of these embryos formed their micromeres in the same direction as that in which the second quartet would normally be formed; however, there were a few cases where the micromeres were given off horizontally or in the opposite direction. Similar isolation experiments have been done on *Patella* [van den Biggelaar and Guerrier, 1979] and *Cerebratulus* [Hörstadius, 1937b; Freeman, 1978]. In these papers no statement is made about the chirality of the next cleavage; however, these experiments were done by investigators who would have been alert to a change in cleavage pattern and noted if it had occurred.* The fact that the outcome of these isolation experiments depends upon the species used suggests that different mechanisms may be operating to generate the spiral cleavage pattern in different forms. The results argue that asymmetrical blastomere rotation only occurs in some groups with spiral cleavage.

The cleavage clock is another mechanism that can be used to explain the alternating direction of cleavage in spiralians. The results of the macromere isolation experiments after first quartet formation in *Cerebratulus*, *Limax*, and *Patella* are consistent with this mechanism. One problem with this mechanism is that the cytoplasmic machinery that controls spindle orientation appears to function with reference to the egg axis (at least in sea urchins). In a variety of spiralians it is possible to shift the axis along which the quartets are given off so that it does not correspond to the animal-vegetal axis of the egg [Conklin, 1917; Freeman, 1978; Guerrier, 1968, 1970a]. In spite of this the alternating directionality of spiral cleavage continues to occur normally. This suggests that the egg uses a set of cues that depend on the location of the first two cleavage planes to orient the spindles that are responsible for generating the quartets. These problems make it clear that the mechanisms that underlie spiral cleavage are still not well understood.

CONCLUSIONS

Cleavage patterns are clearly important developmental phenomena with many interesting ramifications. The bases for these patterns deserve more serious attention than they have received in recent years.

*In *Cerebratulus* [Freeman, 1978] the first quartet cells were removed shortly after they formed long before fourth cleavage spindle formation; a sample of these embryos was examined for the specific purpose of finding out if the direction of the next cleavage was changed by the operation. In every case this cleavage was counterclockwise (Freeman, unpublished observations).

ACKNOWLEDGMENTS

I want to thank Jim Cather, Jonathan Henry, Judy Lundelius, Mark Martindale, and JoAnn Render for discussing cleavage patterns with me. This work was supported by grant GM 20024-09 from the National Institutes of Health.

REFERENCES

Ancel P, Vintemberger P (1948): Recherches sur le déterminisme de la symétrie bilatéral dans l'oeuf des Amphibiens. Bull Biol Fr Belg [Suppl] 31:1–182.

Biggelaar van den J, Guerrier P (1979): Dorsoventral polarity and mesentoblast determination as concomitant results of cellular interactions in the mollusk *Patella vulgata*. Dev Biol 68:462–471.

Clement A (1952): Experimental studies on germinal localization in *Ilyanassa*. I. The role of the polar lobe in determination of the cleavage pattern and its influence in later development. J Exp Zool 121:593–626.

Coe W (1899): The maturation and fertilization of the egg of *Cerebratulus*. Zool Jahrb Abt Anat Ontog 12:425–476.

Conklin E (1902): Karyokinesis and cytokinesis in the maturation, fertilization and cleavage of *Crepidula* and other gastropoda. I. Karyokinesis. II. Cytokinesis. J Acad Nat Sci (Philadelphia, Ser 2) 12:1–121.

Conklin E (1905a): Mosaic development in ascidian eggs. J Exp Zool 2:145–223.

Conklin E (1905b): The organization and cell lineage of the ascidian egg. J Acad Nat Sci (Philadelphia, Ser 2) 13:1–119.

Conklin E (1912): Experimental studies on nuclear and cell division in the eggs of *Crepidula*. J Acad Nat Sci (Philadelphia, Ser 2) 15:503–591.

Conklin E (1917): Effects of centrifugal force on the structure and development of the eggs of *Crepidula*. J Exp Zool 22:311–419.

Costello D (1945): Experimental studies of germinal localization in Nereis. I. The development of isolated blastomeres. J Exp Zool 100:19–46.

Costello D (1961): On the orientation of centrioles in dividing cells and its significance: A new contribution to spindle mechanics. Biol Bull 120:285–312.

Costello D, Henley C (1976): Spiralian development: A perspective. Am Zool 16:277–291.

Crampton H (1894): Reversal of cleavage in a sinistral gastropod. Ann NY Acad Sci 8:167–291.

Dan K (1979): Studies on unequal cleavage in sea urchins. I. Migration of the nuclei to the vegetal pole. Dev Growth Differ 21:527–535.

Dan K, Ikeda M (1971): On the system controlling the time of micromere formation in sea urchin embryos. Dev Growth Differ 13:285–301.

Freeman G (1976): The role of cleavage in the localization of developmental potential in the ctenophore *Mnemiopsis leidyi*. Dev Biol 49:143–177.

Freeman G (1978): The role of asters in the localization of the factors that specify the apical tuft and the gut in the nemertine *Cerebratulus lacteus*. J Exp Zool 206:81–108.

Freeman G (1979): The multiple roles which cell division can play in the localization of developmental potential. In Subtelny S, Konigsberg I (eds): "Determinants of Spatial Organization." New York: Academic Press, pp 53–76.

Freeman G (1980): The role of cleavage in the establishment of the anterior-posterior axis of the hydrozoan embryo. In Tardent P, Tardent R (eds): "Developmental and Cellular Biology of Coelenterates." Amsterdam: Elsevier/North Holland, pp 97–108.

Freeman G, Lundelius J (1982): The developmental genetics of dextrality and sinistrality in the gastropod *Lymnaea peregra*. W Rouxs Arch 191:69–83.

Guerrier P (1968): Origine et stabilité de la polarité animale-végétative chez quelques Spiralia. Ann Embryol Morphol 1:119–139.

Guerrier P (1970a): Nouvelles données expérimentales sur la segmentation et l'organogenése chez *Limax maximus* (Gastéropode pulmoné) Ann Embryol Morphol 3:283–294.

Guerrier P (1970b): Les caractéres de la segmentation et la détermination de la polarite dorsoventrale dans le développement de quelques Spiralia. III. *Pholas dactylus* et *Spisula subtruncata* (Mollusques Lamellibranches). J Embryol Exp Morphol 23:667–692.

Harvey E (1956): "The American Arbacia and Other Sea Urchins." Princeton: Princeton University Press.

Hörstadius S (1937a): Investigations as to the localization of the micromere-, the skeleton-, and the entoderm-forming material in the unfertilized egg of *Arbacia punctulata*. Biol Bull 73:295–316.

Hörstadius S (1937b): Experiments on determination in the early development of *Cerebratulus lacteus*. Biol Bull 73:317–342.

Hörstadius S (1973): "Experimental Embryology of Echinoderms." London: Oxford University Press.

Lillie F (1906): Observations and experiments concerning the elementary phenomena of embryonic development in *Chaetopterus*. J Exp Zool 3:153–268.

Longo F (1973): Fertilization: A comparative ultrastructural review. Biol Reprod 9:149–215.

Lutz D, Inoué S (1982): Colcemid but not cytochalasin inhibits asymmetric nuclear positioning prior to unequal cell division. Biol Bull 163:373–374.

Maruyama Y (1981): Precocious breakdown of the germinal vesicle induces parthenogenetic development in sea cucumbers. Biol Bull 161:382–391.

Meshcheryakov V (1976): Asymmetrical cytotomy and spindle orientation in early isolated blastomeres of gastropod mollusks. Ontogenez 7:558–565.

Meshcheryakov V (1978a): Orientation of cleavage spindles in pulmonate molluscs. I. Role of blastomere shape in orientation of second cleavage spindles. Ontogenez 9:558–566.

Meshcheryakov V (1978b): Orientation of cleavage spindles in pulmonate molluscs. II. Role of structure of intercellular contacts in orientation of the third and fourth cleavage spindles. Ontogenez 9:567–575.

Meshcheryakov V, Beloussov L (1975): Asymmetrical rotations of blastomeres in early cleavage of gastropoda. W Rouxs Arch 177:193–203.

Meshcheryakov V, Veryasova G (1979): Orientation of cleavage spindles in pulmonate molluscs. III. Form and location of mitotic apparatus in binuclear zygotes and blastomeres. Ontogenez 10:24–35.

Morgan T, Spooner G (1909): The polarity of the centrifuged egg. Arch Entmech Org 28:104–117.

Morrill J, Blair C, Larsen W (1973): Regulative development in the pulmonate gastropod *Lymnaea palustris*, as determined by blastomere deletion experiments. J Exp Zool 183:47–56.

Rappaport R (1974): Cleavage. In Lash J, Whittaker J (eds): "Concepts of Development." Stamford, CT: Sinauer, pp 76–98.

Robert A (1902): Recherches sur le developpement des troques. Arch Zool Exp Genet 3^{e} Ser 10:269–538.

Sakai H (1968): Contractile properties of protein threads from sea urchin eggs in relation to cell division. Int Rev Cytol 23:89–112.

Schatten G (1981): Sperm incorporation, the pronuclear migrations, and their relation to the establishment of the first embryonic axis: Time-lapse microscopy of the movements during fertilization of the sea urchin *Lytechinus variegatus*. Dev Biol 86:426–437.

Shimizu T (1981): Cortical differentiation of the animal pole during maturation division in fertilized eggs of *Tubifex* (Annelida, Oligochaeta). Dev Biol 85:65–76.
Shirai H, Kanatani H (1980): Effect of local application of 1-methyladenine on the site of polar body formation in starfish oocyte. Dev Growth Differ 22:555–560.
Tanaka Y (1976): Effects of the surfactants on the cleavage and further development of the sea urchin embryos. I. The inhibition of micromere formation at the fourth cleavage. Dev Growth Differ 18:113–122.
Thompson D (1942): "On Growth and Form, 2nd Ed." London: Cambridge University Press.
Wilson E (1925): "The Cell in Development and Heredity, 3rd Ed." New York: MacMillan.

Time, Space, and Pattern in Embryonic Development, pages 197–220

The Polar Lobe in Eggs of Molluscs and Annelids: Structure, Composition, and Function

M.R. Dohmen

Zoological Laboratory, University of Utrecht, Padualaan 8, 3508 TB Utrecht, The Netherlands

Right from the beginning of the science of experimental embryology, which starts in 1888 with Wilhelm Roux's blastomere deletion experiments on frog eggs, polar lobes have played an important role in theories on the influence of cytoplasmic determinants on development. At Woods Hole, Crampton [1896] demonstrated that the polar lobe of *Ilyanassa* can be removed at the trefoil stage (Fig. 1) without immediate damage to the egg. As a result, the mesentoblast cell 4d fails to appear and, consequently, the mesoderm bands do not develop.

Polar lobes arise by a transient constriction that separates the anuclear vegetal region of the egg cytoplasm more or less completely from the remainder of the egg. The process starts during the maturation divisions, but the constrictions remain rather shallow. During first cleavage the constriction is very deep, resulting in a lobe that is connected by a thin stalk to the egg. This stage is called the trefoil stage (Fig. 1c). The lobe is subsequently resorbed into one of the blastomeres, called the CD blastomere. The process is repeated during second cleavage and in some species at third cleavage as well. In *Littorina* polar lobe formation continues even up until fifth cleavage [Moor, 1973]. In this way the vegetal region of the egg is shunted to the D quadrant of the egg, which will become the dorsal quadrant of the future embryo.

In most species it is unknown whether a polar lobe is formed or not. It is difficult, therefore, to establish a relationship between the occurrence of

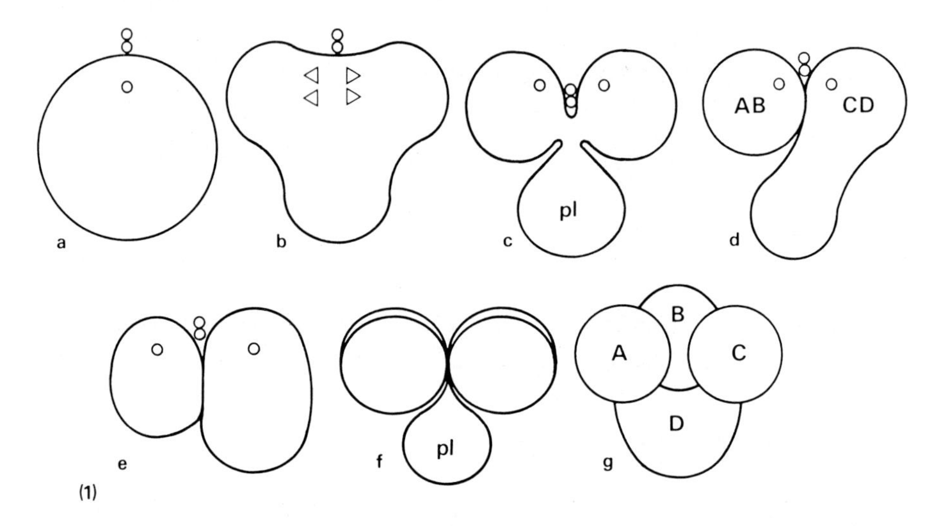

Fig. 1. Schematic representation of polar lobe formation. a. Uncleaved egg. b. Constriction in the vegetal hemisphere, just before first cleavage. c. Trefoil stage: constriction of the polar lobe (pl) is maximal. d. Resorption of the lobe into the CD cell. e. Resorption is completed. f. Lobe formation at second cleavage. g. Four-cell stage after resorption of the second polar lobe. Circles represent nuclei or polar bodies. Triangles are chromosomes during mitosis.

polar lobes and the systematic position of a species [see Fioroni, 1979]. Molluscs have been most extensively studied in this respect. Polar lobes have never been observed in eggs of Polyplacophora. Among the Aplacophora a polar lobe has been described in the species *Epimenia* [Baba, 1951]. In the class of the Scaphopoda *Dentalium* has a polar lobe. No data are available on the subclasses Protobranchia and Septibranchia of the Bivalvia. Many Lamellibranchia have a polar lobe, but several species within this subclass show unequal cleavage without preceding formation of a lobe. Among the Gastropoda the archaeogastropods have equal cleavage without polar lobe formation. Most of the mesogastropods have small polar lobes, whereas the neogastropods have large lobes. In the opisthobranchs and the pulmonates polar lobe formation has never been described [see Verdonk and van den Biggelaar, 1983].

MORPHOGENETIC SIGNIFICANCE

The polar lobe contents appear to be of crucial importance for development, as can be demonstrated by removing the lobe at the trefoil stage. This can be done without injuring the blastomeres. The egg continues cleaving, and subsequent development, although severely defective, is not chaotic but follows a characteristic and regular course. The first effect to be seen after polar lobe deletion is the synchronization of the cleavages in the four quadrants. In normal development the time schedule of the divisions in the D quadrant differs from that of the other quadrants (Fig. 2). These characteristic asynchronies are abolished by removal of the lobe, as has been shown in *Ilyanassa* [Clement, 1952] and *Dentalium* [van Dongen and Geilenkirchen, 1974, 1975].

The polar lobe also affects the pattern of the early cleavages. In normal development, some cells in the D quadrant are smaller than the corresponding cells in the other quadrants, and other cells are larger [Clement, 1952; van Dongen and Geilenkirchen, 1974, 1975]. These differences disappear upon removal of the lobe: the egg becomes radialized. We may conclude, therefore, that the polar lobe contains factors that control the initiation of the cell cycle and the position and orientation of the cleavage planes in the cell lines to which the lobe is segregated during cleavage.

The effects of polar lobe deletion on larval development are remarkably uniform. Both in annelids *(Sabellaria)* and mollusks *(Ilyanassa, Dentalium, Mytilus)* removal of the lobe at first cleavage results in the absence of the apical tuft and of the posttrochal region (Fig. 3) from which the mantle fold, shell, and foot normally develop [Wilson, 1904; Hatt, 1932; Novikoff, 1938; Clement, 1952; Rattenbury and Berg, 1954].

The influence of the lobe on the morphogenesis of adult organs has been investigated in the gastropods *Ilyanassa* [Clement, 1952; Atkinson, 1971]

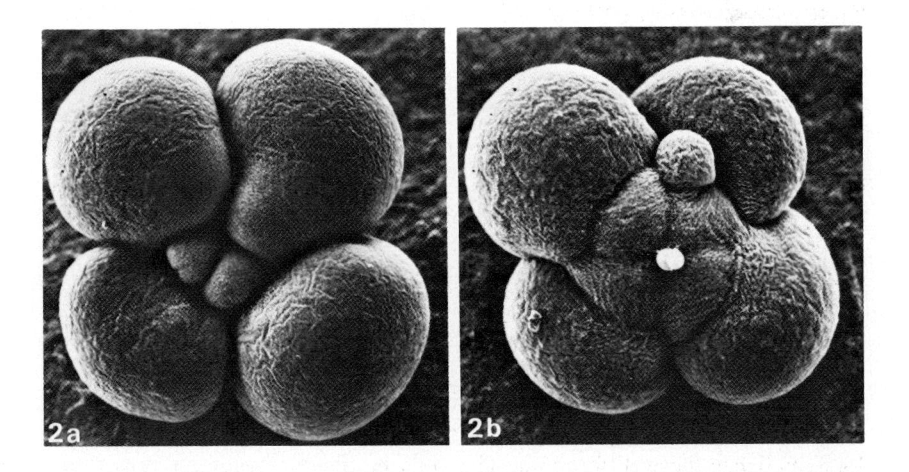

Fig. 2. Asynchronous formation of micromeres in eggs of *Crepidula fornicata*. a. Formation of the first quartet of micromeres: 1d and 1c are already formed. b. Formation of the second quartet: 2d appears first. Magnification ×250.

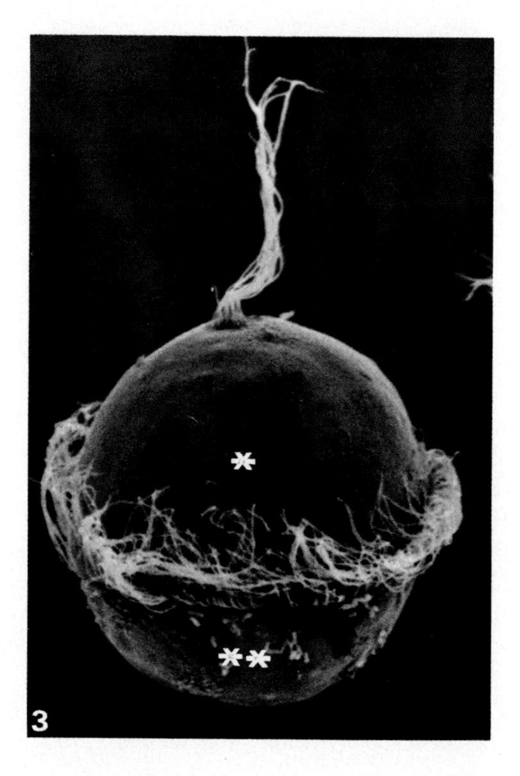

Fig. 3. Trochophore larva of *Sabellaria alveolata,* showing apical tuft, pretrochal region (*), prototroch, and posttrochal region (**).

and *Bithynia* [Cather and Verdonk, 1974] and in the scaphopod *Dentalium* [van Dongen, 1976]. Adult organs missing as a result of polar lobe deletion are shell, foot, operculum, otocysts, eyes, tentacles, intestine, and a beating heart. Part of these structures, such as heart and intestine, are directly lobe dependent, as they originate from the D quadrant, which receives the lobe contents. Others, such as eyes and tentacles, develop from other quadrants, but their appearance depends on an interaction with the D quadrant.

The influence of the polar lobe on cells outside the D quadrant is already evident at cleavage stages. In normal development of *Dentalium,* the time schedule and the pattern of cleavage in some cells of the A and C quadrant deviate from the corresponding cells in the B quadrant. Removal of the polar lobe abolishes these differences. The cells in all four quadrants now divide as they do normally in the B quadrant [van Dongen and Geilenkirchen, 1974, 1975]. The far-reaching influence of the polar lobe has led to the opinion that the vegetal pole plasm confers to the D quadrant the quality of an organizer [Cather, 1971; Clement, 1976].

SEGREGATION AND ACTIVATION OF MORPHOGENETIC DETERMINANTS

There is no evidence that one of the blastomeres at first cleavage is predetermined to receive the polar lobe contents. Both blastomeres appear to be competent to respond properly to the incorporation of polar lobe material. This can be inferred from experiments by Guerrier et al. [1978], who treated eggs of *Dentalium* with cytochalasin B before first cleavage. At certain concentrations, the formation of a polar lobe was inhibited, but the egg cleaved and two equally sized blastomeres were formed, each containing half of the vegetal pole plasm. Both blastomeres behaved as a CD cell normally does, forming a polar lobe at second cleavage. Consequently, a four-cell stage was formed consisting of two large D cells and two small C cells. The two D cells showed all characteristic features normally displayed by the D quadrant, both with respect to the time schedule and to the pattern of cleavage, and during further development a duplication of lobe-dependent structures was observed.

Centrifugation experiments with *Bithynia* eggs also indicate that both blastomeres at first cleavage are equally competent to behave as a CD cell, provided that they receive the polar lobe contents. In *Bithynia* eggs the polar lobe contains a special plasm, the vegetal body (Fig. 4), that can be displaced from its polar position by centrifugation [van Dam et al., 1982]. Polar lobes devoid of the vegetal body can be removed without affecting normal development, thereby demonstrating that the lobe-specific determinants are concentrated in the vegetal body. It may be assumed that the vegetal body is not preferentially displaced into a particular blastomere by centrifugation. As normal development ensues irrespective of the position of the vegetal body, it is obvious that both blastomeres at first cleavage are able to react properly to the presence of the polar lobe determinants.

At second cleavage, the resorption of the polar lobe obeys a strict rule in all lobe-forming species. The daughter cell that is situated in a clockwise direction when viewed on the animal pole always receives the second polar lobe and thus becomes the D cell.

In *Bithynia* and *Dentalium* there is evidence that the polar lobe determinants are not exclusively shunted into the D blastomere. Verdonk and Cather [1973] studied the developmental potential of blastomeres at the four-cell stage in *Bithynia*. Isolated AD and BC combinations developed similarly and both combinations gave rise to lobe-dependent structures such as eyes, tentacles, foot, and shell. Cather et al. [1976] studied the development of the various types of three-quarter embryos, obtained by deletion of one blasto-

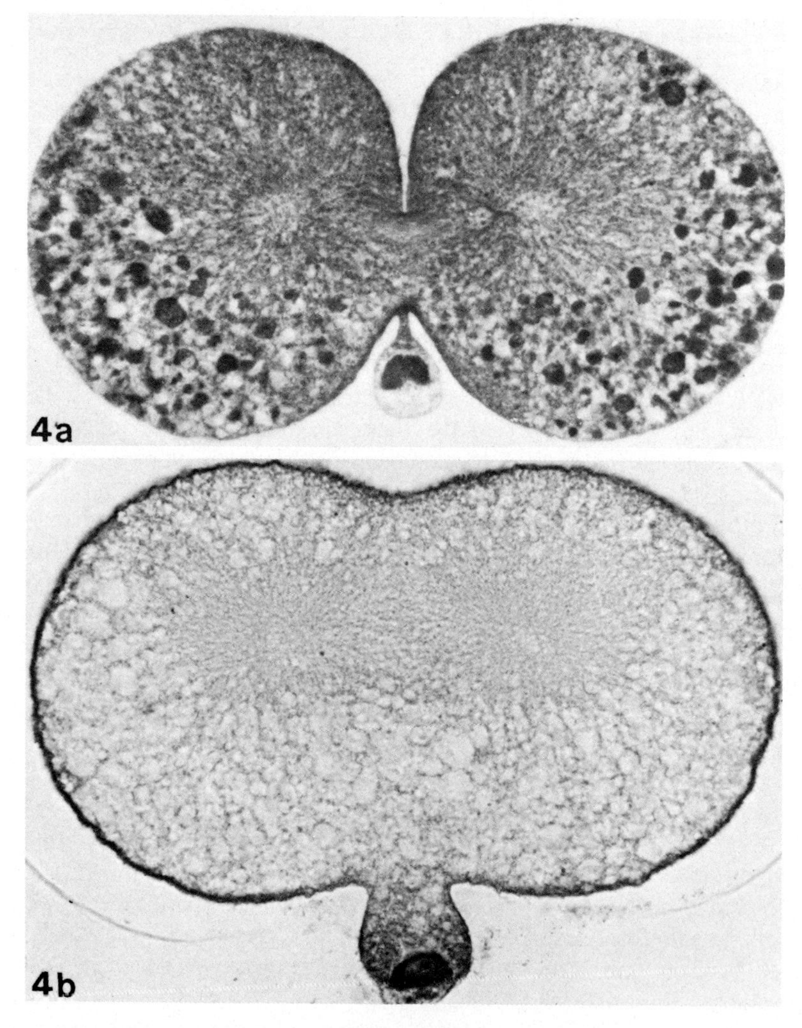

Fig. 4. Sections of *Bithynia* eggs at the trefoil stage. a. Hematoxylin-eosin staining. The polar lobe contains a cup-shaped structure, the vegetal body. b. Pyronin staining. The vegetal body stains intensely.

mere at the four-cell stage. It was found that the ABC combination develops all lobe-dependent structures. Both sets of experiments indicate that the C blastomere receives part of the polar lobe determinants. In *Dentalium,* the determinants for the formation of an apical tuft, present in the first polar lobe, do not become incorporated in the second polar lobe, as can be deduced from the observation that an apical tuft may be formed after deletion of the second polar lobe. At third cleavage the apical tuft determinants must be present at the animal pole of the C and D blastomere as they are segregated into the 1c and 1d micromeres from which the apical tuft develops [Cather and Verdonk, 1979; van Dongen and Geilenkirchen, 1974].

The segregation of polar lobe determinants at later cleavage stages has been studied in *Ilyanassa* [Clement, 1962] and in *Dentalium* [Cather and Verdonk, 1979] by deleting the D-quadrant macromeres at successive cleavages. Deletion of the cells 1D and 2D has the same effect as removal of the second polar lobe. Deletion of 3D results in a well-organized larva that possesses most of the adult structures. Deletion of 4D has no effect at all, but in *Ilyanassa* removal of the mesentoblast 4d results in a veliger lacking heart and intestine [Clement, 1960]. After deletion of 4d in *Dentalium*, no defects in the larvae could be detected [Cather and Verdonk, 1979]. These results indicate that the polar lobe determinants are segregated into the 2D macromere and that their influence is exerted mainly during the interval between fifth and sixth cleavage. Since several adult structures observed in larvae raised from eggs lacking 3D (e.g., eyes, tentacles, and shell) originate from micromeres present already before the formation of 3D, an inductive action is probably exerted by 3D on these micromeres right after fifth cleavage and just before 3D is deleted.

IDENTIFICATION OF MORPHOGENETIC DETERMINANTS

Biochemical Identification

The biochemical nature of the polar lobe determinants has been investigated mainly by looking for possible differences in molecular composition between isolated polar lobes and blastomeres and differences in synthetic activity between normal and lobeless embryos. Research has been concentrated almost exclusively on *Ilyanassa.* The most significant differences in composition that emerged from these efforts are a higher concentration of adenosine triphosphate (ATP) in the polar lobe of *Ilyanassa* [Collier and Garone, 1975] and a higher concentration of guanosine triphosphate (GTP) in the polar lobe of *Nassarius* [van Dongen et al., 1981]. Collier [1983] considers the possibility that ATP might be transferred through gap junctions from the D macromere into specific micromeres and might thus act to increase the cleavage rates in these cells so that the number of stem cells required for the differentiation of an organ would be produced. The morphogenetic deficiencies of the lobeless embryo would thus result from the formation of an insufficient number of stem cells of lobe-dependent organs. Collier does not exclude the possibility that, apart from this nonspecific effect of the polar lobe, there are specific determinants for the differentiation of lobe-dependent organs.

Electrophoretic analysis of RNA synthesis in normal and lobeless embryos of *Ilyanassa* after a 24-hour incubation with ^{3}H-uridine [Koser and Collier, 1976] showed that there are no detectable changes in the pattern of RNA synthesis before gastrulation. After gastrulation there is proportionally less radioactivity in 34S RNA and more in 12–16S RNA in lobeless embryos. The interpretation of these data is that there is a decrease in hnRNA production and an accumulation of untranslated mRNAs in the lobeless embryo. Collier [1975] also measured the proportion of polyadenylated RNA and found that deletion of the polar lobe did not alter the proportion of poly(A)RNA synthesized during the early cleavage stages. These results indicate that the deficiencies of lobeless embryos do not result from a major defect in RNA synthesis. It cannot be ruled out, however, that the lobeless embryo fails to produce a set of specific RNAs that go undetected with the methods used.

The influence of the polar lobe on protein synthesis in *Ilyanassa* has been studied most recently by Collier and McCarthy [1981] and Brandhorst and Newrock [1981] using two-dimensional electrophoresis of ^{35}S-methionine-labeled polypeptides. They found that the intact egg, the lobeless egg, and the isolated polar lobe make identical sets of polypeptides and it appeared that removal of the polar lobe results in some quantitative but no qualitative changes in proteins synthesized by the 29-cell embryo. Collier and McCarthy [1981] also demonstrated that the synthesis of 98% of the 350 polypeptides detected by electrophoresis was insensitive to actinomycin D. This is in agreement with the observation that in *Sabellaria* lobe-dependent structures can develop in the presence of actinomycin D [Guerrier, 1971]. The main conclusion from these data is that there is no evidence for the segregation of a specific set of maternal mRNAs in the polar lobe.

New views on the role of RNA in polar lobes may come from studying the polar lobe of *Bithynia.* In this species the lobe contains a special plasm, the vegetal body, which reacts very strongly with nucleic acid-specific stains such as pyronin (Fig. 4b), acridine orange, ethidium bromide, and Hoechst 33258 [Dohmen and Verdonk, 1974, 1979a]. We do not know which kind of RNA is accumulated in the vegetal body and it remains to be established whether this RNA has the capacity to act as a morphogenetic determinant.

Structural Identification

An experimental procedure that greatly facilitates the identification of structures that are likely to contain lobe-specific determinants has been applied to *Ilyanassa* [Clement, 1968], *Dentalium* [Verdonk, 1968], and *Bithynia* [van Dam et al., 1982]. The procedure consists of removing cytoplasmic components from the polar lobe by centrifugation and assaying the morphogenetic potential of the components that remain in the polar lobe by deleting the lobe, or, alternatively, by removing the part of the egg that contains the components that were displaced from the lobe [Clement, 1968]. A limitation of this procedure is that most eggs are disrupted by high centrifugal forces. Resistance is sufficient, however, to allow the stratification of the common cytoplasmic inclusions such as yolk platelets, lipid droplets, mitochondria, etc. In this way it has been established that, in *Ilyanassa* and *Dentalium*, centrifugation does not displace the morphogenetic determinants from the polar lobe. An ultrastructural study of centrifuged eggs showed that all the easily identifiable inclusions of the polar lobe, such as mitochondria, double-membrane vesicles, yolk, lipid, and endoplasmic reticulum, are readily displaced and thus cannot account for the morphogenetic potential of the lobe [Dohmen and Verdonk, 1979a].

In *Bithynia,* however, the vegetal body (Fig. 4) proved to be exceptionally resistant to centrifugation. Only when forces up to 1,400g were applied did about 50% of the eggs show a displaced vegetal body, which then could be found intact in one of the blastomeres. These properties of the vegetal body were exploited by van Dam et al. [1982] to demonstrate that, when the polar lobe is devoid of the vegetal body, it can be deleted without affecting normal development. These results prove that the vegetal body contains the morphogenetic determinants and that its position in the egg is of minor importance.

Now that a structure has been identified that contains the lobe-specific determinants it should be possible to make some progress toward identifying the molecular nature of the determinants. RNA is a likely candidate as the vegetal body is extremely rich in it. The RNA is likely to be stored in the vesicles that make up the vegetal body (Fig. 5). Similar aggregates of vesicles have been reported in the polar lobes of *Crepidula* [Dohmen and Lok, 1975], *Littorina* [Dohmen and Verdonk, 1979a], and *Buccinum* (Fig. 6).

An interesting feature that emerged from ultrastructural studies of polar lobes of *Crepidula* and *Buccinum* is the occurrence of structures that bear a striking resemblance to the germ plasm of insects and amphibians [Dohmen and Verdonk, 1979b]. Preliminary evidence indicates that in *Crepidula* these plasms segregate into the mesentoblast 4d, from which the gonads develop.

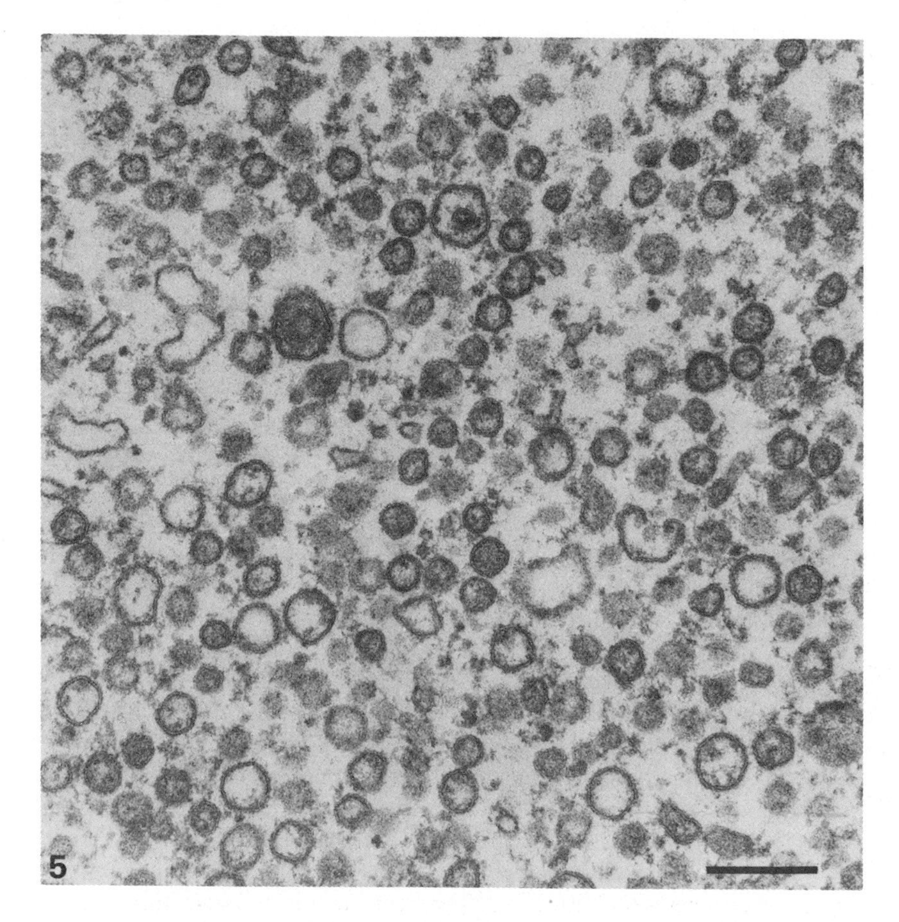

Fig. 5. Detail of the vegetal body of *Bithynia*. Bar represents 0.2 μm.

It should be worthwhile to study the segregation of these plasms during further development in order to see if they become localized in the primordial germ cells. Experimental indication that the gonads or the germ cells are lobe-dependent differentiations is not available. It is doubtful whether lobe deletions will ever allow this issue to be resolved as the gonads develop very late: in *Buccinum* after six weeks, at the late veliger stage [Giese, 1978], and in *Crepidula adunca* after hatching of the young adults [Moritz, 1939]. It is generally not feasible to raise lobeless embryos for such prolonged periods.

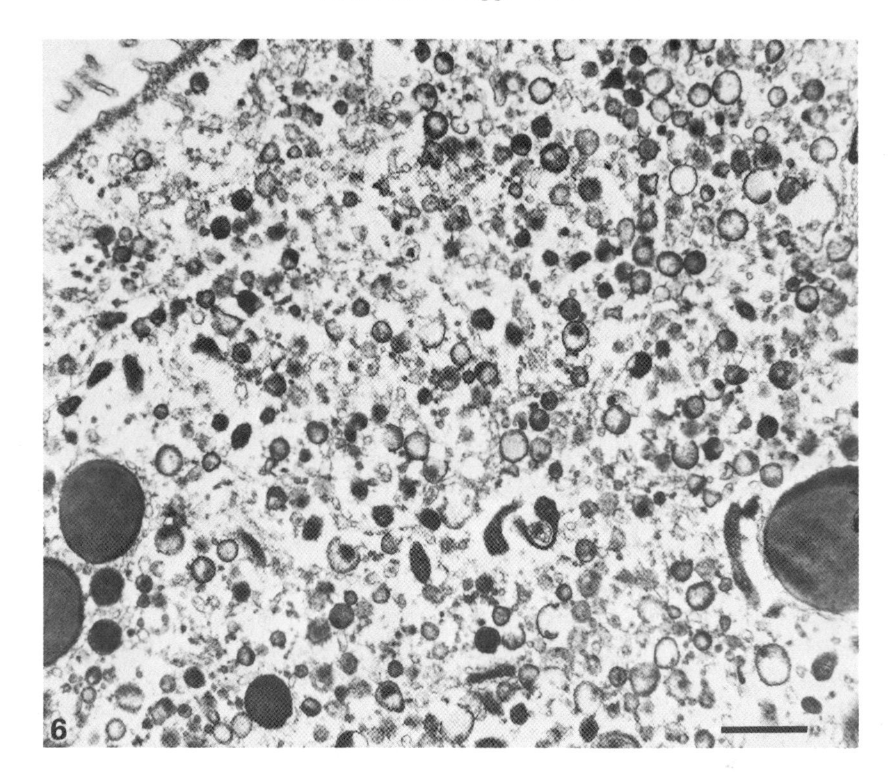

Fig. 6. Detail of vesicle-rich cytoplasm, present in the polar lobe of *Buccinum undatum*. Bar represents 1.0 μm.

THE SURFACE STRUCTURE

The centrifugation experiments of Clement [1968], Verdonk [1968], and van Dam et al. [1982] have demonstrated that the lobe-specific determinants are firmly attached to the vegetal cortex of the egg. In *Bithynia* the localization of the morphogens at the vegetal pole begins during oogenesis. In this species the nascent vegetal body is localized at the basal pole of the oocyte right from its first appearance at the beginning of the vitellogenic phase [Dohmen, 1983]. Indirect evidence for the localization of determinants during oogenesis is provided by experiments in which unfertilized eggs of *Dentalium* were divided in an animal and a vegetal part. After fertilization of both parts animal parts form polar bodies but no polar lobe or lobe-dependent structures. Vegetal parts, however, form a polar lobe and lobe-dependent structures [Wilson, 1904; Verdonk et al., 1971].

The mechanisms that bring about this localization and provide the strong attachment of the determinants to the egg cortex are unknown. Microfilaments may be involved, as can be inferred from the observation that treatment of *Sabellaria* eggs with cytochalasin B throughout meiosis resulted in frequent abnormalities affecting lobe-dependent structures [Peaucellier et al., 1974].

Apart from the question of the mechanics of the localization process, there is also the question of the topography: How is the area selected in which localization is to take place? It has often been argued, on the basis of centrifugation experiments, that the cortex of the egg contains the coordinates for the spatial organization of the egg [see Kühn, 1965]. This implies the existence of regional specializations of the egg surface and this is exactly what can be observed very clearly at the vegetal pole of a number of polar lobe-forming gastropod eggs [Dohmen and van der Mey, 1977].

The eggs of *Nassarius reticulatus* (Fig. 7) and *Bithynia tentaculata* possess a dense array of very large villi at the vegetal pole. In *Crepidula fornicata* (Fig. 8), *Buccinum undatum,* and *Nucella lapillus* (Fig. 9) a characteristic pattern of ridges is present at the vegetal pole. As far as can be judged from studying living eggs with phase optics, these structures are static. They are not comparable to the undulating movements reported to occur on the surface of the eggs of the bivalve *Barnea candida* [Pasteels, 1966]. These movements result from waves of elongation of microvilli. Quite another type of surface differentiation has been found in the eggs of *Dentalium.* In these eggs a tuft of bacteria adheres to the egg surface, exclusively at the vegetal pole [Geilen-

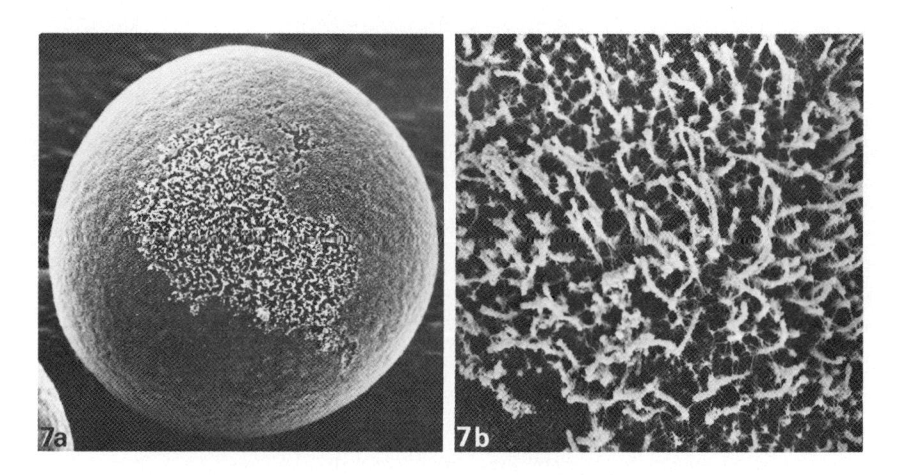

Fig. 7. a. Uncleaved egg of *Nassarius reticulatus,* showing a local surface differentiation at the vegetal pole. b. Detail of the vegetal pole surface. Magnification a. ×300; b. ×1200.

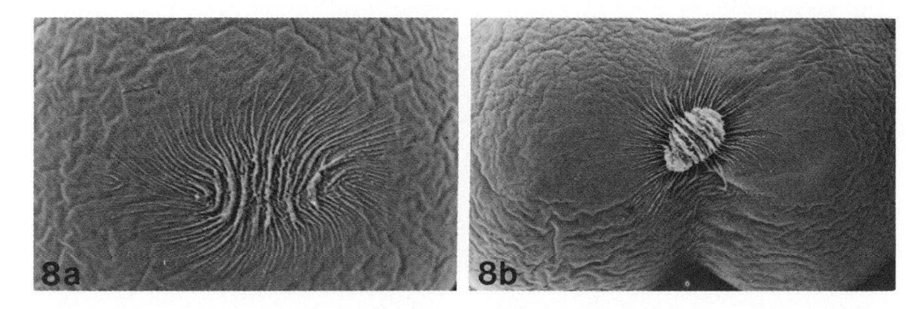

Fig. 8. a. Bilateral symmetrical pattern of folds at the vegetal pole of the uncleaved egg of *Crepidula fornicata*. b. First cleavage with polar lobe. Note the orientation of the pattern of folds with respect to the plane of cleavage. Magnification a. 300×; b. 200×.

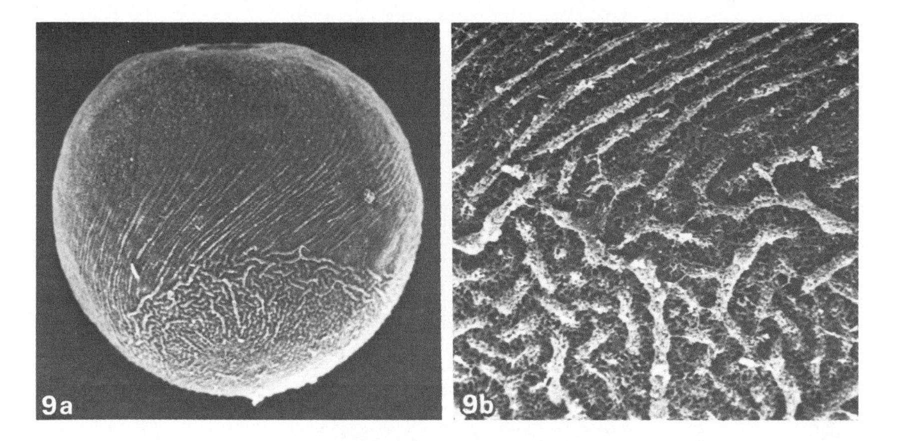

Fig. 9. a. Uncleaved egg of *Nucella lapillus*, showing extensive area of surface folds on the vegetal hemisphere. b. Detail of surface folds, showing the characteristic pattern of parallel folds spiraling upward from the vegetal pole area. Magnification a. ×250; b. ×1100.

kirchen et al., 1971]. The bacteria do not penetrate into the egg cytoplasm; they are attached to the egg surface (Fig. 10). Their exclusive presence at the vegetal pole indicates that in this area the plasma membrane or the extracellular material must possess specific properties.

The regional differentiation of the vegetal egg surface probably arises during oogenesis. This view is strongly supported by the phenomena observed in the eggs of *Nassarius reticulatus*. In this species the oocytes protrude into the lumen of the ovarium tubule, leaving only their basal area in contact with follicle cells. The free surface of the oocyte is covered with

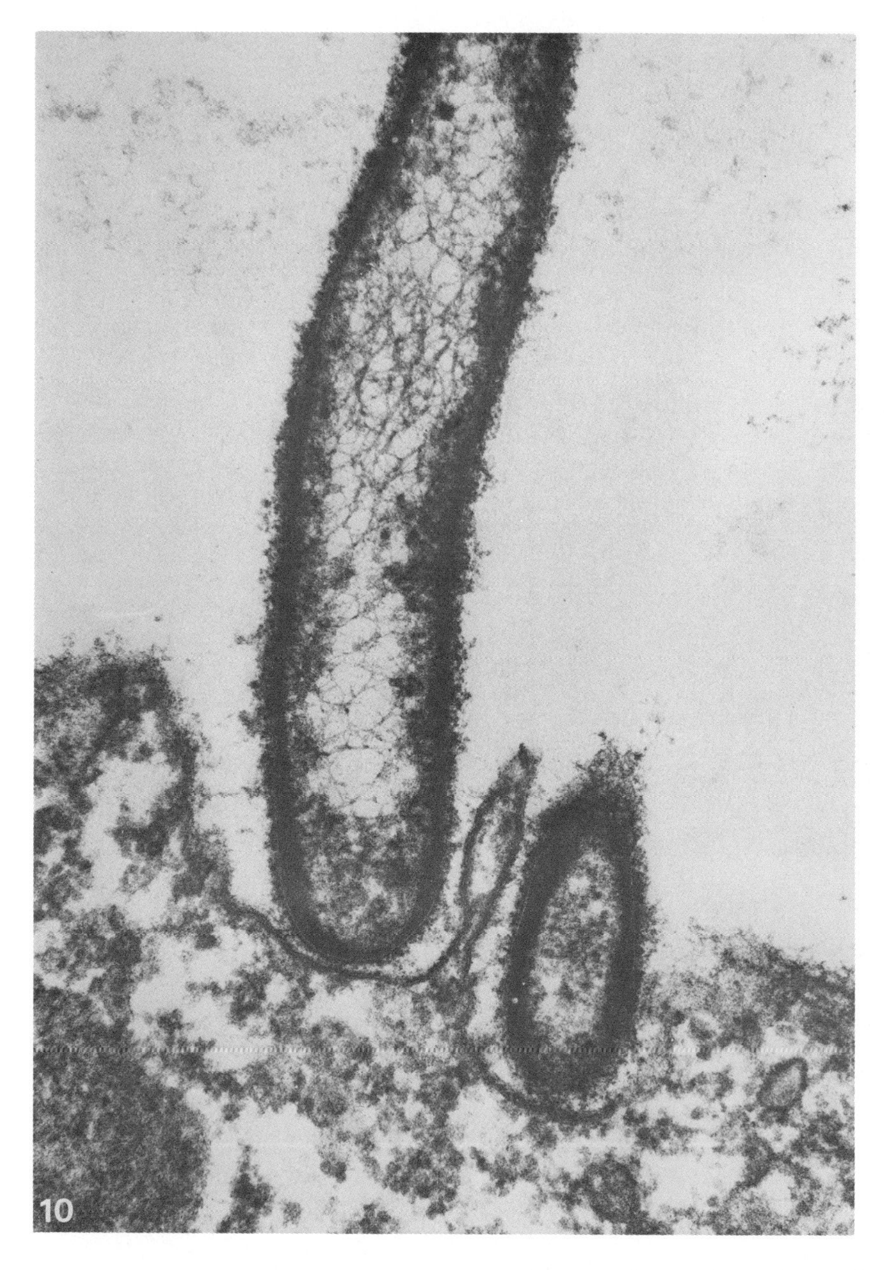

Fig. 10. Detail of the vegetal pole surface of a *Dentalium* egg, showing the attachment of bacteria. Magnification ×18000.

microvilli and bears a carbohydrate coat. Both these features are lacking at the contact area with the follicle cells. Precisely in this contact area, which is smooth initially, a dense array of very long villi develops after oviposition (Fig. 7). The strict correspondence between the original contact area and the differentiated area suggests a causal relationship.

In all the species investigated, the structure of the vegetal pole surface is gradually modulated during early development. Initially, the vegetal pole surface is smooth. During the maturation divisions typical villi or ridges appear. These structures are not temporally coupled to the appearance of the successive polar lobes. They persist and are segregated onto the D macromere at second cleavage (Fig. 11). In some species, such as *Bithynia*, they disappear soon after third cleavage. In other species they persist until gastrulation and are further modulated. In *Nassarius,* for instance, the long microvilli at the vegetal pole disappear during the early cleavages and are replaced by low excrescences, but the differentiated polar area remains sharply demarcated from the remaining egg surface [Dohmen and van der Mey, 1977]. In *Crepidula* the ridges become slightly less pronounced during cleavage, but they are still present at gastrulation.

It remains to be established what significance should be attributed to these regional surface differentiations. They are formed at the vegetal pole of the egg, so they might be instrumental in establishing or maintaining the animal-vegetative polarity of the egg without any direct relationship with the morphogenetic determinants present in the polar lobes. Another view is that they might function in the localization and attachment of these determinants or in the mechanism that regulates their expression. This problem can be ap-

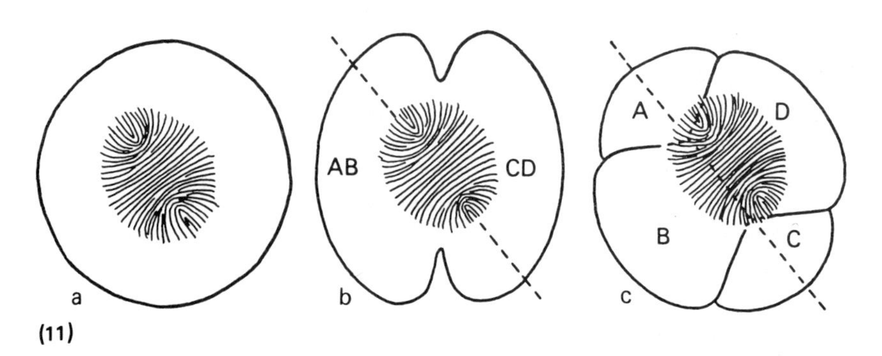

Fig. 11. Schematic drawing of the bilateral symmetrical pattern displayed by the surface folds at the vegetal pole of *Crepidula* eggs. The first cleavage plane is orientated at an angle of 40–60° with respect to the symmetry axis of this pattern. This relationship is retained during subsequent cleavages.

proached in several ways. One of these is to destroy the surface architecture at the vegetal pole and study the consequences for further development. We have done this with eggs of *Nassarius reticulatus*. Treatment of uncleaved eggs with concanavalin A completely removes the large vegetal villi while leaving intact the animal microvilli. Further development of these eggs is highly abnormal, but the original polarity is conserved, as can be seen from the appearance of micromeres at the animal pole. We may conclude, therefore, that the surface architecture at the vegetal pole is not necessary for maintaining the animal-vegetative polarity. The abnormal development resulting from concanavalin A treatment cannot, as yet, be related to malfunctioning of polar lobe determinants. The eggs of *Bithynia* are very suitable objects for studying these problems, as the polar lobe determinants are present in the form of an easily identifiable structure, the vegetal body. This body is synthesized during vitellogenesis and it is localized from the beginning at the basal pole of the oocyte, which corresponds to the future vegetal pole of the egg. This implies that the surface architecture at the vegetal pole is not involved in localizing the vegetal body, as this surface architecture appears after oviposition. In an experiment, described above, the vegetal body was displaced by centrifuging the egg before first cleavage. Deletion of the polar lobe, which is devoid of the vegetal body but still has its normal surface architecture, did not affect further development. This result implies that the surface of the vegetal pole is not necessary for activating the morphogenetic determinants present in the vegetal body. The role of the surface features at the vegetal pole thus remains a puzzling problem.

A common feature of all types of surface differentiation present on polar lobes is that they constitute an increase in surface area. This might provide a clue to their function. Possibly, the increased surface area generates electrical currents. In eggs of *Fucus* and *Xenopus,* and in several other developmental systems, electrical currents have been observed. The function of these currents is still hypothetical [reviewed by Borgens et al., 1979].

In all but one species, the surface features at the vegetal pole are radially symmetrical. Only in *Crepidula* can a bilaterally symmetrical pattern be distinguished (Fig. 11). The plane of first cleavage has a fixed relationship with the axis of this pattern. The angle between them is about 40–60° [Dohmen, unpublished]. This relationship demonstrates that the first cleavage plane is not only predetermined in the sense that it is orientated meridionally. Its position is apparently restricted more rigorously. Probably there is no direct coupling between the surface pattern at the vegetal pole and the first cleavage plane, as the latter originates at the animal pole. Both probably result from an inherent bilaterally symmetrical structure of the cortex of the uncleaved egg.

CYTOKINETIC ASPECTS

For the student of cytoplasmic localization two questions concerning the cytokinetic aspects of polar lobe formation are particularly interesting: Which mechanism triggers the constriction of the lobe and how is the position of the constriction determined?

In normal development the lobes appear in synchrony with the cleavages. The stimulus for furrow formation in cell division in known to be provided by the mitotic apparatus [Rappaport, 1971; Rappaport and Rappaport, 1974]. Lobe formation, however, is not dependent on the mitotic apparatus, since isolated anuclear vegetal parts of an egg are capable of forming polar lobes. This process in isolated vegetal parts is more or less synchronous with cleavage of the isolated animal part of the egg [Wilson, 1904; Morgan, 1935a–c; Clement, 1935; Verdonk et al., 1971]. Synchrony of these cytokinetic events is also observed in egg parts that have been isolated by centrifugation [Morgan, 1935a–c; Clement, 1935]. These data can be explained by assuming that the initiation of both cleavage and polar lobe formation is determined by a single clock mechanism residing in the cortex of the egg.

The position of the polar lobe constriction may vary enormously between different species, and consequently the size of the lobe may vary from very small, less than 1% of the egg volume in *Bithynia* (Fig. 4) and *Crepidula* (Fig. 8), to very large, one third of the egg volume in *Dentalium,* (Fig. 12) and *Ilyanassa.* The polar lobe constriction is always formed perpendicularly to the animal-vegetal axis of the egg. This is even true under experimental conditions. After centrifugation, for instance, the eggs of *Ilyanassa* [Clement, 1968] and *Dentalium* [Verdonk, 1968] often show an oblique or equatorial first cleavage, but the polar lobe always appears in its normal position, irrespective of the direction of first cleavage.

The potential to form a polar lobe is apparently restricted to a limited part of the egg. If the vegetal part of an uncleaved egg of *Dentalium* is removed, the egg will not form a polar lobe at first cleavage. If the deleted volume is less than the volume of a lobe, a smaller lobe than usual will be formed [Verdonk et al., 1971]. An influence of the mitotic apparatus on the position of the lobe constriction is suggested by experiments on *Dentalium* [Wilson, 1904; Verdonk et al., 1971]. Unfertilized eggs were cut equatorially into an animal and a vegetal half and both halves were fertilized. The animal half cleaved without forming a polar lobe. The vegetal half, however, forms a normally proportioned trefoil, implying that the position of the lobe constriction has been adapted. A decisive role of the mitotic apparatus in determining the position of the lobe constriction is strongly suggested by the observation that in the vegetal half of fertilized eggs, in which no nucleus is present, no

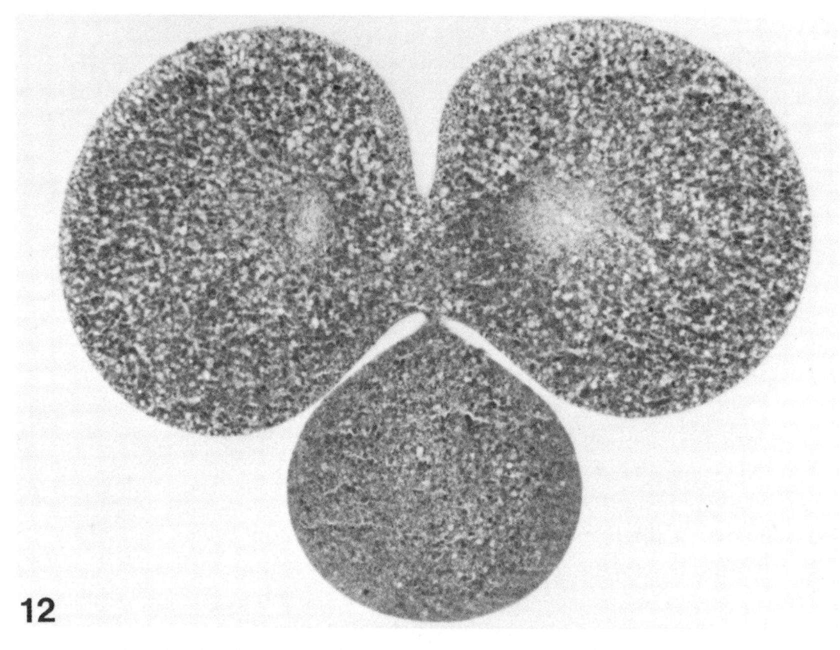

Fig. 12. Section of a trefoil stage of *Dentalium*. Magnification ×350.

adaptation takes place: a polar lobe of normal size is formed. A role of the mitotic apparatus is in agreement with the results obtained by treating eggs with colchicine. Treatment at an early stage, before constriction of the lobe becomes visible, inhibits lobe formation. Treatment at later stages has no effect anymore [Raff, 1972; Conrad and Williams, 1974a]. The observations suggest that the microtubules of the mitotic apparatus are involved in the initiation of lobe formation. Only the vegetal half of the egg is capable of responding to this stimulus. However, in the absence of any mitotic apparatus the vegetal part of the egg is still capable of forming a polar lobe. These conflicting data have not been reconciled.

The constriction of the lobes is executed by a ring of microfilaments [Conrad, 1973; Conrad et al., 1973; Conrad and Williams, 1974a]. Treatment with high concentrations of cytochalasin B inhibits lobe formation as well as cleavage [Raff, 1972; Conrad et al., 1973; Conrad and Willims, 1974a]. Lower concentrations may result in inhibition of lobe formation, but cleavage proceeds normally [Guerrier et al., 1978]. This differential sensitivity for cytochalasin B suggests that the processes involved in lobe constriction and cleavage furrow formation are not quite identical, although both operate by contraction of microfilaments. A similar differential sensitivity is observed when eggs of the annelid *Chaetopterus* are compressed. These eggs do not form a real polar lobe, but only a broad flattened bulge, through which the cleavage plane passes [Tyler, 1930]. The reverse effect, suppression of cleavage and unimpaired lobe formation, has been obtained by treating *Sabellaria* eggs with seawater enriched in KCl. In most of the eggs cleavage is inhibited but a number of them form polar lobes of the same size and at the same time as in the controls [Novikoff, 1940]. Treatment of *Nassarius* eggs with concanavalin A also inhibits cleavage, but the polar lobe is constricted off very well, though not quite normally (Fig. 13). As yet no explanation is available for these differences in sensitivity.

Polar lobelike protrusions can be induced anywhere on the egg by exposing fertilized eggs of *Ilyanassa* to isotonic CaCl after second maturation division [Conrad and Williams, 1974b]. Within ten to 15 minutes protrusions begin to form, usually in the vegetal hemisphere, but also on the animal side. If this treatment is applied at the two-cell stage, a lobelike protrusion is frequently observed on the AB cell, whereas in normal development only the CD cell forms a polar lobe. Similar effects are observed after iontophoretic

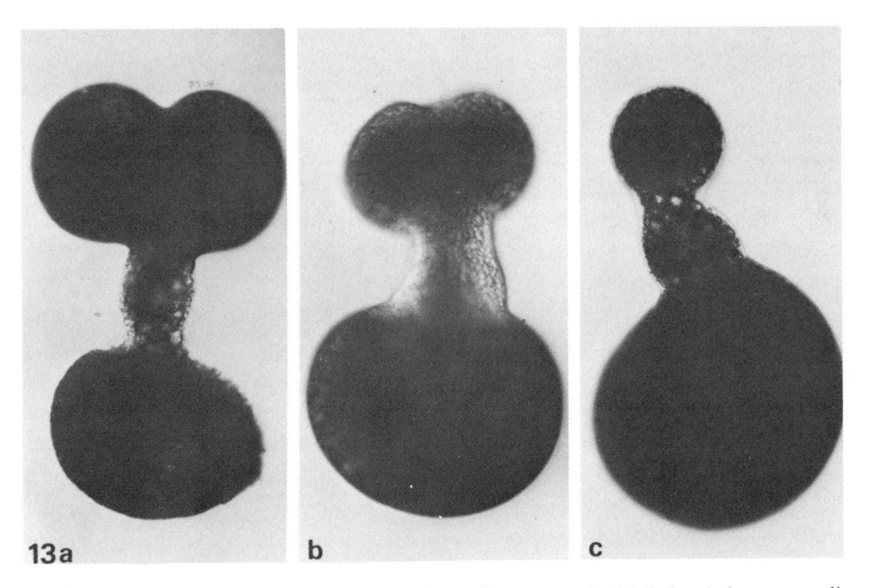

Fig. 13. *Nassarius reticulatus*: various reactions of eggs treated with 0.5 μg/ml concanavalin A for 20 minutes after the second maturation division. Cleavage is suppressed, but the beginning of a furrow may be present temporarily. Polar lobe constriction is well developed but abnormal in varying degrees. The volume of the polar lobe may be increased at the expense of the blastomere volume (b, c). a, b, and c do not represent successive stages, but different ways in which the eggs may react on treatment with concanavalin A. Magnification ×250.

injection of calcium or cyclic adenosine monophosphate (AMP) [Conrad and Davis, 1977] and after treatment with ionophores [Conrad and Davis, 1980]. These data show that the whole egg surface is capable of forming lobelike constrictions. The normal position of the polar lobe at the vegetal pole apparently depends on the correct positioning of the triggering signal. The mitotic apparatus may be instrumental in this process, but its presence is not absolutely required for the appearance of the lobe at the correct time and the correct place.

CONCLUDING REMARKS

Polar lobes are evidently of crucial importance for development. They contain factors that determine the appearance of a large array of adult structures. The potential of large and small polar lobes is the same. Apparently, in small polar lobes the determinants are highly concentrated. The process of polar lobe formation serves to ensure that the vegetal pole plasm is shunted into one blastomere at first cleavage. Which one of the two blastomeres receives the lobe material does not seem to be of great importance, but it is absolutely necessary for normal development that only one blastomere receives the lobe material. In species in which the lobe-specific determinants are concentrated in a single tightly coherent structure, such as the vegetal body of *Bithynia,* this might be achieved without polar lobe formation, since the cleavage furrow is not likely to divide this structure in two parts. Centrifugation experiments demonstrate that, wherever the vegetal body comes to lie, development is normal, most probably because the vegetal body is not fragmented and thus displaced as a whole into one blastomere.

As far as we know, in other species the polar lobe determinants are not contained within a single coherent structure. In these species the cleavage furrow can easily separate the accumulated determinants into two parts, as demonstrated by experiments in which the formation of the polar lobe is suppressed. It is essential, therefore, that the determinants remain strictly confined to the cytoplasmic compartment that will be set apart by the constriction of the polar lobe. Attachment of the determinants to the vegetal cortex seems to be an efficient way to accomplish this confinement. In *Bithynia,* however, the coherence of the vegetal body and the competence of both blastomeres to react properly to its presence anywhere in the cytoplasm render superfluous polar lobe formation and all the processes connected with it, such as the localization of the determinants at the vegetal pole and their attachment to the cortex. These processes might be functional during oogenesis, however, in order to ensure the concentration of the determinants in a single structure. The conservation of the localization and attachment mechanisms in *Bithynia,* although presumably atavistic in some respects, is very fortunate, for it opens up the possibility of investigating these mechanisms in a more direct way than would be possible in other species in which the structures containing the determinants have not yet been identified.

ACKNOWLEDGMENTS

I would like to thank Prof. N.H. Verdonk for his stimulating suggestions and criticisms in the preparation of this chapter.

REFERENCES

Atkinson JW (1971): Organogenesis in normal and lobeless embryos of the marine prosobranch gastropod, *Ilyanassa obsoleta.* J Morphol 133:339–352.

Baba K (1951): General sketch of the development in a solenogastre, Epimenia verrucosa (Nierstrasz). Misc Rep Res Inst Nat Resour (Tokyo) 19–21:38–46.

Borgens RB, Vanable JW Jr, Jaffe LF (1979): Bioelectricity and regeneration. Bioscience 29:468–474.

Brandhorst BP, Newrock KM (1981): Post-transcriptional regulation of protein synthesis in *Ilyanassa* embryos and isolated polar lobes. Dev Biol 83:250–254.

Cather JN (1971): Cellular interactions in the regulation of development in annelids and molluscs. Adv Morphog 9:67–125.

Cather JN, Verdonk NH (1974): The development of *Bithynia tentaculata* (Prosobranchia, Gastropoda) after removal of the polar lobe. J Embryol Exp Morphol 31:415–422.

Cather JN, Verdonk NH, Dohmen MR (1976): Role of the vegetal body in the regulation of development in *Bithynia tentaculata* (Prosobranchia, Gastropoda). Am Zool 16:455–468.

Cather JN, Verdonk NH (1979): Development of *Dentalium* following removal of D-quadrant blastomeres at successive cleavage stages. W Rouxs Arch Dev Biol 187:355–366.

Clement AC (1935): The formation of giant polar bodies in centrifuged eggs of *Ilyanassa.* Biol Bull 69:403–414.

Clement AC (1952): Experimental studies on germinal localization in *Ilyanassa.* I. The role of the polar lobe in determination of the cleavage pattern and its influence in later development. J Exp Zool 121:593–625.

Clement AC (1960): Development of the *Ilyanassa* embryo after removal of the mesentoblast cell. Biol Bull 119:310.

Clement AC (1962): Development of *Ilyanassa* following removal of the D macromere at successive cleavage stages. J Exp Zool 149:193–216.

Clement AC (1968): Development of the vegetal half of the *Ilyanassa* egg after removal of most of the yolk by centrifugal force, compared with the development of animal halves of similar visible composition. Dev Biol 17:165–186.

Clement AC (1976): Cell determination and organogenesis in molluscan development: A reappraisal based on deletion experiments in *Ilyanassa.* Am Zool 16:447–453.

Collier JR (1975): Polyadenylation of nascent RNA during the embryogenesis of *Ilyanassa obsoleta.* Exp Cell Res 95:263–268.

Collier JR (1983): The biochemistry of molluscan development. In Verdonk NH, van den Biggelaar JAM (eds): "Biology of the Mollusca. Development." New York: Academic Press.

Collier JR, Garone LM (1975): The localisation of ATP in the polar lobe of the *Ilyanassa* egg. Differentiation 4:195–196.

Collier JR, McCarthy ME (1981): Regulation of polypeptide synthesis during early embryogenesis of *Ilyanassa obsoleta.* Differentiation 19:31–46.

Conrad GW (1973): Control of polar lobe formation in fertilized eggs of *Ilyanassa obsoleta* Stimpson. Am Zool 13:961–980.

Conrad GW, Williams DC, Turner FR, Newrock KM, Raff RA (1973): Mirofilaments in the polar lobe constriction of fertilized eggs of *Ilyanassa obsoleta.* J Cell Biol 59:228–233.

Conrad GW, Williams DC (1974a): Polar lobe formation and cytokinesis in fertilized eggs of *Ilyanassa obsoleta.* I. Ultrastructure and effects of cytochalasin B and colchicine. Dev Biol 36:363–378.

Conrad GW, Williams DC (1974b): Polar lobe formation and cytokinesis in fertilized eggs of *Ilyanassa obsoleta.* II. Large bleb formation caused by high concentrations of exogenous calcium ions. Dev Biol 37:280–294.

Conrad GW, Davis SE (1977): Microiontophoretic injection of calcium ions or of cyclic AMP causes rapid shape changes in fertilized eggs of *Ilyanassa obsoleta.* Dev Biol 61:184–201.

Conrad GW, Davis SE (1980): Polar lobe formation and cytokinesis in fertilized eggs of *Ilyanassa obsoleta.* III. Large bleb formation caused by Sr^{2+}, ionophores X537A and A23187, and compound 48/80. Dev Biol 74:152–172.

Crampton HE (1896): Experimental studies on gasteropod development. Arch Entwicklungsmech Org 3:1–19.

van Dam WI, Dohmen MR, Verdonk NH (1982): Localization of morphogenetic determinants in a special cytoplasm present in the polar lobe of *Bithynia tentaculata* (Gastropoda). W Rouxs Arch Dev Biol 191:371–377.

Dohmen MR (1983): Gametogenesis. In Verdonk NH, van den Biggelaar JAM (eds): "Biology of the Mollusca. Development." New York: Academic Press, pp 1–48.

Dohmen MR, Verdonk NH (1974): The structure of a morphogenetic cytoplasm, present in the polar lobe of *Bithynia tentaculata* (Gastropoda, Prosobranchia). J Embryol Exp Morphol 31:423–433.

Dohmen MR, Lok D (1975): The ultrastructure of the polar lobe of *Crepidula fornicata* (Gastropoda, Prosobranchia). J Embryol Exp Morphol 34:419–428.

Dohmen MR, van der Mey JCA (1977): Local surface differentiations at the vegetal pole of the eggs of *Nassarius reticulatus, Buccinum undatum,* and *Crepidula fornicata* (Gastropoda, Prosobranchia). Dev Biol 61:104–113.

Dohmen MR, Verdonk NH (1979a): The ultrastructure and role of the polar lobe in development of molluscs. In Subtelny S, Konigsberg IR (eds): "Determinants of Spatial Organization." New York: Academic Press, pp 3–27.

Dohmen MR, Verdunk NH (1979b): Cytoplasmic localizations in mosaic eggs. In Newth DR, Balls M (eds): "Maternal Effects in Development." Cambridge: Cambridge University Press, pp 127–145.

van Dongen CAM (1976): The development of *Dentalium* with special reference to the significance of the polar lobe. VII. Organogenesis and histogenesis in lobeless embryos of *Dentalium vulgare* (da Costa) as compared to normal development. Proc K Ned Akad Wet Ser C 79:421–432.

van Dongen CAM, Geilenkirchen WLM (1974): The development of *Dentalium* with special reference to the significance of the polar lobe. I, II, and III. Division chronology and development of the cell pattern in *Dentalium dentale* (Scaphopoda). Proc K Ned Akad Wet Ser C 77:57–100.

van Dongen CAM, Geilenkirchen WLM (1975): The development of *Dentalium* with special reference to the significance of the polar lobe. IV. Division chronology and development of the cell pattern in *Dentalium dentale* after removal of the polar lobe at first cleavage. Proc K Ned Akad Wet Ser C 78:358–375.

van Dongen CAM, Mikkers FEP, De Bruyn Ch, Verheggen TPEM (1981): Molecular composition of the polar lobe of first cleavage stage embryos in comparison with the lobeless embryo in *Nassarius reticulatus* (Mollusca) as analyzed by isotachophoresis. In Everaerts FM (ed): "Analytical Isotachophoresis." Amsterdam: Elsevier, pp 207–216.

Fioroni P (1979): Zur Struktur der Pollappen und der Dottermakromeren—eine vergleichende Übersicht. Zool Jahrb Abt Anat Ontog Tiere 102:395–430.

Geilenkirchen WLM, Timmermans LPM, van Dongen CAM, Arnolds WJA (1971): Symbiosis of bacteria with eggs of *Dentalium* at the vegetal pole. Exp Cell Res 67:477–479.

Giese K (1978): Zur Embryonalentwicklung von *Buccinum undatum* L. (Gastropoda, Prosobranchia, Stenoglossa [Neogastropoda], Buccinacea). Zool Jahrb Abt Anat Ontog Tiere 100:65–117.

Guerrier P (1971): A possible mechanism of control of morphogenesis in the embryo of *Sabellaria alveolata* (Annelide polychaete). Exp Cell Res 67:215–218.

Guerrier P, van den Biggelaar JAM, van Dongen CAM, Verdonk NH (1978): Significance of the polar lobe for the determination of dorsoventral polarity in *Dentalium vulgare* (da Costa). Dev Biol 63:233–242.

Hatt P (1932): Essais expérimentaux sur les localisations germinales dans l'oeuf d'un annélide (*Sabellaria alveolata* L.). Arch Anat Microsc Morphol Exp 28:81–106.

Koser RB, Collier JR (1976): An electrophoretic analysis of RNA synthesis in normal and lobeless *Ilyanassa* embryos. Differentiation 6:47–52.

Kühn A (1965): "Entwicklungsphysiologie, 2nd Ed." Berlin: Springer-Verlag.

Moor B (1973): Zur frühen Furchung des Eies von *Littorina littorea* L. (Gastropoda Prosobranchia). Ein neues Beispiel von Pollappenbildung. Zool Jahrb Abt Anat Ontog Tiere 91:546–573.

Morgan TH (1935a) Centrifuging the eggs of *Ilyanassa* in reverse. Biol Bull 68:268–279.

Morgan TH (1935b): The separation of the egg of *Ilyanassa* into two parts by centrifuging. Biol Bull 68:280–295.

Morgan TH (1935c): The rhythmic changes in form of the isolated anti-polar lobe of *Ilyanassa.* Biol Bull 68:296–299.

Moritz CE (1939): Organogenesis in the gasteropod *Crepidula adunca* Sowerby. Univ Calif Berkeley Publ Zool 43:217–248.

Novikoff AB (1938): Embryonic determination in the annelid, *Sabellaria vulgaris.* II. Transplantation of polar lobes and blastomeres as a test of their inducing capacities. Biol Bull 74:211–234.

Novikoff AB (1940): Morphogenetic substances or organizers in annelid development. J Exp Zool 85:127–156.

Pasteels JJ (1966): Les mouvements corticaux de l'oeuf de *Barnea candida* (Mollusque Bivalve) étudiés au microscope électronique. J Embryol Exp Morphol 16:311–319.

Peaucellier G, Guerrier P, Bergerard J (1974): Effects of cytochalasin B on meiosis and development of fertilized and activated eggs of *Sabellaria alveolata* L. (Polychaete Annelid). J Embryol Exp Morphol 31:61–74.

Raff RA (1972): Polar lobe formation by embryos of *Ilyanassa obsoleta.* Effects of inhibitors of microtubule and microfilament function. Exp Cell Res 71:455–459.

Rappaport R (1971): Cytokinesis in animal cells. Int Rev Cytol 31:169–213.

Rappaport R, Rappaport BN (1974): Establishment of cleavage furrows by the mitotic spindle. J Exp Zool 189:189–196.

Rattenbury JC, Berg WE (1954): Embryonic segregation during early development of *Mytilus edulis*. J Morphol 95:393–414.

Tyler A (1930): Experimental production of double embryos in annelids and mollusks. J Exp Zool 57:347–408.

Verdonk NH (1968): The effect of removing the polar lobe in centrifuged eggs of *Dentalium*. J Embryol Exp Morphol 19:33–42.

Verdonk NH, Geilenkirchen WLM, Timmermans LPM (1971): The localization of morphogenetic factors in uncleaved eggs of *Dentalium*. J Embryol Exp Morphol 25:57–63.

Verdonk NH, Cather JN (1973): The development of isolated blastomeres in *Bithynia tentaculata* (Prosobranchia, Gastropoda). J Exp Zool 186:47–62.

Verdonk NH, van den Biggelaar JAM (1983): Early development and the formation of the germ layers. In Verdonk NH, van den Biggelaar JAM (eds): "Biology of the Mollusca. Development." New York: Academic Press, pp 91–122.

Wilson EB (1904): Experimental studies on germinal localization. I. The germ regions in the egg of *Dentalium*. J Exp Zool 1:1–72.

Time, Space, and Pattern in Embryonic Development, pages 221–239

Localization and Determination in Embryos of *Caenorhabditis elegans*

William B. Wood, Susan Strome, and John S. Laufer

Department of Molecular, Cellular, and Developmental Biology, University of Colorado, Boulder, Colorado 80309

More than 100 years ago, early European embryologists had posed the two central questions of animal development: First, how is the sameness of cells and organisms maintained during development and reproduction, and what factors transmit this hereditary information? Second, how do the cells of an embryo become different; what factors dictate that a particular cell at a particular time and position becomes committed to a particular developmental pathway? In the intervening century, we have largely answered the first question, acquiring extensive information about the genetic machinery and how it works. By contrast, we have gained little new understanding of the epigenetic process responsible for temporal and positional control of cell determination in embryos. How this process operates remains a central problem of contemporary developmental biology.

EMBRYONIC DEVELOPMENT IN *C. elegans*

In several respects *C. elegans* provides an excellent experimental system with which to approach questions of early embryogenesis. The embryos are transparent and develop rapidly, giving rise to first-stage larvae with only 558 nuclei. Although inconvenient for physical manipulation, the embryos are well suited for analysis by light and electron microscopy. Moreover, the organism is convenient for genetic analysis [Brenner, 1974] so that mutations perturbing embryogenesis can be easily isolated and studied for clues to normal developmental mechanisms.

From microscopic observations of embryogenesis in living specimens, the lineages and developmental fates of all the cells in the embryo are now known [Deppe et al., 1978; Krieg et al., 1978; Sulston et al., 1983]. These relationships are shown for the first few cleavages in Figure 1. The zygote (P_0) undergoes a series of asymmetric divisions, each giving rise to a somatic precursor and a germ line (P) cell. Most of these divisions take place in the direction of the anterior-posterior axis, but several have dorsal-ventral or left-right components. The resulting positions of the cells at each stage, as well as the fates of their progeny, are invariant from embryo to embryo [Deppe et al., 1978; Krieg et al., 1978; Laufer et al., 1980].

How does the same cell in every embryo become committed to the same developmental fate? We can propose two general kinds of mechanisms, either one or both of which could explain these observations. (1) The fate of a cell may be determined by cues from outside that depend upon its position, its interaction with neighboring cells in the embryo, or both. (2) The fate of a cell may be determined in response to lineally transmitted, internally segregating

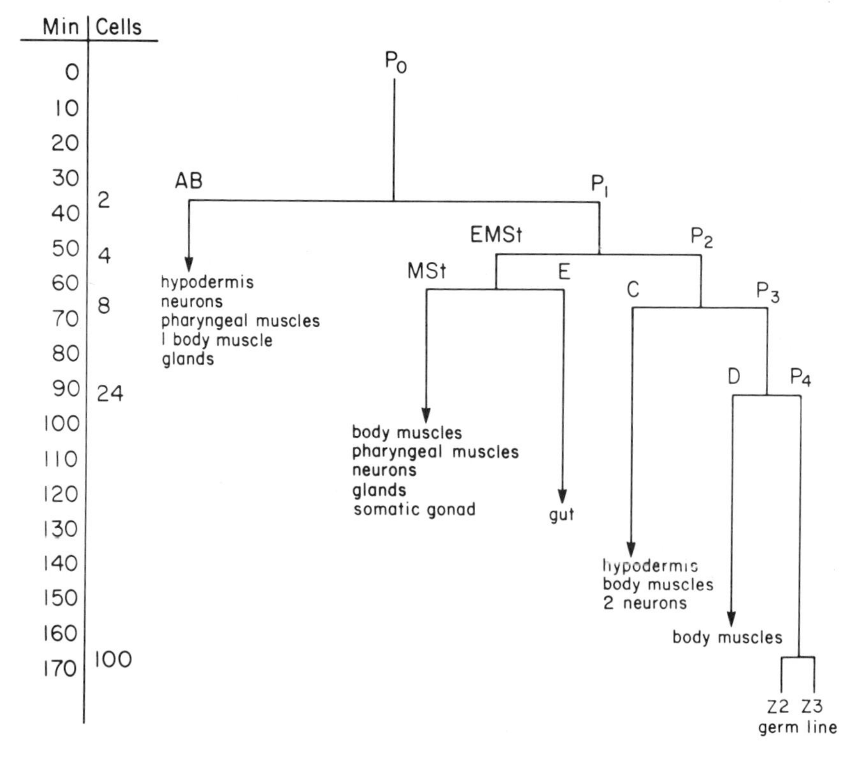

Fig. 1. Lineage relationships, timing, and progeny cell fates in the early cleavage divisions of *C. elegans* embryos at 25°C. Progeny cell fates are based on data from Deppe et al. [1978], Krieg et al. [1978], and Sulston et al. [1983].

determinants. Because the lineages, positions, and fates of cells in the embryo are invariant, causal relationships between these parameters generally cannot be deduced from observations of normal development, but must be investigated by perturbing the system.

MUTATIONAL PERTURBATION OF EARLY CLEAVAGES

Temperature-sensitive mutants of *C. elegans* that show lethal defects in embryogenesis at 25°C but not 16°C were first isolated and characterized by D. Hirsh and collaborators [Hirsh and Vanderslice, 1976; Vanderslice and Hirsh, 1976]. Genetic analysis showed that 21 out of 24 of these mutants define genes that show maternal effects; 11 are strict maternal mutants for which viability of the embryo depends only on the maternal genotype and is independent of the embryonic genotype [Wood et al., 1980]. Similar results have been obtained with other mutants analyzed independently by G. von Ehrenstein and co-workers [Schierenberg et al., 1980; Cassada et al., 1981]. Therefore, in *C. elegans*, as in most organisms, normal embryogenesis depends on many gene products made during maternal gametogenesis before embryonic development begins.

Several of the strict maternal mutants show striking morphological perturbations in the first two cleavages (Fig. 2) but then continue division to 100 cells or more before development arrests and the embryos die [Wood et al., 1980; Wolf and Hirsh, personal communication]. Analysis of the defects in these mutants eventually should provide information about the normal developmental process. However, because the mutant embryos do not develop far enough to allow identification of differentiated tissues by morphology, such analysis requires more sophisticated means of assaying at early stages for developmental potential and for the presence of specific macromolecules as markers for determinative events. In developing and exploiting such assays we also have perturbed embryos chemically and physically, with results that provide information about the nature of at least one determinant for differentiation in a somatic cell lineage.

SEGREGATION OF A SOMATIC DETERMINANT

In experiments with the sea squirt *Ciona intestinalis*, Whittaker [1973] showed that cytochalasin-treated or colchicine-treated early embryos, although arrested in cell division, nevertheless would produce enzymes characteristic of specific differentiated tissues approximately on schedule. Furthermore, production of these marker enzymes was restricted to the correct tissue progenitor cells. Such experiments provide a means of assaying the differentiation potential of cells at early cleavage stages.

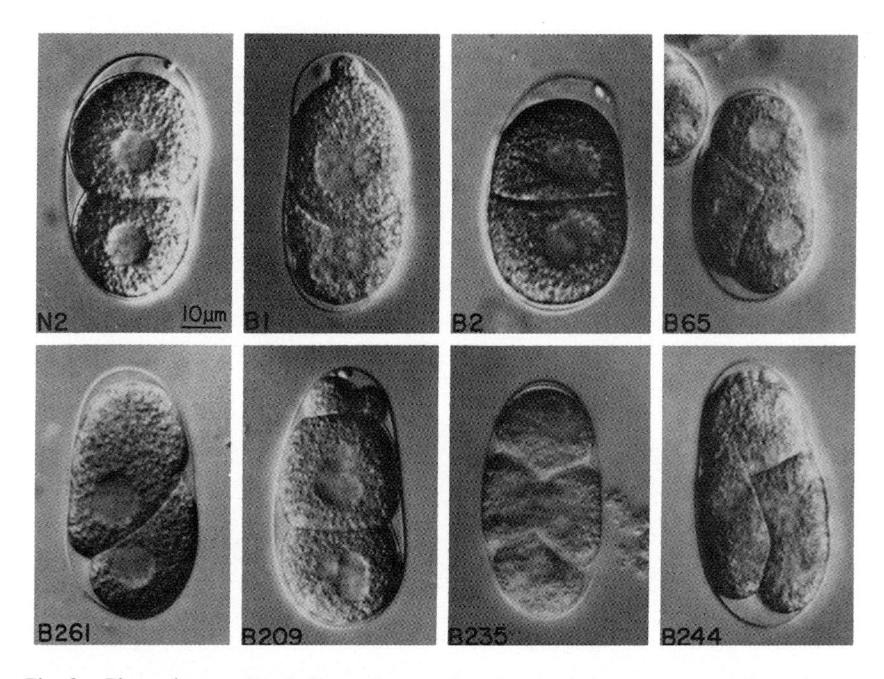

Fig. 2. Photomicrographs of abnormal two-cell embryos from *ts* embryonic-lethal mutants. Embryos oriented with the anterior pole at the top were photographed shortly after first cleavage using a Zeiss Photomicroscope with Nomarski differential interference-contrast optics. The upper left panel shows the normal first cleavage of a wild-type (N2) embryo to form the larger AB and the smaller P1 cell. From Wood et al. [1980]. All subsequent photomicrographs in this article were taken with Nomarski optics unless indicated otherwise.

In similar experiments with *C. elegans* we determined the potential of embryonic cells to express a gut-specific differentiation marker after blocking the cleavage of early embryos with cytochalasin B and colchicine. As a marker we used the so-called rhabditin granules [Chitwood and Chitwood, 1950], fluorescent refractile bodies that contain tryptophan catabolites and that normally appear in gut cells at the 100-cell stage of embryogenesis. All the gut cells in *C. elegans* are derived from the E cell, which arises at the third embryonic cleavage as a descendant of the P1 and EMSt cells (Fig. 1). When embryos were blocked at the two-cell stage, gut granules developed only in the P1 cell. When embryos were blocked at the four-cell and eight-cell stages, granules developed only in the EMSt cell and the E cell, respectively. The same results were observed in cleavage-blocked partial embryos, in which one or more cells were ablated at the two-cell, four-cell, or eight-cell stages (Fig. 3). Therefore, the potential for this gut-specific differentiation depends neither on cell division after the two-cell stage nor on the

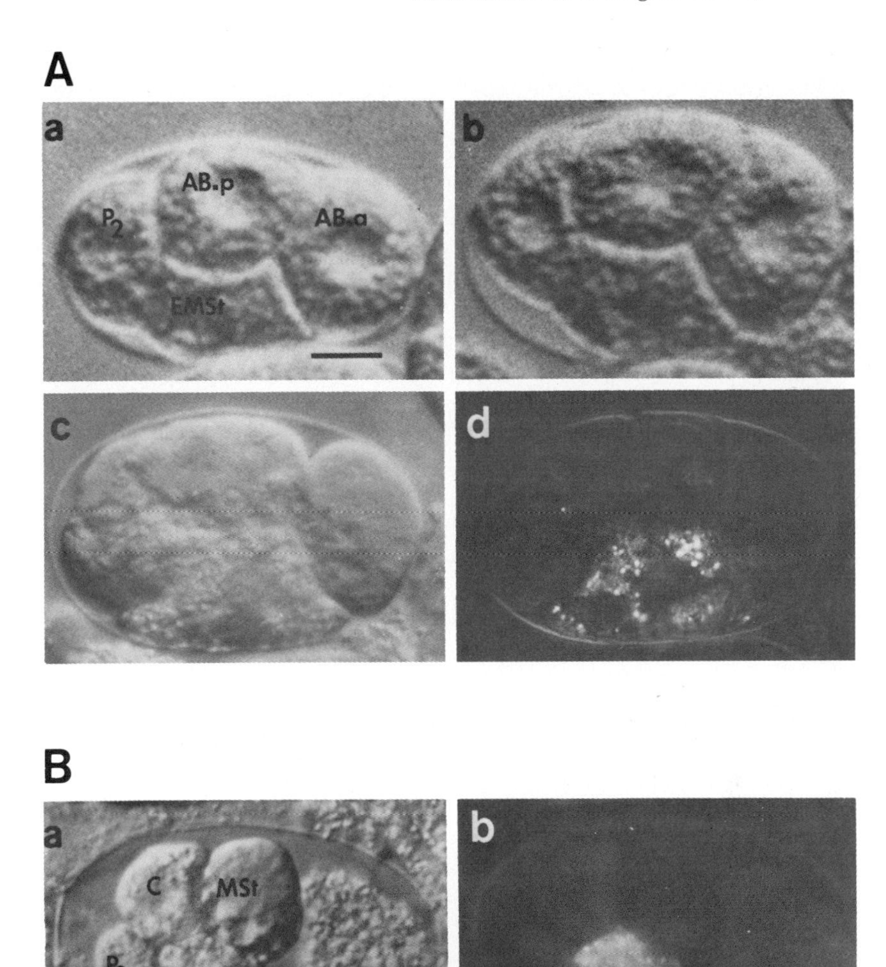

Fig. 3. Formation of rhabditin granules in cleavage-blocked embryos made permeable by rupturing the egg shell with locally applied pressure. Bars = 10 μm. A. A four-cell embryo in medium containing 50 μg/ml colchicine and 25 μg/ml cytochalasin B (a) before rupture, (b) immediately after rupture, (c) after incubation at room temperature for 19 hours, and (d) as in (c) with epifluorescence optics. B. A partial embryo that was ruptured at the two-cell stage with lysis of the AB cell; the remaining P1 cell was allowed to undergo two divisions to yield the cells shown, identified from the cleavage pattern, and then colchicine and cytochalasin were added at the same concentrations as in A. The cleavage-blocked embryo is shown after 16-hour incubation with (a) Nomarksi and (b) epifluorescence optics. The granules in such experiments first appear after about six hours of incubation. From Laufer et al. [1980].

positional relationships or cell contacts characteristic of untreated embryos. Rather, it depends on cell-autonomous, internal determinants, which are partitioned from the P1 cell into the EMSt cell and thence into the E cell and its descendants [Laufer et al., 1980].

Such determinants could be either nuclear or cytoplasmic. However, Laufer and von Ehrenstein [1981] have shown that removal of substantial portions of cytoplasm from early embryos does not prevent normal development. Puncture of the eggshell with a laser microbeam often leads to extrusion of a membrane-bounded bleb of cytoplasm, followed by resealing of the puncture and continued development of the embryo (Fig. 4). They obtained larvae that grew into fertile adult worms from embryos that had lost up to 20% of the cytoplasm from either end of the uncleaved egg or up to 60% of the cytoplasm from any one of several different cells during early cleavage stages. These results suggest that if there are prelocalized cytoplasmic determinants essential for development, they must be anchored in some manner to cell components that are not free to flow with the bulk cytoplasm through a small hole.

This view is strengthened by recent blastomere-fusion experiments carried out in our laboratory to determine whether early embryonic nuclei can

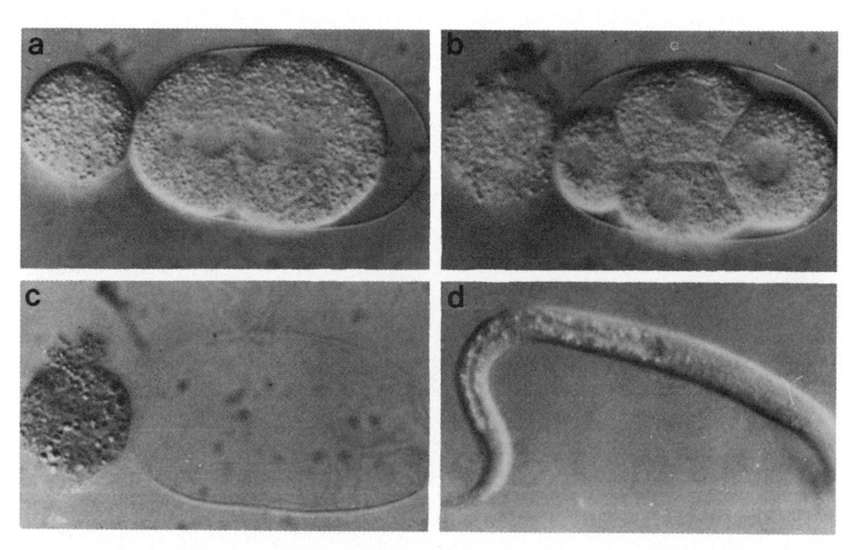

Fig. 4. Development of an embryo to hatching after extrusion of cytoplasm at the one-cell stage. (a) Laser irradiation of the egg shell at the posterior pole and gentle pressure has led to extrusion of about 15% of the cytoplasm as a membrane-bounded bleb, followed by apparent resealing of the cell membrane. Photograph shows embryo completing the first cleavage; (b) the same embryo at the four-cell stage; (c) same embryo just prior to hatching; (d) the resulting hatched larva, which matured into a fertile hermaphrodite.

become differently programmed as a result of exposure to the cytoplasm of a neighboring cell. Neighboring blastomeres in early embryos can be fused by firing a 365-nm beam from a rhodamine dye laser directed through a microscope at the membranes between the two cells (Schierenberg, personal communication). The laser treatment opens a small hole between the cells, clearly visible in the microscope. Several stages in such an experiment are shown in Figure 5.

In rare cases the hole reseals quickly. When resealing occurs, the embryo often develops normally to a fertile adult, indicating that side effects of the laser irradiation are minimal. Usually, the hole enlarges after the initial puncture, and the membrane retracts completely so that the two blastomere nuclei reside in a single cytoplasm (Fig. 5b). The nuclei then approach each other and touch (Fig. 5c). The fused cell subsequently undergoes a tetrapolar division, at a time between the normal division times of its component blastomeres, often giving rise to four cells that are similar in position and subsequent behavior to the corresponding four blastomeres of a normal eight-cell embryo (Fig. 5e–g). Although after blastomere fusion the embryos show grossly abnormal late morphogenesis and never hatch, their early development appears normal in many respects, and they produce rhabditin granules at the normal time in development (Fig. 5h). Using this approach, we have asked whether exposure of a non-gut-precursor nucleus to EMSt cytoplasm can lead to subsequent production of gut granules in cells containing the nongut nucleus or one of its progeny.

The most successful experiments of this kind have been carried out by fusing EMSt and AB.a in four-cell embryos as shown in Figure 5. During the subsequent tetrapolar division, both nuclei form spindles in approximately the same orientation as in a normal embryo. Cleavage yields four cells which, based on their positions, correspond to the E, MSt, AB.aa, and AB.ap cells of a normal eight-cell embryo. Usually, the MSt and AB.ap cells contain two nuclei each, probably because a secondary spindle arises between adjacent poles of the two primary spindles during the tetrapolar cleavage. Occasionally these cells re-fuse with one another and undergo another tetrapolar cleavage. The putative E cell, however, behaves similarly to its counterpart in a normal embryo with regard to migration of its nucleus toward the P3 cell, the temporal sequence of its subsequent cleavages relative to those of other cells, and the movements of its daughter cells during gastrulation.

Two kinds of experiments were undertaken to investigate the segregation of gut granule potential in blastomere-fused embryos. In the first, after tetrapolar cleavage of the fused cell, either the presumed E blastomere or the presumed MSt blastomere was irradiated with a 297-nm UV microbeam. This treatment kills the irradiated cell, which stops dividing and becomes vacuolated, without visibly affecting the rest of the embryo, in which the

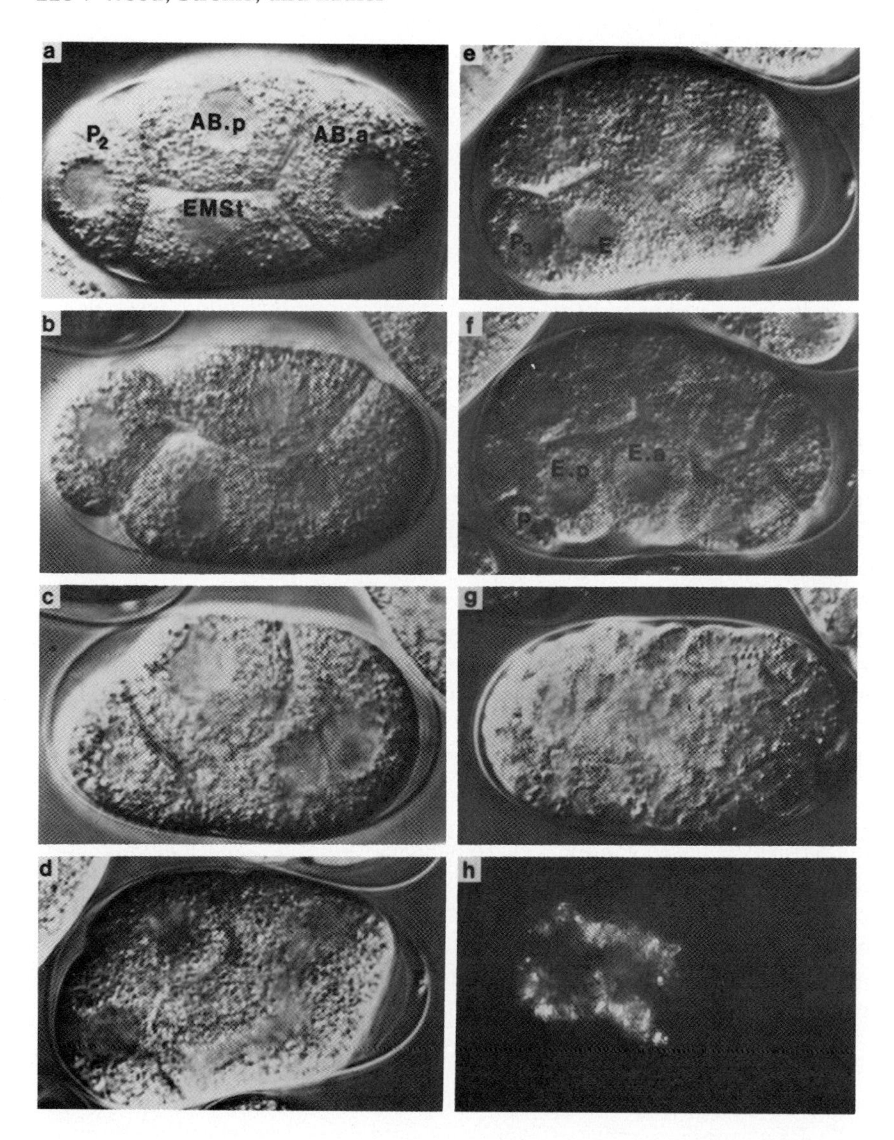

Fig. 5. Development of an embryo after fusion of EMSt and AB.a at the four-cell stage. (a) Before fusion; (b) immediately after fusion; (c) the EMSt and AB.a nuclei approach each other; (d) the nuclei set up discrete spindles; (e) the fused cell divides into four daughter cells, and the "E" nucleus migrates toward P_3; (f) gastrulation takes place; (g) about six hours later a normal-looking gut has formed; (h) same as (g) with polarization optics to show the rhabditin granules. This series includes photographs of three different embryos.

cells continue to divide apparently normally. After ten hours the embryos were scored for production of rhabditin granules. In 15 of 17 experiments in which the presumed E cell was irradiated, no granules appeared (Table I). In two such experiments, however, granules did appear, not in the irradiated E cell, but in more anterior cells presumed to be progeny of the MSt cell.

In the second kind of experiment, embryos were made permeable by laser irradiation of the eggshell after the first tetrapolar cleavage, and subsequent cell division was inhibited with cytochalasin B as described above. The cleavage-blocked embryos were incubated and then scored for the presence and location of rhabditin granules. In 39 of 40 embryos that formed rhabditin granules, the granules were clearly confined to the E cell. The one exceptional embryo had granules in both the E cell and a more anterior cell.

Similar results were obtained following fusions of EMSt with AB.p cells. Again, tetrapolar cleavage yielded four daughter cells which could be identified by their positions as presumed E, MSt, and two AB blastomeres. Of nine embryos cleavage-blocked after the tetrapolar division, all developed granules only in the presumed E cell. Details of these experiments will be published elsewhere.

The foregoing results show that the potential for production of rhabditin granules almost always is restricted to the presumed E blastomere. Only in rare instances does this potential also segregate to other progeny of the fused cell. Such results are not consistent with a freely diffusible cytoplasmic determinant for gut cell determination having the ability to reprogram nuclei that are exposed to it. Rather, these observations suggest that the determinants conferring potential for rhabditin granule synthesis are either nuclear or bound to a cytoplasmic structure that is usually partitioned into the E blastomere.

TABLE I. Formation of Rhabditin Granules in Embryos Following Blastomere Fusion at the Four-Cell Stage, Tetrapolar Cleavage, and Killing of Single Cells by Irradiation With a UV Microbeam

Cells fused	Nucleus irradiated	Embryos with granules/ total embryos
None	E	0/8
None	MSt	8/8
EMSt-AB.a	None	13/18
EMSt-AB.a	E	2/17
EMSt-AB.a	MSt	8/14

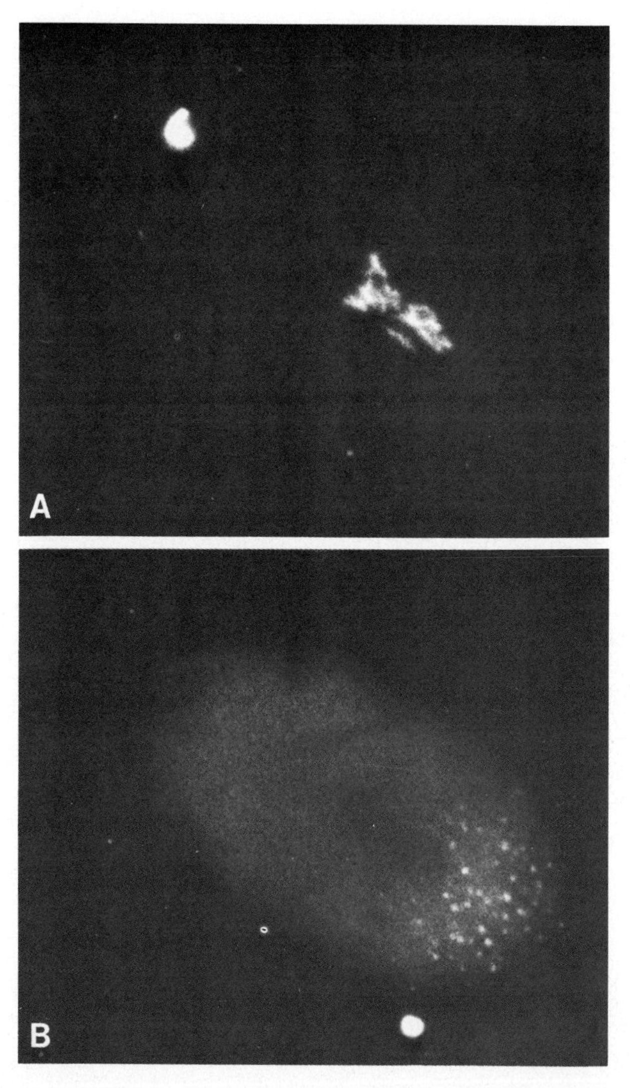

Fig. 6. Zygote showing localization of P granules at the posterior pole just prior to pronuclear fusion. Embryos in this and subsequent figures were fixed, treated with fluorescein-conjugated anti-P-granule antibody, or a nonfluorescent primary antibody and a fluorescein-conjugated secondary antibody, washed, stained with diamidinophenylindole (DAPI) hydrochloride, rinsed, mounted for microscopy, and photographed as described by Strome and Wood [1982]. Embryos in this and all subsequent photographs are shown in opposite orientation to those in Figures 3–5, with anterior left and posterior right. A shows, from left to right, the polar body at the anterior pole, the chromosomes of the maternal pronucleus, and the chromosomes of the paternal pronucleus as seen by DAPI fluorescence. B shows P granules as seen in the same embryo by immunofluorescence.

PRELOCALIZATION AND SEGREGATION OF CYTOPLASMIC COMPONENTS UNIQUE TO THE GERM LINE

Immunofluorescence microscopy provides a powerful method for determining the location of specific macromolecules in cells or tissues. Using the hybridoma technique of Kohler and Milstein [1975] we have generated a collection of monoclonal antibodies from mice immunized with nematode embryo and adult tissue homogenates. These mouse antibodies have been employed, in combination with fluorochrome-conjugated rabbit antimouse antibody, to determine the distribution of several specific antigens in fixed *C. elegans* embryos and larvae (S. Strome, M. Hobbs, and W.B. Wood, unpublished). The most interesting results from these studies so far, however, began with the chance finding that a preparation of fluorochrome-conjugated rabbit antimouse antibody (F-RAM) reacted directly, in the absence of mouse antibodies, with cytoplasmic components specific to germ-line precursor cells. This antibody preparation and others of similar specificity that we have subsequently obtained, including monoclonal antibodies, have allowed us to follow the segregation and distribution during development of these germ-line components, which we have called P granules [Strome and Wood, 1982].

P granules are detectable in the uncleaved zygote initially as dispersed, faintly staining particles, which appear to coalesce and prelocalize at the posterior pole of the embryo prior to pronuclear fusion (Fig. 6). After the first cleavage they are detected only in the P1 cell (Fig. 7). In the subsequent divisions they are progressively segregated to the P2, P3, and P4 cells (Fig. 8a,b). At the 100-cell stage, P4 divides into germ-line precursor cells Z2 and Z3 (Fig. 8c), which remain the only stained cells until the embryo hatches as a first-stage larva.

In each of the first three cleavages, the P-granules become prelocalized, during prophase, to the region of the cytoplasm that will be included in the

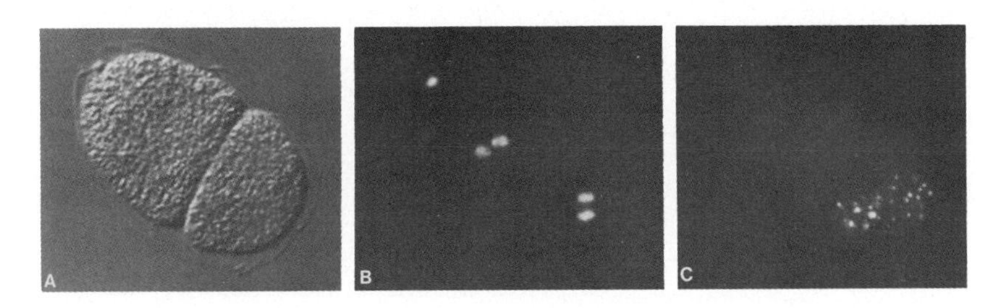

Fig. 7. A two-cell embryo showing localization of the P granules prior to the second cleavage. P1 nucleus (right) is in anaphase. Embryos were prepared and photographed as in Figure 6. A, Nomarski image; B, DAPI image; C, immunofluorescence image.

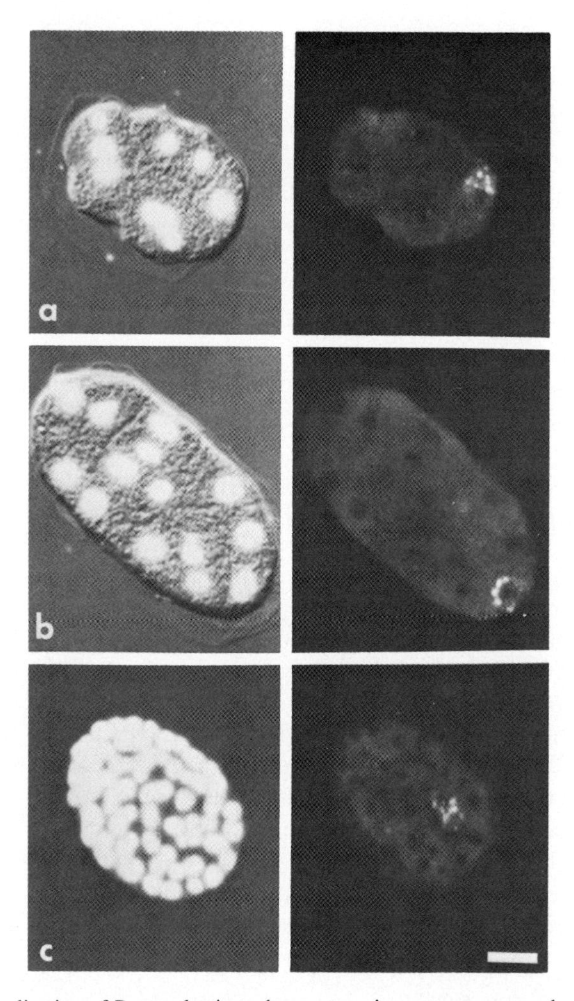

Fig. 8. Localization of P granules in embryos at various stages prepared and photographed as in Figure 6. Nomarski-DAPI images are shown on the left, and immunofluorescence images on the right. (a) Six-cell embryo in the process of EMSt division. The EMSt nucleus is in late anaphase or telophase; the P2 nucleus is in early prophase. Note prelocalization of P granules in the ventral region of the P2 cell. (b) Embryo of about 15 cells, showing perinuclear granules in the P3 cell. (c) Embryo of > 100 cells shortly after P4 division into Z2 and Z3, both of which contain perinuclear granules. Bar = 10 μm. From Strome and Wood [1982].

next P-cell daughter (Figs. 7, 8a). After the second cleavage, they change in size and distribution. In one-cell to four-cell embryos they are numerous and small, located apparently randomly in the cytoplasm during interphase and near the cortex during cell division. However, by about the 16-cell stage, the small granules seem to have coalesced, forming three to five large granules, which are located around the nucleus (Fig. 8b).

This perinuclear organization of the granules persists throughout subsequent embryogenesis and larval development. Granules remain clearly visible in germ cells of the developing larval gonad (Fig. 9) and later in the distal arm of both adult hermaphrodite and male gonads. As oocytes mature in the hermaphrodite, the granules appear to disperse from the nucleus, leading to diffuse cytoplasmic staining (Fig. 10). Sperm are not stained, either in hermaphrodites (Fig. 11) or males (not shown).

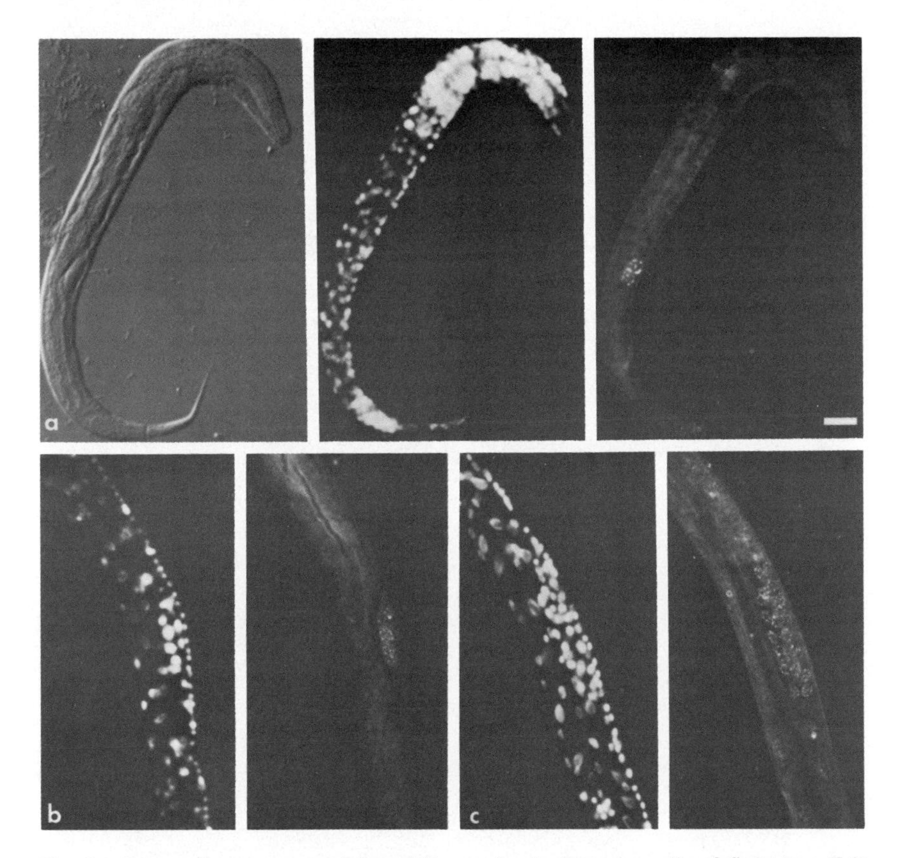

Fig. 9. Immunofluorescence staining of P granules in the germ cells of first stage (L1) larvae. (a) Nomarski (left), DAPI (middle), and immunofluorescence (right) images of a newly hatched L1 showing immunofluorescence staining of only the Z2 and Z3 cells in the four-cell gonad primordium. (b) DAPI (left) and immunofluorescence (right) images of a later-stage L1 after germ-cell proliferation has begun. (c) DAPI (left) and immunofluorescence (right) images of a late-stage L1. No staining is detected in somatic cells of the gonad. Bar = 10 μm. From Strome and Wood [1982].

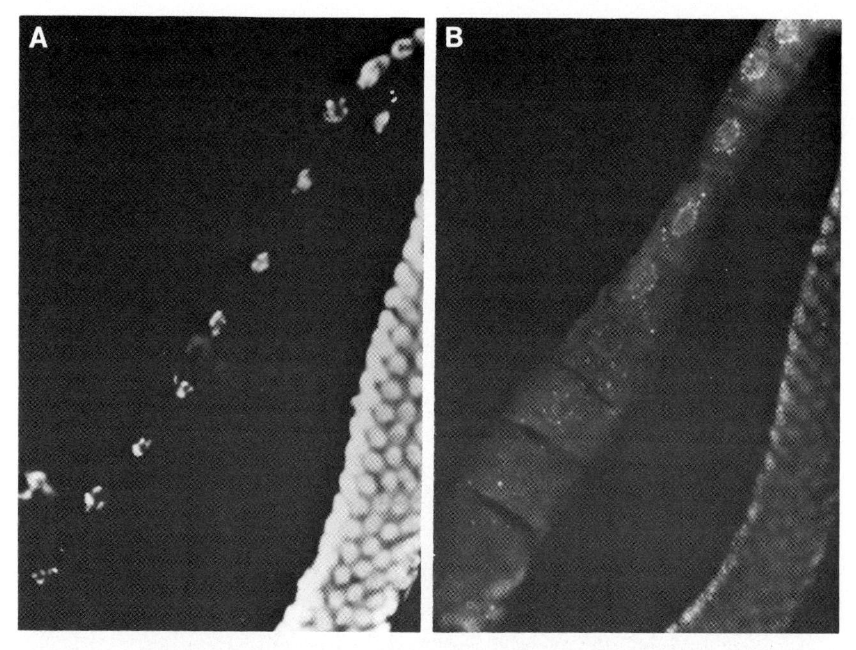

Fig. 10. A dissected adult hermaphrodite gonad, showing perinuclear immunofluorescence staining of P granules in the distal arm to right, and dispersal of P granules into cytoplasm during oocyte maturation in the proximal arm to left. A, DAPI image; B, immunofluorescence image.

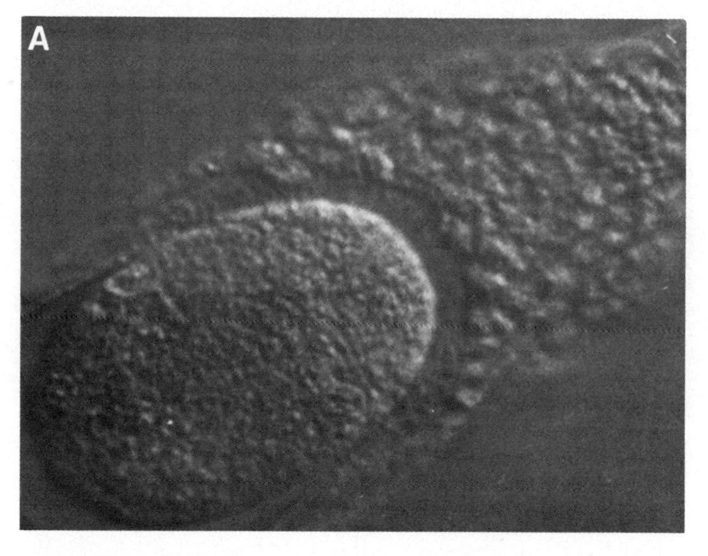

Fig. 11. Newly fertilized embryo in a dissected hermaphrodite gonad just after passage through the spermatheca. A. Nomarski image. B. DAPI image showing condensed nuclei of sperm in spermatheca. C. immunofluorescence image showing no staining of sperm, and faint staining of P granules distributed throughout the oocyte cytoplasm.

The origin of anti-P-granule antibodies in the original F-RAM preparation and other rabbit and goat sera we have subsequently obtained remains unclear. However, these sera, in contrast to anti-P-granule monoclonal antibodies, also stain nongonadal tissues in larvae and adults, as if new antigens recognized by these sera appear after hatching. This finding lends support to the possibility that nematode infections in the animals from which sera were

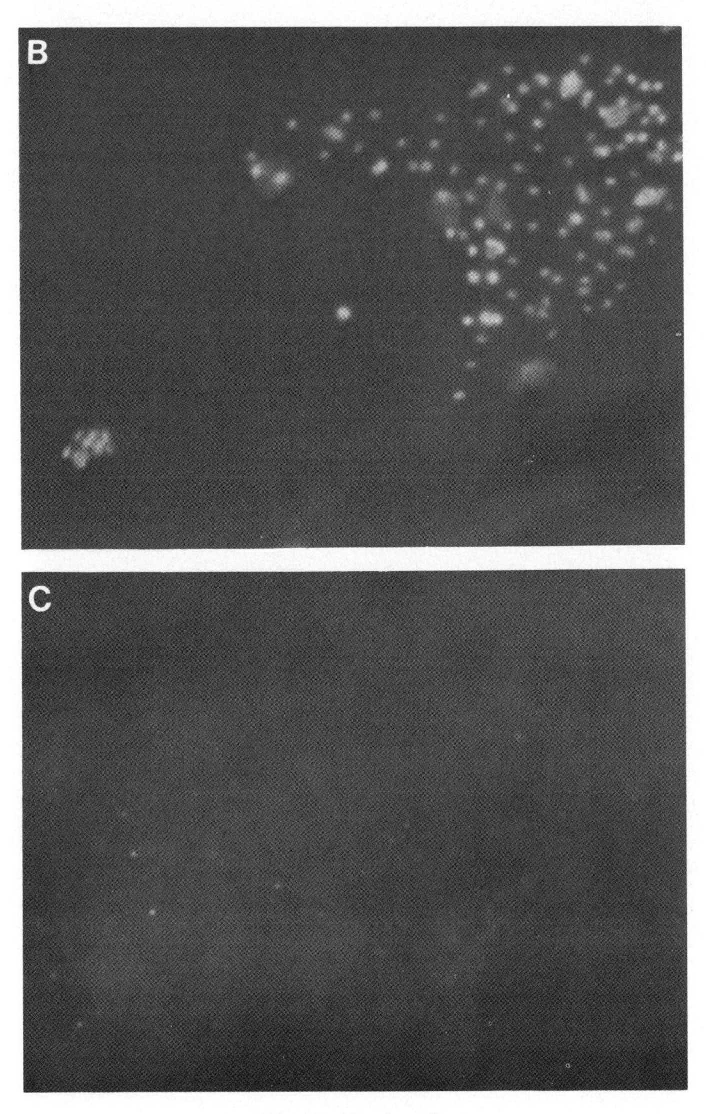

Fig. 11. (Continued)

obtained elicited production of antibodies that cross-react with *C. elegans* antigens.

Three questions currently under investigation using anti-P-granule antibodies are: How are the granules segregated; what is their composition; and what is their functional significance? Preliminary results are consistent with the view that granule movement in early cleavages is not determined by the

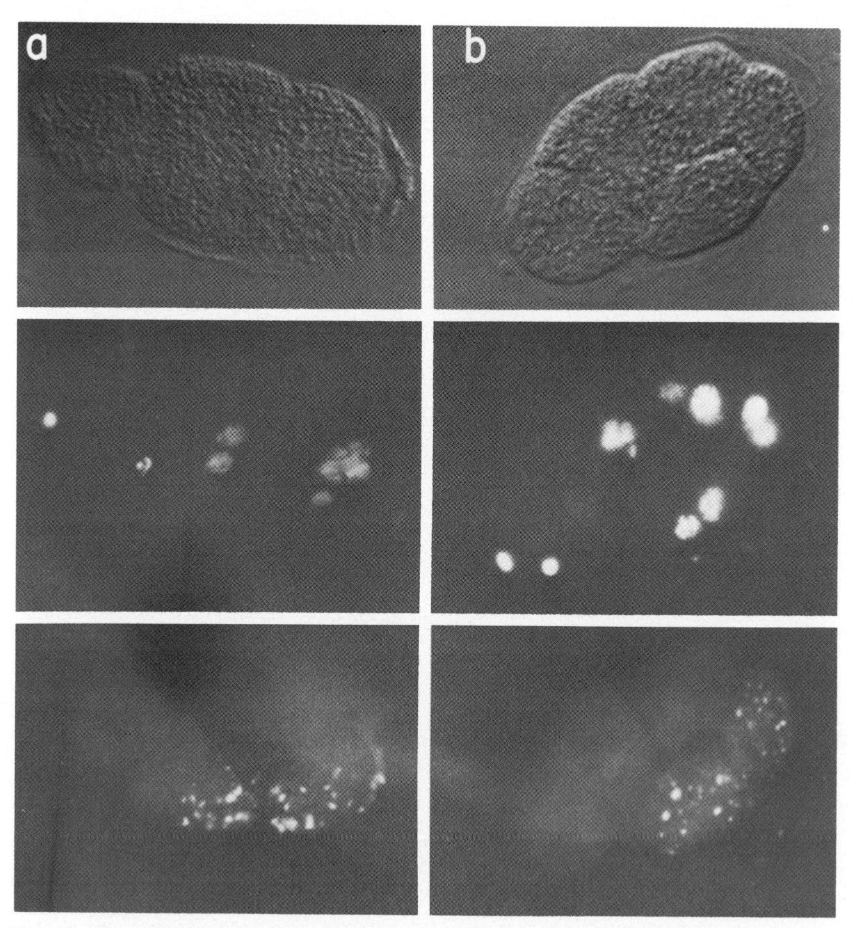

Fig. 12. Two-cell (a) and (b) four-cell embryos of *zyg-9* mutant. Nomarski (upper panel), DAPI (center panel), and immunofluorescence (lower panel) images of each embryo are shown. The embryo in (a) underwent an abnormal longitudinal first cleavage to yield an anterior cytoplast and two elongated cells of about equal size, which rotated before fixation so that one cell lies posterior to the other. Both cells contain P granules, which prelocalized normally at the posterior pole of the zygote despite the abnormal transverse axis of the first-cleavage spindle. The embryo in (b) underwent a similar first cleavage followed by a second asymmetric cleavage in which the P granules segregated to the two posterior daughter cells.

mitotic spindle. In embryos of *zyg-9* mutants, such as *tsB244* (Fig. 2), in which the axis of the first cleavage spindle is usually perpendicular to its normal orientation, the P granules nevertheless appear to migrate normally to the posterior pole of the embryo prior to first cleavage, and then become incorporated into both blastomeres at the two-cell stage. Figure 12a shows such an embryo, in which first cleavage yielded a cytoplast (anterior) and two cells, which rotated so that one now lies posterior to the other; both contain granules. Figure 12b shows a similar embryo at the four-cell stage, with granules in two of the four cells. This result is reminiscent of Boveri's finding that centrifuged *Ascaris* embryos, in which the first division spindle was rotated 90° to give a longitudinal first cleavage, showed no chromosome diminution in progeny of both the resulting blastomeres, as if both cells at first cleavage had acquired germ-line character [Boveri, 1910].

In preliminary drug inhibition experiments, treatment of embryos with cytochalasin B at levels that block cleavage also prevents the asymmetric localization of granules in subsequent mitoses. Further experiments with both mutant embryos and drug inhibition are in progress.

The composition and functional significances of the granules remain unclear. Germ-line-specific cytoplasmic elements, referred to as nuage or germ plasm, have been observed by electron microscopy as electron-dense bodies in a variety of both vertebrate and invertebrate organisms [Eddy, 1975], including *C. elegans* [Krieg et al., 1978]. Like P granules, these elements are restricted to P cells during early cleavage stages in *C. elegans*. Moreover, the number, size, and distribution of these elements change during early cleavages in the same manner and with the same timing as observed for P granules, from numerous small cytoplasmic bodies to fewer larger perinuclear bodies [Wolf et al., 1983]. This correspondence suggests that anti-P-granule antibodies could be reacting with components of germ plasm.

The widespread occurrence of germ plasm suggests that this material may be important in either the determination of germ-line cells during early embryogenesis or the functions of germ-line cells during gametogenesis or both. Consistent with a determinative role are the findings by Boveri, mentioned above, that posterior zygote cytoplasm may act to prevent subsequent chromosome diminution in *Ascaris* embryos [Boveri, 1910], the results of Wakahara [1978] showing that injection of extra *Xenopus* egg vegetal pole cytoplasm into recipient *Xenopus* eggs causes induction of supernumerary primordial germ cells, and the demonstration by Illmensee and Mahowald [1974] that in *Drosophila* embryos, transplanted oocyte cytoplasm containing the nuagelike elements known as polar granules can cause normally somatic embryonic cells to become functional germ-line precursors. The anti-P-granule antibodies provide possible approaches to determining the correspondence between nuage and P granules by immunoelectron microscopy, to isolating P granules for biochemical analysis, and to exploring their role in development of germ-line cells in *C. elegans*.

ACKNOWLEDGMENTS

This research was supported by grants to W.B.W. and S.S. from the National Institutes of Health (HD11762, HD14958) and the American Cancer Society (CD-96). S.S. was also supported by postdoctoral fellowships from the Anna Fuller Fund and National Institutes of Health, and J.S.L. by a predoctoral fellowship from the National Science Foundation.

REFERENCES

Boveri T (1910): Über die Teilung centrifugierter Eier von *Ascaris megalocephala*. Arch Entwicklungsmechanik 30:101–125.

Brenner S (1974): The genetics of *Caenorhabditis elegans*. Genetics 77:71–94.

Cassada R, Isnenghi E, Culotti M, von Ehrenstein G (1981): Genetic analysis of temperature-sensitive embryogenesis mutants in *C. elegans*. Dev Biol 84:193–205.

Chitwood BG, Chitwood MB (1950): "Introduction to Nematology." Reprinted 1974. Baltimore: University Park Press, p 106.

Deppe U, Schierenberg E, Cole T, Krieg C, Schmitt D, Yoder B, von Ehrenstein G (1978): Cell lineages of the embryo of the nematode *Caenorhabditis elegans*. Proc Natl Acad Sci USA 75:376–380.

Eddy EM (1975): Germ plasm and the differentiation of the germ cell line. Int Rev Cytol 43:229–280.

Hirsh D, Vanderslice R (1976): Temperature-sensitive developmental mutants of *Caenorhabditis elegans*. Dev Biol 49:220–235.

Illmensee K, Mahowald AP (1974): Transplantation of posterior polar plasm in *Drosophila*. Induction of germ cells at the anterior pole of the egg. Proc Natl Acad Sci USA 71:1016–1020.

Kohler G, Milstein C (1975): Continuous cultures of fused cells secreting antibody of predefined specificity. Nature 256:495–497.

Krieg C, Cole T, Deppe U, Schierenberg E, Schmitt D, Yoder B, von Ehrenstein G (1978): The cellular anatomy of embryos of the nematode *Caenorhabditis elegans*; analysis and reconstruction of serial section electron micrographs. Dev Biol 65:193–215.

Laufer JS, Bazzicalupo P, Wood WB (1980): Segregation of developmental potential in early embryos of *Caenorhabditis elegans*. Cell 19:569–577.

Laufer JS, von Ehrenstein G (1981): Nematode development after removal of egg cytoplasm: Absence of localized unbound determinants. Science 211:402–405.

Schierenberg E, Miwa J, von Ehrenstein G (1980): Cell lineages and developmental defects of temperature-sensitive embryonic arrest mutants in *C. elegans*. Dev Biol 76:141–159.

Strome S, Wood WB (1982): Immunofluorescence visualization of germ-line-specific cytoplasmic granules in embryos, larvae, and adults of *Caenorhabditis elegans*. Proc Natl Acad Sci USA 79:1558–1562.

Sulston J, Schierenberg E, White J, Thomson N (1983): The embryonic cell lineage of the nematode *Caenorhabditis elegans*. Dev Biol (in press).

Vanderslice R, Hirsh D (1976): Temperature-sensitive zygote-defective mutants of *Caenorhabditis elegans*. Dev Biol 49:236–249.

Wakahara M (1978): Induction of supernumerary primordial germ cells by injecting vegetal pole cytoplasm into Xenopus eggs. J Exp Zool 203:159–164.

Whittaker JR (1973): Segregation during ascidian embryogenesis of egg cytoplasm information for tissue-specific enzyme development. Proc Natl Acad Sci USA 70:2096–2100.

Wolf N, Priess J, Hirsch D (1983): Segregation of germline granules in early embryos of *Caenorhabditis elegans*: An electron microscopic analysis. J Embryol Exp Morphol (in press).

Wood WB, Hecht R, Carr S, Vanderslice R, Wolf N, Hirsh D (1980): Parental effects and phenotypic characterization of mutations that affect early development in *Caenorhabditis elegans*. Dev Biol 74:446–469.

Time, Space, and Pattern in Embryonic Development, pages 241–259

Messenger RNA Localization and Cytoskeletal Domains in Ascidian Embryos

William R. Jeffery

Department of Zoology, University of Texas at Austin, Austin, Texas 78712

> "The sorting out process requires energy and, according to some, even intelligence."
>
> Ross G. Harrison in reference to ooplasmic segregation, 1940.

The role of cytoplasmic determinants in specifying the fates of embyronic cells was established by experimental embryologists in the early part of this century [see Wilson, 1925, for review]. This discovery initiated an experimental period which was focused on determining the identity of the cytoplasmic substances thought to be responsible for embryonic determination. Although the experiments which were conducted during this period were largely unsuccessful, and the identity of these substances is still a significant question in modern developmental biology, they were able to exclude the possibility that large cytoplasmic organelles such as yolk granules, lipid particles, and mitochondria play an instructive role in cytoplasmic localization [Conklin, 1931; Morgan, 1935; Tung et al., 1941; and others].

The discovery of a complex population of maternal messenger RNA (mRNA) molecules which code for most of the proteins synthesized during early development [see Davidson, 1976, for review] has made mRNA the most popular modern candidate for a cytoplasmic determinant [Davidson and Britten, 1971; Whittaker, 1977; see Jeffery, 1983, for review]. Although considerable attention has been devoted to the temporal aspects of mRNA metabolism in recent years, there has been relatively little consideration of the spatial distribution of maternal mRNA. A complete knowledge of the territories occupied by mRNA sequences in the egg cytoplasm and the manner in which they are distributed to the blastomeres during cleavage will almost certainly be crucial to our understanding of the role of maternal mRNA in cytoplasmic localization.

The problem of maternal mRNA localization has been approached in this laboratory by mapping the spatial distribution of mRNA sequences in sections of developing embryos by in situ hybridization with complementary molecular probes. This chapter summarizes our results on the pattern of mRNA distribution and the mechanisms that may be responsible for organizing this pattern during early development of the ascidian embryo.

EARLY DEVELOPMENT OF ASCIDIAN EMBRYOS

The Colored Ooplasms

Ascidian embryos exhibit determinative events very early in their development and consequently provide excellent systems for studies of embryonic determination [see Whittaker, 1979, for review]. Certain genera of these organisms, including *Styela* and *Boltenia*, contain ooplasms of different colors which are indicative of the cellular distribution of cytoplasmic regions exhibiting specific morphogenetic potentials [Conklin, 1905; Whittaker, 1980]. Since the colored ooplasms can be identified in histologically stained sections as well as in the living embryo they can be exploited to study the distribution of mRNA with respect to cytoplasmic localization.

Three different ooplasms, the ectoplasm, the endoplasm, and the myoplasm, are visible in living eggs and sectioned material. They each occupy defined territories in the unfertilized egg. The transparent ectoplasm, derived almost entirely from the sap that escapes from the germinal vesicle (GV) at the time of maturation, is located in the animal hemisphere. The yolky endoplasm surrounds the ectoplasm in the animal hemisphere and almost completely fills the vegetal hemisphere. The myoplasm occupies the entire cortex of the egg and contains pigment granules which are tinted yellow in *Styela* and orange or red in *Boltenia*.

After fertilization the ooplasms are rearranged by a spectacular episode of cytoplasmic movements known as ooplasmic segregation. These movements are considered in detail below (Fig. 1) since they are instrumental in setting up the pattern of mRNA localization in the embryo.

Ooplasmic Segregation

Ooplasmic segregation occurs during the first 30 minutes after fertilization. It can be divided into two parts. The first part accompanies the maturation divisions and begins with the streaming of the cortical myoplasm into the vegetal hemisphere where it collects as an intensely pigmented cap. The myoplasm is quickly followed by the ectoplasm which separates into a

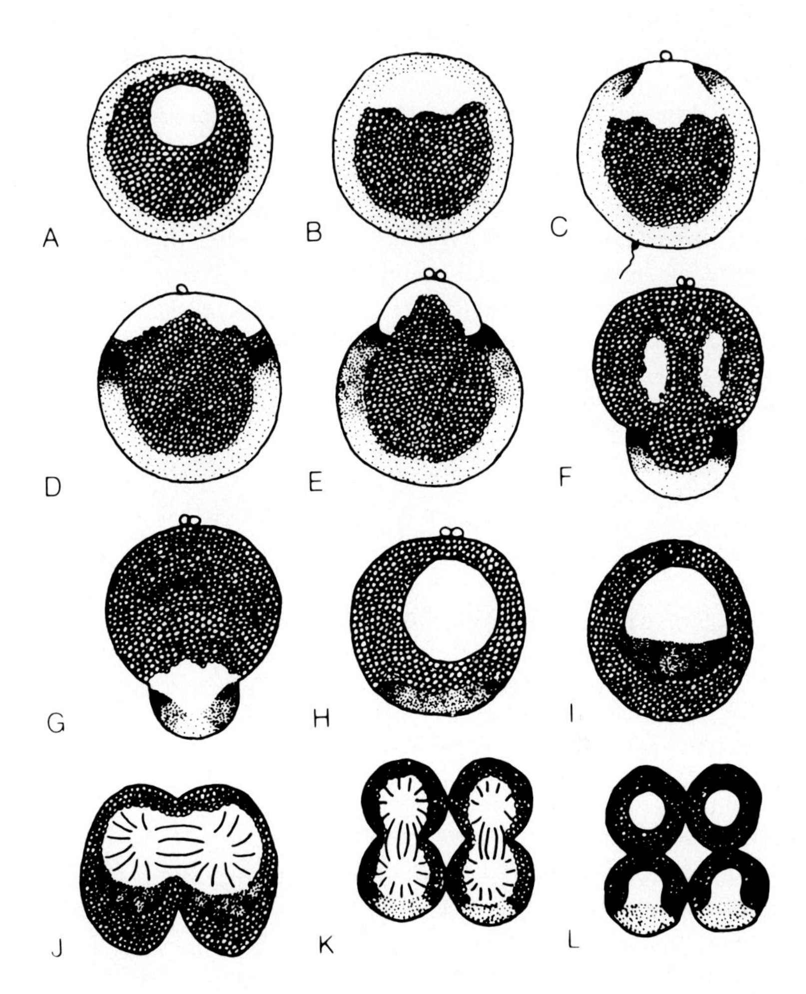

Fig. 1. Ooplasmic segregation and partitioning of ooplasms during cleavage in ascidian eggs. A detailed description of the events shown in this diagram is given in the text. A. Meridional section through a postvitellogenic oocyte containing an intact germinal vesicle (center). B. Meridional section through an unfertilized egg after germinal vesicle breakdown. C–H. Meridional sections through fertilized eggs undergoing the first part of ooplasmic segregation. The sperm enters in the vegetal pole region in C. I,J. Meridional sections through the myoplasmic crescent. K, L. Subequatorial sections through the myoplasmic crescent area of two- and four-cell embryos. The clear areas represent ectoplasm, the finely granular areas represent myoplasm, and the coarse granular areas represent endoplasm. This diagram was constructed from histological studies carried out during early development of *Styela plicata* embryos.

number of protoplasmic islets during its migration and comes to rest immediately above the myoplasmic cap in the vegetal hemisphere (Fig. 1C–G). The movement of myoplasm and ectoplasm into the vegetal hemisphere appears to displace the endoplasm upward into the animal hemisphere (Fig. 1C–G). The movement of cytoplasm through the egg occurs with such force that broad protoplasmic lobes are often transiently protruded in the animal and vegetal regions (Fig. 1E–G) [Zalokar, 1974; Sawada and Osanai, 1981]. The first part of ooplasmic segregation terminates when the ooplasms are stratified along the animal-vegetal axis of the egg (Fig. 1G).

The second part of ooplasmic segregation occurs in concert with the movements of the male pronucleus, which has entered the egg in the vegetal hemisphere (Fig. 1C) [Conklin, 1905]. The myoplasm and ectoplasm migrate up along the vegetal cortex to the future posterior region of the embryo where they are extended into crescent-shaped patches (Fig. 1H-I). The patch of myoplasm is known as the yellow crescent in *Styela* [Conklin, 1905] and the orange crescent in *Boltenia*. Ooplasmic segregation is concluded when the ectoplasm streams into the animal hemisphere, accompanied by the male pronucleus, and the endoplasm returns to its original position in the vegetal hemisphere.

A cytoplasmic pattern appears to be formed as a consequence of ooplasmic segregation which is essential for normal development. When eggs are cut in half prior to fertilization normal larva are obtained from any half containing a nucleus [Reverberi and Ortolani, 1962]. If a similar cut is made after the completion of ooplasmic segregation, however, only nucleated vegetal halves are able to cleave and develop normally [Reverberi, 1937].

Fate of the Ooplasms

The colored ooplasms are unequally segregated to the blastomeres during cleavage. The ectoplasm is largely distributed to the animal hemisphere blastomeres, although some of it surrounds the nucleus in every embryonic cell. Animal blastomeres develop into larval epidermis, palps, and sense organs. The endoplasm is also distributed to all the blastomeres, but a large portion of it enters the anterior pair of vegetal cells that are responsible for gut, spinal chord, and notochord development. The myoplasm enters the posterior vegetal blastomeres which form the larval mesenchyme and muscle cells.

DISTRIBUTION OF mRNA DURING EARLY DEVELOPMENT

If it is assumed that certain mRNA sequences serve as cytoplasmic determinants, one of three different behaviors might be expected with regard to their distribution during early development. First, they might be evenly distributed between the blastomeres but selectively translated in particular cell types because of the differential distribution of specific translational factors. Second, they may be evenly distributed in the egg and either differentially segregated to specific blastomeres or segregated evenly between all the blastomeres but subsequently destroyed in certain cells by specific nucleases. Third, they may be differentially localized in the egg and passively partitioned to different cells during cleavage. To test which of these behaviors occurs during early ascidian development we have subjected sections of eggs and embryos to in situ hybridization with probes complementary to total and specific mRNA sequences [Jeffery and Capco, 1978; Jeffery, 1982a; Jeffery et al., 1983].

Total mRNA

The distribution of total mRNA was determined by in situ hybridization with [^{3}H]-poly(U) [Capco and Jeffery, 1978]. The nucleus and cytoplasm of previtellogenic oocytes hybridized with poly(U) contain equivalent grain concentrations. The concentration of grains in the cytoplasm is gradually diluted during vitellogenesis. Consequently, the GV plasm of postvitellogenic oocytes is characterized by an eight-fold elevation in grain concentration relative to the surrounding endoplasm (Fig. 2A). The grain concentration also varies between the cytoplasmic regions of the oocyte, with the myoplasm exhibiting values five to ten times less than the endoplasm. After GV breakdown the relative grain concentrations of each ooplasm are maintained during ooplasmic segregation (Fig. 2B) and following their partition by cleavage to specific embyronic cells (Fig. 2C, D).

These studies suggest that the ectoplasm is greatly enriched in total mRNA (strictly speaking total poly(A)$^+$RNA) during early development. This enrichment could be related to the origin of the ectoplasm in the oocyte GV. The suitability of the GV plasm for the storage of substances that are subsequently active in early development or are specifically localized in the egg has been recognized for many years [Delage, 1901; Conklin, 1902; Wilson, 1903; Yatsu, 1905; Huff, 1962; Malacinksi, 1974]. In ascidian oocytes, mRNA precursors retained in the nucleus during oogenesis may be selectively modified enabling them to be specifically localized in the ectoplasm after GV breakdown and processed into active mRNA. It is also possible that the mRNA molecules which were stored in the GV represent a unique set of mRNA species which are instrumental in the development of cells which inherit the ectoplasm.

Despite its high concentration of mRNA, the ectoplasm is estimated to contain only 45% of the mass of egg mRNA (Fig. 3). This is due, of course, to its relatively small volume—only about 10% of the egg. The endoplasm, which comprises 80% of the egg volume, accounts for the rest of the mRNA mass. Thus the myoplasm appears to be very poor in mRNA (Fig. 3), a fact

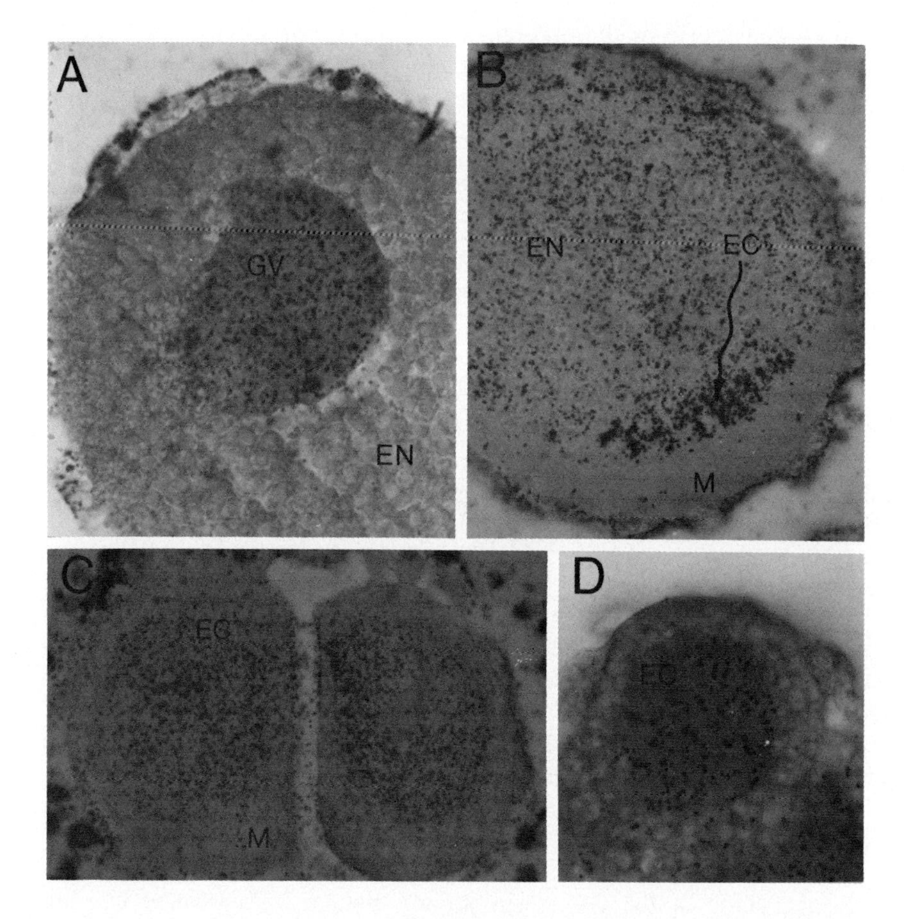

Fig. 2. In situ hybridization with a [^{3}H]-poly(U) probe during early development of *Styela*. A. A post-vitellogenic oocyte with focus on grains concentrated over the germinal vesicle (GV). B. A fertilized egg which has completed the first part of ooplasmic segregation. The ooplasms are stratified along the animal-vegetal axis in the sequence endoplasm-ectoplasm-myoplasm. The grains are concentrated over the ectoplasm (EC). Some grains are also present over the endoplasm (EN), but few grains are seen above the myoplasm (M). C. A two-cell embryo with many grains present over the ectoplasm but few grains present above the myoplasm(M). D. An eight-cell embryo showing grains concentrated over the ectoplasm of an animal hemisphere blastomere. All frames are approximately ×400.

which must be taken into account when assessing the possibility that specific mRNA sequences may be localized in this region.

We wish to strongly emphasize two conclusions from our study of the distribution of mRNA. First, as pointed out above, most of the mRNA molecules are located in the ectoplasm and endoplasm. Second, the ooplasms

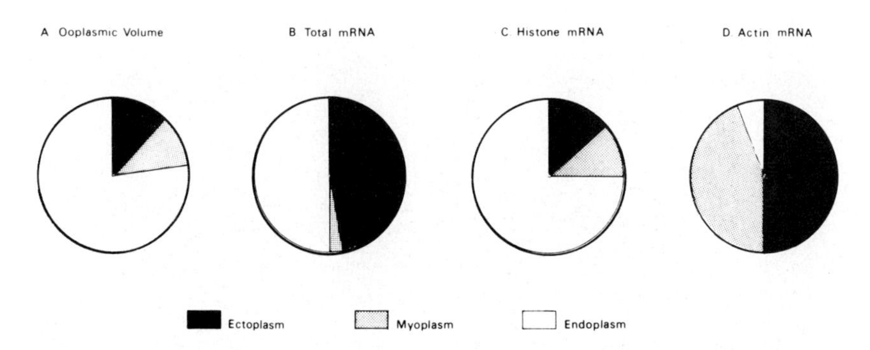

Fig. 3. Relative volume and mRNA content of the three ooplasms of *Styela* eggs. To quantify the distribution of mRNA, the number of grains registered in each ooplasm was multiplied by its volume and this value was expressed as a percentage of the total grains.

retain their characteristic concentrations of mRNA while they are extensively moved through the egg during ooplasmic segregation. The latter observation suggests that mRNA molecules show tenacious affinities for specific ooplasmic constituents, a topic we shall return to later in this chapter.

Specific mRNA

We have examined the distribution of specific mRNA sequences in ascidian embryos by in situ hybridization with cloned DNA probes [Jeffery, 1982a; Jeffery et al., 1983]. At present we have obtained information on the distribution of two classes of mRNA, histone mRNA and actin mRNA.

Histone mRNA. The probe we have used to detect histone mRNA consists of a 4.6-kilobase (kb) restriction fragment containing all five genes of the Dm-500 histone gene complex from *Drosophila melanogaster* [Lifton et al., 1977]. No significant differences were found in the concentration of grains developed over the three ooplasms of oocytes, unfertilized eggs, or embryos after in situ hybridization with a [^{125}I]-labeled histone probe (Fig. 4A, B). Quantification of these results shows that about 75% of the histone mRNA is present in the endoplasm (Fig. 3). Histone mRNA is apparently representative of a class of maternal messages that is evenly distributed in the egg and embryo. This distribution may be required to promote enhanced histone synthesis in each part of the embryo during the period of rapid cleavages which follows fertilization.

Actin mRNA. The probe we have used to detect actin mRNA consists of a 1.8-kb restriction fragment of the Dm-A2 actin gene from *D. melanogaster* [Fryberg et al., 1980]. High concentrations of grains were developed in the ectoplasm and myoplasm at all stages of early development after in situ

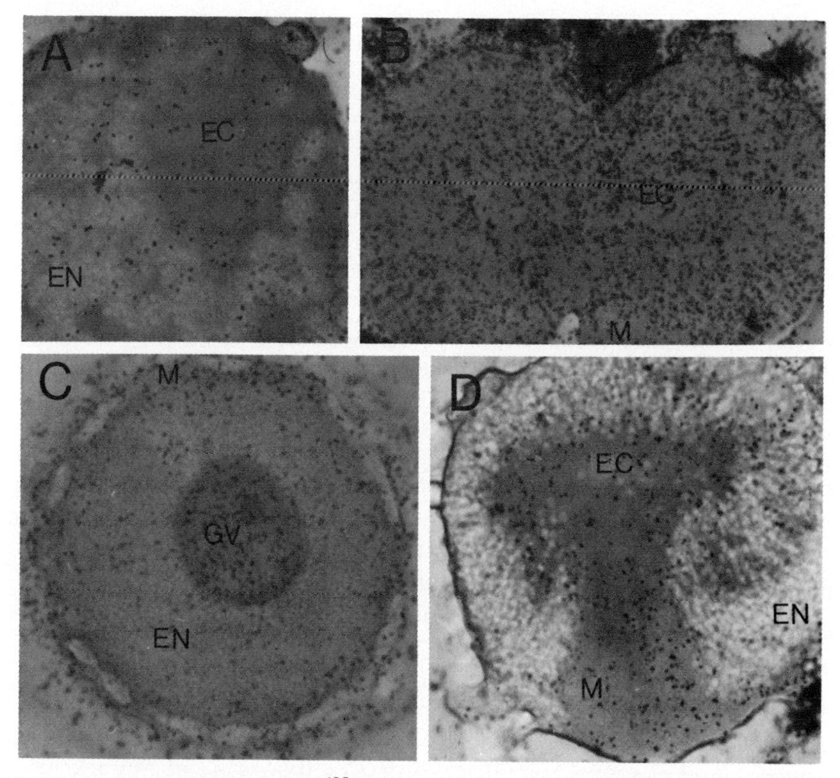

Fig. 4. In situ hybridization with [^{125}I]-labeled histone and actin DNA probes during early development of *Styela*. A. An unfertilized egg probed with histone DNA. Grains are uniformly distributed above all three ooplasms. ×400. B. A two-cell embryo probed with histone DNA. Grains are uniformly distributed above all three ooplasms. The dark patches above the egg represent labeled follicle cells. ×400. C. A postvitellogenic oocyte probed with actin DNA. Grains are concentrated above the germinal vesicle (GV) and the cortical myoplasm(M), ×400. D. One of the blastomeres of a two-cell embryo probed with actin DNA. Grains are concentrated over the ectoplasm (EC) and the myoplasm(M) but not the endoplasm (EN). ×1,000.

hybridization with the [^{125}I]-labeled actin probe (Fig. 4C, D). Quantification of these results shows that the ectoplasm and myoplasm contain about 50% and 42%, respectively, of the actin mRNA molecules (Fig. 3). Two conclusions can be drawn from this study. First, since actin mRNA molecules seem to be concentrated in the ectoplasm, they may be among the mRNA species thought to be localized after accumulation in the GV plasm during oogenesis. Second, a preferential localization of actin mRNA appears to exist in the myoplasm. This conclusion is greatly strengthened by the previous result indicating that the myoplasm is generally very poor in mRNA. Actin mRNA is apparently representative of a class of messages which is localized in the egg and passively partitioned to specific blastomeres during cleavage.

Developmental Role of Localized Actin mRNA

The ascidian egg rapidly develops into a complex larva equipped with a muscular tail. In this relatively small egg it may be more efficient to localize mRNA molecules and use them in specific cell lineages to code for many copies of a protein than to store the proteins themselves or to enlist the capacity to transcribe large amounts of new mRNA. Thus it is possible that a maternal mRNA species which codes for the muscle-specific isoform of actin is stored in the myoplasm and activated in the muscle cells during embyrogenesis. It should be pointed out that this would not be a surprising phenomenon in ascidian eggs since evidence has already been presented that a localized mRNA coding for alkaline phosphatase is segregated to the endodermal cell lineage during early ascidian development [Whittaker, 1980]. The possibility that a muscle-type actin mRNA is localized in the myoplasm is now being tested in this laboratory by in vitro translation and product analysis of egg mRNA and in situ hybridization with DNA probes complementary to nonconserved regions in the *Styela* muscle actin gene.

MECHANISMS OF mRNA LOCALIZATION

The tenacious association of mRNA molecules with specific ooplasms is strongly implied by their remarkable resistance to mixing during ooplasmic segregation. This is not a phenomenon unique to ascidian eggs; maternal mRNA is also known to show a selective affinity for specific cytoplasmic regions in *Chaetopterus* eggs [Jeffery and Wilson, 1983] and *Xenopus laevis* oocytes [Capco and Jeffery, 1982]. Moreover, poly(A)$^{+}$RNA isolated from the vegetal pole region of *Xenopus* eggs is distributed in a vegetal pole to animal pole gradient when labeled in vitro and microinjected into any part of another egg [Capco and Jeffery, 1981].

We can envision three possible mechanisms for the localization of mRNA. First, they could be localized by an affinity for a specific cytoplasmic microenvironment. Examples would include areas distinguished by differences in pH or the concentration of divalent cations. Second, they could be localized by associations with the plasma membrane, regionalized intracellular membrane systems, or membrane-bound organelles. Third, they could be localized by interactions with specific cytoskeletal domains. The latter possibility seems particularly attractive because a number of investigators have described associations between mRNA molecules and cytoskeletal elements [Lenk et al., 1977; Cervera et al., 1981; van Venrooij et al., 1981; Jeffery, 1982a].

To decide between these possibilities ascidian eggs have been extracted with buffers containing the nonionic detergent Triton X-100 and assayed for the presence of poly(A)$^+$RNA in the soluble and insoluble fractions by poly(U) hybridization. About 70% of the total poly(A) is left in the detergent-insoluble fraction under conditions in which more than 85% of the lipid, 75% of the protein, and 70% of the total RNA is extracted by Triton X-100 [Jeffery and Meier, 1983]. These results suggest that the detergent-insoluble fraction contains most of the egg mRNA molecules. They also imply that most of the mRNA molecules are associated with the egg cytoskeleton or its associated components since Triton X-100 would be expected to destroy both cytoplasmic microenvironments and cellular membrane systems. In the following section of this chapter our recent studies of the cytoskeleton of ascidian eggs are described and further experimental evidence for the possible role of this structure in setting up the pattern of mRNA distribution is presented.

THE CYTOSKELETON OF ASCIDIAN EGGS

General Features

Our understanding of the general organization of eukaryotic cells has markedly changed in recent years. It is now known that the cytoplasm is not entirely fluid, but contains an elaborate filamentous network consisting of microtubules, microfilaments, and intermediate filaments. A filamentous structural organization also appears to be present in the nucleoplasm. These structures and their associated organelles, collectively known as the cytoskeletal framework (CF), are left as an insoluble residue after cells are extracted with Triton X-100 [Brown et al., 1976].

The CF of ascidian eggs has been examined by biochemical and microscopic methods after extraction with Triton X-100 [Jeffery and Meier, 1983]. Analysis of proteins from the detergent insoluble fraction by two-dimensional gel electrophoresis shows that the CF contains three major acidic polypeptides. Not surprisingly, one of the major polypeptides is actin. The identities of the other major cytoskeletal polypeptides are currently unknown, but their electrophoretic mobilities are similar to those of intermediate filament proteins. Despite the extraction of most of the cellular proteins and lipids, eggs treated with Triton X-100 retain an amazing amount of their structural detail, including cytoskeletal domains corresponding to their ectoplasm, endoplasm, and myoplasm. The CF seems to be the modern equivalent of what was referred to by the early embryologists as the cytoplasmic ground substance.

Although each of the cytoskeletal domains of the ascidian egg shows unique structural features [Jeffery and Meier, 1983], the myoplasm was chosen for further consideration in this chapter because it contains a localization of actin mRNA and its position at the egg surface permits a straightforward demonstration of its organization by scanning electron microscopy (SEM). As discussed in detail below, the myoplasmic cytoskeletal domain is comprised of at least two distinct elements, a superficial plasma membrane lamina (PL) and a deeper filamentous lattice which connects the pigment granules and possibly other myoplasmic organelles to the PL.

The Myoplasmic Cytoskeletal Domain

The surface of the myoplasmic cytoskeletal domain is shown in Figure 5A. The PL can be seen as a network of filaments organized parallel to the plane of the plasma membrane which is, of course, removed by the detergent. The PL is thought to be a remnant of cytoskeletal elements lying immediately below the plasma membrane in the living egg. Submembrane cytoskeletons of this nature have been previously reported in erythrocytes and other vertebrate somatic cells [Hainfield and Steck, 1977; Boyles and Bainton, 1979; and others] but, to our knowledge, have not been seen before in eggs.

The origin and fate of the PL during early development is closely linked to the history of the myoplasm. In the unfertilized egg it is located over the entire surface of the CF. In the fertilized egg the PL appears to recede into the vegetal hemisphere along with the myoplasm during ooplasmic segregation (Fig. 5B). The average distance between the interstices of the network is also markedly reduced during ooplasmic segregation, as if the entire PL contracted toward a single position on the cell surface (compare Fig. 5A and C). During early embryogenesis the PL is specifically partitioned to the presumptive mesenchyme and muscle cells. This is easily demonstrated by preparing the CFs of ectoplasmic, endoplasmic, and myoplasmic cells separated from a 32-cell embryo. As illustrated in Figure 5D–F, the PL was present only in the CFs of myoplasmic cells.

A more elaborate system of filaments is present in the myoplasmic cytoskeletal domain beneath the PL. It consists of a filamentous lattice coursed with bundles of filaments (Fig. 6A, B) and appears to be attached to both the PL and the underlying pigment granules (Fig. 6C). These connections are probably responsible for the resistance of the myoplasmic pigment granules to Triton X-100 extraction.

Since the cortical regions of sea urchin eggs are known to contain actin microfilaments [Vacquier, 1981] and actin is present in the CF of ascidian eggs, experiments were conducted to determine whether the myoplasmic cytoskeletal domain contains actin filaments. As in other cell types [Hitchcock et al., 1976; Raju et al., 1978], actin is selectively removed from the

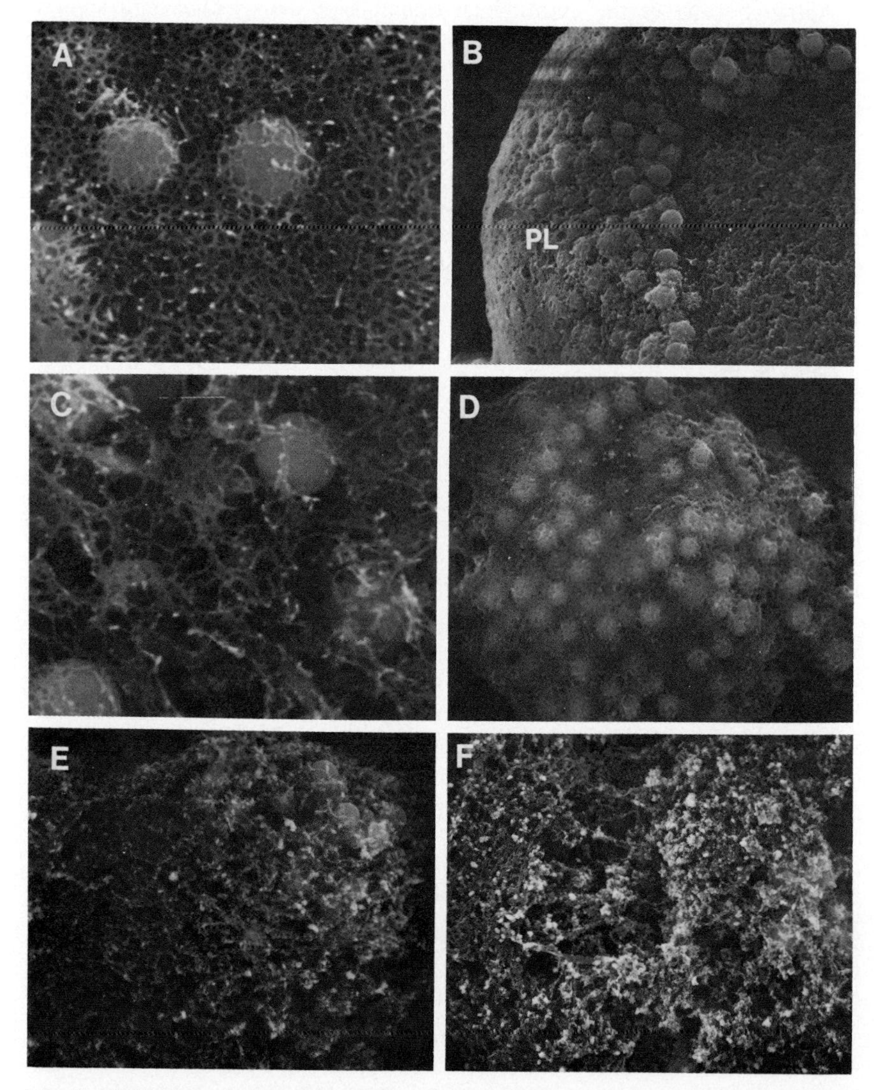

Fig. 5. SEM of the plasma membrane lamina (PL) of Triton X-100-extracted ascidian eggs. A. The surface of the myoplasmic cytoskeletal domain in the cytoskeletal framework of a fertilized *Styela* egg. ×10,000. B. The myoplasmic cytoskeletal domain of the cytoskeletal framework of fertilized *Boltenia* eggs undergoing ooplasmic segregation. ×200. The surface of the myoplasmic cytoskeletal domain in the cytoskeletal framework of an unfertilized *Styela* egg. ×10,000. Note the greater distances between the filaments of the PL in C compared to A. Cytoskeletal frameworks of myoplasmic(D), ectoplasmic(E), and endoplasmic(F) cells separated from a 32-cell *Styela* embryo. ×27,000. Only the myoplasmic cytoskeletal frameworks contain the PL.

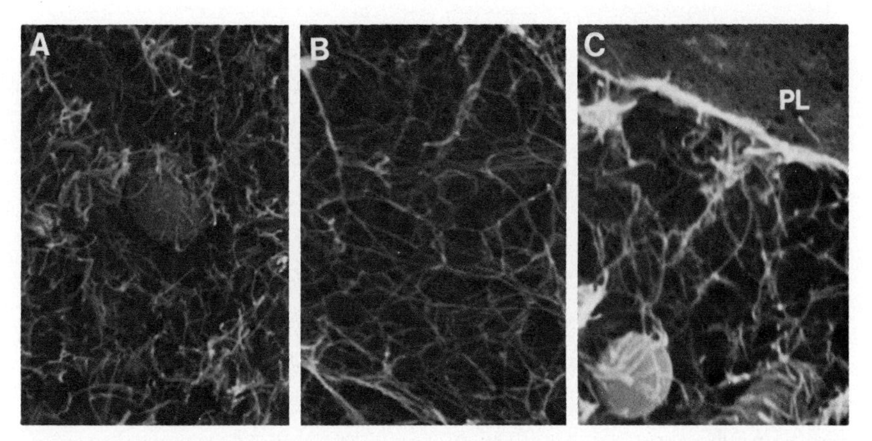

Fig. 6. SEMs showing the internal filamentous organization of the myoplasmic cytoskeletal domain of Triton X-100 extracted *Styela* zygotes. A. Lattice of internal filaments which are associated with a pigment granule observed at a place in the cytoskeletal framework where the PL has been torn away. B. A bundle of filaments can be seen to course the filamentous lattice. C. The lattice of internal filaments is associated with underlying pigment granules and the overlying PL. All frames are ×9,000.

CF of ascidian eggs by DNase I [Jeffery and Meier, 1983]. When the myoplasmic cytoskeletal domain of CFs treated with DNase I were examined most of the internal filaments were present, but the PL was almost entirely removed (Fig. 7). This result suggests that the PL is at least in part composed of actin filaments. The presence of actin filaments also supports the possibility that the PL is a contractile structure.

Distribution of mRNA in the Cytoskeletal Framework

To determine whether the CF is involved in the localization of mRNA we subjected sections of Triton X-100-extracted eggs and embryos to in situ hybridization with some of the probes discussed earlier. The appropriate methods for in situ hybridization of sectioned CFs have been recently developed in this laboratory [Jeffery, 1982a]. The different cytoskeletal domains of the ascidian egg can be distinguished in these histological preparations by their staining properties and, in the case of the myoplasmic cytoskeletal domain, by the presence of embedded pigment granules. In a typical experiment, a population of eggs or embryos is divided into two parts. One part is extracted with Triton X-100 while the other remains untreated. The CFs and the intact specimens are then mixed, immediately fixed, and processed together for in situ hybridization. In this way the distribution of grains in the ooplasms and their cytoskeletal domains can be compared in a quantitative

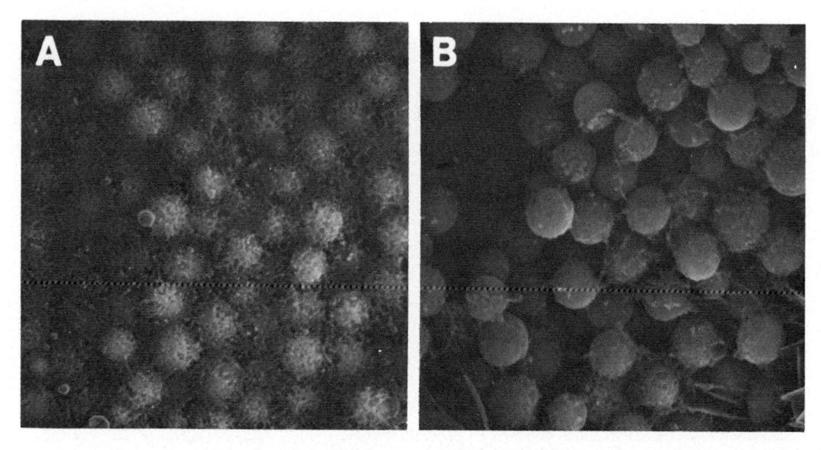

Fig. 7. Removal of the plasma membrane lamina from the surface of the myoplasmic cytoskeletal domain by DNase I in the cytoskeletal framework of *Boltenia* zygotes. A. A control cytoskeletal framework which was treated with bovine serum albumin. The plasma membrane lamina is intact. B. A cytoskeletal framework treated with DNase I. The plasma membrane lamina is missing. Both frames are ×3,400.

fashion. When an experiment of this nature was conducted using the poly(U) and the actin probes it was found that 70% of the total mRNA and 60% of the actin mRNA was present in the CF. More significantly, both total mRNA and actin mRNA molecules are concentrated in precisely the same regions of the CF as they are in the intact egg (Fig. 8). Total mRNA is concentrated in the ectoplasmic cytoskeletal domain, while actin mRNA is present in both the ectoplasmic and the myoplasmic cytoskeletal domains. This result suggests that mRNA localization is based on the association of mRNA molecules with specific cytoskeletal elements or their associated organelles. It will be an exciting challenge to determine the identity of the cytoskeletal structures which interact with mRNA.

ROLE OF CALCIUM IN OOPLASMIC SEGREGATION AND mRNA LOCALIZATION

Ooplasmic segregation and changes in the distribution of mRNA molecules are among the earliest indications of polarity in the ascidian egg. Although we are beginning to understand the cellular basis for mRNA localization, little is known about how localized mRNA molecules and other cytoplasmic constituents find specific positions on the fertilized egg. It has been suggested that ooplasmic segregation and the distribution of cytoplasmic macromolecules may be polarized by an intracellular gradient of calcium

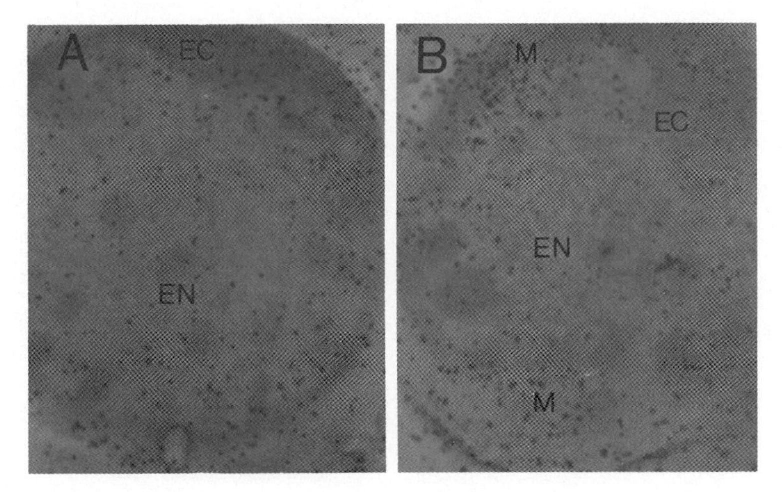

Fig. 8. In situ hybridization with [^{3}H]-poly(U) and [^{125}I]-actin DNA probes of cytoskeletal frameworks from Triton X-100-extracted *Styela* zygotes. A. A cytoskeletal framework probed with poly(U). Grains are concentrated over the ectoplasmic cytoskeletal domain (EC) where it surfaces at several positions in the sectioned cytoskeletal framework. The endoplasmic cytoskeletal domain (EN) contains few grains. B. A cytoskeletal framework probed with actin DNA. The grains are concenteated over the ectoplasmic(EC) and myoplasmic(M) cytoskeletal domains which surface at several sites in the section, but not over the endoplasmic cytoskeletal domain(EN). Both micrographs are ×400.

ions [Jaffe, 1981]. This proposal was particularly attractive to us because calcium ions have also been implicated in contractile responses at the egg surface [Gingell, 1970; Schroeder, 1974], like those that appear to be involved in ooplasmic segregation in ascidian eggs [Zalokar, 1974; Sawada and Osanai, 1981].

The possible role of calcium ions in polarizing ooplasmic segregation and mRNA localization was tested by treating *Boltenia* eggs with the calcium-transporting ionophore A23187 [Jeffery, 1982b]. Ionophore A23187 was previously shown to activate ascidian eggs and promote several partial cleavages [Steinhardt et al., 1974; Bevan et al., 1977] but, since the species investigated lacked eggs with colored ooplasms, no information on ooplasmic segregation was obtained. Most of the *Boltenia* eggs treated with ionophore A23187 in our experiments were induced to undergo ooplasmic segregation and formed an orange cap of myoplasm on one side of the cell. The orange cap induced by ionophore A23187 was shown to contain all the visible organelles, as well as the characteristic localization of actin mRNA, present in the myoplasm of fertilized eggs.

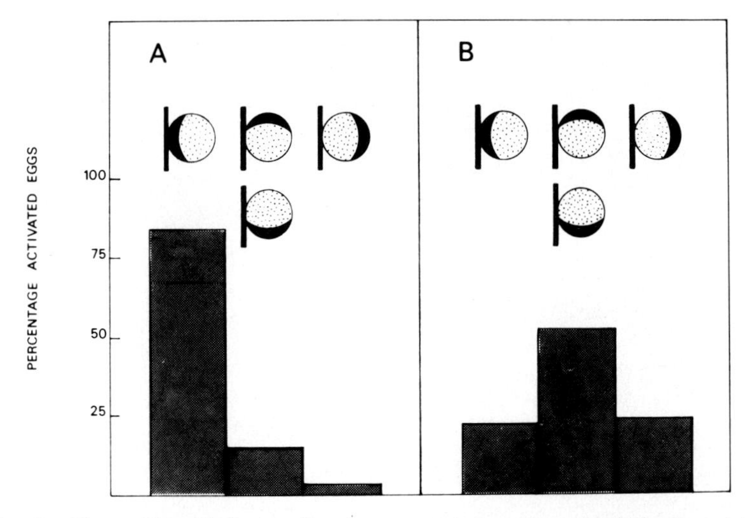

Fig. 9. The position of the myoplasmic crescent in ionophore A23187-activated *Boltenia* eggs aligned against glass rods. A. The direction of orange cap polarization in eggs aligned against the ionophore-coated rod. Most eggs form orange caps on the side nearest the rod. B. The direction of orange cap polarization in eggs aligned against an untreated rod. Ionophore A23187 was present in the surrounding seawater. The orange caps form at random positions in the eggs with respect to the rod.

There are three possibilities for the relation between the site of calcium ion flux and orange cap formation in ionophore A23187-treated eggs. The orange cap may be polarized in the direction of the highest intra-cellular calcium concentration. The orange cap may be polarized in the direction of the lowest calcium concentration. The formation of an orange cap may be induced by the calcium flux but its polarization may be dictated by another element of polarity already present in the egg. To test whether a relation exists between the site of calcium flux and orange cap formation eggs were aligned against a small glass rod which was coated with ionophore [Robinson and Cone, 1980]. A steep gradient of ionophore emanates from the rod. The majority of the eggs which were aligned against the ionophore-coated rod formed an orange cap on the side of the cell facing the rod (Fig. 9A). Thus the orange cap seems to be polarized in the direction of the highest intracellular calcium concentration. These results cannot be explained by a tendency of the glass rod itself to polarize ooplasmic segregation, since orange caps appear at random positions along the circumference of ionophore-activated eggs aligned adjacent to untreated glass rods (Fig. 9B). The simplest interpretation of these results is that an elevation in calcium ion concentration polarizes the direction of ooplasmic segregation and mRNA localization in ascidian eggs. The calcium ion flux may be initiated at the point of sperm entry during normal development.

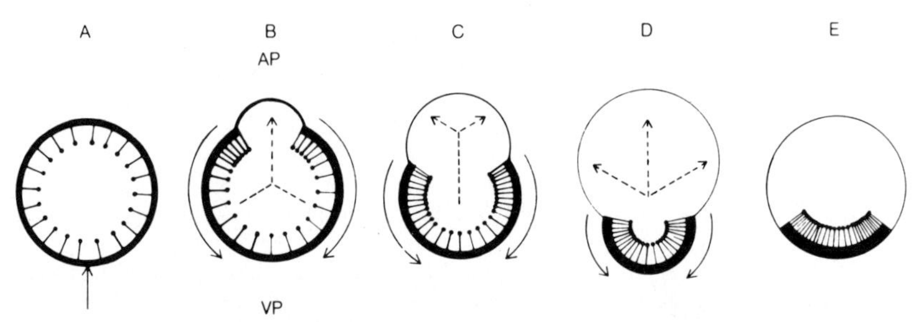

Fig. 10. A model for mRNA localization during myoplasmic segregation. A. An unfertilized egg. The outer arrow indicates the point of sperm entry. B–D. Fertilized eggs in the process of ooplasmic segregation. The arrows represent the directions of ooplasmic movement—arrows with solid lines for the myoplasm and arrows with broken lines for the endoplasm. E. A zygote which has completed the first part of ooplasmic segregation. In each diagram the thick egg boundaries represent parts of the plasma membrane underlied by the plasma membrane lamina. The thin egg boundaries represent parts of the plasma membrane without the plasma membrane lamina. The structures attached to the inside of the plasma membrane lamina in the myoplasm represent the internal filamentous lattice and its associated components, the pigment granules and the localized mRNA molecules of the myoplasmic cytoskeletal domain. AP, animal pole; VP, vegetal pole.

A MODEL FOR mRNA LOCALIZATION DURING MYOPLASMIC SEGREGATION

In Figure 10 we present a model which integrates our results on mRNA localization, the association of localized mRNA with the CF, and the role of calcium ions in polarizing ooplasmic segregation of the myoplasm. We propose that localized mRNA sequences are fixed to the myoplasmic cytoskeletal domain during oogenesis. At the time of fertilization a local calcium ion elevation causes the PL to contract toward the point of sperm entry in the vegetal hemisphere of the egg. The other cytoskeletal elements and associated mRNA sequences of the myoplasmic cytoskeletal domain are linked to the PL and its contraction would therefore provide direction for mRNA localization and a fixed position in the vegetal region of the egg which would insure the subsequent deposition of localized mRNA molecules in the presumptive muscle and mesenchyme cells. This model would predict that the integrity of mRNA-cytoskeletal interactions is essential for normal development, a possiblity that is presently being tested in our laboratory.

ACKNOWLEDGMENTS

Many of the experiments discussed in this chapter were performed in collaboration with my students and colleagues Richard Brodeur, David Capco, Stephen Meier, and Craig Tomlinson. Technical assistance was provided by Lynn Hunter, Priscilla Kemp, Dianne McCoig, and Bonnie Brodeur. Our work is supported by grants from the National Institutes of Health (GM-250119 and HD-13970) and the Muscular Dystrophy Association.

REFERENCES

Bevan SJ, O'Dell DS, Ortolani G (1977): Experimental activation of ascidian eggs. Cell Differ 6:313–318.

Boyles J, Bainton DF (1979): Changing patterns of plasma membrane-associated filaments during the initial phases of polymorphonuclear leukocyte adherence. J Cell Biol 82:347–368.

Brown S, Levinson W, Spudich JA (1976): Cytoskeletal elements of chick embryo fibroblasts revealed by detergent extraction. J Supramol Struct 5:119–130.

Capco DG, Jeffery WR (1978): Differential distribution of poly(A)-containing RNA in the embryonic cells of *Oncopeltus fasciatus*: Analysis by in situ hybridization with a [^{3}H]-poly(U) probe. Dev Biol 67:137–152.

Capco DG, Jeffery WR (1981): Regional accumulation of vegetal pole poly(A)$^+$RNA injected into fertilized *Xenopus* eggs. Nature 294:255–257.

Capco DG, Jeffery WR (1982): Transient localizations of messenger RNA in *Xenopus laevis* oocytes. Dev Biol 89:1–12.

Cervera M, Dreyfuss G, Penman S (1981): Messenger RNA is translated when associated with the cytoskeletal framework in normal and VSV-infected HeLA cells. Cell 23:113–120.

Conklin EG (1902): Karyokinesis and cytokinesis in the maturation, fertilization, and cleavage of *Crepidula* and other Gasteropoda. J Acad Nat Sci (Philadelphia) 12:5–121.

Conklin EG (1905): The organization and cell lineage of the ascidian egg. J Acad Nat Sci (Philadelphia) 13:1–119.

Conklin EG (1931): The development of centrifuged eggs of ascidians. J Exp Zool 60:1–119.

Davidson, EH (1976): "Gene Activity in Early Development." New York: Academic Press, Chap 7, pp 245–318.

Davidson, EH, Britten RJ (1971): Note on the control of gene expression during development. J Theor Biol 32:123–130.

Delage Y (1901): Etudes experimentales chez les Echinoderms. Arch Zool Exp Genet (Ser 3) 9:285–326.

Fryberg EA, Kindle KL, Davidson N, Sodja A (1980): The actin genes of *Drosophila*: A dispersed multigene family. Cell 19:365–378.

Gingell D (1970): Contractile responses at the surface of an amphibian egg. J Embryol Exp Morphol 167:183–186.

Hainfield JF, Steck TL (1977): The sub-membrane reticulum of the human erythrocyte: A scanning electron microscope study. J Supramol Struct 6:301–311.

Hitchcock SE, Carlsson L, Lindberg U (1976): Depolymerization of F-actin by deoxyribonuclease I. Cell 7:531–542.

Huff RE (1962): The developmental role of material derived from the nucleus (germinal vesicle) of mature ovarian eggs. Dev Biol 4:398–422.

Jaffe LF (1981): Calcium explosions as triggers of development. Ann NY Acad Sci 399:86–101.

Jeffery WR (1982a): Messenger RNA in the cytoskeletal framework: Analysis by *in situ* hybridization. J Cell Biol 95:1–7.

Jeffery WR (1982b): Calcium ionophore polarizes ooplasmic segregation in ascidian eggs. Science 216:545–547.

Jeffery WR (1983): Maternal RNA and the embryonic localization problem. In Siddiqui MAQ (ed): "Control of Embryonic Gene Expression." Boca Raton, FL: CRC Press.

Jeffery WR, Capco DG (1978): Differential accumulation and localization of maternal poly(A)-containing RNA during early development of the ascidian, *Styela*. Dev Biol 67:152–166.

Jeffery WR, Meier S (1983): A yellow crescent cytoskeletal domain in ascidian eggs and its role in early development. Dev Biol (in press).

Jeffery WR, Wilson LJ (1983): Localization of messenger RNA in the cortex of *Chaetopterus* eggs and early embryos. J Embryol Exp Morphol (in press).

Jeffery WR, Tomlinson C, Brodeur R (1983): Ooplasmic segregation and localization of actin messenger RNA in the yellow crescent cytoplasm of *Styela* eggs (submitted).

Lenk R, Ransom L, Kaufman Y, Penman S (1977): A cytoskeletal structure with associated polyribosomes obtained from HeLa cells. Cell 10:67–74.

Lifton RP, Goldberg ML, Karp RW, Hogness DS (1977): The organization of the histone genes in *Drosophila melanogaster*: Functional and evolutionary implications. Cold Spring Harbor Symp Quant Biol 42:1047–1051.

Malacinski GM (1974): Biological properties of a presumptive morphogenetic determinant from the amphibian oocyte germinal vesicle nucleus. Cell Differ 3:31–44.

Morgan TH (1935): Centrifuging the egg of *Ilyanassa* in reverse. Biol Bull 68:268.

Raju TR, Stewart M, Buckley IK (1978): Selective extraction of cytoplasmic actin-containing filaments with DNA-ase I. Cytobiology 17:307–311.

Reverberi G (1937): Richerche sperimentali sull'uovo di ascidie. Pubbl Staz Zool Napoli 11:168–193.

Reverberi G, Ortolani G (1962): Twin larvae from halves of the same egg in ascidians. Dev Biol 5:84–100.

Robinson KR, Cone R (1980): Polarization of fucoid eggs by a calcium ionophore gradient. Science 207:77–78.

Sawada T, Osanai K (1981): The cortical contraction related to ooplasmic segregation in *Ciona intestinalis* eggs. W Rouxs Arch Dev Biol 190:208–214.

Schroeder TE (1974): Ionophore, calcium and contractivity in frog eggs. Exp Cell Res 83:139–142.

Steinhardt RA, Epel D, Carroll EJ, Yanagimachi R (1974): Is calcium ionophore a universal activator for unfertilized eggs? Nature 252:41–43.

Tung T, Ku S, Tung Y (1941): The development of the ascidian egg centrifuged before fertilization. Biol Bull 80:153–168.

Vacquier VD (1981): Dynamic changes in the egg cortex. Dev Biol 84:1–26.

van Venrooij WJ, Sillekens PTG, van Eekelen CAG, Reinders RJ (1981): On the association of mRNA with the cytoskeleton in uninfected and adenovirus infected human KB cells. Exp Cell Res 135:79–91.

Whittaker JR (1977): Segregation during cleavage of a factor determining the endodermal alkaline phosphatase development in ascidian embryos. J Exp Zool 202:139–153.

Whittaker JR (1979): Cytoplasmic determinants of tissue differentiation in ascidian eggs. In Subtelny S, Konigsberg IR (eds): "Determinants of Spatial Organization." New York: Academic Press, pp 29–51.

Whittaker JR (1980): Acetylcholinesterase development in extra cells caused by changing the distribution of myoplasm in ascidian embryos. J Embryol Exp Morphol 55:343–345.

Wilson EB (1903): Experiments on cleavage and localization in the nemertine egg. Arch Entwicklungsmech Org 16:411–460.

Wilson EB (1925): "The Cell in Development and Heredity." New York: MacMillan, Chap 14, pp 1035–1121.

Yatsu N (1905): The formation of centrosomes in enucleated egg fragments of *Cerebratulus*. J Exp Zool 2:287–312.

Zalokar M (1974): Effect of colchicine and cytochalasin B on ooplasmic segregation of ascidian eggs. W Rouxs Arch Dev Biol 175:243–248.

Time, Space, and Pattern in Embryonic Development, pages 261–286

Control of Polarity in the Amphibian Egg

J. Gerhart, S. Black, R. Gimlich, and S. Scharf

Department of Molecular Biology, University of California, Berkeley, California 94720

Within the subject of "control of polarity in the amphibian egg," we have chosen to discuss the cytoplasmic localization and stage-specific activity of axial determinants—that is, the agents that normally bias a certain region of the amphibian egg to develop into the dorsal structures of the body axis of the embryo. These axial structures include the central nervous system, notochord, and somites, arranged in an organ complex characteristic not only of the amphibian tadpole but of all vertebrates (Fig. 1C). The neoclassical view of cytoplasmic localizations, deriving largely from work on the eggs of invertebrates and ascsidians, might be summarized as follows: (1) determinants are agents capable of controlling specific gene expression; (2) they are uniformly distributed in the egg or oocyte cytoplasm at an early stage and subsequently are partitioned to specific regions of the cell by intracellular localization mechanisms; (3) localized determinants, once they become contained in the lineages of cells cleaved from those regions, act differentially on totipotent nuclei of the lineages to establish an "embryo in the rough"; (4) the greater the variety of determinants and the more precise the localization machinery of the egg, the more complete will be the patterning accomplished by determinants at these early stages; and (5) details of pattern are subsequently filled in by intercellular rather than intracellular processes, for example by inductions. We cite these idealizations because amphibian eggs may depart substantially from certain of them, as will be discussed.

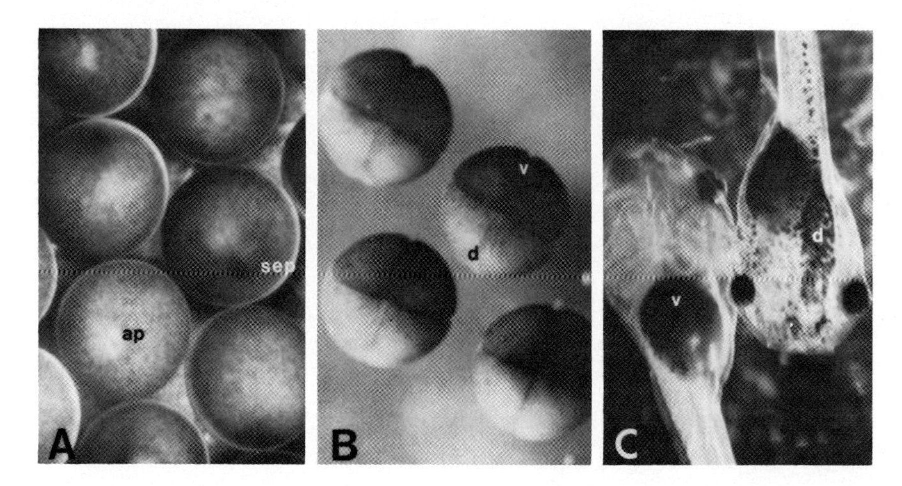

Fig. 1. Developmental stages of *Xenopus laevis*. A. Eggs 30 minutes after fertilization. Note the dark spot of pigment accumulated at the sperm entry point (sep). The eggs have lifted the vitelline membrane from the egg surface and have been dejellied. They have rotated to a vertical orientation, with the lightly pigmented, dense, vegetal hemisphere downward. The animal pole (ap) of an unfertilized egg is visible. B. Early four-cell stage, approximately two hours after fertilization at 22°C. The two cleavages have been vertical, through the animal pole. The lighter pair of blastomeres on each egg will provide material for the dorsal (marked d) embryonic organs, whereas the darker pair of blastomeres will go to ventral (marked v) organs. The sperm entrance point was on the side of the dark blastomeres. C. Tadpoles at four days postfertilization of the egg, at 22°C. Note the dorsal complex (d) of the central nervous system, notochord, and segmental muscle blocks of the trunk and tail, and the ventral (v) gut structures.

THE ANIMAL-VEGETAL AXIS OF THE EGG CORTEX AND ENDOPLASM

The unfertilized egg of *Xenopus laevis,* and of other amphibia, is a highly organized cell in which the contents are layered along an animal-vegetal axis and are grossly divided into two hemispheres. This axis extends from the animal pole, where the meiotic spindle and first polar body reside (Fig. 1A) to the geometrically opposite vegetal pole. The animal hemisphere of the egg contains (1) a layer of pigment granules in the cortex of the cell, just under the plasma membrane; (2) a layer of nuclear sap components in the endoplasm (deep cytoplasm) near the animal pole—these deposited by the enormous germinal vesicle of the oocyte when it broke down in meiosis; (3) a layer of common cytoplasmic constituents such as mitochondria and ribosomes; and (4) some nutrient reserve materials such as glycogen granules, fat droplets, and small yolk platelets. The vegetal hemisphere contains in its endoplasm massive amounts of large yolk platelets, and relatively few cytoplasmic constituents, and in its cortex almost no pigment granules, compared to the animal hemisphere. These characteristics are shown in Figures 2A, 3A, and 4C. The interface of the two hemispheres defines the equatorial level of the egg. The equator is easy to distinguish externally as the boundary of the dark and light hemispheres.

The developmental fates of the hemispheres are known from dye marking experiments. Spots of vital dyes placed on the animal hemisphere surface end up in neural and epidermal derivatives of the embryonic ectoderm, whereas dye spots placed on the vegetal hemisphere surface go to the lining of the digestive tract and other endodermal structures. As Keller [1975, 1976] has shown, mesodermal derivatives in *X. laevis* do not originate from surface regions of the egg, but from internal materials of the equatorial level. Although germ layer assignments can be made in the unfertilized egg, dorsal and ventral fates cannot be predicted until after fertilization, as will be discussed later.

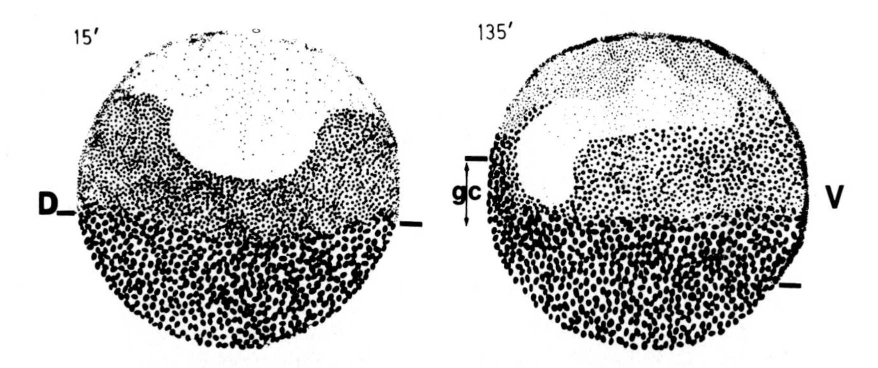

Fig. 2. Internal reorganization of the egg in the period from fertilizaton to first cleavage, as reported by Klag and Ubbels [1975]. Eggs of the frog *Discoglossus pictus* were oriented and sectioned at various times. Midsaggital sections are shown, with the animal pole at the top. The left-hand section represents an egg 15 minutes after fertilization; note the approximate radial symmetry of the egg contents around the vertical animal-vegetal axis, and note the trilayer organization of cytoplasmic contents, with a topmost layer rich in nuclear sap materials from the disrupted germinal vesicle of the oocyte, a middle layer of medium-sized yolk platelets interspersed with mitochondria, ribosomes, and familiar cytoplasmic materials, and a bottom layer of large, densely packed yolk platelets. The right-hand section represents an egg 135 minutes after fertilization, with first cleavage due at approximately 150 minutes. Note the left-hand side of the section where the top and bottom layers have moved into contact, with exclusion of the middle layer. This indicates the reorganization of egg contents on the prospective dorsal (D) side. The gray crescent (gc) is formed at the equatorial surface, the arrow indicating the distance of movement of pigment granules from the crescent area. On the other side of the egg, the contents remain trilayered, as before fertilization; this is the prospective ventral (V) side.

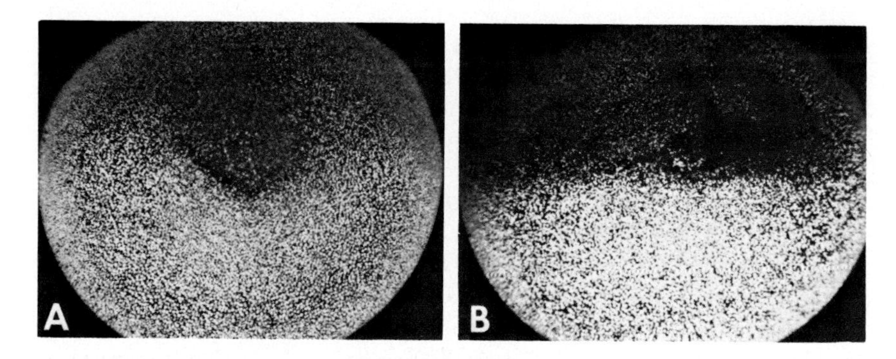

Fig. 3. Reorganization of the internal contents of the *X. laevis* egg. Eggs were fixed in 2% glutaraldehyde-50 mM sodium phosphate, pH 7.2, embedded in JB-4, and sectioned at 5 μm. Unstained sections were observed by epifluorescence microscopy (optimized for fluorescein), so that autofluorescence by the yolk platelets dominates the image. A. Fertilized egg at 0.4 (that is, 40% of the time interval from fertilization to first cleavage). The animal pole is at the top. The sperm had entered on the left side. Note the symmetric cup-shape distribution of yolk platelets. The dark central area had been occupied by the germinal vesicle in the oocyte. Small yolk platelets are found in the animal hemisphere. B. Fertilized egg at 0.9, animal pole at the top and sperm entry point on the left. Note the general movement of platelets from the animal hemisphere, giving a sharp boundary of the animal and vegetal hemispheres at the equatorial level. There remains a trail of platelets on the right side, the prospective dorsal side, that is, the region of the gray crescent. A slight downward shifting of platelets has occurred on the left side, the prospective ventral side.

The inversion experiments of Pasteels [1941] give interesting information about the distribution of determinants in the hemispheres of the newly fertilized egg. Pasteels centrifuged the dense vegetal components of that hemisphere into the pigmented animal hemisphere, at the same time displacing the animal hemisphere endoplasm into the nonpigmented vegetal hemisphere. In this way, he reversed the arrangement of the endoplasmic components of the animal-vegetal axis relative to the cortical components, at least as reflected by the location of pigment granules. These eggs developed well, and derived their ectodermal structures from the light hemisphere and their endodermal structures from the surface of the pigmented hemisphere, the exact reverse of normal. Thus, Pasteels could conclude that at this stage the egg does not have determinants localized in its cortex, whereas moveable endoplasmic components of the animal-vegetal axis invariably dictate the developmental fate of germ layers. From this experiment, though, it is not possible to say whether determinants reside in both the animal and vegetal endoplasm of the egg, or just in one of them.

The origin of the animal-vegetal axis in oogenesis has long fascinated embryologists, and C.M. Child early this century proposed that the naive

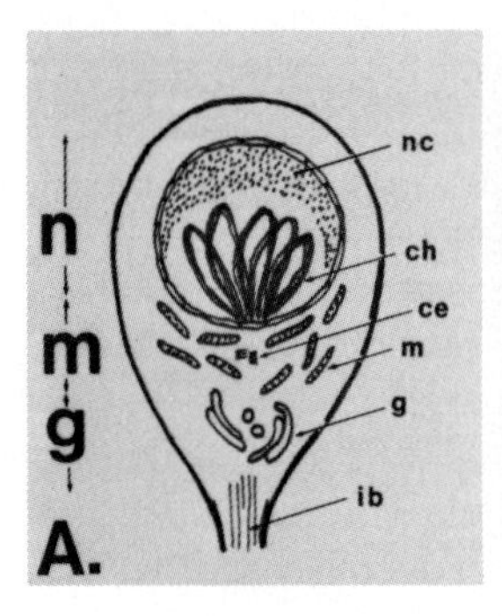

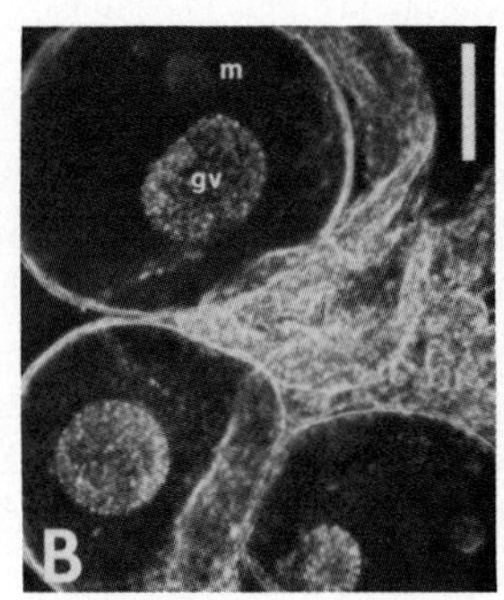

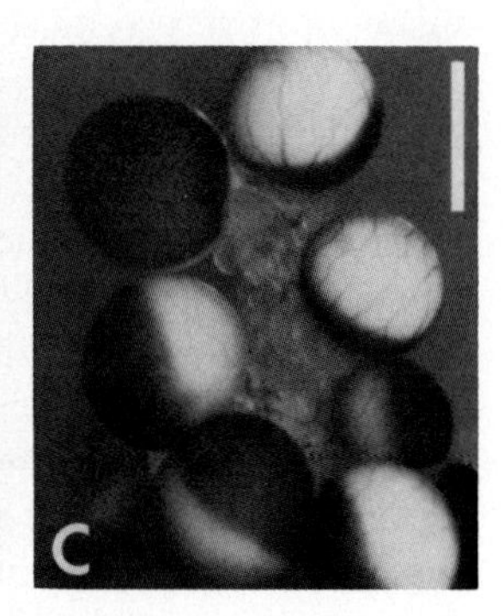

Fig. 4. Polarity of the oocytes of *X. laevis*. The pachytene stage (diameter approximately 25 μm), summarized from Tourte et al. [1981], Coggins [1973], and Billet and Adam [1976]. The axis of the cell is divided into three zones: nuclear zone (n), mitochondrial zone (m), and Golgi apparatus zone (g). The nucleus contains a nucleolar cap (nc) which at this stage consists of rDNA and partially formed nucleoli, and contains meiotic prophase chromosomes (ch) in the bouquet configuration. The mitochondria and Golgi apparatus are oriented around a centriole pair (ce) beneath the nucleus. At the base of the cell is an intercellular bridge (ib) with remnant microtubules from the last mitotic division of the oogonium. B. Late previtellogenic oocytes (bar length: 100 μm). The nucleus, also called the germinal vesicle (gv), contains approximately 1,500 nucleoli. The mitochondrial mass (m) is still visible, although it has moved away from the nucleus and will soon disperse. Follicle cells surround each oocyte, although these are not visible here. C. Vitellogenic oocytes (bar length: 1 mm). A portion of ovary was removed from an adult frog; the small clear oocytes are previtellogenic and equivalent to those shown in B. Note the extensive capillary system covering each large oocyte, through which vitellogenin arrives from the liver, to be pinocytosed by the oocyte. Note the light-pigmented vegetal hemisphere and the dark-pigmented animal hemisphere. The animal-vegetal axis of the oocyte connects the poles of these hemispheres. The equator is defined as the interface of these hemispheres.

oocyte gains its polarity from the asymmetric environment of its follicle. A.W. Bellamy [1919] reported, in fact, that *Rana pipiens* oocytes are oriented in the follicle with the pigmented animal hemisphere toward the single stalk through which capillaries bring oxygen and nutrients (as shown in Fig. 4C), as if the more vital animal hemisphere arises in the more stimulating environment. However, several researchers since then [Van Gansen and Weber, 1972; Brønsted and Meijer, 1950] have found that oocytes are not aligned in the follicle relative to the stalk, and also are random with respect to the frog's body axis and to the gravitational vector; therefore the notion of the environmental induction of polarity has lost support.

The alternate idea is that animal-vegetal polarity of the egg originates from prior internal structure of the oogonium or earliest oocyte, particularly as an elaboration of the nucleus-centrosome axis characterizing not only the oogonium, but also most metazoan cell types. Coggins [1973], Tourte et al. [1981], Billet and Adam [1976], and others examined the earliest stages of amphibian oogenesis and found a remarkable polarity already present, as schematized in Figure 4A. At early pachytene stage, the cell is tapered, connected at the base to one or more of its 16 clonal neighbors by an incomplete division bridge, and contains (1) a nucleus at the wide end, (2) a mass of proliferating mitochondria surrounding a centrosome (a centriole pair and microtubule organizing center) at the midsection, and (3) an extensive Golgi system in the narrow part, near the bridge. Even the nuclear contents are polarized, with the replicating rDNA at one end and the bouquet-stage chromosomes oriented toward the centrosome at the other end. At later times, the intercellular bridges seal off, and follicle cells surround the individual oocytes.

This polarity of the early oocyte may be amplified directly into the animal-vegetal axis of the egg; but since this relationship has not yet been demonstrated, we can only review speculations about a few main steps of the process. First, the elaboration of the vegetal aspect of the egg's axis concerns massive yolk platelet formation and localization, a process of "vitellogenesis" accompanying the enlargement of the oocyte from approximately 300 μm to 1.3 mm in diameter in the case of *X. laevis*. As Wallace, Ansari, and others have shown [reviewed by Wallace and Bergink, 1974], yolk proteins are synthesized in the liver as a large precursor (vitellogenin) and circulated to the oocyte where they are actively internalized by pinocytosis into membrane-bounded vesicles. These vesicles fuse to each other and to extant platelets; then the precursor is cleaved proteolytically to lipovitellin and phosvitin, which precipitate as a crystalline complex in the platelet. The amount of membrane surrounding the platelets of the full-grown oocyte exceeds by roughly 100-fold the amount of membrane in the plasmalemma from which it derives during pinocytosis. Thus, the oocyte engages in a high level of internalization and replacement of the plasma membrane during vitellogenesis.

With regard to the mechanism of localization of large, closely packed platelets in the vegetal hemisphere during vitellogenesis, there are two speculations. First, Wittek [1952] suggests that the nucleus of the oocyte releases a large halo of materials into the animal hemisphere, and as a consequence the platelets are displaced into the vegetal hemisphere. She also suggests (1) that the oldest platelets of the oocyte are the largest ones, due to the continuous addition of new contents and (2) that they occupy the centralmost position of the vegetal hemisphere of a large oocyte because younger, smaller platelets are added from the surface. While the nucleus is undoubtedly important in furnishing the animal hemisphere with materials such as ribosomes, the position and large size of platelets in the vegetal half may reflect the active organizing ability of that hemisphere itself—for example, based on the asymmetrically placed Golgi system of the oocyte. This system, which seems to occupy the centrosome region of other types of cells as well, is known to function in pinocytotic vesicle processing and in membrane recycling—the very activities of vitellogenesis [Farquhar and Palade, 1981]. In the movement of vesicles to various parts of the oocyte, the centrosomal aster may provide direction, and the functional coupling of microtubular organelles and vesicle-membrane systems is suggested by the finding that colchicine inhibits vitellogenin uptake very effectively in the oocyte [Wallace and Ho, 1972].

Second, the asymmetric location of the Golgi system might explain the animal-vegetal difference in pigmentation of the cortex. It is known that pigment granules form and embed in the cortex of small oocytes during a short period at the start of vitellogenesis, and are not formed thereafter [Dumont, 1972]. At first, pigment is distributed over the entire surface of the small oocyte, and only later does a light hemisphere appear gradually as the vitellogenic oocyte enlarges. Now, if the Golgi terminus of the cell were a site of active plasma membrane insertion, then a new pigmentless surface might originate from one pole, while the old and less expanding surface might gradually recede with all the pigment granules. Thus, although the facts are lacking, a plausible scheme can be made for the genesis of the specialized polarity of the egg from a set of inherent, general, polarized cell functions. Fortunately, some of these possibilities for the genesis of polarity can be tested directly now that Wallace and his colleagues have succeeded in culturing vitellogenic oocytes in vitro [Wallace et al., 1981].

NEW POLARITY AFTER FERTILIZATION

A single sperm enters the frog egg at a position of random contact in the animal hemisphere, and a small dark spot of accumulated pigment identifies the point of entry (Fig. 1A). We have tested the topographic relation of the sperm entry point (SEP) to the eventual position of the embryonic axis, and find as expected that dorsal axial structures originate from equatorial regions of the egg most distant from the SEP. Reciprocally, ventral embryonic structures originate from the side of sperm entry [Kirschner et al., 1980; Gerhart et al., 1981]. Since the sperm can enter anywhere in the animal hemisphere, we conclude that the entire equatorial circumference has the potential to produce dorsal or ventral structures and that "determination" in one direction or the other depends on the sperm. In this sense, the sperm can be considered a source of determinants, although one would have to define this term in a broad sense. It is thought, but not proven, that the centrosome or centriole of the sperm is the actual organelle indicating sperm location in the egg, and relative to which the dorsal-ventral aspects of the fate map are oriented [Manes and Barbieri, 1977]. The ultimate test of the centrosome as a determinant of the orientation of the fate map could now be made with localized injections of purified centrioles (Mitcheson and Kirschner, unpublished), which are known to support parthenogenesis [Heidemann and Kirschner, 1975; Maller et al., 1976].

Historically, much more emphasis has been placed on the gray crescent than on the sperm centrosome as an axial determinant, because: (1) the gray crescent appears at the equatorial region from which the blastopore lip will first appear, from which the Spemann organizer derives, and from which dorsal axial structures will arise [Elinson, 1980], whereas the SEP correlates with ventral nonaxial structures; (2) egg fragments thought to contain the gray crescent can develop both dorsal and ventral embryonic structures whereas the noncrescent (SEP) portions develop only ventral structures [Spemann, 1902; Ruud, 1925; Fankhauser, 1948]; (3) the crescent is enduring in its location, whereas the centrosome migrates, replicates, and partitions to all cells; and (4) the centrosome is not usually thought of as an agent controlling differential gene expression, whereas the gray crescent might be. The gray crescent forms at a specific time after fertilization, namely, in the period from one half to three quarters of the interval from fertilization to first cleavage (abbreviated as 0.5–0.75 on a normalized scale), as shown in Figure 5 [Manes and Elinson, 1980]. In Figure 1B an *X. laevis* egg at the four-cell stage is shown when the pigmentation differences related to the gray crescent are particularly clear. Since the gray crescent does regularly form in response to the position of sperm entry, it is assumed that the sperm centrosome in some way controls crescent formation. For example, in the period from 0.15 to 0.5, the centrosome organizes a large aster which moves toward the center of the egg, extending microtubules across the egg diameter [Subtelney and Bradt, 1963; Stewart-Savage and Grey, 1982], and perhaps contacting the cortex of the far side. However, the centrosome is probably not required for gray crescent formation since artifically activated eggs, lacking a sperm centrosome, nonetheless derive a normal-appearing gray crescent at approximately the correct time, though at an unpredictable locus of the equator. Thus, one is left with the impression that the egg itself is scheduled to form a gray crescent at a certain time, and that the sperm centrosome acts only as a transient spatial cue for positioning the crescent. We will ask later whether other determinants in the egg also act only as spatial cues.

Dalcq and Pasteels [1937] have outdone other embryologists in interpreting the role of the gray crescent in the determination of dorsal and axial development [reviewed by Pasteels, 1964]. They devised a double gradient hypothesis, of which one gradient originated from the gray crescent and declined across the egg surface whereas the other came from the yolk mass

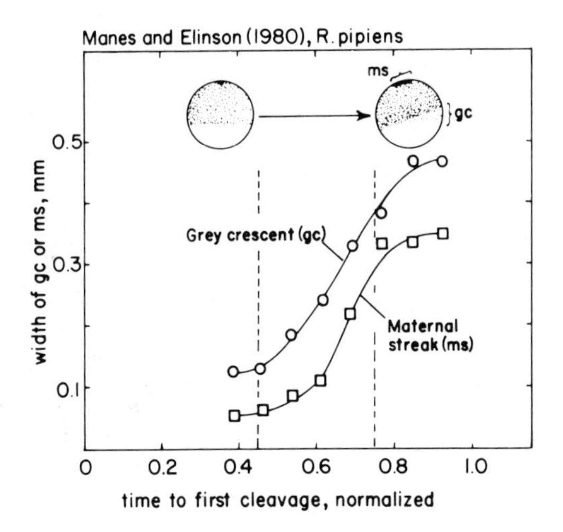

Fig. 5. Time of formation of the gray crescent in *Rana pipiens* [adapted from Manes and Elinson, 1980]. The width of the gray crescent (abbreviated gc) was measured in approximately 20 eggs at various times, indicated on a normalized decimal scale, with fertilization at 0.0 and first cleavage at 1.0. First cleavage occurred at approximately 150 minutes. Movement of the polar body spot at the animal pole was also recorded; the spot elongates into a streak, the "maternal streak," abbreviated ms. Note that the equatorial movement of the pigment at the gray crescent is approximately equal in magnitude to the movement in the maternal streak, as if the whole cortex moves relative to deeper materials.

of the vegetal hemisphere and declined vertically along the animal-vegetal axis. The first was a cortical gradient of one kind of material (abbreviated C for cortical), and the second was an endoplasmic gradient of another kind of material (abbreviated V for vegetal or vitelline). Every point of the bilateral sides of the egg surface received a unique amount of the two materials, as shown in Figure 6A. Then, as cells were cleaved from the egg material, each cell performed two computations with its gradient values: one gave a product ($C \times V$), defining the anterior-posterior level of the differentiation it undertook. For example, the highest value of $C \times V$ would become the dorsal lip of the blastopore and the region of head mesoderm. The second was a ratio of C/V defining the dorsal-lateral-ventral level of differentiation, as indicated in Figure 6B. Without dwelling on details, we can say that Dalcq and Pasteels considered it likely that extensive final determination takes place at the time of gray crescent formation—that is, even before first cleavage, based on the quantitative uniqueness to each egg region. Subsequent steps of development would merely express this complex pattern. Although Dalcq and Pasteels could avoid invoking a great variety of *kinds* of determinants, they still had

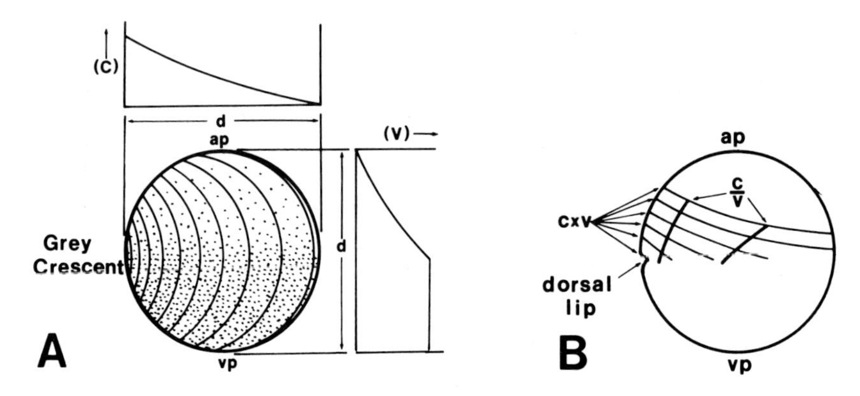

Fig. 6. The double gradient interpretation of the amphibian fate map, as proposed by Dalcq and Pasteels [1937]. A. The egg is shown shortly after gray crescent formation, with the animal pole (ap) at the top, the vegetal pole (vp) at the bottom, and the crescent on the left. The crescent is the center of a gradient of material C located in the egg cortex. The gradient attenuates from left to right across the egg, and this attenuation is plotted at the top as concentration of C vs. distance (d) across the egg. A second gradient is generated in the endoplasm in relation to the yolk mass and is composed of material V (for vitellin or vegetal). This gradient attenuates vertically, as shown on the right where the concentration of (V) is plotted vs. distance(d) along the animal-vegetal axis of the egg. B. The egg is oriented as in A. The family of lines across the egg are contour lines of various values of the multiple C × V; the closer the line is to the animal pole, the lower is the value of the multiple. On the left the C term dominates in the multiple and on the right the V term dominates; therefore as one follows the contour in the right-hand direction, it comes closer to the vegetal hemisphere where V is most concentrated. The multiple C × V is thought to represent a product dependence in a bimolecular reaction; the product was called "organisine" by Dalcq and Pasteels. At the highest C × V value, the blastopore first forms and dorsal midline structures originate. A second family of thicker and more vertical lines are contour lines of values of the ratio C/V, the ratio perhaps representing opposing reactions. The values of the contours decrease with lines ever more to the right. Note how the C × V contours and C/V contours cross one another in an array roughly approximating the amphibian fate map, where the leftmost stack of zones would represent anterior to posterior regions of chordamesoderm (both prechordal plate and notochord), the middle stack of zones would represent anterior to posterior somites, and the rightmost stack would be lateral plate mesoderm.

to invoke spatially precise quantitative detail in order to generate a complete fate map in one step. In that sense, they may have created a numerical homunculus. Their evidence for a cortical gradient came largely from (1) their inability to move the site from which dorsal and axial structures develop in eggs centrifuged after the time of gray crescent formation, and (2) from the reasonable assumption that unmovable materials must be cortical materials.

The process of gray crescent formation affects endoplasmic components of the egg in addition to those of the cortex, as shown by the cytology of Klag and Ubbels [1975] on *Discsoglossus* eggs. As shown in Figure 2B, the animal hemisphere layer of nuclear sap is brought into contact with the most vegetal layer of large yolk platelets on the gray crescent side, whereas the opposite side retains the endoplasmic organization of the unfertilized egg. In *Xenopus* eggs, there are related but less clear endoplasmic movements in the first cell cycle, as shown in Figure 3B. Thus, the period of gray crescent formation is a period of widespread reorganization of both the cortical and endoplasmic contents of the egg. The actual mechanism of reorganization is not known, but two published speculations are (1) that the entire cortical layer of the egg rotates relative to the endoplasm, moving up on the gray crescent side and down on the sperm entry side, or (2) that the cortical layer contracts toward the sperm entry point, pulling away from the gray crescent side [Elinson, 1980]. By either view, the gray crescent forms where pigmented cortex is withdrawn from the equator on one side, leaving a less-pigmented zone. It seems plausible that actin-dependent contractile forces are generated between the cortex and endoplasm, to accomplish a unitary and directional displacement of one compartment relative to the other. Since colchicine blocks gray crescent formation in *R. pipiens,* microtubules may also be involved [Manes et al., 1978]. The direction of this displacement is somehow biased by the sperm centrosome, acting as a spatial cue. Whatever the details of mechanism, the effect is a regional cytoplasmic differentiation of the egg's contents at the equatorial level, to give a new polarity not possessed by the oocyte. While it is true that a new polarity is in principle discernible as soon as the sperm enters the egg at one point (not the animal pole), regional cytoplasmic differences probably remain small until the major shifts of the period of gray crescent formation. When reorganization is complete, important determinants for axial development have probably been brought into place on the gray crescent side or have been activated locally on that side. These early steps are summarized in Figure 7, parts 1 and 2. Subsequent steps will be discussed later.

EXPERIMENTAL TESTS OF DETERMINANTS IN THE FIRST CELL CYCLE

We will summarize here experiments on blocking the egg's ability to localize determinants of axial development and on artificial means to direct the localization of these determinants.

(1) Fertilized eggs exposed briefly to cold shock (1°C, four minutes), hydrostatic pressure (8,000 psi, four minutes) or UV-irradiation prior to 0.8 of the first cell cycle give apparently normal late blastulae, which gastrulate symmetrically instead of starting with a dorsal lip, and which fail to neurulate

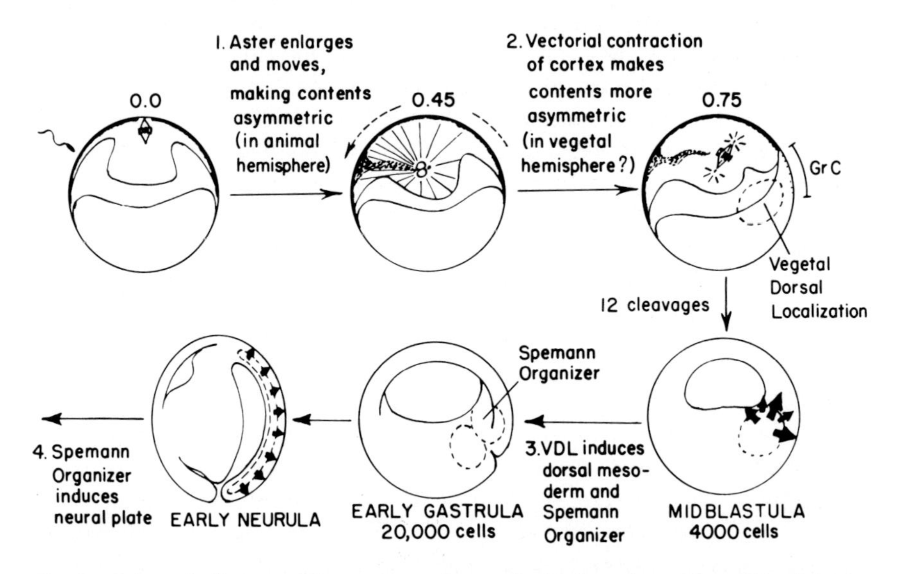

Fig. 7. Schematic diagram of the sequence of events in the formation of the embryonic body axis in anurans. The scheme is largely speculative, based on interpretations of several embryologists, as referenced in the text. In the upper left-hand portion of the figure the radially symmetric unfertilized egg at the time of fertilization is shown, with the sperm entering in the animal hemisphere randomly at one point on the left. In the period until 0.45 (45% of the time interval to first cleavage), the sperm centrosome (the centriole and microtubule organizing center [MTOC]) forms a large aster and moves to the center of the egg, coordinating the migration of the sperm and egg pronuclei, which complete DNA synthesis and meet at approximately 0.45. It is suggested that aster formation and movement from one side of the egg cause a slight reorganization of the animal hemisphere contents and perhaps a slight regional difference in the build-up of the animal hemisphere cortex (step 1). In the period 0.45–0.75, the cortex shifts relative to the deep contents of the egg by a rotation or directional contraction toward the sperm entry side, perhaps driven by microfilaments of the cortex. In the course of this displacement, the gray crescent emerges as a less-pigmented region of the cortex. Deep contents adhering to the cortex are drawn up on the gray crescent side. It is suggested that this cortex-driven reorganization of the deep contents activates certain "dorsalizing determinants" in the vegetal hemisphere, just on one side, to give a "vegetal dorsalizing localization," abbreviated VDL (step 2). This region of cytoplasm is inherited by vegetal cells arising in the next 12 cleavages. At some time in the early cleavages, the vegetal cells induce adjacent animal hemisphere cells to become prospective mesoderm, with the VDL inducing the prospective dorsal mesoderm of the Spemann organizer region (step 3). Then, during gastrulation, the Spemann organizer cells (prospective dorsal mesoderm) migrate up the blastocoel wall and induce the neural plate (step 4), which in subsequent steps will form definitive dorsal parts of the nervous systems.

[Grant and Wacaster, 1972; Malacinski et al., 1977; Scharf and Gerhart, 1980, 1983]. The embryos develop a simple three-layered anatomy with only a few cell types such as ciliated epidermis, red blood cells, and gut cells. They lack any trace of a body axis and seem to derive their organization directly from the animal-vegetal organization of the oocyte. They resemble strikingly the "ventral embryos" or "belly pieces" produced by Fankhauser [1948] and Spemann [1902] from ligated crescent-deficient fragments of newt eggs (see comparison in Fig. 8A–E). We think these embryos represent "limit forms" which have failed to reorganize their cytoplasmic contents and therefore have failed to localize or to activate regionally the agents needed to originate axial and dorsal development. These treatments probably inactivate *not* the determinants themselves, but the *process* by which they are localized or locally activated. Related to this interpretation, for example, cold and pressure are known to interrupt microtubule-dependent processes by depolymerizing the tubules to monomers. Furthermore, Manes and Elinson [1980] report that UV-irradiation blocks gray crescent formation in *R. pipiens* eggs, and Scharf (unpublished) has observed a reduced movement of pigment granules in irradiated *X. laevis* eggs.

(2) The sensitive period of eggs to these treatments ends by the time 0.8 [Scharf and Gerhart, 1980, 1983; Youn and Malacinski, 1980]. The sharp transition to resistance suggests that the egg really does accomplish an important event in axis determination in a critical period prior to 0.8, whereas previous morphological observations of crescent formation and of internal reorganization could only imply but not assess the importance of the change. However, we can't conclude from these results that axis-determining processes are in any way final by this time, since the sensitive process affected here may be just the first of a series of coupled processes occurring over many stages. In fact, we favor this view, as discussed later. The beginning of the critical period is more difficult to define, but probably occurs at approximately 0.4, since this is the time when cold and pressure can first affect axis determination [Scharf and Gerhart, 1983]. It is noteworthy that this sensitive period of 0.4–0.8 corresponds well with the observed period of gray crescent formation and internal reorganization (Fig. 5). Apparently the accomplishments of the egg before 0.4, as the centrosome organizes an aster, are not sufficient to launch axial development on their own.

(3) Treatment of the egg with D_2O, an agent known to stabilize microtubules, protects eggs from the axis-impairing effects of UV and cold. This is further circumstantial evidence that the reorganization mechanism requires microtubules. But futhermore, D_2O on its own causes excessive development of anterior axial structures (Scharf, unpublished). In extreme cases, the embryo has a radial sucker and extra eyes, as shown in Figure 8F and G. Thus, it may be possible for the reorganization mechanism of the egg to "overdo" the localization or activation of determinants.

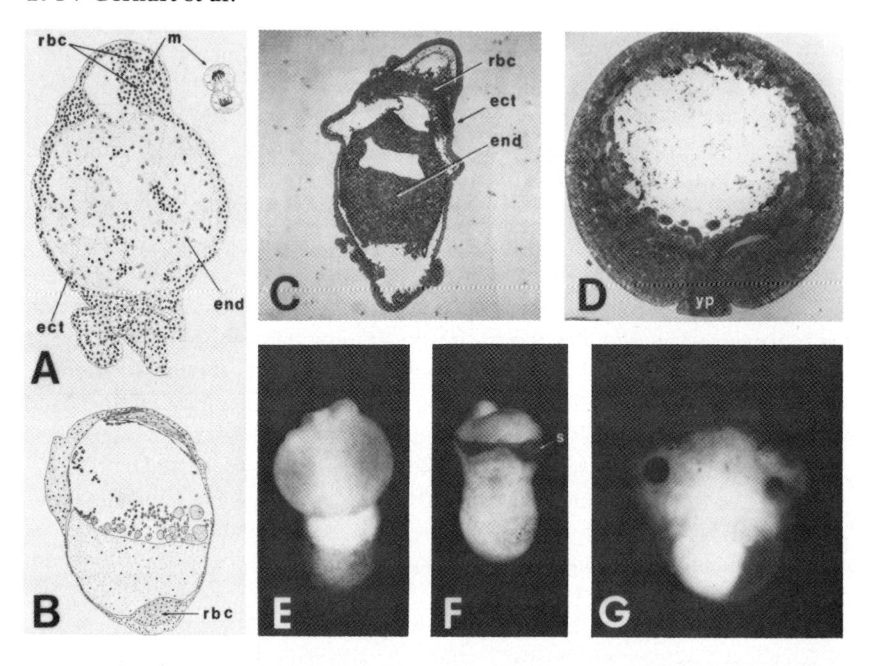

Fig. 8. Experimental inhibition and exaggeration of axis determination in the first cell cycle of the fertilized egg of *X. laevis*. A. Drawing of a section of a fixed, stained "belly piece" obtained by Spemann [1902] from a blastomere removed at the time of first cleavage and allowed to develop for a time sufficient for the heartbeat stage to be reached by controls. Spemann speculated that the blastomere lacked the gray crescent, although this could not be seen by him in the newt species used for these experiments. The animal pole is toward the top. rbc, red blood cells; end, endodermal mass of cells; ect, ectodermal surface sheet of cells; m, mitotic cell, shown at two magnifications. B. Drawing of a section of a "ventral embryo" obtained by Frankhauser [1948] from a fragment ligated from the egg before first cleavage. The gray crescent of this newt species was also not visible. In this case, the animal pole is toward the top and the red blood cells are toward the bottom. C. Section of a grade 5 embryo obtained by Scharf and Gerhart [1983] from an egg of *X. laevis* cold shocked (0°C, four minutes) at 0.62 in the first cell cycle, and allowed to develop for a time sufficient for control embryos to reach stage 33 (heartbeat stage). The animal pole is toward the top and the closed blastopore toward the bottom. The abbreviations are the same as in A. D. Cross section of a midgastrula of a prospective grade 5 embryo, prepared as described for C. The blastopore is toward the bottom, with a small yolk plug (yp) protruding. Note the radial symmetry, with left and right slices of the circular blastopore lips looking identical and both like ventral lips of a normal blastopore. The liquid space is the blastocoel. The archenteron is small and circular, just adjacent to the lips. E. External view of a grade 5 embryo, similar to the one sectioned for C. Length approximately 2 mm. F. A grade "minus 5" embryo obtained from an egg exposed to 70% D_2O for ten minutes starting at 0.25 (Scharf, unpublished) and allowed to develop for a time sufficient for controls to reach stage 33. Notice the sucker which has formed around the entire circumference; this is a very anterior structure. This embryo had produced a circular blastopore lip precociously and an exaggerated extension of involuting material. G. An embryo equivalent to that of F, allowed to develop further. Notice the eyes and head structures. These actually comprise part of a janus twin, from which trunk and tail structures are lacking.

(4) Partial treatment of the egg with cold, pressure, or UV in the sensitive period leads to partial loss of the embryonic body axis, in a regular anterior-posterior progression. That is, low doses cause a truncation of the axis at the head level (trunk and tail still being normal), medium doses cause truncation at the trunk level, and high doses at the tail level [Scharf and Gerhart, 1980, 1983]. Thus, we think the quantity of reorganization is related to the quantity of localization of determinants or to the quantity of local activation of them. As a speculation, we suggest that the extent of the egg's dislocation of cortical constituents relative to endoplasmic ones might have "cell-biological" consequences for the efficiency with which the cellularized units of that cytoplasmic region later carry out various processes such as exocytosis, endocytosis, and membrane recycling, all off which require cortical-endoplasmic interactions.

(5) Artificial rearrangement of the egg's contents can be achieved by gravity or centrifugal force (e.g., 10g, four minutes) acting at an angle to the animal-vegetal axis, as shown by Born [1885], Schultze [1898], Penners and Schleip [1928a,b], Ancel and Vintemberger [1948, 1949], and Pasteels [1938, 1946, 1948], and reviewed by Gerhart [1980]. For example, the egg can be held 90° off-axis so that the equator is vertical, whereupon the dense yolk mass of the vegetal hemisphere, driven by gravitational force, slips downward relative to the cortex. This artificial reorganization works amazingly well in causing the localization or regional activation of axial determinants, as demonstrated by the rescue experiment of Scharf and Gerhart [1980, 1983]: UV-irradiated, or cold or pressure treated eggs are obliquely oriented (90° off-axis) for 60 minutes after the impairment and are found to develop to completely normal tadpoles, whereas they otherwise would develop only as "ventral embryos" lacking body axes. When impaired eggs are given shorter periods of forced reorganization, such as 15 or 30 minutes, the degree of rescue is partial and dose dependent, progressing in a posterior to anterior direction, the reverse of the impairment progression. Tail structures are rescued first, then trunk, and finally head. Thus, we think the artificial rearrangement of the egg's contents by gravity can completely substitute for the endogenous "natural" rearrangements driven by contractile mechanisms of the cortex.

As perceived by Pasteels [1946] and diagrammed in Figure 9, the artificial and natural processes have much in common as far as the rearrangements they cause, namely, in both cases a dislocation of the cortex relative to the endoplasm. Thus, the egg seems to have a definite subcellular organization which specifies the type of reorganization it can undergo, whereas the particular force used to drive that reorganization can be as "unstructured" as gravity itself. In this regard, we consider it plausible that the cortical contraction used normally by the egg to move its contents is itself not spatially precise in the pattern of forces it generates, but just provides a sufficient force and "evokes" the egg's precise response.

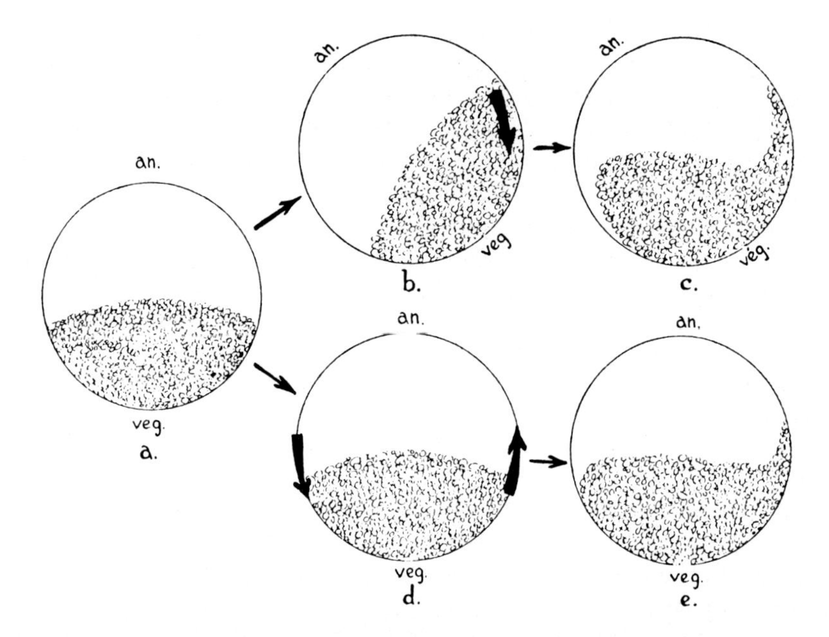

Fig. 9. Interpretation of the effects of oblique orientation of the fertilized, uncleaved egg on alignment of the prospective dorsal-ventral axis. Modified slightly from Pasteels [1948]. In (a) an egg with radially symmetric contents of the animal and vegetal hemispheres is shown. In (d) and (e), the lower portion of the figure, the cortex moves relative to the deep contents, in an upward direction on the right side where the gray crescent is formed; deep contents on the right side are dragged upward near the cortex. This is the prospective dorsal side. In (b) and (c), the egg is tipped 45° off the vertical axis so that the dense vegetal yolk mass is no longer in gravitational equilibrium. These dense materials start to move downward, driven by gravity, leaving a layer of yolk platelets adsorbed to the cortex on the uppermost region of the animal-vegetal interface. This will become the embryonic dorsal region. Notice that the reorganization of the internal materials of the egg looks rather equivalent whether the cortex moves relative to the deep materials, perhaps driven by actomyosin contractility [(d) and (e)], or whether the deep contents move relative to the cortex, driven by gravity [(b) and (c)]. Thus, the egg contents reorganize in a prescribed way, irrespective of the motive force, presumably due to the prior organization of the egg cell—for example, according to the depth to which deep contents adhere to the cortex.

(6) Artificial rearrangement of the egg's contents in an experimentally controlled direction can completely override any axis-orienting influences of the sperm. For example, eggs held off axis (90°) with the SEP side uppermost for 20 minutes form dorsal and axial structures on the *same* side as the SEP, not on the opposite side, as is normally the case [Kirschner et al., 1980; Gerhart et al., 1981]. In fact, whatever region of the equator is uppermost for this brief period becomes the site of dorsal and axial development. Thus,

the position of the SEP can be experimentally uncoupled from the orientation of the axis, indicating that the SEP does not provide fixed, required, irreplaceable determinants. Brief lateral centrifugation (10g, four minutes) has the same effect. The egg is most sensitive to experimental control of its axis position in the 0.4–0.8 period, and by 0.9 becomes quite reistant to 1g or 10g [Black and Gerhart, 1983a]. However, even at 0.9, some eggs reverse their axis orientation at forces of 30g or 50g. This shows that even the position of the gray crescent can be experimentally uncoupled from the final orientation of the axis in contrast to the expectations of Dalcq and Pasteels [1937]. Resistance to forced rearrangement seems to reflect cell-cycle dependent changes in cytoplasmic architecture, and not the creation of an unmovable cortical locus of dorsal determinants. By 0.8 the egg has entered metaphase, and it is noteworthy that during its previous metaphase in the meiotic unfertilized egg, the endoplasmic contents had also resisted movement. We think (1) that the crescent simply reflects the direction of cortical displacement but is not itself an important product of the displacement and (2) that the endoplasmic contents moved normally by the cortex or artificially by centrifugal or gravitational force are the seat of determinants needed for axial development.

(7) Forced reorganization of the egg by gravity or centrifugation provokes twinning at high frequencies, as shown in Figure 10. For example, eggs are centrifuged laterally at 30g for four minutes at 0.4 and then at a later time centrifuged in the opposite direction; these often give 80%–100% twins with complete, opposed body axes oriented according to the centrifugal force vectors [Black and Gerhart, 1983b]. The second centrifugation evokes the second axis and is effective even at times after first cleavage. This is evidence that our artificial means of localizing determinants or activating determinants can be accepted by the egg long after its own reorganization process is over. As related evidence, UV-irradiated, cold, or pressure-impaired eggs can be rescued by centrifugation well after first cleavage [Scharf and Gerhart, 1980]. Presumably the egg confines its own reorganization to the 0.4–0.8 period of the first cell cycle simply because that is the period when its cortical contractile activity is scheduled and when its cytoplasm is most movable [Gerhart et al., 1983], and not because that is the only period of receptivity to new polarities.

We were surprised at the ease and completeness of twinning and find this hard to explain by a strict localization scheme where uniformly distributed axial determinants become entirely concentrated in one equatorial region of the egg. If this were the case, the second centrifugation would have to move half the determinants from the original concentrated locus and put them on the opposite side. Also, since axes always arise from the *centripetal* side of the egg, we would have to say axial determinants are less dense than the

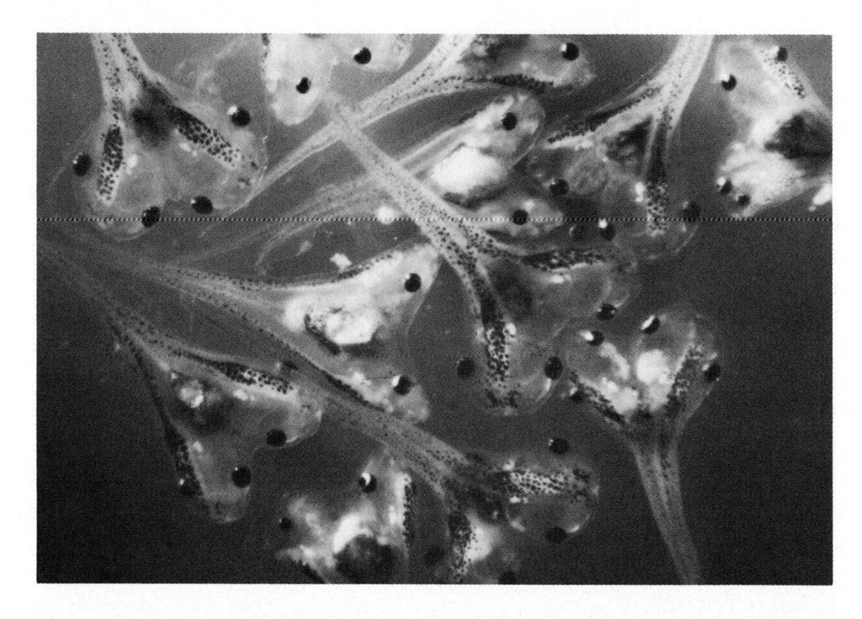

Fig. 10. Twins produced by centrifugation of eggs. Fertilized, dejellied eggs were embedded in gelatin [Black and Gerhart, 1983b] and centrifuged at 30g for four minutes with the centrifugal force directed 90° to the animal-vegetal axis. The dish of eggs was removed from the rotor and placed flat in the incubator, in which position gravity was acting to move the egg contents back to their original positions. Maximum twinning was found for eggs centrifuged at 0.4 in the interval from fertilization to first cleavage. At the late neurula stage, embryos were removed from gelatin by placing the dish in 32°C water for three minutes to melt the gelatin, at which time embryos were quickly pipetted into 20% amphibian Ringer solution at 22°C, in which they were subsequently raised. In this batch, 100% twins was obtained. Note the complete double axes which join in some cases at the posterior trunk or tail level. All twins share ventral structures.

cytosol and float when centrifuged. Instead, we prefer to think that natural and artificial reorganization of the egg's contents causes, not localization, but localized *activation* of uniformly distributed determinants. Thus, when the first centrifugation activates determinants on one side, it does not deplete them elsewhere; thereafter, the second centrifugation can activate them on the opposite site.

Also, the ease of twinning by gravity led us to doubt the validity of the cortical transplantation results of Curtis [1960, 1962], the main evidence for cortical determinants of axis formation. We think it possible that the operation conditions, involving holding eggs in off-axis positions, provoked twinning, and that the essentiality of the gray crescent graft for the effect was not proven by adequate controls [Gerhart et al., 1981].

In summary, we may suggest the following picture of axis determination in the first cleavage period: (1) the egg has a force-generating mechanism by which it shifts its cortex in one direction relative to its endoplasm; this is scheduled to occur in the 0.4–0.8 period; (2) normally this mechanism is oriented spatially by some influence from the sperm aster, but without that influence it will occur anyway, although not in a predictable direction; (3) the dislocation of the cortex achieves an activation of determinants in some region of the endoplasm (to be discussed later); (4) artificial forced dislocation, by gravity or centrifugation, of the endoplasm relative to the cortex can also activate determinants; (5) determinants can be activated at times outside the 0.4–0.8 period, by forced shifts; and (6) we do not know if the activation occurs at the same time as the shift or at a later time.

DETERMINANTS IN THE BLASTULA

Cytoplasmic materials of the vegetal hemisphere probably continue long-distance rearrangements until the 64–128-cell stage when paratangential furrows finally divide the vegetal hemisphere in a horizontal direction [see Ballard, 1955, for the movement of vegetal dye marks; and Züst and Dixon, 1975, for the movement of germ plasm granules]. These movements may or may not be important for axis determination. Shortly thereafter, at the 256–512-cell stage, equatorial blastomeres on the gray crescent side of the blastula acquire a new developmental capacity to generate dorsal mesodermal structures of the body axis when cultured for several days as small explants [Nakamura et al., 1970; Asashima, 1980]. Before that time, explanted blastomeres of this same region can only form ciliated epidermis. There is little information about this change of determination of dorsal equatorial cells, except what can be extrapolated from the results of Nieuwkoop and his colleagues [Nieuwkoop, 1973, 1977] concerning determinative interactions of cells in the midblastula (1,000–8,000 cells) stages. We will review these important findings and then return to the 256–512-cell stage.

Nieuwkoop removed the equatorial cells from a midblastula of *X. laevis* or *A. mexicanum* and tested the ability of the remaining animal hemisphere cap of cells, and the remaining vegetal hemisphere base of cells (prospective for subblastoporal endoderm), to form mesoderm autonomously in culture. Neither portion alone could do so, a confirmation of the fact that only the equatorial cells (now removed) have this potentiality at this stage. Then Nieuwkoop recombined the cap and base portions, suitably marked with ^{3}H-thymidine, and found that the healed combination *could* form mesodermal structures, all of which derived from the cells of the animal hemisphere cap. Many of the recombinates also formed well-proportioned body axes, including neural tube derivatives, implying that even the Spemann organizer region of dorsal mesoderm had originated from animal hemisphere cells after they had been in contact with vegetal base cells. When the cap was rotated 180° relative to the base, dorsal mesoderm arose in the animal hemisphere at positions specified by the dorsalmost cells of the vegetal hemisphere, not of the animal hemisphere. Thus, the vegetal cells must exert an instructive induction on the animal hemisphere cells, promoting them to form mesoderm of dorsal, lateral, and ventral types. Furthermore, fragments of the vegetal hemisphere caused different inductions: the most dorsal piece (that closest to the gray crescent) inducing animal hemisphere cells to differentiate dorsal mesodermal derivatives including notochord and somites (characteristic of the Spemann organizer mesoderm); whereas lateral pieces induced lateral mesodermal structures, such as pronephric tubules; and ventral pieces induced ventral mesoderm. Thus, at least by the midblastula stage, the prospective subblastoporal vegetal cells carry a rather mosaic pattern of inductive abilities, whereas the animal hemisphere cells appear to be indifferent, multipotent, and uniformly responsive to regional induction by their vegetal neighbors. It is plausible that this mechanism of induction operates not just in recombinates but also in the intact early or midblastula and controls the inevitable emergence of mesoderm at the equatorial interface of animal and vegetal hemispheres.

The implications of these results for our discussion of determinants are the following: (1) At the blastula stage, the vegetalmost cells are the ones with strong regional "determination" for controlling subsequent axial development—that is, they are the ones carrying localized or regionally activated determinants; (2) the vegetal cells on the gray crescent side comprise a "vegetal dorsalizing localization" that is later responsible for the induction of the Spemann organizer and axial dorsal mesoderm; (3) since the vegetal cells themselves do not become axial dorsal mesodermal or neural derivatives, but probably instead give rise only to ventral gut structures, it would be incorrect to think in a literal sense of *axial* determinants in the frog egg as ones entering directly the lineages of blastomeres destined for axial structures; these dorsal vegetal cells only provide a spatial cue for axis formation (in a sense the way the sperm centrosome provided a cue); and (4) mesoderm itself is patterned at the blastula stages not by prelocalized determinants but by an induced imprint of the regional pattern of the subjacent vegetal cells. These results of Nieuwkoop and his colleagues require a substantial departure from classical notions of axial determinants in amphibian eggs; true axial determinants probably do not exist in frog eggs, unless one says that the animal hemisphere as a whole receives "axial" determinants giving its cells "competence" to form axial mesoderm if they are lucky enough to be induced by vegetal cells. But this is a different use of the concept of determinants than is usually applied. Nonetheless, classical determinants still have a place in amphibian development in that the egg does regionalize something needed by vegetal cells for their ability to induce axial mesoderm in their neighbors.

More recently, Gimlich and Scharf (unpublished) have devised an important extension of the Nieuwkoop experiments, in an effort to link conceptually the determinative events of the first cell cycle to those of the midblastula. As shown in Figure 11, they took UV-irradiated eggs as hosts at the 64-cell stage

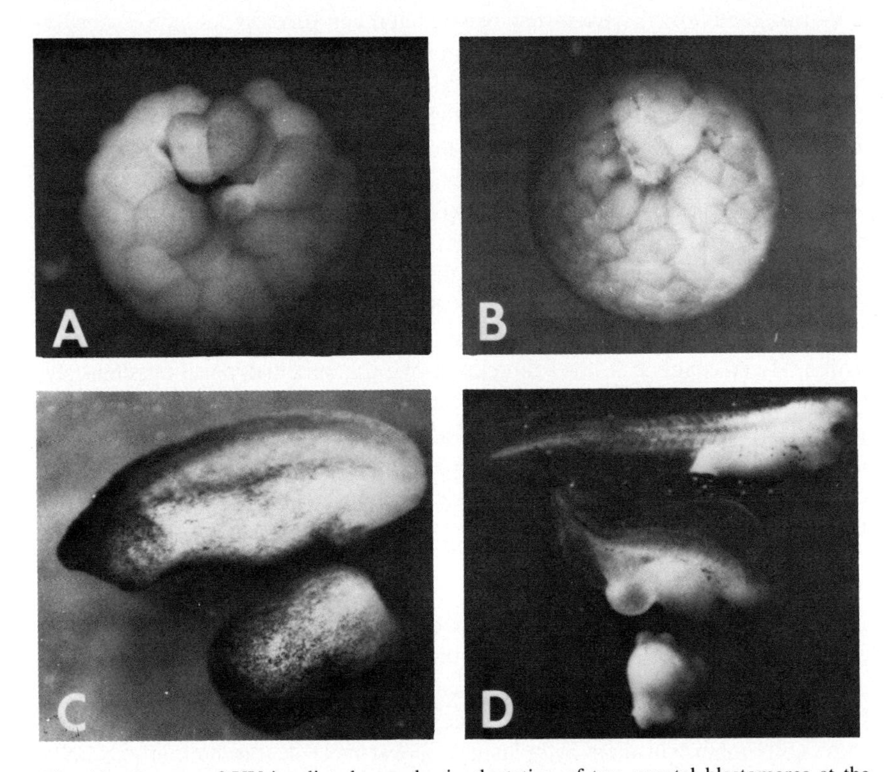

Fig. 11. Rescue of UV-irradiated eggs by implantation of two vegetal blastomeres at the 64-cell stage (Gimlich and Scharf, unpublished). A. The vegetal hemisphere of a 64-cell egg which had been irradiated at 0.4 of the first cell cycle. The vegetal pole is directly in the center. Two vegetal blastomeres have been removed and replaced with two corresponding blastomeres from an *un*irradiated control embryo. These occupy a prospective subblastoporal location. They are loosely settled in place, but not yet healed in. B. Approximately 30 minutes later than A; the implanted blastomeres have healed in and another cleavage has occurred. C. The lower embryo derives from an irradiated egg which has *not* received a blastomere replacement, whereas the upper embryo is from an irradiated egg which has. The upper one has neurulated and shows much more extension of a body axis, although the head region is somewhat reduced in size, indicating incomplete rescue. D. Three embryos of equivalent age; the upper one is almost fully rescued, representing approximately a grade 1 rescue from grade 5. The middle embryo shows partial rescue from grade 5 to grade 3. The lower embryo is a nontransplanted irradiated control, demonstrating an unrescued grade 5.

and performed blastomere exchanges with nonirradiated donors of the same age. This host was chosen because it cannot establish dorsal mesoderm and axial structures on its own, due to the UV-impairment of its localization mechanism during the first cell cycle, as discussed earlier. However, when two vegetal blastomeres are removed from the prospective subblastoporal level of the 64-cell stage and are replaced with two corresponding blastomeres taken from the prospective dorsal side of a nonirradiated control, the UV-irradiated host is extensively rescued and can produce an almost-normal embyro (Fig. 11D). The axial structures arise in the host on the side of the transplantation. Thus, these two vegetal dorsal blastomeres must contain the developmental information to set off the patterning mechanisms of the other 62 blatomeres from which essentially all of the dorsal and axial structures of the rescued embryo are developed.

Now we can return to the observation, cited at the beginning of this section, that equatorial blastomeres (the ones fated for mesoderm formation) only acquire at the 256–512-cell stage the capacity for autonomous mesoderm formation in small explants. That is to say, these blastomeres seem to become "determined" to form mesoderm at this stage. These blastomeres are located in the 64-cell stage at a level directly above the subblastoporal vegetal cells transplanted by Gimlich and Scharf. In putting these two experiments together, we propose that the vegetal cells begin inducing their overlying neighbors to acquire mesodermal determinaton at these extremely early cleavage stages, long before gastrulation when these neighbors finally express their mesodermal potentiality.

If induction does occur at the 256–512-cell stage, we have an interesting dilemma in that gene expression in *X. laevis* is detectable only at the 4,000-cell stage. Therefore, we cannot expect that determinants in these vegetal cells are affecting gene expression to make them inductively active. It is of course possible that these cells receive special mRNA molecules allowing them to produce unique proteins needed for activity. It is also possible that these cells inherit in their cytoplasm, not just mRNA, but certain organelle systems localized from the start of oogenesis in the vegetal part of the egg, and as a consequence, these cells precociously take on activities related to these systems. For example, since these cells originate from the most vegetal part of the egg cytoplasm, they may receive an extensive Golgi system of the sort occupying one end of the pachytene oocyte during oogenesis (Fig. 3A). Consequently, the vegetal cells might engage in Golgi-regulated activities of exocytosis, endocytosis, and membrane turnover before their less-endowed animal hemisphere counterparts are able to do so. The question remains, though, of how the vegetal blastomeres differ from one another in their inductive abilities in a dorsal-ventral dissection, and how the reorganization of the egg cytoplasm in the first cell cycle created this difference. We can only speculate that the unidirectional dislocation of cortex and endoplasm in the first cell cycle has local consequences for the operation of endocytotic-cytotic systems, since these involve both the cell cortex (surface) and endoplasm. Thus, the cell structure of the oocyte, when modified slightly in the egg, might dictate aspects of organismal structure at multicellular stages before gastrulation. Future experiments can identify the time of induction by transplanting "dorsal" vegetal blastomeres into UV-irradiated eggs at the 64-cell stage, and then explanting adjacent equatorial blastomeres at various times thereafter to establish when they gain the capacity for autonomous mesoderm differentiation. The grafted vegetal blastomeres can also be inhibited in various cell activities such as cell division, gene expression, endocytosis-exocytosis, or junction formation, to see what is needed for their inductive activity. In this way, we might clarify what cell functions of the vegetal blastomeres are affected by the determinants they received.

SUMMARY

With respect to the axial determinants of frog eggs, we may suggest that they depart greatly from the idealized description given in the beginning of this chapter:

(1) They probably do not pass directly from the egg cytoplasm to the lineages of axial precursor cells, but instead enter into neighboring vegetal cells which will become gut. This conclusion is fairly well supported by current experiments. Thus, axial determinants in the strict sense probably do not exist in amphibian eggs, although "axis orienting" determinants of the vegetal cells do exist.

(2) They may not be concentrated from a uniform distribution into one region of the egg by a localization mechanism, but might be already located throughout the vegetal hemisphere in an inactive state and then locally activated; this is a reasonable interpretation of experimental data but is not directly affirmed.

(3) They may not affect gene expression of the vegetal cells they enter, but may act at a more immediate cytoplasmic level of cellular function needed for cell interactions. This is conjecture at present and requires more study.

(4) The level of patterning accomplished by determinants in the frog egg may be very rudimentary and concern only the level of inductive activity of subblastoporal vegetal cells. All the rest of the pattern of the embryo, including axis formation, may originate from intercellular mechanisms triggered by inductions; this is a general impression, not a presently quantifiable fact.

(5) Patterning of axial structures is probably built up gradually through a series of coupled processes, each setting the temporal and spatial conditions of the next, and each exploiting different aspects of cell structure and function. Patterning probably does not occur in a single step—for example, in the process of gray crescent formation.

Of course, it should be kept in mind that this outline concerns the frog embryo, which as a vertebrate embryo, may emphasize cell interactions in patterning and determination, and deemphasize localized determinants of the egg. This may stand in contrast to the situation for various invertebrate and ascidian species.

ACKNOWLEDGMENTS

The research and preparation of this article were supported by USPHS grant GM 19363 to J.G.; S.B. and R.G. were NIH predoctoral fellows supported by training grants GM 07127 and GM 07232, respectively. S.S. was a University of California Regents predoctoral fellow.

REFERENCES

Ancel P, Vintemberger P (1948): Recherches sur le déterminisme de la symétrie bilaterale dans l'oeuf des Amphibiens. Bull Biol Fr Belg [Suppl] 31:1–182.

Ancel P, Vintemberger P (1949): La rotation de symétrisation, facteur de la polarisation dorso-ventrale des ébauches primordiales, dans l'oeuf des Amphibiens. Arch Anat Microsc Morphol Exp 38:167–183.

Asashima M (1980): Inducing effects of the grey crescent region of early developmental stages of Ambystoma mexicanum. W Rouxs Arch 188:123–126.

Ballard W (1955): Cortical ingression during cleavage of amphibian eggs, studied by means of vital dyes. J Exp Zool 129:77–97.

Bellamy AW (1919): Differential susceptibility as a basis for modification and control of early development in the frog. Biol Bull 37:312–361.

Billet FS, Adam E (1976): The structure of the miotchondrial cloud of Xenopus laevis oocytes. J Embryol Exp Morphol 36:697–710.

Black SD, Gerhart JC (1983a): Reversal of the dorsal-ventral axis by controlled centrifugation of Xenopus laevis eggs. (in preparation).

Black SD, Gerhart JC (1983b): High frequency twin embryos from centrifuged eggs of Xenopus laevis. (in preparation).

Born G (1885): Über den Einfluss der Schwere auf das Froschei. Arch Mikrosk Anat 24:475–545.

Brønsted H, Meijer H (1950): Does a correlation exist between the egg axis and the egg attachment in the ovary of Rana temporaria? Vidensk Medd Dansk Naturhist Foren Kjobenhavn 112:253.

Coggins LW (1973): An ultrastructural and autoradiographic study of early oogenesis in the toad, Xenopus laevis. J Cell Sci 12:71–86.

Curtis ASG (1960): Cortical grafting in Xenopus laevis. J Embryol Exp Morphol 8:163–173.

Curtis ASG (1962): Morphogenetic interactions before gastrulation in the amphibian, Xenopus laevis—The cortical field. J Embryol Exp Morphol 10:410–422.

Dalcq A, Pasteels J (1937): Une conception nouvelle des bases physiologiques de la morphogénèse. Arch Biol 48:669–710.

Dumont JN (1972): Oogenesis in Xenopus laevis (Daudin). I. Stages of oocyte development in laboratory maintained animals. J Morphol 136:153–180.

Elinson R (1980): The amphibian egg cortex in fertilization and early development. Symp Soc Dev Biol 38:217–234.

Fankhauser G (1948): The organization of the amphibian egg during fertilization and cleavage. Ann NY Acad Sci 49:684–702.

Farquhar MG, Palade GE (1981): The Golgi apparatus (complex)—(1954–1981)—from artifact to center stage. J Cell Biol 91:77s–106s.

Gerhart JC (1980): Mechanisms regulating pattern formation in the amphibian egg and early embryo. In Goldberger R (ed): "Biological Regulation and Development." New York: Plenum Press, Vol 2, pp 133–316.

Gerhart J, Black S, Scharf S (1983): Cellular and pancellular organization of the amphibian embryo. In McIntosh R (ed): "Modern Cell Biology Volume 2: Spatial Organization of Eukaryotic Cells." New York: Alan R. Liss, Inc., pp 483–508.

Gerhart JC, Ubbels G, Black S, Hara K, Kirschner M (1981): A reinvestigation of the role of the grey crescent in axis formation in Xenopus laevis. Nature 292:511–516.

Grant P, Wacaster JF (1972): The amphibian grey crescent—a site of developmental information? Dev Biol 28:454–471.

Heidemann SR, Kirschner MW (1975): Aster formation in eggs of Xenopus laevis: Induction by isolated basal bodies. J Cell Biol 67:105–117.

Keller RE (1975): Vital dye mapping of the gastrula and neurula of Xenopus laevis. I. Prospective areas and morphogenetic movements of the superficial layer. Dev Biol 42:222–241.

Keller RE (1976): Vital dye mapping of the gastrula and neurula of Xenopus laevis. II. Prospective areas and morphogenetic movements of the deep layer. Dev Biol 51:118–137.

Kirschner M, Gerhart JC, Hara K, Ubbels GA (1980): Initiation of the cell cycle and establishment of bilateral symmetry in Xenopus eggs. Sym Soc Dev Biol 38:187–216.

Klag JJ, Ubbels GA (1975): Regional morphological and cytological differentiation of the fertilized egg of Discoglossus pictus (Anura). Differentiation 3:15–20.

Malacinski GM, Brothers AJ, Chung H-M (1977): Destruction of components of the neural induction system of the amphibian egg with ultraviolet irradiation. Dev Biol 56:24–39.

Maller J, Poccia D, Nishioka D, Kidd P, Gerhart J, Hartman H (1976): Spindle formation and cleavage in Xenopus eggs injected with centriole-containing fractions from sperm. Exp Cell Res 99:285–294.

Manes ME, Barbieri FD (1977): On the possibility of sperm aster involvement in dorso-ventral polarization and pronuclear migration in the amphibian egg. J Embryol Exp Morphol 40:187–197.

Manes M, Elinson R (1980): Ultraviolet light inhibits grey crescent formation in the frog egg. W Rouxs Arch 189:73–76.

Manes ME, Elinson RP, Barbieri FD (1978): Formation of the amphibian grey crescent: Effects of colchicine and cytochalasin B. W Rouxs Arch 185:99–104.

Nakamura O, Takasaki H, Mizohata T (1970): Differentiation during cleavage in Xenopus laevis. I. Acquisition of self-differentiation capacity of the dorsal marginal zone. Proc Jpn Acad 46:694–699.

Nieuwkoop PD (1973): The "organization center" of the amphibian embryo: Its origin, spatial organization, and morphogenetic action. Adv Morphogenet 10:1–39.

Nieuwkoop PD (1977): Origin and establishment of embryonic polar axes in amphibian development. Curr Top Dev Biol 11:115–132.

Pasteels J (1938): Recherches sur les facteurs initiaux de la morphogénèse chez les amphibiens anoures. I. Resultats de l'experience de Schultze et leur interpretation. Arch Biol 49:629–667.

Pasteels J (1941): Recherches sur les facteurs initiaux de la morphogénèse chez les Amphibiens anoures. V. Les effects de la pesanteur sur l'oeuf de Rana fusca maintenu en position anormale avant la formation du croissant gris. Arch Biol 52:321–339.

Pasteels J (1946): Sur la structure de l'oeuf insegmente d'axolotl et l'origine des prodromes morphogénétiques. Acta Anat 2:1–16.
Pasteels J (1948): Les bases de la morphogénèse chez vertébrés anamniotes en function de la structure de l'oeuf. Folia Biotheor (Leiden) 3:83–108.
Pasteels J (1964): The morphogenetic role of the cortex of the amphibian egg. Adv Morphogenet 3:363–388.
Penners A, Schleip W (1928a): Die Entwicklung der Schultzeschen Doppelbildungen aus dem Ei von Rana fusca. Teil I–IV. Z Wiss Zool 130:305–454.
Penners A, Schleip W (1928b): Die Entwicklung der Schultzeschen Doppelbildungen aus dem Ei von Rana fusca. Teil V und VI. Z Wiss Zool 131:1–156.
Ruud G (1925): Die Entwicklung isolierter Keimfragmente frühester stadien von Triton taeniatus. W Rouxs Arch 105:209–293.
Scharf SR, Gerhart JC (1980): Determination of the dorsal-ventral axis in eggs of Xenopus laevis: Complete rescue of UV-impaired eggs by oblique orientation before first cleavage. Dev Biol 79:181–198.
Scharf SR, Gerhart JC (1983): Axis determination in eggs of Xenopus laevis: A critical period before first cleavage, identified by the common effects of cold, pressure, and UV-irradiation. Dev Biol (in press).
Schultze O (1894): Die künstliche Erzeugung von Doppelbildungen bei Froschlarven mit Hilfe abnormer Gravitation. W Rouxs Arch 1:160–204.
Spemann H (1902): Entwicklungsphysiologisches Studien am Tritonei. II. W Rouxs Arch 16:552–631.
Stewart-Savage J, Grey RD (1982): The temporal and spatial relationships between cortical contraction, sperm trail formation, and pronuclear migration in fertilized Xenopus eggs. W Rouxs Arch 191:241–245.
Subtelny S, Bradt C (1963): Cytological observations on the early developmental stages of activated Rana pipiens eggs receiving a transplanted blastula nucleus. J Morphol 112:45–60.
Tourte M, Mignotte F, Mounolou J-C (1981): Organization and replication activity of the mitochondrial mass of oogonia and previtellogenic oocytes in Xenopus laevis. Dev Growth Differ 23:9–21.
Van Gansen P, Weber A (1972): Determinisme de la polarité de l'oocyte de Xenopus laevis: Etude aux microscopes électroniques à balayage et à transmission. Arch Biol (Liège) 83:215–232.
Wallace RA, Bergink EW (1974): Amphibian vitellogenin: Properties, hormonal regulation of hepatic synthesis and ovarian uptake, and conversion to yolk proteins. Am Zool 14:1159–1175.
Wallace RA, Ho T (1972): Protein incorporation by isolated amphibian oocytes. II. A survey of inhibitors. J Exp Zool 181:303–330.
Wallace RA, Misulovin Z, Etkin LD (1981): Full-grown oocytes from Xenopus laevis resume growth when placed in culture. Proc Natl Acad Sci USA 78:3078–3082.
Wittek M (1952): La vitellogénèse chez les Amphibiens. Arch Biol 63:134–198.
Youn BW, Balacinski GM (1980): Action spectrum for ultraviolet irradiation inactivation of a cytoplasmic component(s) required for neural induction in the amphibian egg. J Exp Zool 211:369–378.
Züst B, Dixon KE (1975): The effect of UV irradiation of the vegetal pole of Xenopus laevis eggs on the presumptive primordial germ cells. J Embryol Exp Morphol 34:209–220.

Time, Space, and Pattern in Embryonic Development, pages 287–312

Cytoplasmic Localizations and Cell Interactions in the Formation of the Mouse Blastocyst

Martin H. Johnson and Hester P.M. Pratt

Department of Anatomy, University of Cambridge, Cambridge CB2 3DY, England

During development a cell is confronted with a sequence of options and its response to each of these options is governed by both its current cell associations and its cellular inheritance. The relative importance of these two influences has been debated for many years by developmental biologists, in discussions which have been conditioned to a great extent by the experimental organism studied. In this review, we consider what is known about the development of the mammalian blastocyst (Fig. 1) and conclude that both previous events and current influences contribute to cell diversification in the morula and blastocyst. Earlier attempts to consider either influence as being exclusive or preeminent have been embodied in two opposing types of hypotheses: a segregation hypothesis, which maintains that the differential segregation of cytoplasmic and/or membrane factors is the most important influence initiating and maintaining cell diversification, and a positional hypothesis, which stresses a dominant role for cell interactions. We suggest that exclusive adherence to either viewpoint has distorted our perception of the mechanisms underlying cell differentiation in the preimplantation mouse embryo and indeed in other types of embryo and we have attempted to reconcile the differing views in the polarization hypothesis [Johnson et al., 1981].

A variety of staining techniques applied to fixed whole-mount or sectioned early embryonic material has demonstrated zones of cytoplasmic or membrane staining during cleavage that could persist to give rise to the discrete staining patterns observed in the two different and committed tissues of the

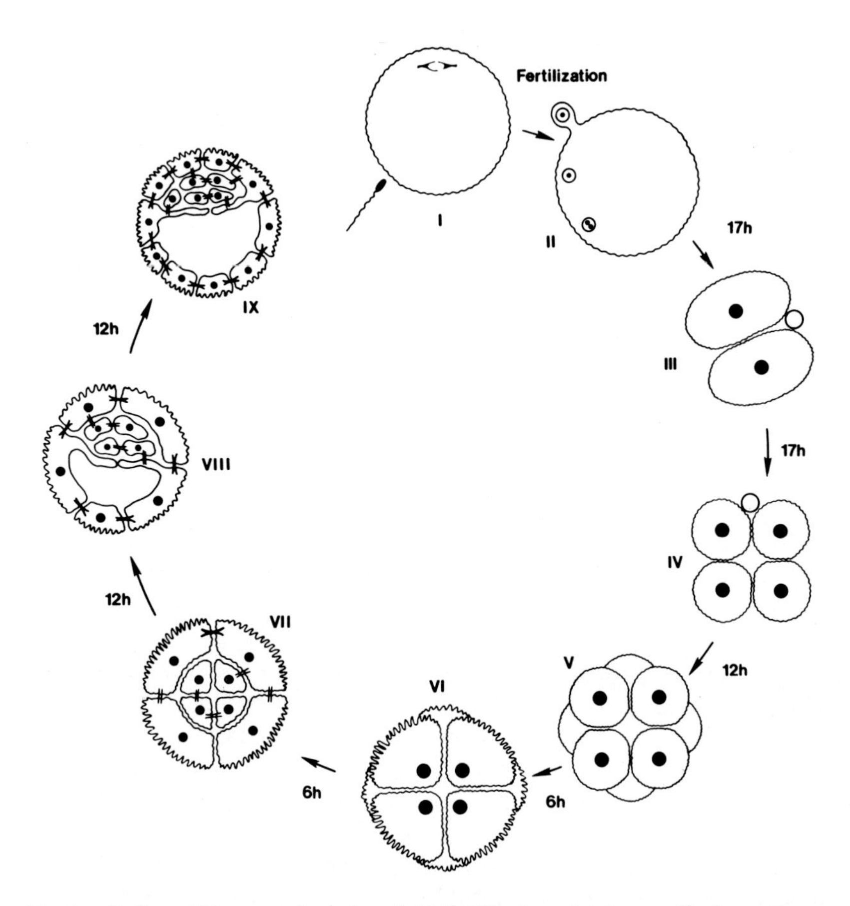

Fig. 1. Outline of blastocyst formation. I. Unfertilized, ovulated egg with chromosomes on the second metaphase spindle; it is surrounded by the zona pellucida (not shown) which persists until the blastocyst stage. II. At fertilization, the second meiotic division is completed leading to the formation of a polar body and the female pronucleus; the male pronucleus forms. III. Pronuclei come together and chromosomes line up on the spindle for the first cleavage division. IV. Cleavage to four cells and (V) eight cells. At the eight-cell stage, the embryo undergoes compaction (VI) in which individual cells polarize to become epitheliallike and flatten on each other, thereby maximizing cell contacts. VII. Specialized focal tight and gap junctions form and the cells divide to give a compact mass of 16 cells which consist of two phenotypically distinct subpopulations, occupying inside and outside positions. These cells in turn divide to yield a 32-cell morula (VIII) in which fluid is secreted internally to form a small blastocoele. The inside cells are covered by processes which extend from the outside cells. The apical tight junctions of the outer cells become zonular. IX. By the 64-cell stage the blastocoele is enlarged and surrounded by trophectoderm cells and contains an eccentrically placed inner cell mass (ICM). The trophectoderm gives rise to the trophoblast and chorionic ectoderm whereas the ICM yields all endoderm, mesoderm, and embryonic ectoderm derivatives.

later blastocyst stage. The concept of membrane and/or cytoplasmic continuity that has arisen from these observations has been applied to the rat and mouse [Dalcq, 1962, 1965; Mulnard, 1955; Izquierdo, 1977], the rabbit [Denker, 1976], the cat [Denker et al., 1978], and the human [Avendano et al., 1975]. However, some of these reports recorded considerable variation between embryos, and, perhaps partly for this reason, have proved difficult to repeat on occasion [Rode et al., 1968; Denker, 1972; Solter et al., 1973]. Further support for the notion that localization, followed by differential inheritance, might be important for the differentiation of the blastocyst came from cell isolation experiments, in which killing one blastomere in early cleaving embryos of rabbit or mouse resulted in the development of some incomplete blastocyst forms [Seidel, 1960; Tarkowski, 1959]. The interpretation placed on these observations was that "zones of information" existed, that were either specifically instructive or generally directive, and these conclusions were incorporated in a *classical segregation hypothesis* [summarized by Denker, 1976].

This idea that continuity of localized material might be important in the formation of the mammalian blastocyst has not received wide support. Apart from the variability in the cytochemical results, the requirement that cell subpopulations must be fixed histologically before they could be identified precluded the direct testing of this hypothesis experimentally. Furthermore, detailed cell isolation experiments demonstrated that, at least for the mouse, each blastomere at the 4-cell, and probably also 8-cell, stage could generate both trophectoderm (TE) and inner cell mass (ICM) tissue [Kelly, 1977]. Indeed, many experiments in which whole embryos, or their constituent cells have been aggregated in a variety of spatial arrays under different experimental conditions have demonstrated considerable flexibility in the developmental potential of most cells in mouse embryos upto, and probably beyond, the 16-cell stage [Mintz, 1965; Tarkowski, 1965; Johnson et al., 1977; Handyside, 1978; Hogan and Tilly, 1978; Spindle, 1978; Johnson, 1979a; Rossant and Lis, 1979; Rossant and Vijh, 1980; Ziomek et al., 1982a]. In some experiments of this general character, it has been observed that the position within the aggregate in which a cell, or group of cells, is placed, appears to affect its subsequent fate [Hillman et al., 1972; Ziomek and Johnson, 1982]. Taken together, the observations on developmental lability and positional effects led to the hypothesis that recognition of relative cell position is the single most important element in the generation of cell subpopulations. This notion has been embodied in the so-called *inside:outside hypothesis* of Tarkowski and Wroblewska [1967], in which the internal or external position of cells within the morula are thought to direct their development along the path of inner cell mass (ICM) or TE, respectively [Herbert and Graham, 1974]. The way in which internal and external positions are recognized, and the nature of the response to these recognition signals, have been the subject of speculation, but only rarely of experimentation [Pedersen and Spindle, 1980; Johnson et al., 1979], and then with somewhat inconclusive results.

The emphasis placed by the proponents of the inside:outside hypothesis on cell interaction and developmental flexibility has led to the neglect, or even rejection, of differential segregation as a potential contributory device for generating cell diversity in the early mouse embryo. Yet it is acknowledged that the observation of lability within a developing system does not of itself preclude an important role for prelocalization in the system [e.g., Wilson, 1925; Waddington, 1956; Davidson, 1976; Graham and Wareing, 1976; Denker, 1976; Dohmen and Verdonk, 1979; Johnson, 1981a; Reeve, 1983]. It merely precludes the *exclusive* control of development by continuity of cytoplasmic or membrane components. Recently, models for early mouse development have been proposed that rehabilitate an important, but not exclusive, role for localization and inheritance and the models have been tested experimentally [Graham and Deussen, 1978; Johnson et al., 1981; Izquierdo and Ebensperger, 1982]. As a result, the mouse embryo can be accommodated within the same conceptual framework of development as, for example, molluscan, insect, or amphibian embryos [Johnson et al., 1983], and we elaborate on the similarities among these systems toward the end of this chapter. First, however, we will describe the experiments that have led us to propose that a combination of inheritance and interaction direct the generation of cell diversity in the early mouse embryo.

EVIDENCE THAT CELL INHERITANCE INFLUENCES CELL FATE

Zones of cytoplasm or cells, marked genetically or physically in the early cleaving embryo, are identifiable as discrete entities which show little evidence of mixing until well after implantation. Growth of cells is coherent and clonal in situ [Wilson et al., 1972; Gardner and Johnson, 1975; Garner and McLaren, 1974; Graham and Deussen, 1978; Kelly, 1979; Ziomek, 1982 Ziomek and Johnson, 1982; Balakier and Pedersen, 1982]. Thus, although mobility of some cytoplasmic elements has been observed [Graham and Deussen, 1978; Reeve, 1981a], a basic and conserved cytoplasmic framework exists within the embryo which survives cytokinesis, cytodifferentiation, and the early phases of growth. Such a framework could provide a basis for the differential distribution of material amongst the cells of the cleaving embryo.

A careful analysis by Graham and his colleagues [Graham and Deussen, 1978; Kelly et al., 1978; Graham and Lehtonen, 1979; Lehtonen, 1980] has provided a detailed description of the patterns underlying this coherent clonal allocation of cytoplasm during the transition from the two-cell to 32-cell stages. The studies revealed that the patterns of cell contacts among blastomeres influence the disposition of their progeny after cytokinesis. The more contacts a cell had, and therefore the "deeper" within the cluster of cells it was located, the more likely were its progeny to contribute to the inside cells of the developing embryo. The number and nature of the contact relationships observed were not immutable but were affected by the presence of the zona pellucida, the division order of the cells, and experimental rearrangements of blastomeres. Thus, at any point in development the disposition of cytoplasm was determined by the previous sequence of contact relationships. Qualitative analyses of the cells involved, particularly during the eight-cell stage and later, have revealed special features of the cell surface that explain the nature of some of the contact relationships observed by Graham and the mechanisms by which they direct cells to internal or external positions. These analyses also point to cytoplasmic and/or membrane segregation as providing an important component of the process by which cell diversity is generated.

The early eight-cell embryo consists of approximately spherical, symmetrical, and overtly similar cells. During the life of each eight-cell blastomere, the surface of the cell is reorganized fundamentally to yield a polar phenotype with an outward apical pole of microvillous membrane and an inner basolateral membrane largely devoid of microvilli (Fig. 2). This structural reorganization may be visualized at both electron and light microscope levels [Calarco and Brown, 1969; Calarco and Epstein, 1973; Ducibella and Anderson, 1975;

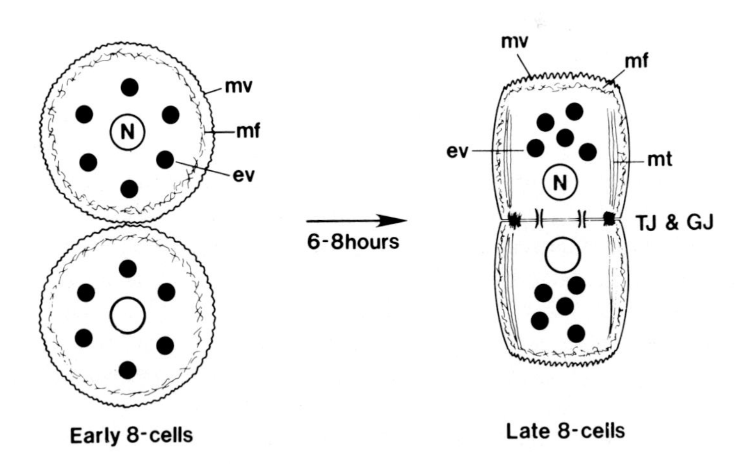

Fig. 2. Polarization of mouse blastomeres during the eight-cell stage. Each cell in a pair of early eight cells cultured in contact for six to eight hours develops an axis of polarity normal to the point of contact. The nuclei (N) move basally, microvilli (mv) are restricted to the apical region, a microfilamentous mesh (mf) is absent from regions of cell contact at which tight and gap junctions (TJ and GJ) form, microtubules (mt) align along lateral surfaces, and endocytotic vesicles (ev) accumulate between apical pole and basal nucleus.

Ducibella et al., 1977; Handyside, 1980; Ziomek and Johnson, 1980; Reeve and Ziomek, 1981] and is accompanied by a major cytoplasmic reorganization [Reeve, 1981b; Reeve and Kelly, 1983; Ducibella et al., 1977]. The properties of the apical and basolateral membranes differ. The apical microvillous membrane is relatively nonadhesive [Johnson and Ziomek, 1981a] and appears to lack the capacity to form specialized junctions [Goodall and Johnson, 1982]. In contrast, the basolateral surfaces are highly adhesive, flattening on each other in situ [Lewis and Wright, 1935; Ducibella, 1977; Lehtonen, 1980; Pratt, 1978], on and around other inner nonmicrovillous cells from later embryos [Rossant, 1975; Kimber and Surani, 1981] and forming specialized junctions [Ducibella, 1977; Magnuson et al., 1978; Lo and Gilula, 1979].

The polarized surface phenotype assumed by each eight-cell blastomere is stable and persists after isolation of single cells and during the division of eight-cell blastomeres to the 16-cell stage either in isolation or in situ [Johnson and Ziomek, 1981b; Reeve, 1981a]. Thus, the 16-cell embryo is characterized by two subpopulations of cells which are distinguishable by both phenotype and position, and which arise by the division of polarized eight cells. The inner cells are derived from the basolateral portions of the polarized eight cells, are sparsely microvillous, highly adhesive, and flatten on each other to yield a compact mass of cells. The outer cells incorporate the apical regions of the polarized eight cells, and therefore inherit a polar cap of microvilli and a basolateral surface which flattens on and leads to the envelopment of the compact clusters of apolar cells [Handyside, 1981; Reeve and Ziomek, 1981; Ziomek and Johnson, 1981; Kimber and Surani, 1981]. The cleavage plane during division of the eight-cell blastomeres is somewhat eccentric in most cases, reflecting the basal position of the nucleus [Reeve and Kelly, 1983], and thereby generates larger outer cells and smaller inner cells [Johnson and Ziomek, 1981a; Randle, 1982; Kimber and Surani, 1981]. Moreover, the apparently random orientation of the cleavage plane results in bisection of the microvillous pole in approximately one out of eight cells. This yields an average of nine outer and seven inner cells in most 16-cell embryos [Johnson and Ziomek, 1981b; Handyside, 1981; Ziomek et al., 1982b]. These observations suggest that the cellular organization of individual parent eight-cell blastomeres lays the foundation for the two cell subpopulations of the 16-cell stage.

There is evidence that these two cell subpopulations are the progenitors of the inner cell mass and the trophectoderm of later stages. Thus, the 32-cell late morula and early blastocyst are also characterized by two cell subpopulations, an outer group of polar cells and an inner group of nonpolar, highly adhesive cells [Wiley and Eglitis, 1981; Johnson and Ziomek, 1982]. Moreover, there are on average 18–20 outer polar cells and 12–14 inner apolar cells, suggesting that the ratio of 9:7 outer:inner cells at the 16-cell stage is maintained throughout division to the 32-cell stage [Handyside, 1978]. Direct evidence in support of the proposition that inner cells yield ICM and outer cells yield TE has come from experiments in which inner or outer cells at the 16-cell stage were marked with a short-term lineage marker, recombined in various numerical and spatial arrays with nonlabelled cells, and their fate followed. For example, a couplet consisting of a labelled apolar cell with a nonlabelled polar cell developed in most cases to yield a quartet of 32-cell blastomeres in which two nonlabelled trophectodermlike cells surrounded two labelled ICM-like cells—a "mini-blastocyst" (Fig. 3) [Johnson and Ziomek, 1983]. When the labelled apolar cell was placed on the inside of an aggregate of 15 other nonlabelled cells, it contributed in most cases only to the ICM, whereas a labelled polar cell placed on the outside contributed almost exclusively to the trophectoderm [Ziomek and Johnson, 1982]. Randle [1982], using a different short-term lineage marker system, has achieved similar results. There is thus quite a substantial body of data to support the notion that properties transmitted by cells from the eight- to 16- and then to 32-cell stages are involved in generating and maintaining cell divergence.

There is, however, one important series of observations which appears to contradict this conclusion. Balakier and Pedersen [1982] used horseradish peroxidase (HRP) iontophoresed into outside cells as a short-term lineage marker in the morula and concluded that a large proportion of outside cells could generate ICM as well as or even instead of trophectoderm tissue. This result, which contrasts with that of both Randle [1982] and Ziomek and

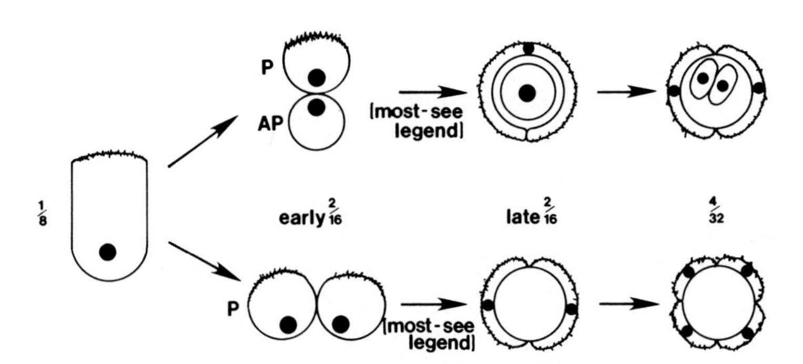

Fig. 3. Scheme of natural cell lineage from eight- to 32-cell stage illustrated for a single polarized eight-cell blastomere. Division may be parallel to (lower) or normal to (upper) the axis of polarity. In the lower case, two polar 16-cell blastomeres generally yield four trophectoderm cells (but see Fig. 4). If division is normal to the axis of polarity, the polar 16-cell blastomere so formed normally envelops the apolar, basally derived cell and a miniblastocyst with two inner and two trophectoderm cells results.

Johnson [1982], requires confirmation and explanation. It is clear that iontophoresis of HRP in many cases affects cell viability and/or ability to undergo division, since there is a deficiency of labelled cells in many blastocysts. Also, the outside cells at the time of labelling were identified purely on the basis of their external position and not also by their polar phenotype. As the embryos were decompacted prior to injection, thereby allowing inside cells to pop out to the surface [Reeve and Ziomek, 1981], it is possible that some injected cells were exposed apolar cells, rather than truly external polar cells. Clearly it will be of the greatest importance to confirm this result under conditions in which the injected cell is identified by both position and phenotype.

In conclusion, there is considerable evidence that the process of differential inheritance is of crucial importance in generating cell diversity in the early mouse embryo.

EVIDENCE THAT CELL INTERACTION INFLUENCES CELL FATE

The rehabilitation of differential inheritance should not lead us to abandon cell interaction as having only a trivial influence on cell fate in the morula. The results of Graham and his colleagues described earlier show that the organization of cells is influenced by preceding patterns of cell contact. Thus, the number and organization of cell interactions sets up a pattern and this pattern predicts the subsequent disposition of cells. Similarly, the qualitative changes in the cell surface observed at polarization also arise as a result of cell interaction. Thus, isolated eight-cell blastomeres do not polarize [Ziomek and Johnson, 1980]; neither do eight-cell blastomeres that are totally surrounded by other blastomeres [Johnson and Ziomek, 1981a]. Only when an eight-cell blastomere engages in asymmetric cell contact with one [Ziomek and Johnson, 1980] or more [Johnson and Ziomek, 1981a] other blastomeres does polarity develop and the axis of that polarity is determined by the location of the point (or points) of contact. These results imply that cell interaction is important in generating the polar phenotype of the late eight-cell blastomere.

The nature of the interaction is not yet defined. It does not seem to be a consequence of nonspecific cell contact, since the capacity of cells to induce polarity is absent prior to and shortly after fertilization and develops progressively during cleavage [Johnson and Ziomek, 1981a]. Furthermore, agents which reduce or modify the *extent* of physical interaction between adjacent eight-cell blastomeres do not seem to interfere with the induction of polarity [Pratt et al., 1982], suggesting that extensive areas of cell contact are not required. The possibility that secreted extracellular matrix glycoproteins might play a role in the induction of polarity seems unlikely, since most reports suggest that their secretion commences only at the 16-cell stage [Leivo et al., 1980; Zetter and Martin, 1978; Adamson and Ayers, 1979; Sherman et al., 1980]. Furthermore, Cooper and MacQueen [1983] have shown by metabolic labelling that all three subunits of laminin, the earliest of the matrix glycoproteins to be secreted, are first synthesized coordinately at the late eight- to 16-cell stage. Gap junctions form for the first time between adjacent cells of eight-cell embryos [Lo and Gilula, 1979] and do so over the first three to four hours contemporaneously with the induction of polarity [Goodall and Johnson, 1982]. However, the formation of gap junctions is not essential for the induction process. For example, agents which inhibit gap junction formation do not inhibit development of polarity [Pratt et al., 1982; H. Goodall, unpublished data], and there are cells that can induce polarity in eight-cell blastomeres but that cannot form gap junctions with them [Johnson and Ziomek, 1981a; Goodall and Johnson, 1982, and unpublished data].

Thus, a pattern begins to emerge. A process of cell interaction induces a major cytoplasmic reorganization which leads to the generation of asymmetries in the embryo. These asymmetries are conserved during division to generate two distinct cell subpopulations, thereby founding the two different lineages of the blastocyst. Recent evidence suggests that continuing and specific cell interactions are required for the maintenance of these discrete lineages. In the absence of these specific cell interactions, individual cells within the two subpopulations cease to differentiate and instead express their totipotentiality by generating complementary cell types.

When polar and apolar 16-cell blastomeres are aggregated together, either as pairs or in larger clusters of 16 cells, they undergo a characteristic interaction in which polar cells surround and envelop apolar cells (Fig. 3). Their inherited surface properties reinforce their relative positional differences and, as was pointed out earlier, these positional differences persist during division to the 32-cell stage [Ziomek and Johnson, 1981, 1982; Randle, 1982; Johnson and Ziomek, 1983]. However, if the two subpopulations of cells are isolated from each other, and cultured as single cells, pairs of similar cells, or clusters of 16 similar cells, then some of the cells express their totipotency. Apolar cells reveal their capacity to generate trophectoderm and polar cells their capacity to generate ICM [Ziomek et al., 1982a; Johnson and Ziomek, 1983]. How do polar and apolar cells express their totipotency when the cellular interactions between them are disrupted?

A polar cell, deprived of the opportunity to wrap around a cluster of compact apolar cells, retains a plump morphology. As a result, when it divides, the cleavage plane is more likely to pass parallel to the nonextended pole of microvilli than to bisect it (Fig. 4). In this case, a polar trophectoderm and an apolar ICM cell will form rather than two polar trophectodermal

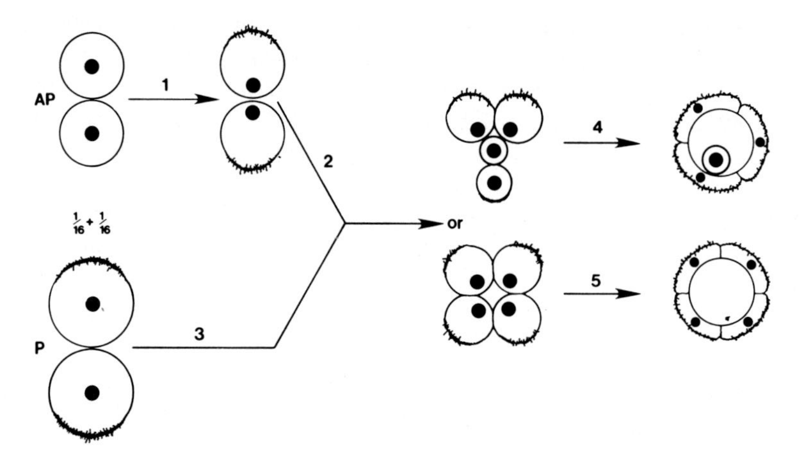

Fig. 4. Expression of totipotentiality in apolar and polar 16-cell blastomeres deprived of mutual interaction. Two apolar cells induce polarity in each other due to their asymmetry of contacts (1), thereby coming to resemble two polar cells. In this way, the apolar cell can generate trophectoderm. In the intact embryo this would not happen as no asymmetry of cell contacts would occur, the apolar cells being enveloped by the polar cells (see upper part of Fig. 3). A couplet of polar cells behaves in one of two ways (2,3). Both cells may divide parallel to the axis of polarity, in which case a vesicle of trophectoderm forms (5). This appears to be invariably the case in the intact normal embryo, or when the polar cell is associated with an apolar cell (see upper part in Fig. 3). However, when a polar cell is deprived of contact with apolar cells, and thus cannot expand to envelop them, the chances of the polar cell dividing normal to the axis of polarity and thereby generating one apolar and one polar cell (4) increases. By this device, a polar cell can contribute progeny to the ICM. In the normal embryo, this would be unlikely to occur due to the attenuated phenotype of the enveloping polar cell.

cells. Therefore it seems that in the normal undisturbed embryo the capacity of polar cells to give rise to apolar cells (i.e., express their totipotency) is suppressed by restricting the orientation of their cleavage planes. This restriction is evidently mediated through their interaction with apolar cells [Johnson and Ziomek, 1983].

When apolar cells are deprived of the encircling and symmetric contact of enveloping polar cells, they are exposed to an asymmetry of contacts with other apolar cells instead. Just as the early nonpolar eight-cell blastomeres responded to asymmetric cell contact by polarizing, so apolar 16-cell blastomeres do likewise (Fig. 4) [Ziomek and Johnson, 1981; Johnson and Ziomek, 1983]. Here again, the normal in situ interactions between polar and apolar cells will prevent asymmetric cell contacts from forming and thereby ensure that the totipotentiality of the cells is suppressed.

Therefore, the inherited cell surface properties determine the interaction pattern that directs the developmental fate of both cell types. However, these inherited cell surface properties can also permit the expression of totipotency if such interaction does not occur.

This totipotentiality of the apolar and polar 16-cell blastomeres probably persists until at least the early to mid 32-cell stage. Careful analysis of late morulae and early blastocysts has revealed that clusters of inside apolar cells which have entered their sixth cell cycle (i.e., number 12–14 in toto) can, if isolated, generate trophectoderm [Handyside, 1978; Spindle, 1978; Rossant and Lis, 1979]. This capacity is lost at some time late in the sixth cell cycle or early in the seventh cell cycle (unpublished data with P. Warren and T. Fleming). Since these totipotent ICMs from the early blastocyst are located eccentrically within the blastocoele, it may reasonably be asked why those cells adjacent to the blastocoele, and thereby apparently exposed to asymmetric cell contact, do not generate trophectoderm cells in situ? In fact, during these early stages of blastocyst expansion, and well into the seventh cell cycle of the inside cells, the ICM is not exposed to the blastocoele to any great extent. Rather, trophectodermal processes reach out and interpose between the blastocoele and the ICM cells (Fig. 1), thereby enveloping them and preserving their symmetry of contacts in a manner which may restrict their expression of totipotency (unpublished data with P. Warren and T. Fleming). For the outer, polar cells, the timing of commitment to a trophectodermal lineage is less clearly defined but probably also occurs during the sixth cell cycle when zonular tight junctions first form between adjacent cells, and trophectodermal cells become attenuated as they expand around the blastocoele, thereby reducing the chances of any tangential, and therefore differentiative, divisions.

ANCESTRY AND INTERACTION—DISSOCIABLE PROCESSES?

Our analysis demonstrates that early mouse development is under the combined influence of cell interaction and differential inheritance. These processes act in concert to regulate normal development and can provide mechanisms that explain in simple cellular terms the progressive restrictions on developmental potential as well the retention of a regulative capacity. Our observations lead us to suggest that it is fundamentally misguided to attempt to ascribe relative importance to interaction or inheritance for any developmental decision. The two phenomena are complementary rather than antithetical and represent different parameters of a common developmental process. The properties of a cell are defined by its history and it is this inheritance which provides the substrate upon which cell interactions can operate. In embryonic terms, time can be represented as a sequential program of developmental options [Johnson, 1980]. Each option, serially selected, places constraints on the capacity of a cell to respond whether through sequential modifications to chromatin, through changes to the overall macromolecular structure of the cell, or most probably through a combination of both. Hence the options available to a cell at any point in development are defined and limited by its developmental history. Our analysis of the early mouse embryo leads us to suggest that the particular option implemented is chosen as a consequence of the pattern of interactions between cells of different phenotype within the developing embryo. The embryonic genome need not necessarily be directly involved in this process and differential gene activity might only play a secondary role in the further elaboration of phenotypic asymmetries once they have become established.

IS THE MOUSE EMBRYO UNIQUE?

The demonstration that early mouse embryogenesis is not only regulatory but also mosaic invites more detailed comparison with other embryos, traditionally regarded as being quite distinct from the mammal in their use of early cytoplasmic localizations to determine later patterns of development. From our studies on the mouse we draw certain general conclusions, which may have wider application, and we suggest that the mouse embryo has features that make it particularly appropriate for formulating some underlying embryological principles.

Our general conclusions are as follows:

(1) Regional localizations are common to most or all early embryos, but the point in development relative to fertilization at which these localizations develop may vary.

Cytoplasmic localizations are established during oogenesis in the *Caenorhabditis* egg and appear to exert fixed determinative influences after cleavage and parcelling out of cytoplasm [Hirsch, 1979]. Other mosaic eggs [e.g., ascidians: Conklin, 1905; *Amphioxus:* Conklin, 1932; and some annelids: Lasserre, 1975] also show overt cytoplasmic localizations during oogenesis, but in these cases it is not clear whether the mosaicism is determinative in the sense observed in *Caenorhabditis* or, as more frequently seems to be the case, simply reflects an underlying animal-vegetal axis. For example, in molluscs the vegetal pole of the egg, and the vegetal body of *Bithynia* located at this pole, are recognizable in the growing oocyte [Dohmen and Verdonk, 1979; Verdonk et al., 1971], and in *Limnaea* a characteristic pattern of subcortical organelles formed during oogenesis anticipates the dorsoventral cytoplasmic axis that develops as the egg passes along the female genital tract [Raven, 1967]. Similarly, in both echinoderms [Hörstadius, 1939] and amphibia [Hara et al., 1980] elements of an animal-vegetal axis are established by the time of ovulation. Insect eggs also show evidence of anteroposterior and possibly dorsoventral polarity during oogenesis. Physical manipulations to, and genetic influences on or within, the developing oocyte have predictable effects on the subsequent anteroposterior axis of the embryo [e.g., *Smittia:* Kalthoff, 1979; pea-beetle: Van der Meer, 1978, cited in Kalthoff, 1979; leafhopper: Sander, 1976; harlequin fly: Yajima, 1970; *Drosophila:* Bull, 1966]. A specific localization, as well as a general axis, is also observed in insect eggs where polar granules, which ultimately become incorporated into the germ cells, are restricted to the posterior pole of the oocyte late in oogenesis [Mahowald, 1971]. Devices for segregating germ plasm and soma which involve localized oocytic cytoplasm have also been described for leeches [the pole plasm: Needham, 1969] and *Ascaris* [gradients of "diminisher": King and Beams, 1938; see also Beams and Kessel, 1974].

In other embryos, fertilization marks the appearance of developmentally significant localizations. These may be either assembled de novo or built upon axial localizations already established during oogenesis. For example, both *Fucus* and *Pelvetia* appear to be without an axis until fertilization [Whittaker, 1940; Jaffe, 1968; Nuccitelli, 1978], and in some oligochaetes fertilization marks the separation of two polar plasms from the endoplasm and thus the appearance of an animal-vegetal axis [Penners cited in Lasserre, 1975]. Eggs of ascidians, *Amphioxus*, molluscs, and amphibians, in which an animal-vegetal axis may already be present, establish their dorsoventral axis at fertilization. A dramatic and directed streaming of organelles and coloured granules results from fertilization of ascidian eggs [Conklin, 1905]. However, if the unfertilized eggs are bisected in equatorial, meridional, or oblique planes each fragment can, on fertilization, generate a normal localization pattern and give rise to essentially normal embryos [Reverberi and Ortolini, 1962], thereby demonstrating that fertilization is the critical event for organization of the cytoplasm. Likewise, grey crescent formation and the establishment of the principal axes of the amphibian egg are activated at fertilization [Pasteels, 1964; Brachet, 1977].

During cleavage also, localizations may occur either de novo, e.g., the mouse [Johnson et al., 1981], or as an elaboration of preexisting patterns, as for example, in ctenophores, in which the oral:aboral axis is established at fertilization but the localization of comb-forming potential does not occur until after the first cleavage [Freeman, 1977]. In the hydrozoan embryo, *Phialidium gregarium*, the anteroposterior axis of the planula larva is determined by the site of the first cleavage initiation. If the site of cleavage initiation is relocated, the axis is shifted accordingly [Freeman, 1981a, b]. Similarly, the complex sequence of absorption and regeneration of the polar lobe in molluscs [Dohmen and Verdonk, 1979], the early phases of segmentation in insects [Nusslein-Volhard and Wieschaus, 1980], the irreversible establishment of dorsoventrality in echinoderms [Hörstadius, 1939], and the establishment of "handedness" in some spiralians [see Waddington, 1956] all involve cytoplasmic reorganizations that occur during early cleavage.

These few examples appear to support three important generalizations. First, the device of cytoplasmic localization as a means for determining subsequent developmental organization is widespread amongst vertebrates, invertebrates, and plants. Second, the time at which developmentally important localizations can first be detected may vary from quite early in oogenesis to late in cleavage. Third, these localizations appear to accumulate sequentially.

(2) The pattern of regional localization is established by processes involving cell interaction.

Both Conklin [1932] and Waddington [1956, p 106] hinted that cell interaction within the ovary might influence axes of oocyte polarity. They pointed out that the site of attachment of the oocyte in the ovary was at the developing vegetal pole in the majority of invertebrates and at the animal pole in *Amphioxus* and ascidians and speculated that the attachment site might relate to the reversal of dorsoventral axis between vertebrates and invertebrates. Some circumstantial evidence supports this speculation. For example, Dohmen and van der May [1977] have described a distinct microvillous area at the vegetal pole (and subsequently in the polar lobe) of the *Nassarius* egg where the oocyte previously made contact with the ovarian follicle cells. More remarkable is the *Limnaea* oocyte [Raven, 1963, 1967; Ubbels et al., 1969], in which six discrete subcortical accumulations of a dense cytoplasmic matrix and granules correspond almost exactly to the asymmetric cuplike pattern of contacts with the six follicle cells in the inner follicular layer of the gonad. The distribution of these subcortical accumulations, and their subsequent behaviour, suggests that they are involved in establishing the dorsoventral axis of the embryos. More direct observations to suggest that patterns of contact within the ovary determine axes within the oocyte come from the *Drosophila* mutant dicephalic [Lohs-Schardin and Sander, 1976]. In the wild-type follicle, the nurse cells are restricted to the anterior pole of the oocyte but in the mutant dicephalic, nurse cells are present at the posterior as well as the anterior pole and a "double-headed" embryo forms. In the silk moth, *Hyalophora*, a mechanism by which nurse cell contacts might generate axes has been described [Woodruff and Telfer, 1980]. The asymmetrically placed nurse cells generate a polarity in the oocyte that is mediated via electrical gradients.

Where localizations appear at fertilization, the site of sperm entry may influence the spatial pattern that develops. In *Limnaea, Amphioxus,* and ascidians the cytoplasmic streaming that occurs at fertilization is focussed around the point of sperm entry. In the ascidian egg for example, the orange crescent forms in relation to the site of sperm entry [Conklin, 1905], and if two spermlike signals are given an orange crescent forms at each site [Jeffery, 1982]. Sperm penetration of amphibian eggs also results in a localized cortical contraction [Elinson, 1975] that may be orientated by the sperm centriole [Gerhart et al., 1981]. The asymmetries thereby induced lead to grey crescent formation and establishment of the dorsoventral axis [Gerhart et al., 1981].

During cleavage, the dispositon of cytoplasmic localizations is influenced by cell interactions, some of which may be inductive [e.g., *Ilyanassa:* Clement, 1962]. Wherever a regulative capacity is retained, as manifested by the capacity to compensate for loss or damage of individual cells or groups of cells, then continuing cell interaction is implicated in the establishment of natural patterns of localization [Hörstadius, 1939; Roux, 1883; Freeman, 1981b]. The amphibian example is instructive. Deletion of one blastomere at the two-cell stage nonetheless permits development of a complete embryo. However, if the blastomere is killed but not removed then the embryo that develops is incomplete [Roux, 1883]. As discussed earlier, the mouse embryo has been demonstrated to rely directly on cell interaction for initiating localization patterns as well as for segregating them to cellular progeny.

These examples, drawn from a variety of phyla, are limited, but remarkably few experimental investigations on the way in which localization patterns are induced have been undertaken. The mouse embryo is particularly suitable for this type of study since its primary mosaicism develops in the free-living embryo rather than in the ovarian-bound oocyte. Although the other examples are not very numerous they nevertheless constitute a reasonable case for the notion that developmentally important localizations arise as a consequence of cell interactions.

(3) A localization pattern need not assemble in an all-or-none (quantal) fashion, but it is more likely to develop progressively through a series of reversible intermediate stages which culminate finally in the "committed" stabilized pattern.

Demonstration of a regulative capacity during or after the development of an overt localization may be taken as evidence in favour of the progressive, rather than quantal, acquisition of that pattern, as for example in experiments where centrifugation or surgery disturbs patterns and yet embryos regulate to develop normally [*Bithynia:* Dohmen and Verdonk, 1979; *Limnaea:* Raven, 1967; *Dentalium:* Verdonk et al., 1971].

More direct and controlled experimental evidence comes from altering the pattern-inducing stimulus at various times after its initial application. In *Fucus* zygotes, for example, polarity develops over the eight-hour period leading to rhizoid formation, but changing the orientation of the polarizing stimulus at any time during the first five hours changes the orientation of the rhizoid that develops [Quatrano, 1972, 1973]. Similarly the labile period in *Pelvetia* is about seven hours [Nuccitelli, 1978]. Studies on the establishment of the dorsoventral axis of the frog egg suggest that there is a limited period during which gravitational forces can reverse the sperm-induced axis and after this period the pattern becomes progressively more refractory to such disturbances [Gerhart et al., 1981]. In the *Drosophila* mutant bicaudal, the anteroposterior axis develops normally during oogenesis, as evidenced by the polarity of egg shell. However, the axis developed in the oocyte does not persist in the egg unless the temperature is manipulated appropriately after the egg is laid [Nusslein-Volhard, 1979], again suggesting that the initial phases of polarity are labile. In the mouse, the induction of polarity in the eight-cell mouse blastomere takes between three and five hours and after this period the axis of polarity is stable. Prior to this, however, a change in the location of the inducing signal will result in a predictable change in the final axis of polarity achieved [Johnson and Ziomek, 1981a]. Indeed, there is evidence that the earliest signs of contact-induced polarization in mouse blastomeres may occur as early as the four-cell stage [Johnson, 1981a; Johnson and Ziomek, 1981a; Pratt et al., 1982; Reeve and Kelly, 1983; Goddard and Pratt, 1983] but that throughout this period the polarity that is induced is labile. Similarly, as was described earlier, the two distinct cell phenotypes generated at the 16-cell stage require a period of 12–24 hours during which continuing cell interactions progressively reinforce and ultimately stabilize their phenotypes as that of trophectoderm or inner cell mass. A similar progressive restriction on lability of cells has been observed in ascidian, nemertean, and echinoderm cleavage stages [Reverberi, 1948; Hörstadius, 1937, 1939], in which blastomeres, although clearly differentiating divergently as judged by progressive segregation of properties, nonetheless retain for a limited period a degree of potentiality in excess of their prospective fate.

These examples lead us to believe that the phenomenon of progressive stabilization of organization is not confined to the mouse embryo, but is widespread, characterizing even those embryos traditionally considered to have a mosaic pattern of development. Viewed in this context, the differences in developmental strategies employed by so-called "mosaic" and "regulative" embryos become superficial and reflect mainly the length of time required to stabilize each developmental asymmetry and the developmental age at which this is achieved. The mouse embryo develops slowly, cleavage to 32-cells taking 3½ days (Fig. 1), and each major developmental decision takes about 27 hours to finalize [Johnson, 1981b]. This makes it an attractive embryo for studying how developmental decisions are made, or, in the terminology adopted earlier, how an option is selected. As the sequence of events from fertilization to blastocyst expansion is defined with more precision, it becomes clear that if there are "quantal" or "switch-throwing" events in early mouse development, they are more likely to be the natural and terminal consequences of an incremental series of changes rather than abrupt and causal events which initiate a series of changes [Johnson, 1979b]. A "switch" of the former type would fix the cell as a product of its recent history, rendering its newly and progressively acquired phenotype irreversible and enabling the cell to confront its next choice of developmental options. This idea of a progressive and plastic moulding of the cell during each phase of development is in direct contrast to concepts invoking biochemical "determinants" as specific instructional molecules which cause discrete and discontinuous qualitative changes in phenotype. Whilst these ideas have developed from our work on the mouse we feel that their application may be much wider and even extend to organisms like *Drosophila* where each binary decision is assumed to be the consequence of a quantal event rather than leading to one. Indeed, as Nusslein-Volhard [1979] points out with reference to *Drosophila* "a gradient in development may lead to . . . an unequal distribution of determinants." In other words, a flexible primary response can provide a framework upon which more stable asymmetries can be assembled subsequently.

(4) Regional localizations in the early embryo are generally concerned with a limited range of developmental specifications.

The components inherited by the early embryo can influence its future development in two main ways—by supporting its basic metabolism and by providing information specifically concerned with development [McLaren, 1979; Cohen, 1979]. Clearly the former function is most obviously fulfilled by the cytoplasmic endowment of mitochondria, ribosomes, etc., as well as transporting and metabolizing enzymes. Indeed, many of the maternal effects described for amphibians [Malacinski and Spieth, 1979], insects [Okada et al., 1974], and possibly even mammals [Verussio et al., 1968; Gärtner and Baunack, 1981] can be ascribed to defects in such "housekeeping" components.

The developmental information inherited by the embryo seems to be of a fairly general kind and it is difficult to find solid evidence in favour of specific instructional molecules ("determinants") which are concerned with directing or regulating specific developmental decisions. One such molecule might be associated with the operation of the o (ova deficient) mutant of the axolotl [Brothers, 1976] in which an oocyte-derived factor interacts with embryonic DNA to yield a stable heritable change. However, this factor requires confirmation and identification [Malacinski and Spieth, 1979]. Almost all other localizations studied appear to have less specific consequences—for example, segregating the somatic and germ lines [Eddy, 1975], separating vegetative or nutritional areas from animal or embryonic areas [e.g., Elinson, 1975], affecting planes and rates of cell division [molluscs: Dohmen and Verdonk, 1979; amphibia: Hara et al., 1980], or establishing embryonic axes and patterns of segmentation [*Limnaea:* Raven, 1976; *Lytechinus*: Schatten, 1981; *Drosophila*: Lohs-Schardin and Sanders, 1976; Nusslein-Volhard and Wieschaus, 1980; molluscs: Dohmen and Verdonk, 1979; amphibia: Gerhart et al., 1981]. In the preimplantation mouse embryo there is no evidence to support a distinct germ line [Heath, 1981]. However, the overall radial symmetry established at the eight-cell stage is perpetuated through the interactions between constituent cells to generate the embryonic (or animal) ICM and the nutritional (or vegetal) trophectoderm. Since mammalian eggs inherit little yolk from the ovary the developmental strategy appears to have been to generate a yolk-gathering tissue, the trophectoderm, which is capable of transporting the yolk equivalent derived from the uterine secretion or decidua to the developing ICM [Johnson and Everitt, 1980]. Hence differentiation of trophectoderm could be viewed as the mammalian equivalent of developing separate animal and vegetal poles.

MATERNAL INFLUENCES ON MOUSE DEVELOPMENT

Our foregoing discussion leads us to conclude that the mouse embryo does not differ fundamentally from embryos in other phyla. The only important difference is one of timing. The cell interactions that first induce cytoplasmic localizations in the mouse embryo occur relatively late, as does the time of final commitment to a stabilized localization. In addition, the absolute time scale in the mouse is much slower than that of other embryos. These features may represent an evolutionary consequence of the greater environmental protection afforded to the mammalian embryo by the female genital tract. Any pressure to attain an advanced free-living state rapidly in order to adapt to environmental perturbation is lessened. Additionally, a slower developmental rate may be imposed upon the embryo as a consequence of the dual role of the female reproductive tract as the conveyor of spermatozoa as well as the site of implantation. The optimal conditions for each of these roles differ, which necessitates cyclic changes in the state of the reproductive tract. It is the amount of time taken by the genital tract to carry out these transitions that may dictate the rate at which the embryo undergoes its preimplantation development [Johnson and Everitt, 1980]. The observed delay in establishing the primary radial embryonic axis may be a secondary consequence of this slower developmental timetable.

In other phyla that have been studied biochemically, developmental axes are regulated by maternally derived instruction whether expressed in the oocyte, at fertilization, or during cleavage [e.g., Denny and Tyler, 1964; Brachet et al., 1963; Smith and Ecker, 1965; Gross, 1968; Sargent and Raff,

1976]. The embryo's own genes are largely irrelevant to the process. In the mammal, however, the timetable of development is so slow that the establishment of a stable radial axis in the embryo occurs over a period of some 60 or so hours after fertilization. This raises the possibility that, unlike other embryos, the mouse controls this primary level of organization with its own genes not with those of its mother. Indeed, a substantial body of evidence points to an active role for the embryonic genome itself, as distinct from solely maternally derived gene products, at an early developmental stage. The evidence suggesting that activation of the embryonic genome occurs during the two-cell stage and that the gene products are immediately both translated and essential to development has been reviewed in detail elsewhere [Johnson, 1981b]. More recently evidence of the precise timing of this event and the fate of maternally derived mRNA has been published.

Evidence of some transcriptional activity as early as the late one-cell stage has been presented recently [Clegg and Piko, 1982]. Of the RNA produced, some may be mRNA but the bulk was characterized as heterodisperse, high molecular weight, poly(A)$^-$RNA which was turned over rapidly. The significance of this transcriptional activity for development is not clear, since there is no evidence available to suggest any effect of the embryonic genome on either polypeptide synthetic profile or development prior to the two-cell stage. During this period development is regulated postranscriptionally by modification of existing mRNAs [Clegg and Piko, 1982] or posttranslationally by protein modification [Van Blerkom, 1981]. Once the two-cell stage is reached, however, two bursts of transcriptional activity occur [Flach et al., 1982]. The first occurs immediately after cleavage to two cells, the second immediately after DNA replication. Both transcriptional events are manifested in the appearance of new proteins [Flach et al., 1982; Piko and Clegg, 1982] including one polypeptide (at least) which is coded for by paternally derived genes [Sawicki et al., 1981]. Thus, the embryo intervenes to influence its own development at an early developmental stage (two-cell), although still in comparative terms, at a late developmental age (18–20 hours).

Surprisingly, there is rapid loss of maternal mRNA during the two- and four-cell stages. Indeed, it is difficult to be sure that a significant level of functional maternal mRNA persists after this time [Flach et al., 1982; Piko and Clegg, 1982]. This means that any persistent maternal effects inherited via the cytoplasm of the oocyte may be due to either the expression of small subsets of atypically stable maternal message or the activity of proteins translated from maternal mRNA during the one- and two-cell stages and surviving later into cleavage.

Since developmental localization within the mouse embryo first becomes stabilized at the eight-cell stage, does this mean that the cell interactions leading to polarization occur under embryonic rather than maternal control?

This question cannot yet be answered unambiguously. However, it does seem likely that transcriptional activity is *not* required during the four-cell and eight-cell stages for either the induction of polarity or for the polarizing response. Remarkably, both these properties are observed at the normal developmental time in embryos placed in an effective dose of α-amanitin at the late two-cell stage even though cleavage is blocked (unpublished data). Thus, any mRNAs required for the events of polarization must be present by this time. Whether these mRNAs are maternal or newly transcribed and embryonic is under investigation. Were they to prove maternal in origin, the readmission of the mammalian embryo to the developmental fold would be complete.

ACKNOWLEDGMENTS

We wish to thank Caroline Hunt and Tracy Kelly for their efficient preparation of the manuscript, and Raith Overhill and Roger Liles for preparation of art and photographic work. The research reported in this chapter was funded by grants to the authors from the Cancer Research Campaign and the Medical Research Council.

REFERENCES

Adamson ED, Ayers SE (1979): The localisation and synthesis of some collagen types in developing mouse embryos. Cell 16:953–965.

Avendano S, Croxatto HD, Pereda J, Croxatto HB (1975): A seven-cell human egg recovered from the oviduct. Fertil Steril 26:1167–1172.

Balakier H, Pedersen RA (1982): Allocation of cells to inner cell mass and trophectoderm lineages in preimplantation mouse embryos. Dev Biol 90:352–362.

Beams HW, Kessel RG (1974): The problem of germ cell determinants. Int Rev Cytol 39:413–479.

Brachet J (1977): An old enigma: The gray crescent of amphibian eggs. Curr Top Dev Biol 11:133–186.

Brachet J, Ficq A, Tencer R (1963): Amino acid incorporation into proteins of nucleate and anucleate fragments of sea urchin eggs: Effect of parthenogenetic activation. Exp Cell Res 32:168–170.

Brothers AJ (1976): Stable nuclear activation dependent on a protein synthesised during oogenesis. Nature 260:112–115.

Bull AL (1966): Bicaudal, a genetic factor which affects the polarity of the embryo in *Drosophila melanogaster*. J Exp Zool 161:221–242.

Calarco PG, Brown EH (1969): An ultrastructural and cytological study of preimplantation development of the mouse. J Exp Zool 171:253–283.

Calarco PG, Epstein CJ (1973): Cell surface changes during preimplantation development in the mouse. Dev Biol 32:208–213.

Clegg KB, Piko L (1982): RNA synthesis and cytoplasmic polyadenylation in the one-cell mouse embryo. Nature 295:342–345.

Clement AC (1962): Development of *Ilyanassa* following removal of the D macromere at successive cleavage stages. J Exp Zool 149:193–215.

Cohen J (1979): Maternal constraints on development. In Newth DR, Balls M (eds): "Maternal Effects in Development." Cambridge: Cambridge University Press, pp 1–28.

Conklin EG (1905): Mosaic development in Ascidian eggs. J Exp Zool 2:146–223.

Conklin EG (1932): The embryology of *Amphioxus*. J Morphol 54:69–151.

Cooper A, MacQueen HA (1983): Synthesis of laminin subunits by the early mouse embryo is not coordinate. Dev Biol (in press).

Dalcq AM (1962): Evolution de l'organisation morphogénetique dans l'oocyte chez le Rat et la Souris. Verh Anat Ges Anat Anz [Suppl] 109:373–382.

Dalcq AM (1965): Cytochimie des premiers stades du développement chez quelque mammiferes. Ann Biol 4:129–155.

Davidson EH (1976): "Gene Activity in Early Development, 2nd Ed." New York: Academic Press.

Denker HW (1972): Furchung beim Säugetier: Differenzierung von Trophoblast- und Embryonal-knoten zellen. Verh Anat Ges Anat Anz [Suppl] 130:267–272.

Denker HW (1976): Formation of the blastocyst: Determination of trophoblast and embryonic knot. Curr Top Pathol 62:59–76.

Denker HW, Eng LA, Mootz U, Hamner CE (1978): Studies on the early development and implantation in the cat. 1. Cleavage and blastocyst formation. Anat Anz 144:457–468.

Denny PC, Tyler A (1964): Activation of protein biosynthesis in non-nucleate fragments of sea-urchin eggs. Biochem Biophys Res Commun 14:245–249.

Dohmen MR, van der Mey JCA (1977): Local surface differentiations at the vegetal pole of the eggs of *Nassarius reticulatus, Buccinum undatum,* and *Crepidula fornicata* (Gastropoda, Prosobranchia). Dev Biol 61:104–113.

Dohmen MR, Verdonk NH (1979): The ultrastructure and role of the polar lobe in development of Molluscs. In Subtelny S, Konigsberg IR (eds): "Determinants of Spatial Organization." New York: Academic Press, pp 3–27.

Ducibella T (1977): Surface changes of the developing trophoblast cell. In Johnson MH (ed): " Development in Mammals." Amsterdam: Elsevier/North-Holland Biomedical Press, vol 1, pp 5–30.

Ducibella T, Anderson E (1975): Cell shape and membrane changes in the eight-cell mouse embryo: Prerequisites for morphogenesis of the blastocyst. Dev Biol 47:45–58.

Ducibella T, Ukena T, Karnovsky M, Anderson E (1977): Changes in cell surface and cortical cytoplasmic organization during embryogenesis in the preimplantation mouse embryo. J Cell Biol 74:153–167.

Eddy EM (1975): Germ plasm and the differentiation of the germ cell line. Int Rev Cytol 43:229–275.

Elinson RP (1975): Site of sperm entry and a cortical contraction associated with egg activation in the frog *Rana pipiens*. Dev Biol 47:257–268.

Flach G, Johnson MH, Braude PR, Taylor RAS, Bolton V (1982): The transition from maternal to embryonic control in the 2-cell mouse embryo. EMBO J 1:681–686.

Freeman G (1977): The establishment of the oral-aboral axis in the ctenophore embryo. J Embryol Exp Morphol 42:237–260.

Freeman G (1981a): The cleavage initiation site establishes the posterior pole of the hydrozoan embryo. W Rouxs Arch Dev Biol 190:123–125.

Freeman G (1981b): The role of polarity in the development of Hydrozoan Planula Larva. W Rouxs Arch Dev Biol 190:168–184.

Gardner RL, Johnson MH (1975): Investigation of cellular interaction and deployment in the early mammalian embryo using interspecific chimaeras between the rat and mouse. In

"Cell Patterning." CIBA Foundation Symposium. Amsterdam: Elsevier, Vol 29, pp 183–200.
Garner W, McLaren A (1974): Cell distribution in chimaeric mouse embryos before implantation. J Embryol Exp Morphol 32:495–503.
Gärtner K, Baunack E (1981): Is the similarity of monozygotic twins due to genetic factors alone? Nature 292:646–647.
Gerhart J, Ubbels G, Black S, Hara K, Kirschner M (1981): A reinvestigation of the role of the grey crescent in axis formation in *Xenopus laevis*. Nature 292:511–516.
Goddard M, Pratt HPM (1983): Control of events during early cleavage of the mouse embryo: An analysis of the '2-cell block'. J Embryol Exp Morphol (in press).
Goodall H, Johnson MH (1982): The use of carboxy-fluorescein diacetate to study the formation of permeable channels between mouse blastomeres. Nature 295:524–526.
Graham CF, Deussen ZA (1978): Features of cell lineage in preimplantation mouse development. J Embryol Exp Morphol 48:53–72.
Graham CF, Lehtonen E (1979): Formation and consequences of cell patterns in preimplantation mouse development. J Embryol Exp Morphol 49:277–294.
Graham CF, Wareing PF (1976): "The Developmental Biology of Plants and Animals." Oxford: Blackwell Scientific Publications.
Gross PR (1968): Biochemistry of differentiation. Annu Rev Biochem 37:631–660.
Handyside AH (1978): Time of commitment of inside cells isolated from preimplantation mouse embryos. J Embryol Exp Morphol 45:37–53.
Handyside AH (1980): Distribution of antibody- and lectin-binding sites on dissociated blastomeres from mouse morulae: Evidence for polarization at compaction. J Embryol Exp Morphol 60:99–116.
Handyside AH (1981): Immunofluorescence techniques for determining the numbers of inner and outer blastomeres in mouse morulae. J Reprod Immunol 2:339–350.
Hara K, Tydeman P, Kirschner M (1980): A cytoplasmic clock with the same period as the division cycle in *Xenopus* eggs. Proc Natl Acad Sci USA 77:462–466.
Heath JK (1981): "Experimental Studies of the Mammalian Germ Line." DPhil Thesis, Oxford University.
Herbert MC, Graham CF (1974): Cell determination and biochemical differentiation of the early mammalian embryo. Current topics Dev Biol 8:151–178.
Hillman NM, Sherman MI, Graham CF (1972): The effect of spatial arrangement on cell determination during mouse development. J Embryol Exp Morphol 28:263–278.
Hirsch D (1979): Temperature sensitive maternal effect mutants of early development in *Caenorabditis elegans*. In Subtelny S, Konigsberg IR (eds): "Determinants of Spatial Organization." New York: Academic Press, pp 149–166.
Hogan B, Tilly R (1978): In vitro development of inner cell masses isolated immunosurgically from mouse blastocysts. I. Inner cell masses from 3.5 day p.c. blastocysts incubated for 24 h before immunosurgery. J Embryol Exp Morphol 45:93–105.
Hörstadius S (1937): Experiments on determination in the early development of *Cerebratulus lacteus*. Biol Bull 73:317–342.
Hörstadius S (1939): The mechanics of sea-urchin development studied by operative methods. Biol Rev 14:132–179.
Izquierdo L (1977): Cleavage and differentiation. In Johnson MH (ed): "Development in Mammals." Amsterdam: North-Holland Vol 2, pp 99–118.
Izquierdo L, Ebensperger C (1982): Cell membrane regionalisation in early mouse embryos as demonstrated by 5′ nucleotidase activity. J Embryol Exp Morphol 69:115–126.
Jaffe LF (1968): Localization in the developing *Fucus* egg and the general role of localizing currents. Adv Morphol 7:295–328.

Jeffery WR (1982): Calcium ionophore polarises ooplasmic segregation in ascidian eggs. Science 216:545–547.
Johnson MH (1979a): Molecular differentiation of inside cells and inner cell masses isolated from the preimplantation mouse embryo. J Embryol Exp Morphol 53:335–344.
Johnson MH (1979b): Intrinsic and extrinsic factors in preimplantation development. J Reprod Fertil 55:255–265.
Johnson MH (1980): Position in the embryo. In Johnson MH (ed): "Development in Mammals." Amsterdam: North-Holland Biomedical Press, Vol 4, pp 1–2.
Johnson MH (1981a): Membrane events associated with the generation of a blastocyst. Int Rev Cytol [Suppl] 12:1–37.
Johnson MH (1981b): The molecular and cellular basis of preimplantation mouse development. Biol Rev 56:463–498.
Johnson MH, Everitt BJ (1980): "Essential Reproduction." Oxford: Blackwell Scientific Publications.
Johnson MH, Ziomek CA (1981a): Induction of polarity in mouse 8-cell blastomeres: Specificity, geometry and stability. J Cell Biol 91:303–308.
Johnson MH, Ziomek CA (1981b): The foundation of two distinct cell lineages within the mouse morula. Cell 24:71–80.
Johnson MH, Ziomek CA (1982): Cell subpopulations in the late morula and early blastocyst of the mouse. Dev Biol 91:431–439.
Johnson MH, Ziomek CA (1983): Cell interactions influence the fate of mouse blastomeres undergoing the transition from the 16- to the 32-cell stage. Dev Biol 95:211–218.
Johnson MH, Handyside AH, Braude PR (1977): Control mechanisms in early mammalian development. In Johnson MH (ed): "Development in Mammals." Amsterdam: North-Holland Vol 2, pp 67–97.
Johnson MH, Chakraborty J, Handyside AH, Willison K, Stern P (1979): The effect of prolonged decompaction on the development of the preimplantation mouse embryo. J Embryol Exp Morphol 54:241–261.
Johnson MH, Pratt HPM, Handyside AH (1981): The generation and recognition of positional information in the preimplantation mouse embryo. In Glasser SR, Bullock DW (eds): "Cellular and Molecular Aspects of Implantation." New York: Plenum Press, pp 55–74.
Johnson MH, Ziomek CA, Reeve WJD, Pratt HPM, Goodall H, Handyside AH (1983): The mosaic organisation of the preimplantation mouse embryo. In Van Blerkom J, Motta P (eds): "Current Topics in Ultrastructure Research. Ultrastructure of Reproduction and Early Development." The Hague: Martinus Nijhoff B.V. (in press).
Kalthoff K (1979): Analysis of a morphogenetic determinant in an insect embryo. In Subtelny S, Konigsberg IR (eds): "Determinants of Spatial Organization." New York: Academic Press, pp 97–126.
Kelly SJ (1977): Studies of the developmental potential of 4- and 8-cell stage mouse blastomeres. J Exp Zool 200:365–376.
Kelly SJ (1979): Investigations into the degree of cell mixing that occurs between the 8-cell stage and the blastocyst stage of mouse development. J Exp Zool 207:121–130.
Kelly SJ, Mulnard JG, Graham CF (1978): Cell division and cell allocation in early mouse development. J Embryol Exp Morphol 48:37–51.
Kimber SJ, Surani MAH (1981): Changing cell adhesiveness in preimplantation embryos. Cell Biol Int Rep [Suppl A] 5:29.
King RL, Beams HW (1938): An experimental study of chromatin diminution in Ascaris. J Exp Zool 77:425–443.

Lasserre P (1975): Clitellata. In Giese AC, Pearse JS (eds): "Reproduction of Marine Invertebrates, Vol III. Annelids and Echiurans." New York: Academic Press, pp 215–275.

Lehtonen E (1980): Changes in cell dimensions and intercellular contacts during cleavage-stage cell cycles in mouse embryonic cells. J Embryol Exp Morphol 58:231–249.

Leivo I, Vaheri A, Timpl R, Wartiovaara J (1980): Appearance and distribution of collagens and laminin in the early mouse embryo. Dev Biol 76:100–114.

Lewis WH, Wright ES (1935): On the early development of the mouse egg. Contrib Embryol Carnegie Inst 148:115–143.

Lo CW, Gilula NB (1979): Gap junctional communication in the preimplantation mouse embryo. Cell 18:399–409.

Lohs-Schardin M, Sander K (1976): A dicephalic monster embryo of *Drosophila melanogaster.* W Rouxs Arch Dev Biol 179:159–162.

Malacinski GM, Spieth J (1979): Maternal effect genes in the Mexican axolotl *(Ambystoma mexicanum).* In Newth DR, Balls M (eds): "Maternal Effects in Development." Cambridge: Cambridge University Press, pp 241–268.

Magnuson T, Jacobson JB, Stackpole CW (1978): Relationship between intercellular permeability and junction organization in the preimplantation mouse embryo. Dev Biol 67:214–224.

Mahowald AP (1971): Polar granules of *Drosophila.* III. The continuity of polar granules during the life cycle of *Drosophila.* J Exp Zool 176:329–344.

McLaren A (1979): The impact of pre-fertilization events on post-fertilization development in mammals. In Newth DR, Balls M (eds): "Maternal Effects in Development." Cambridge: Cambridge University Press, pp 287–320.

Mintz B (1965): Experimental genetic mosaicism in the mouse. In Wolstenholme GEW, O'Connor M (eds): "Preimplantation Stages of Pregnancy." CIBA Foundation Symposium. London: Churchill, pp 194–207.

Mulnard J (1955): Contribution à la connaissance des enzymes dans l'ontogénèse. Les phosphomonoestérases acide et alkaline dans le développement du Rat et de la Souris. Arch Biol 66:525–685.

Needham AE (1969): Growth and Development. In Florkin M, Scheer BT (eds): "Chemical Zoology." New York: Academic Press, Vol 4, pp 377–441.

Nuccitelli R (1978): Ooplasmic segregation and secretion in the *Pelvetia* egg is accompanied by a membrane-generated electrical current. Dev Biol 62:13–33.

Nusslein-Volhard C (1979): Maternal effect mutations that alter the spatial coordinates of the embryo of *Drosophila melanogaster.* In Subtelny S, Konigsberg IR (eds): "Determinants of Spatial Organisation." New York: Academic Press, pp 185–211.

Nusslein-Volhard C, Wieschaus E (1980): Mutations affecting segment number and polarity in Drosophila. Nature 287:795–801.

Okada M, Kleinman IA, Schneiderman HA (1974): Repair of a genetically-caused defect in oogenesis in *Drosophila melanogaster* by transplantation of cytoplasm from wild type eggs and by injection of pyrimidine nucleosides. Dev Biol 37:55–62.

Pasteels JJ (1964): The morphogenetic role of the cortex of the amphibian egg. Adv Morphogenet 3:363–388.

Pedersen RA, Spindle AI (1980): Role of the blastocoele microenvironment in early mouse differentiation. Nature 284:550–552.

Piko L, Clegg KB (1982): Quantitative changes in total RNA, total poly(A), and ribosomes in early mouse embryos. Dev Biol 89:362–378.

Pratt HPM (1978): Lipids and transitions in embryos. In Johnson MH (ed): "Development in Mammals." Amsterdam: North-Holland, Vol 3, pp 83–129.

Pratt HPM, Ziomek CA, Reeve WJD, Johnson MH (1982): Compaction of the mouse embryo: An analysis of its components. J Embryol Exp Morphol 70:113–132.

Quatrano RS (1972): An ultrastructural study of the determined site of rhizoid formation in *Fucus* zygotes. Exp Cell Res 70:1–12.

Quatrano RS (1973): Separation of processes associated with differentiation of two-celled *Fucus* embryos. Dev Biol 30:209–213.

Randle B (1982): Cosegregation of monoclonal reactivity and cell behaviour in the mouse preimplantation embryo. J Embryol Exp Morphol 70:261–278.

Raven CP (1963): The nature and origin of the cortical morphogenetic field in *Limnaea.* Dev Biol 7:130–143.

Raven CP (1967): The distribution of special cytoplasmic differentiations of the egg during early cleavage in *Limnaea stagnalis.* Dev Biol 16:407–437.

Reeve WJD (1981a): The distribution of ingested horseradish peroxidase in the 16-cell mouse embryo. J Embryol Exp Morphol 66:191–207.

Reeve WJD (1981b): Cytoplasmic polarity develops at compaction in rat and mouse embryos. J Embryol Exp Morphol 62:351–367.

Reeve WJD (1983): The generation of cellular differences in the preimplantation mouse embryo. In Harrison RJ, Navaratnam V (eds): "Progress in Anatomy." The Anatomical Society of Great Britain and Ireland. Cambridge: Cambridge University Press, Vol 3, (in press).

Reeve WJD, Kelly FP (1983): Nuclear position in the cells of mouse morulae. J Embryol Exp Morphol (in press).

Reeve WJD, Ziomek CA (1981): Distribution of microvilli on dissociated blastomeres from mouse embryos: Evidence for surface polarization at compaction. J Embryol Exp Morphol 62:339–350.

Reverberi G (1948): Nouveaux résultats et nouvelles vues sur le germe des Ascidies. Folia Biotheor 3:59–82.

Reverberi G, Ortolani G (1962): Twin larvae from halves of the same egg in Ascidians. Dev Biol 5:84–100.

Rode B, Damjanov I, Skreb N (1968): Distribution of acid and alkaline phosphatase activity in early stages of rat embryos. Bull Sci Cons Acad RSF Yougoslavie Sect A 13:304.

Rossant J (1975): Investigation of inner cell mass determination by aggregation of isolated rat inner cell masses with mouse morulae. J Embryol Exp Morph 36:163–174.

Rossant J, Lis WJ (1979): Potential of isolated mouse inner cell masses to form trophectoderm derivatives in vivo. Dev Biol 70:255–261.

Rossant J, Vijh KM (1980): Ability of outside cells from preimplantation mouse embryos to form inner cell mass derivatives. Dev Biol 76:475–482.

Roux W (1883): "Uber die Bedentung der Kernteilungsfiguren. Eine hypothetische Erörterung." Leipzig: Engelmann.

Sander K (1976): Specification of the basic body pattern in insect embryogenesis. Adv Insect Physiol 12:125–238.

Sargent TD, Raff RA (1976): Protein synthesis and messenger RNA stability in activated, enucleate sea urchin eggs are not affected by actinomycin D. Dev Biol 48:327–335.

Sawicki W, Magnuson T, Epstein C (1981): Evidence for expression of the paternal genome in the two-cell mouse embryo. Nature 294:450–454.

Schatten G (1981): Sperm incorporation, the pronuclear migrations and their relation to the establishment of the first embryonic axis: Time lapse video microscopy of the movements during fertilization of the sea urchin *Lytechinus variegatus.* Dev Biol 86:426–437.

Seidel F (1960): Die Entwicklungsfähigkeiten isolierter Furchungszellen aus dem Ei des Kaninchens *Oryctolagus cuniculus.* W Rouxs Arch Entwicklungsmech Organismen 152:43–130.

Sherman MI, Gay R, Gay S, Miller EJ (1980): Association of collagen with preimplantation and peri-implantation mouse embryos. Dev Biol 74:470–478.

Smith LD, Ecker RE (1965): Protein synthesis in enucleated eggs of *Rana pipiens*. Science 150:777–779.

Solter D, Damjanov I, Skreb N (1973): Distribution of hydrolytic enzymes in early rat and mouse embryos—a reappraisal. Z Anat Entwick Gesch 139:119–126.

Spindle AI (1978): Trophoblast regeneration by inner cell masses isolated from cultured mouse embryos. J Exp Zool 203:483–489.

Tarkowski AK (1959): Experiments on the development of isolated blastomeres of mouse eggs. Nature 184:1286–1287.

Tarkowski AK (1965): Embryonic and postnatal development of mouse chimaeras. In Wolstenholme GEW, O'Conner M (eds): "Preimplantation Stages of Pregnancy." CIBA Foundation Symposium. London: Churchill, pp 183–193.

Tarkowski AK, Wroblewska J (1967): Development of blastomeres of mouse eggs isolated at the 4- and 8-cell stage. J Embryol Exp Morphol 18:155–180.

Ubbels GA, Bezem JJ, Raven CP (1969): Analysis of follicle cell patterns in dextral and sinistral *Limnaea peregra*. J Embryol Exp Morphol 21:445–466.

Van Blerkom J (1981): Structural relationship and posttranslational modification of stage-specific proteins synthesised during early preimplantation development in the mouse. Proc Natl Acad Sci USA 78:7629–7633.

Verdonk NH, Geilenkirchen WLM, Timmermans LPM (1971): The localization of morphogenetic factors in uncleaved eggs of *Dentalium*. J Embryol Exp Morphol 25:57–63.

Verussio AC, Pollard DR, Fraser FC (1968): A cytoplasmically transmitted diet-dependent difference in response to the teratogenic effects of 6-aminonicotinamide. Science 160:206–207.

Waddington CH (1956): "Principles of Embryology." London: Allen & Unwin Ltd.

Whitaker DM (1940): Physical factors of growth. Growth [Suppl] 75–88.

Wiley LM, Eglitis MA (1981): Cell surface and cytoskeletal elements: Cavitation in the mouse preimplantation embryo. Dev Biol 86:493–501.

Wilson EB (1925): "The Cell in Development and Heredity, 3rd Ed." New York: Macmillan.

Wilson IB, Bolton E, Cutler RH (1972): Preimplantation differentiation in the mouse egg as revealed by microinjection of vital markers. J Embryol Exp Morphol 27:467–479.

Woodruff RL, Telfer WH (1980): Electrophoresis of proteins in intercellular bridges. Nature 286:84–86.

Yajima H (1970): Study of the development of the internal organs of the double malformations of *Chironomus dorsalis* by fixed and sectioned materials. J Embryol Exp Morphol 24:287–303.

Zetter BR, Martin GR (1978): Expression of a high molecular weight cell surface glycoprotein (LETS protein) by preimplantation mouse embryos and teratocarcinoma stem cells. Proc Natl Acad Sci USA 75:2324–2328.

Ziomek CA (1982): The use of fluorescein isothiocyanate (FITC) as a short-term cell lineage marker in the peri-implantation mouse embryo. W Rouxs Arch Dev Biol 191:37–41.

Ziomek CA, Johnson MH (1980): Cell surface interaction induces polarization of mouse 8-cell blastomeres at compaction. Cell 21:935–942.

Ziomek CA, Johnson MH (1981): Properties of polar and apolar cells from the 16-cell mouse morula. W Rouxs Arch Dev Biol 190:287–296.

Ziomek CA, Johnson MH (1982): The roles of phenotype and position in guiding the fate of 16-cell mouse blastomeres. Dev Biol 91:440–447.

Ziomek CA, Johnson MH, Handyside AH (1982a): The developmental potential of mouse 16-cell blastomeres. J Exp Zool 221:345–355.

Ziomek CA, Pratt HPM, Johnson MH (1982b): The origins of cell diversity in the early mouse embryo. In Finbow ME, Pitts JD (eds): "Functional Integration of Cells in Animal Tissues." Br Soc Cell Biol Symp No 5. Cambridge: Cambridge University Press, pp 149–165.

Time, Space, and Pattern in Embryonic Development, pages 313–348

Cytoplasmic Determinants in Dipteran Eggs

Klaus Kalthoff

Department of Zoology, University of Texas at Austin, Austin, Texas 78712

After a late entry [Seidel, 1926] into the world of experimental embryology, insects are currently receiving much attention by investigators of cytoplasmic localization and cell determination. This is due largely to the highly developed genetics of *Drosophila melanogaster*. Female sterile and homoeotic mutants of this species have considerably boosted the search for localized cytoplasmic determinants and genes involved in determination. Another advantage of insects is their wealth of appendages and cuticular structures, which provide a detailed body pattern even during embryonic stages. Finally, there is an almost unlimited diversity of insects from which the investigator can choose the system most suitable to an experimental design. This diversity, however, also restricts the ability to transfer results and conclusions obtained among different groups of insects. This chapter is for the most part limited to dipterans; for more broadly based reviews, readers are referred to articles of Krause [1939, 1981], Counce [1973], and Sander [1976].

Dipteran eggs follow the superficial mode of cleavage as do most other insects (Fig. 1). During an extended period of "intravitelline cleavage," nuclei undergo several mitotic divisions within the yolk-rich endoplasm of the egg. Most cleavage "energids," i.e., nuclei with jackets of cytoplasm around them, move toward the superficial periplasm, while a few remain in the yolk-rich endoplasm as primary vitellophages. The energids at the egg surface become arranged as a dense monolayer while mitotic divisions continue. Eventually, infoldings of the plasma membrane enclose each superficial nucleus as a cell [Fullilove and Jacobson, 1971]; the resulting columnar

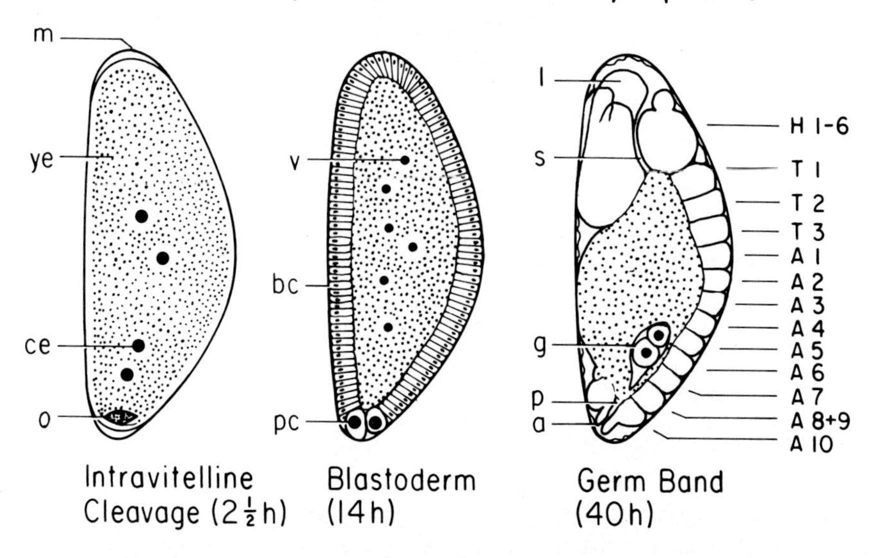

Fig. 1. Embryonic development of *Smittia* sp. Anterior pole up, dorsal side to the left. a, anal papilla; bc, blastoderm cell; ce, cleavage energid; g, gonad; l, labrum; m, micropyle; o, oosome; p, proctodeum; pc, pole cell; s, stomodeum; v, vitellophage; ye, yolk endoplasm; H1–H6, head segments; T1–T3, thoracic segments; A1–A10, abdominal segments. Length of egg: 200 μm.

epithelium is called the "blastoderm." During the process of cellularization, changes in the pattern of RNA synthesis are observed in *Drosophila* and other embryos. The synthetic rate of nonnucleolar, polyadenylated RNA increases, while nucleolar transcription also becomes prominent [McKnight and Miller, 1976; Lamb and Laird, 1976]. The blastoderm cells remain connected to the yolk-rich endoplasm by cytoplasmic bridges, some of which persist until extension of the germ anlage [Rickoll, 1976; Rickoll and Counce, 1980]. These bridges provide for an extended period of electrical and humoral communication among blastoderm cells.

A particular group of cells known as pole cells are segregated early at the posterior pole of the embryo. The nuclei that become included in the pole cells move precociously into the periplasm (Fig. 1). Pole cells can be distinguished from blastoderm cells by their larger size and round shape; they retain no cytoplasmic bridges with the endoplasm, and their divisions are out of synchrony with those of the blastoderm cells. At least some of the pole cells become primordial germ cells. In accord with this fate, they retain the full genetic complement in species that eliminate a number of chromosomes from their somatic cells. Pole cells also differ from blastoderm cells in their scheduling of transcriptional activation [Lamb and Laird, 1976; Zalokar, 1976] and protein synthesis [Pietruschka and Bier, 1972; Lundquist and Emanuelsson, 1980].

Most blastoderm cells contribute to the unsegmented "germ anlage" while the rest form amnion and serosa. The germ anlage, through processes including gastrulation and segmentation, is transformed into the segmented "germ band" which already reflects the basic insect body plan (Fig. 1). The anterior portion of the head containing eyes and brain is thought to consist of three segments (procephalon, H1–H3). Of the following three head segments (gnathocephalon, H4–H6) each has characteristic mouth parts. The thoracic segments (T1–T3) will carry legs, wings, and halteres in the adult. The number of abdominal segments originally formed in many dipteran germ bands is ten plus a telson. Abdominal segments visible in late embryos and larvae are often reduced in number to eight or nine by fusions among posterior abdominal segments [Kalthoff and Sander, 1968; Turner and Mahowald, 1977].

Fate maps of *Drosophila* embryos at the blastoderm stage have been constructed by observation of living and sectioned embryos, and by killing or removing blastoderm cells [see Underwood et al., 1980a]. With 0% egg length (EL) and 100% EL corresponding to the posterior and anterior pole, respectively, the approximate longitudinal positions of cells giving rise to larval epidermis (hypoderm) were mapped as follows: head 75–60% EL, thorax 60–50% EL, abdomen 50–15% EL. Blastoderm cells located anterior to 75% or posterior to 15% give rise to brain, stomodeum, anterior midgut, proctodeum, and posterior midgut. Given a consistent wobble in fate mapping data, they indicate that in a population of embryos the topographical position of a cell restricts the fate of its progeny to within a few segments. However, a clone of hypoderm cells derived from a single marked blastoderm cell in an individual embryo is confined to a single segment or compartment thereof [Steiner, 1976; Wieschaus and Gehring, 1976; Lawrence et al., 1978; Szabad et al., 1979].

CELL DETERMINATION IN EARLY EMBRYOS

Not only are the fates of blastoderm cells restricted, the cells are also "determined" to contribute to certain embryonic segments and not to others. This was first shown by Chan and Gehring [1971], who dissociated and reaggregated genetically marked blastoderm cells. While cells from anterior halves of embryos gave rise to adult head and thoracic structures, cells from posterior halves produced thoracic and abdominal structures. A more detailed analysis was provided by Illmensee [1978], who transplanted single blastoderm cells to blastoderm recipients distinguished by different genetic markers. After homotopic transplantations (e.g., anterior to anterior, ventral to ventral), transferred cells became integrated for the most part into the adult structures formed by the surrounding recipient cells. After extremely heterotopic transplantations (e.g., anterior to posterior, posterior to anterior), the transplanted cells segregated from the host cells and developed according to the position they had occupied in the donor. For instance, anterior blastoderm cells transplanted among posterior host cells developed into antennal structures. Moderately heterotopic transplantations (e.g., ventral to posterior) led to either integration or segregation. Thus, blastoderm cells seem to be generally determined for anterior vs. posterior structures, but not necessarily restricted to certain fates within those areas. Similar experiments by Kauffman [1980] indicate that semistable anterior vs. posterior nuclear commitments are already present at the "syncytial blastoderm" stage, i.e., before cellularization.

How do embryonic cells acquire and maintain their determined states? External factors such as gravity, light, or the geometry of the egg shell do not seem to have determinative effects in insect embryos. Rather, the earliest determinative steps are controlled by localized cytoplasmic factors, which are therefore called cytoplasmic determinants. The first to be discovered was basophilic granular material, also known as the polar disk or oosome, located at the posterior pole of many Coleoptera, Diptera, and Hymenoptera [Anderson, 1972; Eddy, 1975]. Hegner [1908], as others before him, had observed that these components were included in the pole cells. He concluded that the oosome was a germ cell determinant or a visible marker associated with such a determinant. In support of this view, he found that after puncture or cauterization of the posterior pole, no germ cells were formed in the adult. The most conclusive evidence that posterior pole plasm of *Drosophila* contains pole cell determinants came from a heterotopic transplantation experiment (below).

A different cytoplasmic determinant which is also localized at the posterior pole has been observed in the egg of the leaf hopper, *Euscelis plebejus*. Sander [1960, 1961] has used a cluster of symbiotic bacteria at the posterior pole as a visible marker in experiments involving the displacement of posterior pole material. By combining such translocations with egg ligations, Sander was able to demonstrate long-range effects of posterior pole material on the specification of the segment pattern. After ligation during cleavage, anterior fragments of *Euscelis* eggs produced only procephalic or no embryonic structures at all. Such fragments became capable, however, of forming complete embryos if posterior pole material was shifted anteriorly before ligation. In cases in which embryos in anterior fragments were less than complete, supernumerary segments were "added on" to form, e.g., a complete head and the first thoracic segment. These results suggest that the posterior pole material does not necessarily determine the character of a particular segment but may act to complete the segment pattern in a stepwise manner.

Many important questions surrounding the concept of cytoplasmic determinants are unresolved. There is no agreement on how many determined states they can evoke in cells. The oosome case suggests that at least some determinants are *bidirectional*—i.e., their presence vs. absence appears to direct a decision between two determined states. Other determinants, like the posterior pole material in *Euscelis*, have been viewed as *multidirectional*, i.e., graded levels of their activity are thought to release more than two determined states. Multidirectional determinants showing steady decline along a body axis are known as *gradients*. Another important question is whether determined states of cells are controlled by *absolute* activity levels of determinants or by *relative* activity levels of antagonistic pairs of determinants. We also know little about the origin and maintenance of the localization of cytoplasmic determinants, and hardly anything about the way they interact with chromatin. Obviously, the analysis of any of these problems would be greatly facilitated if the molecular nature of cytoplasmic determinants were known. Evidence is accumulating in a few cases that cytoplasmic ribonucleoprotein particles act as determinants. In the following, I will review data on three sets of cytoplasmic determinants in dipteran eggs, i.e., pole cell determinants, determinants of anteroposterior polarity, and determinants of dorsoventral polarity.

POLE CELL DETERMINANTS

Germ cell determination in animals is a basic and well-reviewed subject [Beams and Kessel, 1974; Eddy, 1975; Nieuwkoop and Sutasurya, 1981]. A most detailed series of investigations has been carried out on pole cell determination in *Drosophila* [see Mahowald et al., 1979]. *Drosophila* mutants showing abnormal germ line development, e.g., *grandchildless*, will be discussed elsewhere (Mahowald, this volume). I will therefore give only a short account of the work on *Drosophila* and provide more coverage of data obtained with other dipterans.

Relationship Between Pole Cells and Germ Cells

Pole cells are defined by a set of characteristics including their origin at the posterior pole, their round shape and larger size in comparison to blastoderm cells, and their early segregation and slow rate of mitotic divisions. Germ cells are defined as cells whose sole fate is to become spermatogonia or oogonia. Among the dipterans, the Nematocera have few pole cells, all of which appear to become included in the gonads. The Cyclorrhapha tend to have large numbers of pole cells, e.g., *Drosophila melanogaster* has about 50. Less than half become included in the gonads [Underwood et al., 1980b], and probably a smaller fraction actually contributes to gamete formation [Illmensee, 1978]. What happens to pole cells not included in the gonads is controversial.

A number of investigators [see Anderson, 1972] have proposed that *Drosophila melanogaster* pole cells contribute to the formation of secondary vitellophages and midgut epithelium. On the other hand, Underwood et al. [1980b] maintain that those pole cells which do not find their way into gonads either become trapped in neighboring tissues, degenerate in the yolk, or are excreted. They refute the validity of data which led earlier authors to propose that pole cells can also give rise to somatic structures. In my opinion the earlier view is still supported by two independent lines of evidence. First, Poulson and Waterhouse [1960] prevented pole cell formation by UV-irradiation of the posterior pole. As a result, not only were the gonads devoid of germ cells but also the cuprophilic midgut cells were missing or reduced in number. Underwood et al. [1980b] argue that the UV-irradiation might have damaged some adjacent regions determined to form cuprophilic midgut cells. However, the existence during early cleavage of a special anlage for cuprophilic midgut cells is unlikely. Second, cells derived from a single genetically marked pole cell may contribute to both germ line and midgut [Illmensee et al., 1976]. These observations indicate that pole cells are in a preliminary state of determination which includes the capability of producing germ cells but does not exclude contributions to somatic structures. Therefore, I prefer the term "pole cell determination" to "germ cell determination" in cases in which pole cells may contribute to somatic structures.

Pole Cell Determination in *Drosophila*

A key experiment in the analysis of pole cell determination in *Drosophila* has been the heterotopic transplantation of posterior pole plasm to the anterior pole of a host embryo during early intravitelline cleavage [Illmensee and Mahowald, 1974]. The transplantation caused, in an otherwise somatic region, the formation of cells with the morphological characteristics of pole cells. These cells contained posterior pole plasm from the donor and a nucleus with the genetic markers of the host. After transplantation to the posterior pole of a second host carrying different genetic markers, the cells became integrated into the gonads of the second host. There they gave rise to functional gametes, as shown by the progeny, some of which carried the genetic markers of the first host. Such mixed progeny were not derived from second hosts which had been control injected with normal anterior blastoderm cells. Another important control experiment was carried out by Okada et al. [1974] and Warn [1975], who "rescued" *Drosophila* embryos sterilized by

UV. The capabilities of forming pole cells and normal gonads were restored by microinjection of posterior but not anterior pole plasm.

Histochemical and electron microscopical observations on polar granules have provided important clues to pole cell determination in *Drosophila*. Whether polar granules actually function as pole cell determinants or serve only as visible markers is unresolved. Their ultrastructure and association with other organelles change during the life cycle of *Drosophila* [see Mahowald et al., 1979], although heterospecific transplantations show that they are probably not self-replicating [Mahowald et al., 1976]. Localization of polar granules at the posterior pole is already visible in stage 10 oocytes, but the capacity of posterior pole plasm to determine pole cells was not found before stage 13 of oogenesis [Illmensee et al., 1976]. This has been interpreted in terms of ooplasmic maturation before stage 13 [Mahowald et al., 1979]. However, it should be borne in mind that pole cell determinants might not be a permanent component of polar granules but only be temporarily associated with them. Histochemical evidence indicates that polar granules contain RNA and protein. However, the RNA moiety seems to disappear from the granules upon pole cell formation [see Mahowald et al., 1979]. Because polar granules become surrounded with polysomes when their RNA content is decreasing, Mahowald has proposed that polar granules release mRNA coding for germ-cell-determining proteins.

Recently, Ueda and Okada [1982] had prepared, from whole *Drosophila* eggs, a 27,000 g precipitate containing ribosomes, other RNP particles, and possibly polar granule material. This fraction, upon injection to the posterior pole, restored the ability of forming pole cells to UV-irradiated embryos. However, those pole cells did not develop into germ cells. Moreover, the fraction failed to induce pole cell formation ectopically, in contrast to unfractionated posterior cytoplasm [Illmensee and Mahowald, 1974]. The same limited "rescue" was also obtained with RNA extracted from the 27,000 g precipitate [Togashi and Okada, 1982]. These observations indicate that pole cell formation alone requires at least one UV-sensitive and one UV-resistant factor. At least one other factor seems to be necessary for germ cell development, and each of these factors appears to be localized at the posterior pole.

UV Inhibition of Pole Cell Formation in *Smittia*

With eggs of the chironomid midge *Smittia* sp., we have recently obtained experimental evidence that pole cell formation and gonad development require the activity of a nucleic acid-protein complex localized near the posterior pole [Brown and Kalthoff, 1983]. Both pole cell formation and gonad development were inhibited by UV-irradiation of the posterior pole without a reduction in the viability of the embryos. The UV effect was highly region

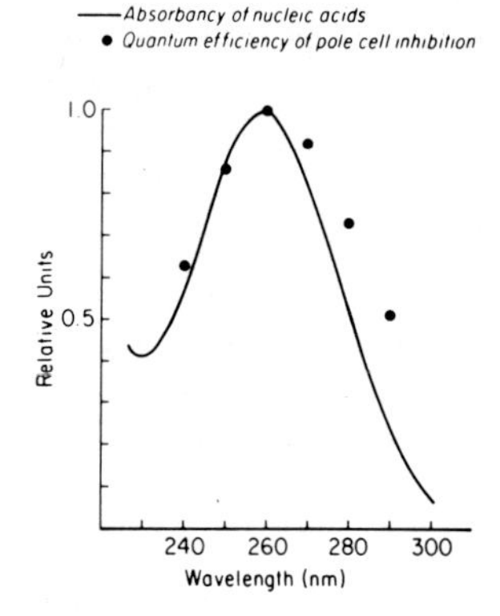

Fig. 2. Action spectrum for UV inhibition of pole cell formation in eggs of *Smittia* sp. Circles represent relative quantum efficiencies, after correction for wavelength dependence of quantum energy and transmittance of chorion. Curve represents nucleic acid absorption spectrum.

specific, since microbeam-irradiation of the oosome region inhibited pole cell formation, whereas irradiation of other regions did not. There was also a considerable stage specificity. Until one hour after deposition, i.e., until 1.5 hours before the first pole cell buds off in unirradiated controls, small UV doses caused significant pole cell inhibition. Thereafter, the UV dose required for the same level of inhibition increased steeply. The action spectrum for UV inhibition of pole cell formation in *Smittia* (Fig. 2) follows closely the absorption spectrum of nucleic acids, suggesting that a nucleic acid moiety acts as a primary absorber in the effective targets. The somewhat increased quantum efficiencies between 270 and 290 nm indicate an association of the nucleic acid moiety with a slightly photosensitizing protein(s). Independent evidence for the involvement of a nucleic acid is provided by the fact that UV inhibition of pole cell formation and gonad development in *Smittia* is photoreversible. Exposure to light of 380 or 440 nm wavelength after UV inhibition partly restored both pole cell formation and gonad development [Brown and Kalthoff, 1983]. Such photoreversible UV effects in *Smittia* eggs have been ascribed to inactivation of RNA by UV-induced pyrimidine dimers [see Kalthoff and Jäckle, 1982].

Chromosome Elimination

Posterior pole material also protects the germ line cells from chromatin loss in certain dipterans, which eliminate a number of chromosomes from their somatic cells. This phenomenon has been observed in species from three families, i.e., Chironomidae [Bauer and Beermann, 1952], Cecidomyidae, and Sciaridae [reviewed by Beams and Kessel, 1974]. Elimination of the germ-line-limited chromosomes (termed "extra" or E chromosomes) occurs during intravitelline cleavage. The E chromosomes are left behind close to the metaphase plate while the somatic or S chromosomes become included in the daughter nuclei. The mechanisms and control factors in this process are not known, but elimination is prevented if some posterior pole material is present at the time when elimination would otherwise occur.

Geyer-Duszynska [1959] has analyzed the effects of various experimental treatments on chromosome elimination and subsequent development of germ line cells in *Wachtliella persicariae*. By temporary ligation, nuclei were barred from access to posterior pole plasm during the fourth cleavage when chromosome elimination occurred. As a result, all nuclei underwent chromosome elimination. When such nuclei populated the posterior pole upon removal of the ligature, pole cells formed which Geyer-Duszynska describes as normal except that they contained only S chromosomes. Similar results were obtained with *Mayetiola destructor* by Bantock [1970], who also studied gonad development after UV-irradiation or temporary ligation. In both males and females, only the germ-line-derived portions of the gonads were affected, whereas the somatically derived parts were normal. Thus, the function of the E chromosomes seems indispensable for normal gametogenesis. Kunz et al. [1970] have presented evidence that mRNA is transcribed from the E chromosomes in the oocyte nucleus. An experimental dissection of pole cell determinants from the agent protecting against chromosome elimination was attempted by Geyer-Duszynska [1959]. However, evidence crucial to her argument could not be reproduced by Wolf [1969].

DETERMINANTS OF ANTEROPOSTERIOR POLARITY

The insect embryo develops from an ovoid of blastoderm cells and is thus two-dimensional. One dimension corresponds to the anteroposterior axis. The second extends laterally from the ventral midline of the early embryo. As its flanks grow around the yolk, the mediolateral direction is bent ventrodorsally. I will refer to it as the dorsoventral axis. The anteroposterior and dorsoventral axes of the embryo are foreshadowed by the egg architecture. The anterior pole of the chorion is marked by an opening, the micropyle, which is used in sperm entry. The ventral side of the egg is often more convex than the dorsal side (Fig. 1). Thus, the typical dipteran egg, like most other insect eggs, has a bilateral symmetry [Anderson, 1972]. We will now examine how the longitudinal body pattern of the embryo develops from the anteroposterior polarity of the egg.

The Gap Phenomenon

The gap phenomenon has become a hallmark of experimental insect embryogenesis [see Sander, 1976]. Insect embryos can be separated into anterior and posterior fragments by ligation or by pinching between razor blades. If both fragments develop into partial germ bands, the anterior one will always include procephalic structures, whereas the posterior one will include the most posterior abdominal segments. If fragments are separated at cellular blastoderm or later stages, the resulting partial germ bands complement each other so that together they form a virtually complete pattern. However, if the operation is carried out during intravitelline cleavage, some intermediate segments are missing from both partial germ bands, constituting a gap in the longitudinal body pattern.

While this phenomenon may be ascribed in part to local damage inflicted by the separation procedure [Vogel, 1977], the gap also reflects a disruption of interactions between anterior and posterior fragments which seem necessary for a complete body pattern to be formed. This is shown most clearly in cases in which a gap is avoided when a second operation follows early separation. Such an example is the combined ligation and translocation experiment of Sander [1960, 1961] with *Euscelis* embryos. In another experiment, temporary pinching of *Drosophila* embryos with a razor blade caused separation of the fragments by two plasma membranes, along which one or two layers of blastoderm cells were formed [Newman and Schubiger, 1980]. The development of this barrier could be prevented by puncturing the plane of separation with a fine glass needle, allowing formation of complete larvae, whereas pinching alone produced a large gap in the segment pattern [Schubiger et al., 1977]. The transverse layer of plasma membranes and/or blastoderm cells may thus prevent interactions between anterior and posterior fragments which are necessary for development of the complete segment pattern. Newman and Schubiger [1980] have applied a fate-mapping procedure to both posterior fragments and normal embryos. Aside from the considerable scatter in their data, the results indicate that blastoderm cells around 40% EL, which normally contribute to the third or fourth abdominal segment, become part of the fifth to seventh segment in posterior fragments. No corresponding shift in the fate of blastoderm cells was found in anterior fragments.

Anterior Determinants in Chironomid Embryos

Four distinct body patterns. For an analysis of cytoplasmic determinants involved in the determination of anteroposterior polarity, chironomid eggs are especially suitable. Following various experimental interferences they give rise to four distinct body patterns, i.e., normal embryos, inverted embryos, double cephalons, and double abdomens (Fig. 3). The normal body pattern consists of a head, three thoracic segments, and nine abdominal segments (Figs. 1, 3a). The inverted embryos (Fig. 3b) look exactly like the normal ones except that their anteroposterior polarity is reversed and their pole cells are sometimes trapped in head structures [Yajima, 1978, 1983].

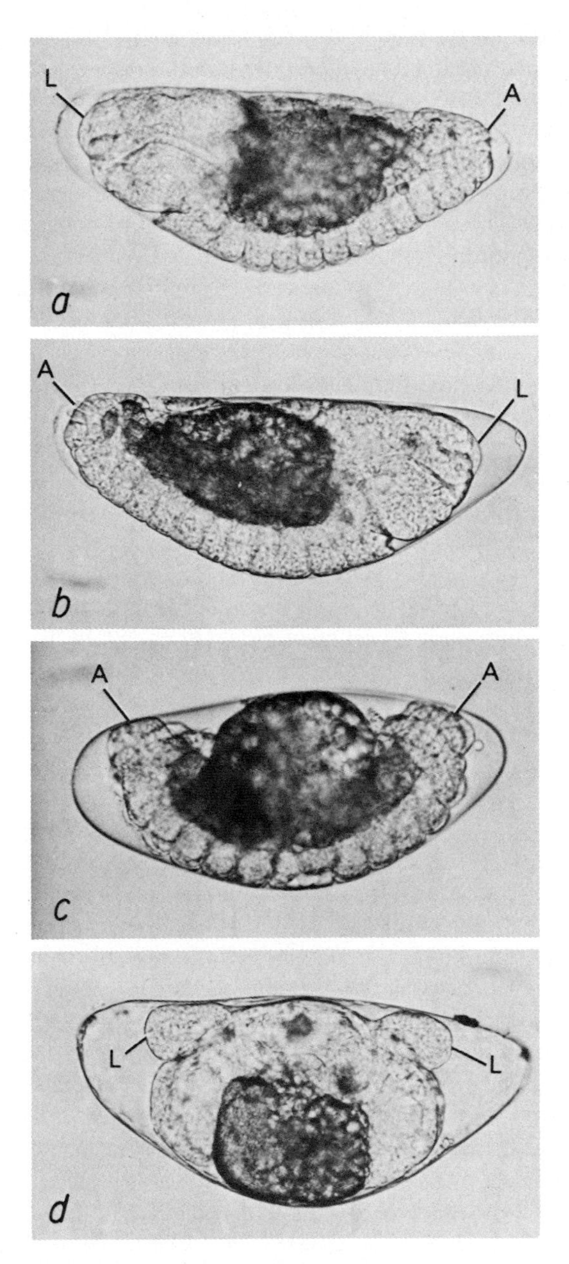

Fig. 3. Four basic body patterns generated in *Smittia* embryos after centrifugation during intravitelline cleavage (6,600g, posterior pole centrifugal). Dorsal side up, anterior pole to the left. L, labrum; A, anal papillae. a. Normal embryo. b. Inverted embryo. c. Double abdomen. d. Double cephalon. From Rau and Kalthoff [1980].

The inverted embryos develop in reverse orientation relative to the egg shell and to the yolk endoplasm [Rau and Kalthoff, 1980]. In the double abdomens (Fig. 3c) head and thorax are replaced by a mirror-image duplication of the abdomen. In *Smittia* sp., most double abdomens are symmetrical in their external and internal structures except for pole cells and gonads, which are found only in the posterior abdomen [Gollub, 1970; Sander, unpublished]. Double cephalons show a mirror-image duplication of anterior head structures including labrum, eyes, and stomodeum, while thorax and abdomen are missing altogether (Fig. 3d). Pole cells are found in the posterior but not the anterior half [Yajima, 1970], or in neither half [Lohs-Schardin, 1982].

The body patterns described above have been observed in the suborder Nematocera, including the families Chironomidae, Culicidae [see Kalthoff, 1979], and Sciaridae [Perondini et al., 1982], and in *Drosophila*, i.e., a representative of the suborder Cyclorrhapha. Double abdomens have also been obtained in the pea beetle, *Callosobruchus* [Van der Meer, 1978], and in the leaf hopper, *Euscelis* [Sander, 1960, 1961]. Mechanisms underlying the generation of the normal and abnormal body patterns are probably common to dipterans, and perhaps shared to some extent by other insects. Spontaneous occurrence of double abdomens has been observed in laboratory colonies of the pitcher plant mosquito, *Wyeomyia smithii* [Price, 1958], and of *Chironomus* sp.. In the latter case, the abnormal body pattern is associated with a change in the appearance of the fourth chromosome (Wacker and Kalthoff, unpublished). Moreover, chironomid midges seem most suitable for generating abnormal body patterns by various types of epigenetic interference (Fig. 4). Double cephalons and double abdomens, along with inverted and normal embryos, develop after centrifugation [Yajima, 1960, 1983; Kalthoff et al., 1977, 1982]. Either double cephalons or double abdomens can also be produced by UV-irradiation of posterior or anterior pole regions, respectively [Yajima, 1964, 1983; Kalthoff and Sander, 1968]. While the yield of double cephalons after posterior UV-irradiation of various chironomids has been less than 1% in our hands, double abdomen yields after anterior UV-irradiation are virtually 100% with *Smittia* sp. [see Kalthoff, 1979] and *Chironomus* sp. (Kalthoff, unpublished). Double abdomens in *Smittia* are also formed after puncture at [Schmidt et al., 1975] or application of RNase to [Kandler-Singer and Kalthoff, 1976] the anterior pole. The induction of double abdomens by anterior UV-irradiation is photoreversible, i.e., the UV effect is counteracted by subsequent exposure to light of longer wavelengths. This phenomenon is ascribed to light-dependent repair of UV damage to cytoplasmic RNA [see Kalthoff and Jäckle, 1982]. The sensitive period during which abnormal body patterns can be induced, and photoreverted after UV, extends from egg deposition to early blastoderm formation [see Kalthoff, 1979; Kalthoff et al., 1982].

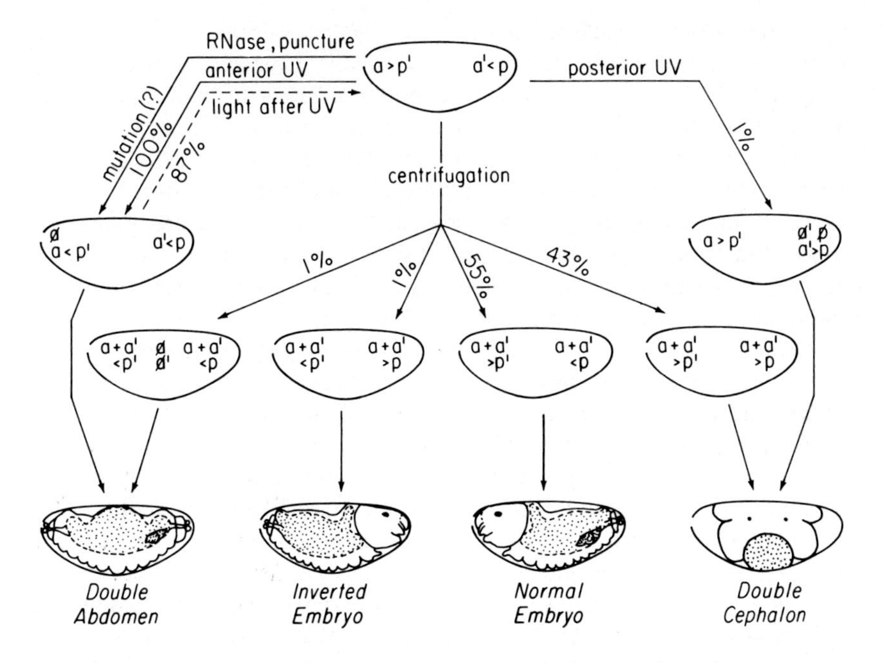

Fig. 4. Model of anteroposterior determination in *chironomid* embryos. Dorsal side up, anterior pole to the left. a, a' = anterior determinants; p, p' = posterior determinants. Crossed letters indicate partial inactivation or loss of determinant. Determination for anterior vs. posterior development is thought to occur independently in each egg half, according to the relative strength of anterior and posterior determinants (symbolized by $<$ or $>$). Application of UV or RNase to the anterior pole, which causes double abdomen formation with high yield, is thought to inactivate mainly a, rendering p' dominant. Exposure to visible light after UV appears to reactivate a. Posterior UV irradiation, which causes double cephalon formation with low yield, is thought to inactivate a' and p to a similar extent, rendering a' dominant only in exceptional cases. The four body patterns found after centrifugation are double cephalons, normal embryos, inverted embryos, and double abdomens. They are ascribed to a variable redistribution of a and a' after centrifugation. Percentages indicate frequency of body pattern obtained with *Smittia* under certain experimental conditions. From Kalthoff et al. [1982].

Anteroposterior decisions in both egg halves. The development of the four basic body patterns in chironomid embryos has been studied by observation of living embryos, microcinematography, and histological sections [Kalthoff and Sander, 1968; Yajima, 1970; Kalthoff, 1975; Rau, 1980; Rau and Kalthoff, 1980]. The way the germ anlage forms is characteristic of each body pattern, which can clearly be distinguished thereafter. However, embryos programmed to form any of these patterns all look alike during the blastoderm stage. Double cephalons and double abdomens do not result from the inability to form posterior or anterior, respectively, blastoderm cells. Rather, the ability to form either head or abdomen is inherent to both the anterior and the posterior half of the embryo. Even an isolated anterior fragment, after UV-irradiation, can develop into an abdomen with reversed polarity [Ritter, 1976; Sander, unpublished, reviewed in Kalthoff, 1979].

In the following model (Fig. 4) I propose that the four basic body patterns observed in chironomid embryos reflect anteroposterior decisions made independently in both anterior and posterior halves. Each half is thought to contain anterior (a,a') and posterior (p,p') cytoplasmic determinants. These are viewed as mutually repressive, and the prevailing determinant is thought to cause a general determination for anteriorness or posteriorness (anteroposterior decision). In a normal embryo, a is thought to prevail over p' in the anterior half, while p prevails over a' in the posterior half. The data suggest that the activity levels of p and p' are similar, whereas the activity of a is thought to exceed the activity of a' by a relatively small margin (about 20%). This would explain why slightly abnormal redistributions of a and a' after centrifugation can lead to the development of four different body patterns in the same batch of centrifuged embryos. On the other hand, inactivation of a small portion of a, e.g., by UV-irradiation of a small anterior target area, would be sufficient to produce double abdomens. Anterior UV-irradiation appears to inactivate a more efficiently than p', thus rendering p' prevalent in virtually every case. However, posterior UV-irradiation seems to inactivate p and a' with similar efficiency, rendering a' prevalent only in rare cases. The different sensitivities of a, a', p, and p' to inactivation by UV are ascribed to differences in shielding and/or association with photosensitizing components. Such differences would also explain species-specific variations, e.g., in double abdomen yield after irradiation with a standard UV dose.

The model outlined above represents a modification of an earlier version [Kalthoff, 1978, 1979] in which the determination for anterior vs. posterior was ascribed to the *absolute* level of anterior determinant activity. In the new version, the anteroposterior decision is controlled by the *relative* strength of anterior and posterior determinants. This modification has been made to accommodate new data, some of which are shown in Figure 5. UV-irradiation of the posterior pole region before centrifugation caused a significant increase in the yields of double cephalons and inverted embryos relative to control embryos that were only centrifuged (Fig. 5b). This result was unexpected on the basis of the old model but is explained by the new version. Irradiation of the posterior pole region is assumed to reduce both p and a' so that a small excess of $a + a'$ resulting from centrifugation is more likely to prevail over the remaining p. Figure 5a shows a similar experiment except

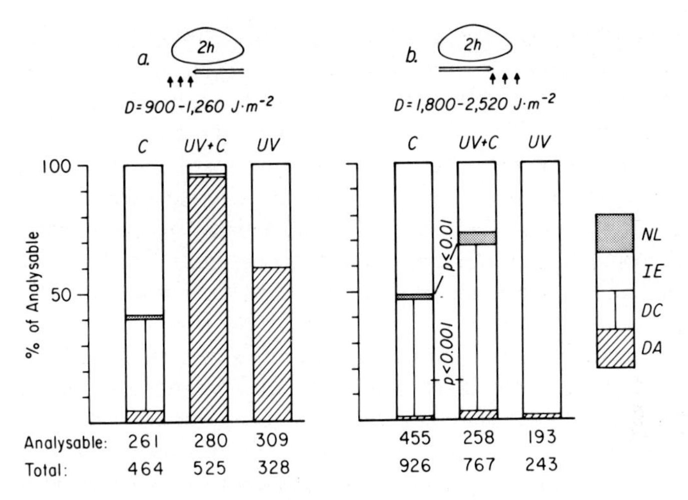

Fig. 5. Modifying effects of UV-irradiation before centrifugation. Diagrams show effects of perpendicular irradiation of the anterior (a) or posterior (b) quarter. D, UV dose. Columns marked UV + C represent embryos which were UV-irradiated at two hours and centrifuged at 3.5 hours. Columns marked C and UV represent the control embryos that were only centrifuged or only UV-irradiated. Portions of columns represent the frequency of normal larvae (NL), inverted embryos (IE), double cephalons (DC), and double abdomens (DA) among the analysable embryos. *P* indicates levels of statistical significance according to chi-square test. From Kalthoff et al. [1982].

that the anterior pole region was irradiated, leading to an increase in double abdomens at the expense of the other three body patterns requiring anterior determinant activity. Comparison with the UV-irradiation control also shows a synergistic effect of UV-irradiation and centrifugation in producing double abdomens. These and other results [Kalthoff et al., 1982] corroborate the basic idea behind the new (and old) model that the same anterior determinants that are inactivated by UV-irradiation are also shifted and redistributed after centrifugation.

Characterization of anterior determinants in *Smittia*. The anterior determinants in *Smittia* eggs have been characterized as cytoplasmic ribonucleoprotein (RNP) particles, based on action spectrum data [Kalthoff, 1973], localized application of RNase and other enzymes [Kandler-Singer and Kalthoff, 1976], and microbeam irradiation of stratified components in centrifuged eggs [Kalthoff et al., 1977]. This characterization is also supported by the fact that double abdomen induction by UV is photoreversible [see Kalthoff, 1979; Ripley and Kalthoff, 1981]. UV-irradiation of *Smittia* embryos causes pyrimidine dimer formation in RNA, while photoreverting light leads to the disappearance of dimers and restores the translatability of mRNA in

UV-irradiation embryos [see Kalthoff and Jäckle, 1982]. Moreover, our UV action spectra indicate that in vivo generation of pyrimidine dimers in *Smittia* RNA involves associated proteins or other photosensitizing components [see Kalthoff and Jäckle, 1982]. This is in line with the different sensitivities to UV of anterior and posterior determinants implied in our model (Fig. 4). Association with proteins or other components might modulate not only the sensitivity to UV but also the biological activity of the RNA sequence involved. An important ramification of this possibility is that the activity of such a determinant does not necessarily reflect the number of RNA copies present.

A gene regulatory function of the anterior determinants in *Smittia* embryos is indicated by the identification of two specific proteins foretelling the development of either head or abdomen (Fig. 6). A posterior indicator protein (designated PI_1, $M_r \sim 50{,}000$ daltons, IEP ~ 5.5) is synthesized in posterior but not in anterior halves of normal embryos during early blastoderm stages. An anterior indicator protein (designated AI_1, $M_r \sim 35{,}000$ daltons, IEP ~ 4.9) is synthesized in anterior but not in posterior halves of normal embryos during late blastoderm and early germ anlage stages. The synthesis of these

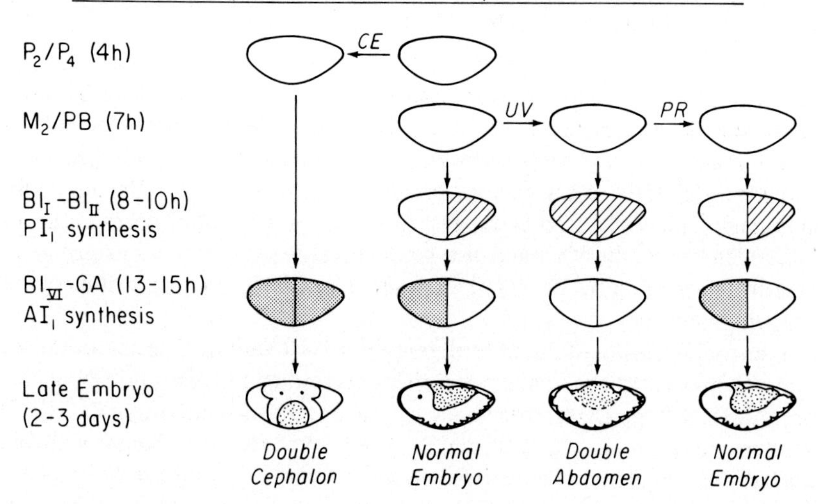

Fig. 6. Effects of UV-irradiation (UV), photoreversal (PR), and centrifugation (CE) on the synthesis of indicator proteins (AI_1 and PI_1). In normal embryos synthesis of PI_1 and AI_1 is region-specific at early blastoderm ($B1_I/B1_{II}$) and late blastoderm to germ anlage ($B1_{VI}$/GA) stages, respectively. Synthesis of AI_1 (PI_1) in a given embryonic half is indicated by hatching (stippling). Note that synthesis of AI_1 always foretells head development, whereas abdomen formation is always preceded by PI_1 synthesis.

proteins was monitored in embryos programmed for double cephalon, double abdomen, or normal development by centrifugation, UV-irradiation, and photoreversal, respectively. In each case, abdomen development in a given embryonic half is foreshadowed by PI_1 synthesis during early blastoderm and by the lack of AI_1 synthesis during late blastoderm and early germ anlage. Conversely, head development is correlated with the lack of PI_1 synthesis and the occurrence of AI_1 synthesis during the respective stages [Jäckle and Kalthoff, 1981]. These data (Fig. 6) cannot be explained by the assumption that double cephalons and double abdomens result from shifting or inactivating unspecific components such as ribosomes. The data are compatible with the idea that anterior and/or posterior determinants are maternal messenger RNP particles. However, other types of gene regulatory RNP particles cannot be excluded.

The topographical localization of anterior determinants within the anterior half is difficult to assess. The shape of the dose-response curves indicates a multiplicity of targets effective in UV-irradiation of double abdomens [Kalthoff, 1971]. The efficiency of UV-irradiation decreases as the target area is shifted posteriorly. For instance, induction of a given yield of double abdomens by microbeam irradiation of target area 1 in Figure 7 requires a smaller UV dose than is necessary with target areas 2, 3, 4. Also, the double abdomen yields resulting from microbeam irradiation of given target areas show a major stage dependence (Fig. 7). Irradiation of target area 1 is effective early and declines during nuclear migration, whereas irradiation of target areas 2, 3, and 4 becomes more effective during later stages. While this shift can be interpreted several ways, a most straightforward explanation would be a shift of the anterior determinants from a concentrated localization in target area 1 to a more dispersed distribution over the anterior surface (Ripley and Kalthoff, unpublished). This interpretation then raises the question of which forces move the anterior determinants.

Asymmetrical double abdomen patterns. While double abdomens are almost invariably symmetrical in *Smittia*, asymmetrical segment patterns with terminal abdominal structures on both ends are frequently observed in *Chironomus*. The typical double abdomen pattern of *Smittia* sp. is shown in Figure 3d. There are seven abdominal segments on both sides of an intermediate complex which is typically wider than one segment but does not exceed the length of two normal segments. Quite often, the intermediate complex shows irregularities such as small indentations. This pattern, with six, seven, or eight segments on either side of the intermediate complex, was found in more than 80% of the double abdomens induced by UV-irradiation under various experimental conditions [Bathe, 1977]. Less than 10% were symmetrical without a clearly recognizable intermediate complex, and less than 10% were asymmetrical, with either the anterior or the posterior abdo-

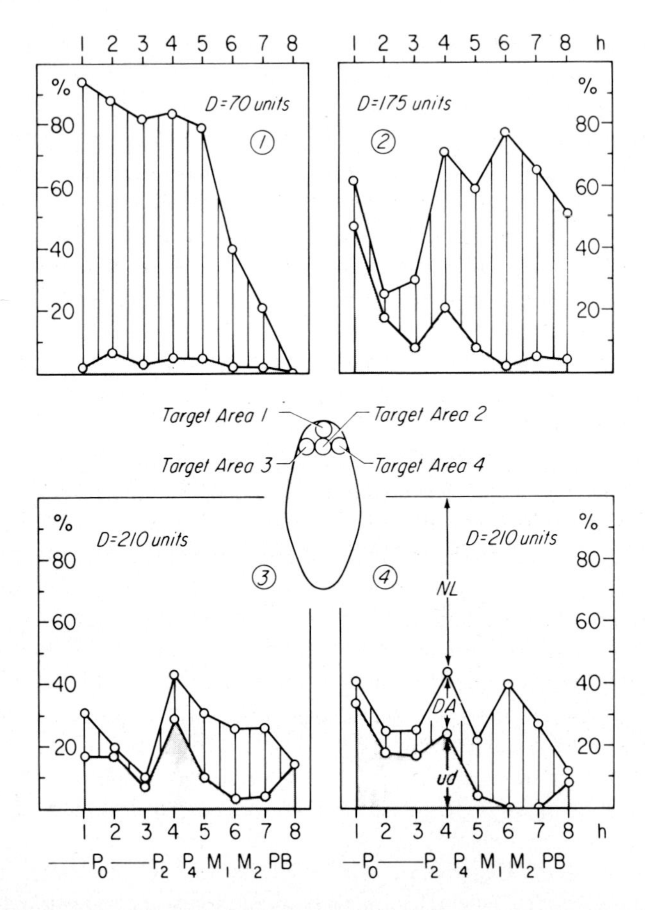

Fig. 7. Microbeam irradiation of different target areas at successive stages of development. Target areas 1, 2, 3, and 4 (focal diameter 20 μm) are indicated in outline of egg. Abscissa: age of embryos in hours after deposition, and corresponding morphological stages at 20°C. P_0, P_2, P_4: zero, two, or four pole cells. M_1, M_2: nuclear migration into periplasm. PB, preblastoderm. BL_I, beginning blastoderm formation. Ordinate, frequency of normal larvae (NL), double abdomens (DA), and undifferentiated results (ud). D, UV dose in relative units. Note that frequency of double abdomens after irradiation of target areas 2, 3, and 4 increases as irradiation of target area 1 becomes less effective.

men having one less segment. Embryos in which either the anterior or the posterior abdomen was missing were observed after irradiation of large areas, with high UV doses, and toward the end of the sensitive period, i.e., at early blastoderm stages [Ripley and Kalthoff, 1981]. One *Smittia* double abdomen with a major asymmetry was found after centrifugation. It had the normal series of thoracic and abdominal segments, with a small number of abdominal segments including the anal papillae attached to the prothorax with reversed polarity [Rau, 1980].

Asymmetrical double abdomens were obtained after oblique centrifugation of *Chironomus dorsalis* [Yajima, 1960], and after anterior UV irradiation of *Chironomus* sp. (Kalthoff, unpublished). These can be grouped tentatively as follows: (1) embryos truncated anterior to the prothorax, (2) embryos with normal thoracic and abdominal segments plus another set of terminal abdominal structures attached to the prothorax with reversed polarity, and (3) embryos appearing normal from abdomen to gnathocephalon but with procephalic structures replaced by terminal abdominal ones. Such asymmetrical double abdomens were also formed spontaneously, although with low frequency, in clusters of an apparently mutant *Chironomus* strain (Wacker and Kalthoff, unpublished). Body patterns with juxtaposed segments that are not normally adjacent are of particular interest. They are difficult to reconcile with models involving continuous gradients of multidirectional determinants. Rather, the asymmetrical patterns suggest that segment characters might be determined sequentially starting at each end.

Drosophila Mutants Mimicking Chironomid Double Abdomens and Double Cephalons

Two *Drosophila melanogaster* mutants with major effects on the longitudinal body pattern have been isolated. Both are maternal effect mutants, i.e., the phenotype of the embryo is controlled by the genotype of the mother but is independent of the genotype of the sperm. Each of the two mutants produces a range of normal body patterns rather than one distinct phenotype.

The mutant *bicaudal* was originally isolated and described by Bull [1966] and has been reinvestigated by Nüsslein-Volhard [1977, 1979]. The mutation is recessive and hypomorphic, i.e., the phenotypes seem to result from a lower than normal amount of the *bicaudal* gene product. In addition to the genetic constitution of the mother, her age and the ambient temperature were found to have major effects on the frequency of *bicaudal* embryos. Temperature shift experiments suggest a sensitive period at the onset of vitellogenesis. The phenotypes of *bicaudal* embryos have been classified as bicaudal, headless, or short [Nüsslein-Volhard, 1977]. In the bicaudal class, anterior structures were replaced by duplications of posterior structures with reversed polarity. Most frequent within this class were symmetrical bicaudal embryos which showed a mirror image duplication of the most posterior one to five (usually three) segments. These were symmetrical in their external and internal structures except that pole cells were only in the normal (posterior) half. Since the segment length was not much enlarged, bicaudal embryos were considerably shorter than normal. Other embryos, referred to as headless bicaudals, consisted of a complete abdomen terminating at the anterior end with a pair of spiracles. Between these and the symmetrical bicaudal types, a range of asymmetrical phenotypes was found. Attempts to phenocopy the *bicaudal* mutant syndrome by anterior UV-irradiation of *Drosophila* wild-type eggs or embryos have in rare cases produced the headless bicaudal phenotype [Bownes and Kalthoff, 1974; Bownes and Sander, 1976]. Symmetrical bicaudal phenotypes could not be phenocopied by UV-irradiation. Why *Drosophila* and *Chironomids* react differently to UV is unclear, but anterior determinants in *Drosophila* may be better shielded from, or more sensitive to, UV so that only a small fraction can be activated by sublethal doses.

The *Drosophila* mutant *dicephalic* [Lohs-Schardin and Sander, 1976; Lohs-Schardin, 1982] has been characterized as semidominant since some heterozygous females produced offspring with the mutant phenotype. The penetrance and expressivity were affected by genetic background and by temperature during oogenesis. In follicles of the *dicephalic* phenotype, the 15 nurse cells were not all in their normal position anterior to the oocyte, but occurred at both poles of the oocyte. Usually, the nurse cells were divided into clusters of six and nine, or seven and eight, with the larger cluster located either anteriorly or posteriorly. The mutant egg shell showed micropyles at both ends. Embryos had no pole cells, two cephalic furrows, and two pairs of largely normal head lobes. Few embryos developed sufficiently to secrete a cuticle. Among 86 of these, 68 cases showed anterior structures at both ends with a plane of polarity reversal between. Some of these anterior duplications included thoracic and even anterior abdominal segments. In 15 out of 86 cases, the segment pattern revealed only a single polarity but segments were misarranged or fused. Finally, in three cases, posterior structures occurred at both ends with a plane of polarity reversal in the fifth, sixth, or seventh abdominal segment.

The *Drosophila bicaudal* and *dicephalic* mutant phenotypes have obvious similarities to the double abdomens and double cephalons produced epigenetically in chironomids. In both groups, there are mirror-image duplications of anterior or posterior segments and similar ranges of asymmetrical double abdomens. However, the numbers of duplicated segments differ considerably. While chironomid double abdomens tend to have 12 or more segments, *Drosophila bicaudals* have ten or less. Conversely, the *Drosophila dicephalic* embryos include thoracic and even abdominal segments in contrast to the chironomid double cephalons which are duplications of less than a complete head.

In order to explain the *bicaudal* and *dicephalic* phenotypes on the basis of the model introduced above (Fig. 4), I suggest that anterior determinants are produced in the nurse cells and transferred to the oocyte during oogenesis.

This hypothesis is compatible with the following observations and conclusions: (1) Double abdomens can be induced in *Smittia* eggs as early as ten minutes after deposition, indicating that the effective targets, i.e., the anterior determinants, have been synthesized earlier. (2) Anterior determinants have been characterized as cytoplasmic ribonucleoproteins, which are known to be transferred from nurse cells to oocytes [see Berry, 1982]. (3) The preferred localization of anterior determinants in the anterior pole region could result from influx at the anterior pole of the oocyte. (4) As abnormal body patterns in *Smittia* are ascribed to faulty redistributions of anterior determinants after centrifugation (Fig. 4), the *dicephalic* phenotypes in *Drosophila* appear to result from abnormal distributions of the nurse cells, considered to be the sources of anterior determinants. Note that the ratio of double cephalons over double abdomens was 68:3 among the *dicephalic* embryos [Lohs-Schardin, 1982] and 1,341:55 among centrifuged *Smittia* embryos [Rau and Kalthoff, 1980]. (5) The *bicaudal* mutant is considered deficient in the synthesis or transfer of anterior determinants, in accord with its maternal effect type and hypomorphic character.

Sectorial Restrictions of Anteroposterior Polarity Reversals

In most of the pattern abnormalities discussed so far, the character of a given segment seems to be identical for its entire dorsoventral circumference. However, exceptions have been observed [Van der Meer, 1978]. In the pea beetle, *Callosobruchus*, unilateral double abdomens were produced which had a double abdomen pattern with polarity reversal on one side, whereas the normal pattern of cephalic, thoracic, and abdominal segments was shown on the contralateral side. In other cases, polarity reversals of the double abdomen type were restricted not to a lateral half, but to an even smaller sector of the circumference. Similar types of sectorial restrictions have been observed in *Drosophila*. In some *dicephalic* embryos, the levels of polarity reversal were found to be in different segments for the right half and the left half of the body [Lohs-Schardin, 1982]. In some of the *bicaudal* phenotypes, only the dorsalmost posterior structures, i.e., the spiracles, were found to be duplicated at the dorsal anterior end of the embryo [Nüsslein-Volhard, 1977].

The cases of sectorial restrictions in polarity reversal are of considerable interest because they are difficult to reconcile with models of pattern specification that rely on concentration gradients of diffusible determinants. The sectorial restrictions involve juxtapositions of cuticular structures which do not normally occur in the same segment or in neighboring segments. How discontinuities in the concentration of a diffusible substance could originate and be maintained in insect embryos before cellularization would seem very difficult to explain.

Embryonic Mutations Affecting the Longitudinal Body Pattern

The longitudinal body pattern of insects is characterized by a partitioning into segmental elements, by segment-specific characters of each element, by the sequence in which these elements are arranged, and by the polarities of each element and of the sequence. All are evidently affected in a coordinate fashion by the maternal effect mutations and experimental manipulations described above. On the other hand, there are embryonic (zygotic) mutations which alter only one or two pattern characteristics while leaving the others unaffected. The properties of these mutants indicate that the embryonic genome is involved in several partially independent processes which together generate the longitudinal body pattern.

Segment boundaries in dipteran embryos first become visible after gastrulation as a series of ectodermal bulges and furrows. In the larval hypodermis and/or cuticula, these boundaries are visible as furrows or constrictions. Cells at a larval segment boundary are assumed to be derived from cells at the corresponding embryonic boundary. Gynandromorph maps tracing larval segments back to the embryonic blastoderm indicate a degree of clonal restriction between precursor cells of different segments [Szabad et al., 1979]. Electrical coupling between cells across segment boundaries is similar to coupling within segments, whereas the transfer of dyes was found to be restricted at segment boundaries [Warner and Lawrence, 1982] and intrasegmental compartment boundaries [Weir and Lo, 1982]. Several mutants of *Drosophila melanogaster* alter the normal partitioning of the larva into segmental units and, in some of the cases, the polarity of parts of the segments. Fifteen loci of this type have been identified by Nüsslein-Volhard and Wieschaus [1980]. In six, a defined band from each segment was replaced by a mirror-image duplication of the remainder. Another group of mutants was characterized by a pair-rule. Homologous portions of segments were skipped in every other segment. Because the deleted portions included segment boundaries, the resulting larvae showed enlarged composite segments. Sander et al. [1980] have described the embryonic development of a *paired* mutant, clearly showing its double-segment character at the germ band stage. Kornberg [1981] has found similar phenotypes among 58 mutations of the *engrailed* locus. The mutant larvae also showed disoriented and fused denticle belts, indicating an inability to maintain segment boundaries.

A class of mutants, known as *homoeotic*, leaves the grid of segment boundaries unaffected but changes segmental characters. Homoeotic genes, and the *engrailed* gene, have been related through the compartment hypothesis to clonal restrictions and cell determination [Garcia-Bellido, 1975; Morata and Lawrence, 1977]. Homoeotic mutations cause the replacement of normal structures by structures normally found elsewhere, e.g., the substitution of an antenna by a leg (see Kaufman, this volume). Thus, homoeotic mutations have profound effects on intrasegmental patterns, including size and shape of a segment and its appendages, arrangement of bristles, tracheal openings, and other external and internal structures. They affect the larval and adult body pattern alike, and may be expressed in single segments, segmental subunits, or in several segments. However, homoeotic mutations do not alter segment polarity or the partitioning of the body into segmental units. An orderly subdivision into segments is maintained even in mutants in which the same intrasegmental pattern is serially repeated several times [Lewis, 1978; Struhl, 1981].

The regulation of homoeotic gene activity is of critical importance to understanding how the anteroposterior polarity of eggs and early embryos is translated into the complex pattern of distinct segments appearing later. In *Drosophila melanogaster*, homoeotic genes in the *bithorax* and *Antennapedia* complexes appear to be under negative and positive control of two genes called *Polycomb* and *Regulator of bithorax*, respectively [see Duncan and Lewis, 1982; Kaufman, this volume]. These regulator genes have both zygotic and maternal effects [Struhl, 1981; Duncan and Lewis, 1982]. Lewis [1978]has proposed that the activity of the *Polycomb* gene decreases posteriorly along the longitudinal axis of the embryo. Genes in the *bithorax* complex are thought to have different affinities for the *Polycomb* gene product that represses them. Thus, low concentrations of this product would inhibit only the *bithorax* genes with the highest affinities, whereas high concentrations of the *Polycomb* product would also repress those *bithorax* genes that have lower affinities.

Embryonic mutants showing abnormal segmental polarity include *engrailed* alleles. The posterior compartments in thoracic and other segments of these mutants are replaced with structures resembling mirror images of the corresponding anterior compartments [see Morata and Lawrence, 1977]. Thus, the normal anteroposterior polarity is reversed in the transformed compartments. Other polarity mutants affecting the orientation of hairs and bristles in several body regions have been described by Gubb and Garcia-Bellido [1982]. The mutant phenotypes were expressed autonomously in large but not in small homozygous clones.

The properties of the embryonic mutants described above suggest the following conclusions: (1) Although segment boundaries normally occur where cells of different segmental characters are juxtaposed, such juxtapositions are not necessary for segment boundaries to form. This is demonstrated by their occurence between repetitive segments of the same character. (2) Conversely, different segmental characters can be established without intervening segment boundaries, as indicated by the composite segments in *paired* mutants. (3) Finally, there is a conspicuous difference in the aberrations of the longitudinal body pattern caused by embryonic vs. maternal effect mutations. The latter, i.e., *bicaudal* and *dicephalic*, have fundamental effects on the overall body pattern, but leave the mesh of segmentation, segmental characters, and polarity intact. On the other hand, this mesh is disrupted by embryonic mutations which may alter segmental characters but conserve segmentation, or vice versa. Taken together, the conclusions suggest that embryonic genes involved in segmentation, specification of segment characters, and polarity can function independently of each other but are under common control of cytoplasmic determinants laid down during oogenesis [Sander et al., 1980].

Maternal Specification Vs. Embryonic Interaction

Behind the concept of cytoplasmic determinants is the basic idea that uneven distributions of cytoplasmic factors, directly or indirectly, cause differential gene activity in a spatial order. Versions of the general concept differ as to how much detail of the embryonic body pattern is ascribed to a maternally controlled "prepattern" of cytoplasmic determinants and how much is left to interactions between embryonic nuclei or cells. According to one set of models of longitudinal pattern formation in insect embryos, many determined states are encoded maternally, either by many bidirectional determinants (mosaic models), or by multiple local levels of one or two multidirectional determinants (gradient models). A different group of models requires only a terminal polarity, specifying, e.g., anteriorness vs. posteriorness, from cytoplasmic determinants and ascribes the rest to interactions between embryonic cells or energids (cell counting and sequential induction models). Some of these and other models have also been reviewed by Kauffman [1981] and Sander [1981].

Mosaic models postulate that the larval and adult body patterns develop under the control of a "mosaic" of many qualitatively different cytoplasmic determinants. Such a mosaic would have to be maintained despite considerable ooplasmic streaming in oocytes and early embryos [Gutzeit and Koppa, 1982; Fullilove et al., 1978]. So far the oosome is the only cytoplasmic component confined to a small egg area and associated with a strictly local determinative effect. Genetic screens for maternal effect embryonic lethals have yielded mutants showing localized defects in blastoderm formation [Rice and Garen, 1975] but no defects that could be associated with particular segments or small body regions [Gans et al., 1975; Mohler, 1977]. Among experimental manipulations of preblastoderm embryos, some produce the overall changes in body pattern described above. Smaller abnormalities seem to reflect a local disturbance in the formation of blastoderm cells rather than a change in the determination of surviving cells [Bownes and Sander, 1976]. None of these results supports the idea of a mosaic of qualitatively different determinants.

A model of antagonistic anterior and posterior gradients was proposed by Sander [1960, 1961] on the basis of his experiments with *Euscelis* embryos. Sander [1976, 1981] has used this idea as a unifying formal concept to explain a large body of experimental data but has also pointed out the difficulty in explaining discontinuities in asymmetrical and unilateral double abdomens on the basis of gradient models involving diffusible substances. These are also at variance with data of Vogel [1978, 1982b], who observed the development of normal segment patterns in middle fragments isolated from the presumed polar sources and in eggs with their anterior and posterior poles juxtaposed by double invagination. A gradient model based on two diffusible substances and lateral inhibition has been developed by Gierer and Meinhardt [1972] and applied to insect embryogenesis [Meinhardt, 1977]. This model, which has the virtue of being spelled out in two differential equations with numerical constants, can in fact explain several sets of experimental data. However, other results are at variance with major features of the model [Kalthoff, 1978; Vogel, 1978, 1982a,b]. Nüsslein-Volhard [1979: Fig. 6] has interpreted the *bicaudal* phenotypes in terms of Meinhardt's model. According to her diagram, the anterior terminal structures of headless *bicaudal* embryos originate between 90 and 100% EL. It is difficult to see how the normal topographical relationship to endoderm and gut ectoderm can be maintained under these conditions. Addressing the discontinuity problem in asymmetrical *bicaudals*, Nüsslein-Volhard suggests that segments might be skipped because the number of blastoderm cells available for the anterior abdomen could be too small to realize each segmental level in the steep anterior arm of the hypothetical gradient. If this were the case, segments should be skipped where the gradient is steepest, i.e., at the terminal end. However, this is not reported; the terminal structures seem always included if duplication occurs.

According to a "cell counting" model proposed by Wolpert [1969], the cell on one end of a body axis would acquire a determined state A_0, the next cell a different state A_1, its neighbor A_2, and so forth. Except for cell A_0, the determined state of a cell would depend on its rank among its neighbors. Thus, cytoplasmic determinants would merely establish a terminal polarity by specifying state A_0 at one end, and possibly another terminal state at the other end of an embryonic axis. In this respect, the cell counting model is fundamentally different from both mosaic and gradient models which, for all cells, rely on interactions between nuclei and their local cytoplasmic microenvironments.

In an attempt to fit the cell counting model to the specification of the longitudinal body pattern in dipteran embryos, one could assume that cells building the first cephalic and the tenth abdominal segment are determined by the local balance between anterior and posterior determinants (Fig. 4). Adjacent strips about three to four cells wide would be "counted off" for the second cephalic and ninth abdominal segments and so forth. This hypothesis would best explain why, in abnormal body patterns resulting from mutation or epigenetic interference, the terminal structures seem present and the first to appear in duplications. This dominance of terminal characters was demonstrated most dramatically by Yajima [1960], who removed, by puncture, graded amounts of cytoplasm from centrifuged *Chironomus* embryos. This caused the disappearance of a proportional number of middle, but not terminal, segments. The cell counting model would explain the symmetry observed in most chironomid double cephalons and double abdomens. However, asymmetrical cases could also be accommodated by assuming a slower rate of counting in the shorter portion, as proposed by Sander [1976, 1981], or by a delayed anteroposterior decision. Either of these assumptions could also account for differences in the number of segments between normal and various abnormal patterns. A problem with this hypothesis is how the counting can proceed from both ends in normal or inverted embryos without producing gaps or supernumerary segments in the middle. To avoid this difficulty, some proportioning or disproportion-correcting mechanism would seem desirable.

Mathematical models based on the sequential induction of locally exclusive states have been developed by Meinhardt and Gierer [1980]. Local exclusion is achieved by mutual repression or autocatalysis coupled with a common repressor. Sequential induction results from mutual enhancement via rapidly diffusing products. Such systems are most suitable for modeling intercalary regeneration because a gap in a sequence of determined states is filled by redetermination of cells next to the gap. By virture of the same properties, such systems could also specify a series of determined states in an embryo. In order to generate a segment pattern, only one or two terminal segments would have to be specified by localized cytoplasmic determinants. This application would avoid major shortcomings of both gradient and cell-counting models. The size regulatory properties of sequential induction should help to fill gaps that might occur in normal embryos with mere counting from both ends. Discontinuities in asymmetrical double abdomen patterns could be explained by a delay in the shorter end, which might leave inadequate time for regulation. Sectorially restricted polarity reversals could be ascribed to anisotropic transmission of signals between cells, which seems at least easier to conceive of than anisotropic diffusion within plasmodium. Unfortunately, the gap phenomenon is not accommodated by the sequential induction models of Meinhardt and Gierer [1980]. Separation of anterior from posterior fragments, according to these models, would result in twinning rather than missing segments. However, the problem might be overcome by introducing a rule about the last determined state that can be induced in a fragment. A rule of this type would also seem to be necessary for generating the segment pattern in the normal embryo. A series of inductive interactions in the specification of the longitudinal body pattern in insect embryogenesis has also been postulated on the basis of experimental data [Vogel, 1978, 1982a]. Vogel proposes that three cytoplasmic determinants (anterior, posterior, and thoracic) become localized during oogenesis and that the additional determined states result from short-range interactions.

THE DORSOVENTRAL POLARITY

Data on the specification of the transverse body pattern are sparse, and experimental studies on this topic have not made much use of dipteran eggs [see Sander, 1976]. The leaf hopper, *Euscelis plebejus*, is therefore used again as a paradigm to illustrate a fundamental difference in the specification of the transverse vs. the longitudinal body pattern. As described above, separation perpendicular to the longitudinal axis results in gap formation. By contrast, separation along the longitudinal axis results in twinning, i.e., complete embryos are formed in lateral, dorsal, and ventral halves [Sander, 1971]. In this respect, the longitudinal axis in insect embryos is similar to the animal-vegetal axis in sea urchin and other eggs.

In *Drosophila melanogaster*, maternal effect mutations have been found which alter the transverse body pattern. Published reports include a description of the mutant *dorsal* [Nüsslein-Volhard, 1979], which has a recessive and a dominant phenotype. The former is fully penetrant while the latter is temperature sensitive and depends on the genetic background of the female. The progeny lacks structures normally derived from ventral blastoderm cells. Neither phenotype forms a ventral furrow. In the recessive phenotype the dorsoventral polarity of the embryo seems entirely lacking. Blastoderm cells in *dorsal* dominant embryos have been mapped with a UV microbeam [Nüsslein-Volhard et al., 1980]. The results indicate that the absence of ventrally derived structures did not result from failure of cell formation or cell death. Rather, the fate of both ventral and lateral cells was altered so that they formed structures derived from more dorsal cells in normal embryos. The fates of the cells were not shifted in anteroposterior direction. The authors conclude that the mutation affects a continuous property, such as a morphogenetic gradient along the dorsoventral axis of the developing embryo. Nüsslein-Volhard [1979] ascribes the different mutant phenotypes to corresponding reductions in the height of a hypothetical gradient with a maximum along the ventral midline. However, Meinhardt [1981] notes that the transverse body pattern in *Drosophila* embryos would be better modelled by sequential induction than by a gradient based on autocatalysis and lateral inhibition.

SYNOPSIS

I have discussed cytoplasmic components which are involved in the determination of pole cells, the anteroposterior polarity, and the dorsoventral polarity in dipteran eggs. Evidence for the existence of these determinants is derived mainly from maternal effect mutations in *Drosophila melanogaster* and from experimental translocation, inactivation, and repair of cytoplasmic components in chironomid embryos. I will now review proposals which question the individual identity of these determinants, e.g., by postulating that pole cell determination and specification of the longitudinal body pattern are controlled by the same determinant, and then I will compare and summarize the properties of the above determinants (Table I).

An attempt to explain both the segregation of pole cells and the specification of the longitudinal body pattern by a single mechanism was made by Deak [1980]. His hypotheis represents an extension of Meinhardt's [1977] lateral inhibition model. Deak's additional provision is that during nuclear migration the posterior pole region acts as a sink for the diffusible "inhibitor" of Meinhardt's model. Deak proposes that the resulting low inhibitor concentration causes both pole cell determination and the autocatalytic buildup of an "activator" peak [see Meinhardt, 1977]. The ensuing generation of a posteroanterior inhibitor gradient would then serve to specify the somatic body pattern. Deak [1980] predicted that transfer of enough posterior pole plasm to the anterior end of the egg should result in the formation of double abdomen embryos. So far, such transplantations have led to the formation of ectopic pole cells but not double abdomens. Similarly, one should expect that inactivation by UV of the hypothetical inhibitor at the anterior pole should generate double abdomens with two sets of pole cells. However, UV-induced double abdomens had pole cells only in the posterior, i.e., normal, abdomen [Yajima, 1970; Gollub, 1970]. Generally, maternal mutations and experimental manipulations disrupting pole cell formation leave the somatic body pattern unaffected and vice versa. These observations are best accommodated by the assumption of separate determinants for pole cells and for the somatic body pattern.

A model that has the potential to unify specification of the longitudinal and the transverse body patterns was proposed by Kauffman [reviewed 1981]. The model is based on successive eigenfunction patterns of reaction-diffusion systems. These would cause transverse along with longitudinal subdivisions. In reviewing the weaknesses of this model, Kauffman [1981] points out its

TABLE I. Cytoplasmic Determinants in Dipteran Eggs

Determinants	Pole cell	Anterior	Ventral
Structures determined	Pole cells	Anterior body pattern	Ventral derivatives
Antagonists	(Factors causing chromosome elim.)	Posterior determinants	?
Mutations in *Drosophila*	Grandchildless	Bicaudal, dicephalic	Dorsal
Localization			
when	Oogenesis	(Oogenesis and embryogenesis)	?
where	Posterior pole	Anterior half	(Ventral)
Association with organelles	Polar granules, oosome	?	?
Molecular nature	(RNP particle)	RNP particle	?
Mode of action	Bidirectional	?	?

sensitivity to changes in egg shape. However, we have observed that longitudinal compression of *Smittia* eggs to about 75% of their normal length did not affect the development of either normal or double abdomen segment patterns [Bathe, 1977]. In the leaf hopper, *Euscelis plebejus*, the development of embryos with normal segment patterns was observed after much more dramatic alterations of the egg shape [Vogel, 1978, 1982b]. Moreover, a fundamental difference in the specification of the longitudinal and transverse body patterns is indicated by the observation of gaps vs. twinning after separation of fragments.

Thus, a minimum set of cytoplasmic determinants in dipteran eggs should include pole cell determinants and determinants of the anteroposterior and dorsoventral body axes. The polarity of the anteroposterior axis appears to be determined by antagonistic determinants in either egg half independently. Whether the dorsoventral body axis is also determined by an antagonistic pair of cytoplasmic components remains to be seen. The *Drosophila* mutant *dorsal*, which has been described only recently, suggests the existence of a ventral determinant, i.e., an ooplasmic component required for the development of ventral embryonic anlagen. There are no obvious antagonists to pole cell determinants, unless one wants to view as such the unknown components causing chromosome elimination from somatic cells in some species.

Localization of polar granules in *Drosophila* at the posterior pole is established in stage 10 oocytes. However, transplantable pole cell determinant activity is not localized at the posterior pole until stage 13 of oogenesis. No pole cell determinant activity has been detected in anterior egg cytoplasm. Anterior determinant activity is present in newly deposited chironomid eggs, suggesting that they become localized during oogenesis. During nuclear migration, anterior determinants seem to spread out from a concentrated localization near the anterior pole to the periphery of the anterior half. The posterior egg half appears to contain anterior determinants of subthreshold activity, since centrifugation is frequently followed by head development in both halves. There is no evidence for localization of posterior determinants. The ventral determinants deduced from the phenotypes of *dorsal* mutants of *Drosophila* should have some degree of ventral localization, since the apparent shift in the fates of blastoderm cells is strongest ventrally. The timing of this hypothetical localization is unknown.

The association of pole cell determinants with distinct polar granules might be intrinsic or fortuitous but in any case has greatly enchanced studies on pole cell determination. Oosomelike aggregates of electron-dense material have been found outside the posterior pole region in various insect eggs and oocytes [see Zissler and Sander, 1982], but no associations with cytoplasmic determinants have been established.

The molecular nature of anterior determinants in *Smittia* has been characterized as ribonucleoprotein (RNP) particles. The nature of their antagonists, the posterior determinants, might be the same since their inactivation by UV is photoreversible. At least some pole cell determinants might also be RNP particles, based on "rescue" experiments, the action spectrum for UV inhibition of pole cell formation, and the histochemical and ultrastructural observations on polar granules.

The mode of action of cytoplasmic determinants might be bidirectional or multidirectional; in the latter case, they are usually thought of as forming gradients. Pole cell determinants seem to be bidirectional because we observe only two determined states, i.e., pole cells vs. blastoderm cells. The anteroposterior polarity is determined by localized anterior determinants and antagonists acting as posterior determinants. In each egg half, the relative strength of these two determinants appears to control binary anteroposterior decisions, leading to the formation of either anterior or posterior terminal structures. In conjunction with sequential inductions or other interactions between embryonic cells, an anteroposterior decision in each egg half would seem sufficient to generate the four basic longitudinal body patterns observed in dipteran embryos. Alternatively, the anterior determinants might act in gradient fashion themselves, or be superimposed on other gradients specifying sequences from terminal to middle segments.

ACKNOWLEDGMENTS

I wish to thank Janet Young for drawing the illustrations, and Dr. Helen Pianka, Lewis Patterson, and Kris Wacker for help with preparing the manuscript. Our current research is supported by grant AI 15046 TMP from the National Institutes of Health.

REFERENCES

Anderson DT (1972): The development of holometabolous insects. In Counce SJ, Waddington CH (eds): "Development Systems: Insects." New York: Academic Press, Vol I, pp 165–242.

Bantock CR (1970): Experiments on chromosome elimination in the gall midge, Mayetiola destructor. J Embryol Exp Morphol 24:257–286.

Bathe MM (1977): Einflüsse auf die Zahl der Segmente in UV-induzierten "Doppelabdomina" bei Smittia spec. (Chironomidae, Diptera). Staatsexamensarbeit, Fakultät für Biologie der Universtät Freiburg.

Bauer H, Beerman W (1952): Der Chromosomencyclus der Orthocladiinen (Nematocera, Diptera). Z Naturforsch 7b:557–563.

Beams HW, Kessel RG (1974): The problem of germ cell determinants. Int Rev Cytol 39:413–479.

Berry SJ (1982): Maternal direction of oogenesis and early embryogensis in insects. Ann Rev Entomol 27:205–227.

Bownes M, Kalthoff K (1974): Embryonic defects in *Drosophila* eggs after partial UV-irradiation at different wavelengths. J Embryol Exp Morphol 31:1–17.

Bownes M, Sander K (1976): The development of *Drosophila* embryos after partial UV-irradiation. J Embryol Exp Morphol 36:394–408.

Brown PM, Kalthoff K (1983): Pole cell formation in Smittia spec. (Chironomidae, Diptera): Action spectrum for UV-inhibition and photoreversibility. Dev Biol 97 (in press).

Bull AL (1966): Bicaudal, a genetic factor which affects the polarity of the embryo in *Drosophila melanogaster*. J Exp Zool 161:221–242.

Chan LN, Gehring W (1971): Determination of blastoderm cells in *Drosophila melanogaster.* Proc Natl Acad Sci USA 68:2217–2221.

Counce SJ (1973): Causal analysis of insect embryogenesis. In Counce SJ, Waddington CH (eds): "Developmental Systems: Insects." New York: Academic Press, Vol II, pp 1–156.

Deak II (1980): Embryogenesis in *Drosophila*: Can a single mechanism explain the developmental segregation of germ-line and somatic cells and their subsequent determination? W Rouxs Arch 188:179–185.

Duncan I, Lewis EB (1982): Genetic control of body segment differentiation in *Drosophila.* In Subtelny S, Green PB (eds): "Developmental Order: Its Origin and Regulation." New York: Alan R. Liss, Inc., pp 533–554.

Eddy EM (1975): Germ plasm and the differentiation of the germ line. Int Rev Cytol 43:229–281.

Fullilove SL, Jacobson AG (1971): Nuclear elongation and cytokinesis in *Drosophila montana.* Dev Biol 26:560–577.

Fullilove SL, Jacobson AG, Turner FR (1978): Embryonic development: Descriptive. In Ashburner M, Wright TRF (eds): "The Genetics and Biology of *Drosophila*." London: Academic Press, Vol 2c, pp 105–227.

Gans M, Audit C, Masson M (1975): Isolation and characterization of sex-linked female sterile mutants in *Drosophila melanogaster.* Genetics 81:683–704.

Garcia-Bellido A (1975): Genetic control of wing disc development in *Drosophila.* In: "Cell Patterning." CIBA Foundation Symposium 29. Amsterdam: Associated Scientific Publishers, pp 241–263.

Geyer-Duszynska J (1959): Experimental research on chromosome elimination in Cecidomyidae (Diptera). J Exp Zool 141:391–447.

Gierer A, Meinhardt H (1972): A theory of biological pattern formation. Kybernetik 12:30–39.

Gollub G (1970): Zur Verteilung der Urgeschlechtszellen in Doppelabdomina von Smittia parthenogenetica (Chir.). Staatsexamensarbeit, Fakultät für Biologie der Universität Freiburg.

Gubb D, Garcia-Bellido A (1982): A genetic analysis of the determination of cuticular polarity during development in *Drosophila melanogaster.* J Embryol Exp Morphol 68:37–57.

Gutzeit HO, Koppa R (1982): Time lapse film analysis of cytoplasmic streaming during late oogenesis of *Drosophila.* J Embryol Exp Morphol 67:101–111.

Hegner RW (1908): Effects of removing the germ cell determinants from the eggs of some chrysomelid beetles. Preliminary report. Biol Bull 16:19–26.

Illmensee K (1978): *Drosophila chimeras* and the problem of determination. In Gehring W (ed): "Genetic Mosaics and Cell Differentiation." New York: Springer, pp 51–69.

Illmensee K, Mahowald AP (1974): Transplantation of posterior polar plasm in *Drosophila.* Induction of germ cells at the anterior pole of the egg. Proc Natl Acad Sci USA 71:1016–1020.

Illmensee K, Mahowald AP, Loomis MR (1976): The ontogeny of germ plasm during oogenesis in *Drosophila.* Dev Biol 49:40–65.

Jäckle H, Kalthoff K (1981): Proteins foretelling head or abdomen development in the embryo of *Smittia* spec. (Chironomidae, Diptera). Dev Biol 85:287–298.

Kalthoff K (1971): Position of targets and period of competence for the UV-induction of the malformation "double abdomen" in the egg of *Smittia* spec. (Diptera, Chironomidae). W Rouxs Arch 168:63–84.

Kalthoff K (1973): Action spectra for UV-induction and photoreversal of a switch in the developmental program of the egg of an insect (*Smittia*). Photochem Photobiol 18:355–364.

Kalthoff K (1975): *Smittia* spec. (Diptera) Normale Embryonalentwicklung. Aberration des Segmentmusters nach UV-Bestrahlung. Göttingen: Encyclop Cinematograph Film E 2158/1974.

Kalthoff K (1978): Pattern formation in early insect embryogenesis: Data calling for revision of a recent model. J Cell Sci 29:1–15.

Kalthoff K (1979): Analysis of a morphogenetic determinant in an insect embryo. In Subtelny S, Konigsberg I (eds): "Determinants of Spatial Organization." New York: Academic Press, pp 97–126.

Kalthoff K, Sander K (1968): Der Entwicklungsgang der Missbildung Doppelabdomen im partiell UV-bestrahlten Ei von Smittia parthenogenetica (Dipt., Chironomidae). W Rouxs Arch 161:129–146.

Kalthoff K, Hanel P, Zissler D (1977): A morphogenetic determinant in the anterior pole of an insect egg (*Smittia* spec., Chironomidae, Diptera). Localization by combined centrifugation and UV irradiation. Dev Biol 55:285–305.

Kalthoff K, Rau KG, Edmond JC (1982): Modifying effects of ultraviolet irradiation on the development of abnormal body patterns in centrifuged insect embryos (*Smittia* spec., Chironomidae, Diptera). Dev Biol 91:413–422.

Kalthoff K, Jäckle H (1982): Photoreactivation of pyrimidine dimers generated by a photosensitized reaction in RNA of insect embryos (Smittia spec.). In Helene C, Charlier M, Montenay-Garestier Th, Laustriat G (eds): "Trends in Photobiology." New York: Plenum, pp 173–188.

Kandler-Singer I, Kalthoff K (1976): RNase sensitivity of an anterior morphogenetic determinant in an insect egg (*Smittia* spec., Chironomidae, Diptera). Proc Natl Acad Sci USA 73:3739–3743.

Kauffman SA (1980): Heterotopic transplantation in the syncytial blastoderm of *Drosophila*: Evidence for anterior and posterior nuclear commitments. W Rouxs Arch 189:135–145.

Kauffman SA (1981): Pattern formation in the *Drosophila* embryo. Philos Trans R Soc Lond [Biol] 295:567–594.
Kornberg T (1981): Engrailed: A gene controlling compartment and segment formation in *Drosophila.* Proc Natl Acad Sci USA 78:1095–1099.
Krause G (1939): Die Eitypen der Insekten. Biol Zbl 59:495–536.
Krause G (1981): Homology studies on insect systems. Fortsch Zool 26:307–333.
Kunz W, Trepte HH, Bier K (1970) On the function of the germ line chromosomes in the oogenesis of *Wachtliella persicariae* (Cecidomyidae). Chromosoma 30:180–192.
Lamb MM, Laird ChD (1976): Increase in nuclear poly(A)-containing RNA at syncytial blastoderm in *Drosophila melanogaster* embryos. Dev Biol 52:31–42.
Lawrence PA, Green SM, Johnston P (1978): Compartmentalization and growth of the *Drosophila* abdomen. J Embryol Exp Morphol 43:233–245.
Lewis EB (1978): A gene complex controlling segmentation in *Drosophila.* Nature 276:565–570.
Lohs-Schardin M (1982): Dicephalic: A *Drosophila* mutant affecting polarity in follicle organization and embryonic patterning. W Rouxs Arch 191:28–36.
Lohs-Schardin M, Sander K (1976): A dicephalic monster embryo of *Drosophila melanogaster.* W Rouxs Arch 179:159–162.
Lundquist A, Emanuelsson H (1980): Polar granules and pole cells in the embryo of *Calliphora erythrocephala*: Ultrastructure and (^{3}H)-leucine labelling. J Embryol Exp Morphol 57:79–93.
Mahowald AP, Illmensee K, Turner FR (1976): Interspecific transplantation of polar plasm between *Drosophila* embryos. J Cell Biol 70:358–373.
Mahowald AP, Allis CD, Karrer KM, Underwood EM, Waring GL (1979): Germ plasm and pole cells of *Drosophila.* In Subtelny S, Konigsberg IR (eds): "Determinants of Spatial Organization." New York: Academic Press, pp 127–146.
McKnight SL, Miller OL Jr (1976): Ultrastructural patterns of RNA synthesis during early embryogenesis of *Drosophila melanogaster.* Cell 8:305–319.
Meinhardt H (1977): A model of pattern formation in insect embryogenesis. J Cell Sci 23:117–139.
Meinhardt H (1981): Pattern formation and the activation of particular genes. Fortsch Zool 26:163–174.
Meinhardt H, Gierer A (1980): Generation and regeneration of sequences of structures during morphogenesis. J Theor Biol 85:429–450.
Mohler JD (1977): Developmental genetics of the *Drosophila* egg. I. Identificaiton of 59 sex-linked cistrons with maternal effects on embryonic development. Genetics 85:259–272.
Morata G, Lawrence PA (1977): Homoeotic genes, compartments and cell determination in *Drosophila.* Nature 265:211–216.
Newman SM Jr, Schubiger G (1980): A morphological and developmental study of *Drosophila* embryos ligated during nuclear multiplication. Dev Biol 79:128–138.
Nieuwkoop PD, Sutasurya LA (1981): "Primordial Germ Cells in the Invertebrates." Cambridge: Cambridge U Press.
Nüsslein-Volhard C (1977): Genetic analysis of pattern formation in the *Drosophila melanogaster* embryo. Characterization of the maternal effect mutant bicaudal. W Rouxs Arch 183:249–268.
Nüsslein-Volhard C (1979): Maternal effect mutations that alter the spatial coordinates of the *Drosophila* embryo. In Subtelny S, Konigsberg IR (eds): "Determinants of Spatial Organization." New York: Academic Press, pp 185–211.
Nüsslein-Volhard C, Wieschaus E (1980): Mutations affecting segment number and polarity in *Drosophila.* Nature 287:795–801.

Nüsslein-Volhard C, Lohs-Schardin M, Sander K, Cremer C (1980): A dorso-ventral shift of embryonic primordia in a new maternal-effect mutant of *Drosophila.* Nature 283:474–476.

Okada M, Kleinman IA, Schneiderman HA (1974): Restoration of fertility in sterilized *Drosophila* eggs by transplantation of polar plasm. Dev Biol 37:43–54.

Perondini ALP, Gutzeit H, Mori L, Sander K (1982): Induktion von "Doppelabdomen" und die Differenzierung anterioren Polplasmas in Embryonen von Sciara. Verh Dtsch Zool Ges 1982.

Pietruschka F, Bier K (1972): Autoradiographische Untersuchungen zur RNS- und Proteinsynthese in der fruehen Embryogenese von Musca domestica. Wilh Rouxs Arch 169:56–69.

Poulson DF, Waterhouse DF (1960): Experimental studies on pole cells and midgut differentiation in Diptera. Aust J Biol Sci 13:541–567.

Price RD (1958): Observations on a unique monster embryo of *Wyeomia smithii* (Coquillett) (Diptera: Culicidae). Ann Entomol Soc Am 51:600–604.

Rau KG (1980): Experimente zur räumlichen Musterbildung in Embryonen von Chironomiden und Käfern. Staatsexamensarbeit, Fakultät für Biologie der Universität Freiburg.

Rau KG, Kalthoff K (1980): Complete reversal of antero-posterior polarity in a centrifuged insect embryo. Nature 287:635–637.

Rice TB, Garen A (1975): Localized defects of blastoderm formation in maternal effect mutants of *Drosophila.* Dev Biol 43:277–286.

Rickoll WL (1976): Cytoplasmatic continuity between embryonic cells and the primitive yolk sac during early gastrulation in *Drosophila melanogaster.* Dev Biol 49:304–310.

Rickoll WL, Counce SJ (1980): Morphogenesis in the embryo of *Drosophila melanogaster*: Germ band extension. W Rouxs Arch 188:163–177.

Ripley S, Kalthoff K (1981): Double abdomen induction with low UV-doses in Smittia spec. (Chironomidae, Diptera): Sensitive period and complete photoreversibility. W Rouxs Arch 190:49–54.

Ritter W (1976): Fragmentierungs- und Bestrahlungsversuche am Ei von Smittia spec. (Chironomidae, Diptera). Staatsexamensarbeit, Fakultät für Biologie der Universität Freiburg.

Sander K (1960): Analyse des ooplasmatischen Reaktionssystems von Euscelis plebejus Fall. (Cicadina) durch Isolieren und Kombinieren von Keimteilen. II. Mitteilung: Die Differenzierungsleistungen nach Verlagern von Hinterpolmaterial. W Rouxs Arch 151:660–707.

Sander K (1961): New experiments concerning the ooplasmatic reaction system of *Euscelis plebejus* (Cicadia). In: "Symposium on Germ Cells and Development." Inst Intern d'Embryologie Pallanza, pp 338–353.

Sander K (1971): Pattern formation in longitudinal halves of leaf hopper eggs (*Homoptera*) and some remarks on the definition of "embryonic regulation." W Rouxs Arch 167:336–352.

Sander K (1976): Specification of the basic body pattern in insect embryogenesis. Adv Insect Physiol 12:125–238.

Sander K (1981): Pattern generation and conservation in insect ontogenesis: Problems, data and models. Fortsch Zool 26:101–119.

Sander K, Lohs-Schardin M, Baumann M (1980): Embryogenesis in a *Drosophila* mutant expressing half the normal segment number. Nature 287:841–843.

Schmidt O, Zissler D, Sander K, Kalthoff K (1975): Switch in pattern formation after puncturing the anterior pole of *Smittia* eggs (Chironomidae, Diptera). Dev Biol 46:216–221.

Schubiger G, Moseley RC, Wood WJ (1977): Interaction of different egg parts in determination of various body regions in *Drosophila melanogaster.* Proc Natl Acad Sci USA 74:2050–2053.

Seidel F (1926): Die Determinierung der Keimanlage bei Insekten. I. Biol Zentralblatt 46:321–343.

Steiner E (1976): Establishment of compartments in the developing leg imaginal discs of *Drosophila melanogaster.* W Rouxs Arch 180:9–30.

Struhl G (1981): A gene product required for correct initiation of segmental determination in *Drosophila.* Nature 293:36–41.

Szabad J, Schüpbach T, Wieschaus E (1979): Cell lineage and development in the larval epidermis of *Drosophila melanogaster.* Dev Biol 73:256–271.

Togashi S, Okada M (1982): Restoration of pole cell forming ability to UV-sterilized *Drosophila* embryos by injection of an RNA fraction extracted from eggs. In Akao H et al. (eds): "The Ultrastructure and Functioning of Insect Cells." Soc fo Insect Cells, Japan, pp 41–44.

Turner FR, Mahowald AP (1977): Scanning electron microscopy of *Drosophila melanogaster* embryogenesis. II. Gastrulation and segmentation. Dev Biol 57:403–416.

Ueda R, Okada M (1982): Induction of pole cells in sterilized *Drosophila* embryos by injection of subcellular fraction from eggs. Proc Natl Acad Sci USA 79:6946–6950.

Underwood EM, Turner FR, Mahowald AP (1980a): Analysis of cell movements and fate mapping during early embryogenesis in *Drosophila melanogaster.* Dev Biol 74:286–301.

Underwood EM, Caulton JH, Allis CD, Mahowald AP (1980b): Developmental fate of pole cells in *Drosophila melanogaster.* Dev Biol 77:303–314.

Van der Meer JM (1978): Region specific cell differentiation during early insect development. PhD Thesis, Faculteit der Wiskunde en Natuurwetenschappen, Katholieke Universiteit te Nijmegen.

Vogel O (1977): Regionalisation of segment-forming capacities during early embryogenesis in *Drosophila melanogaster.* W Rouxs Arch 182:9–32.

Vogel O (1978): Pattern formation in the egg of the leaf hopper, Euscelis plebejus Fall. (Homoptera). Developmental capacities of middle egg fragments after isolation from the polar regions. Dev Biol 67:357–370.

Vogel O (1982a): Experimental test fails to confirm gradient interpretation of embryonic patterning in leafhopper eggs. Dev Biol 90:160–164.

Vogel O (1982b): Development of complete embryos in drastically deformed leafhopper eggs. W Rouxs Arch 191:134–136.

Warn R (1975): Restoration of the capacity to form pole cells in UV-irradiated *Drosophila* embryos. J Embryol Exp Morphol 33:1003–1011.

Warner A, Lawrence P (1982) The permeability of gap junctions at the segmental border in insect epidermis. Cell 28:243–252.

Weir MP, Lo CW (1982): Gap junctional communication compartments in the *Drosophila* wing disk. Proc Natl Acad Sci USA 79:3232–3235.

Wieschaus E, Gehring W (1976): Clonal analysis of primordial disc cells in the early embryo of *Drosophila melanogaster.* Dev Biol 50:249–263.

Wolf R (1969): Kinematik und Feinstruktur plasmatischer Faktorenbereiche des Eies von Wachtliella persiariae L. (Diptera). II Teil: Das Verhalten ooplasmatischer Teilsysteme nach Zentrifugierung im 4 Kern-Stadium. W Rouxs Arch 163:40–80.

Wolpert L (1969): Positional information and the spatial pattern of cellular differentiation. J Theor Biol 25:1–48.

Yajima H (1960): Studies on embryonic determination of the harlequin-fly, *Chironomus dorsalis.* I. Effects of centrifugation and of the combination with constriction and puncturing. J Embryol Exp Morphol 8:198–215.

Yajima H (1964): Studies on embryonic determination of the harlequin-fly, *Chironomus dorsalis.* II. Effects of partial irradiation of the egg by UV light. J Embryol Exp Morphol 12:89–100.

Yajima H (1970): Study of the development of the internal organs of the double malformations of *Chironomus dorsalis* by fixed and sectioned materials. J Embryol Exp Morphol 24:287–303.

Yajima H (1978): On the embryo showing reversed polarity induced by the centrifugation of *Chironomus samoensis* eggs. Zool Magazine (Tokyo) 87:343.

Yajima H (1983): Production of longitudinal double formation by centrifugation or by UV-partial irradiation in the *Chironomus samoensis* egg. Entomol Gener 8 (in press).

Zalokar M (1976): Autoradiographic study of protein and RNA formation during early development of *Drosophila* eggs. Dev Biol 49:425–437.

Zissler D, Sander K (1982): The cytoplasmic architecture of the insect egg cell. In King RC, Akai H (eds): "Insect Ultrastructure." New York: Plenum Press, Vol 1, pp 189–221.

Time, Space, and Pattern in Embryonic Development, pages 349–363

Genetic Analysis of Oogenesis and Determination

Anthony P. Mahowald

Developmental Biology Center and Department of Developmental Genetics and Anatomy, Case Western Reserve University, Cleveland, Ohio 44106

Extensive analysis of oogenesis has shown that the oocyte in most species contains abundant stores of cytoplasmic components required for much of early development. In addition, the zygotic genome draws upon yolk reserves (except in mammals) to carry out all the necessary biosynthetic processes needed to produce an autonomously feeding organism. While extensive molecular analyses [cf., Davidson, 1976] have shown the presence of abundant maternal RNA of high complexity, we have only limited information on the nature of the informational content of these RNAs. Experimental manipulations, especially with nucleocytoplasm hybrids, have clearly demonstrated that the ooplasmic constituents play essential roles in early cellular events. In a few instances, such as *Drosophila* germ plasm [Illmensee et al., 1976], the determinative nature of this ooplasmic information has been shown. There are many other examples of early segregations of cytoplasm which are clearly dependent upon ooplasmic components (e.g., gray crescent of amphibians, polar lobe of annelids and moluscs, anterior determinants in *Smittia*). Some years ago [Gans et al., 1975; Rice, 1973; Mohler, 1977] it became apparent that a thorough genetic analysis of oogenesis in *Drosophila melanogaster* might uncover the range of embryological events which are dependent upon the oocyte genome.

While these screens are still in progress in many laboratories, sufficient information is available on some of these mutations to warrant current discussion. In addition to the possibility of saturating the *Drosophila* genome for particular classes of mutations [cf., Nüsslein-Volhard and Wieschaus,

1980], *Drosophila* has a number of other features that make it ideal for a comprehensive genetic analysis of oogenesis. It is possible to obtain 10–100 gm of eggs and early embryos with relative ease [Allis et al., 1977]. It is also possible to produce ovarian chimeras so as to determine critically the cellular basis of each mutation [Illmensee, 1973; Wieschaus et al., 1981; Perrimon and Gans, 1983]. Finally, it is possible to utilize both the genetic and cytogenetic features of *Drosophila* to produce unique combinations of mutations for a complete genetic analysis.

BRIEF OVERVIEW OF OOGENESIS

Detailed descriptions of oogenesis have been presented recently [Mahowald and Kambysellis, 1980] so that only a brief overview is needed here. Oogenesis in *Drosophila* is especially amenable to both genetic and molecular analysis. The ovary is composed of two separate cell lineages: the germ line derived from the pole cells, and somatic components derived from the mesoderm (Fig. 1). The germ line produces 15 nurse cells and one posteriorly located oocyte by a series of four synchronous divisions of the cytoblast at the anterior end of the germarium (Fig. 2). The nurse cells are responsible for producing nearly all of the RNA-containing components (ribosomes, mRNA, RNA, etc.) of the mature oocyte. The oocyte nucleus lacks a typical nucleolus during most of oogenesis and shows only a brief period of RNA synthesis [Mahowald and Tiefert, 1970] during a portion of vitellogenesis. Nevertheless, it expands greatly, achieving a volume of 400 μm^3 in stage 12.

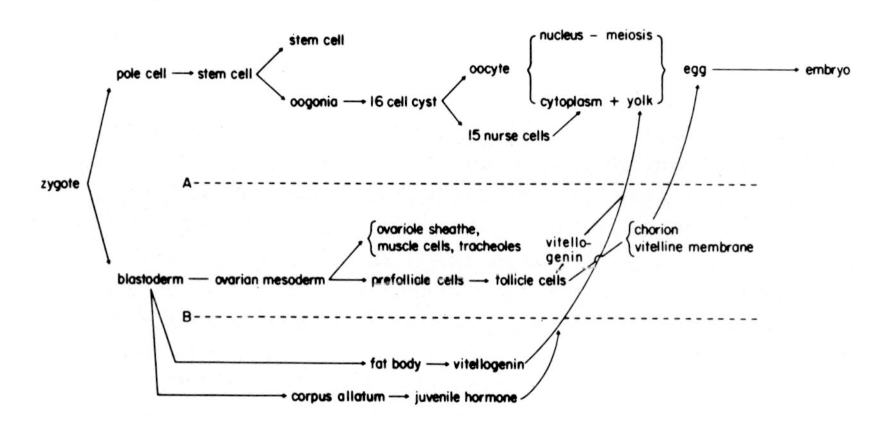

Fig. 1. Cell lineages for the ovary of *Drosophila*. The dotted lines indicate the lineages which can be separated via ovary transplants (B) [King and Bodenstein, 1965] and pole cell transplants or x-ray-induced genetic mosaicism (A). Figure is adapted from Mahowald [1979].

No one has yet determined the nature of the constituents of this expanded nucleus or germinal vesicle. Because of its location at the future anterior dorsal region of the oocyte, concentration of specific oocyte components within the germinal vesicle would be a mechanism for localizing them at the anterior tip.

The somatic components of the gonad produce a series of muscle cells of both the ovariole and ovary sheath. In addition, approximately 80 mesoderm-derived follicle cells become associated with the nurse cell-oocyte composite at the base of the germarium. After four rounds of mitoses producing approximately 1,100 cells, the cells undergo polyploidization to 45C at stage 12 [Mahowald et al., 1979b]. At stages 8–10, the follicle cells become active in synthesis of vitellogenin [Brennan et al., 1982], vitelline membrane [Petri et al., 1976; Fargnoli and Waring, 1982], and finally the chorion [Petri et al., 1976; Spradling and Mahowald, 1979b; Waring and Mahowald, 1979]. Many of the genetic loci for these processes have been identified [Postlethwait and Jowett, 1980; Barnett et al., 1980; Spradling et al., 1980; Griffin-Shea et al., 1980] and the DNA cloned. From an analysis of both the DNA sequence and mutations affecting the expression of these genes [e.g., Spradling and Mahowald, 1981], it should be possible to obtain a thorough understanding of the genetic control of follicular cell function.

CATEGORIES OF MUTATIONS

Gans et al. [1975] have produced a convenient classification of categories of female sterility mutations:

Class I—No or extremely few eggs are laid.

Class II—Mature eggs are produced and laid, but they are morphologically abnormal. Many are flacid, collapsed, and fail to develop.

Class III—Eggs are produced which begin development. Subclass A—Mutant females produce lethal embryos unless the sperm chromosomes include the wild-type allele. In these instances, some or all of the heterozygous embryos survive. Subclass B—Mutants produce lethal embryos which are not affected by the presence of the wild-type allele in the embryos. Subclass C—Mutants produce embryos lacking pole cells so that they produce sterile flies.

Mutations in class I affect early stages of oogenesis. Histological analysis has been carried out on many of these mutations [cf., King and Mohler, 1975], but detailed genetic analysis has been limited. These mutants could be due to failures in establishing either the germ-line derivatives or mesodermally derived components. Use of dominant female steriles, somatic mosaicism, or germ cell transplants will provide information concerning the cellular basis for the ovarian abnormality.

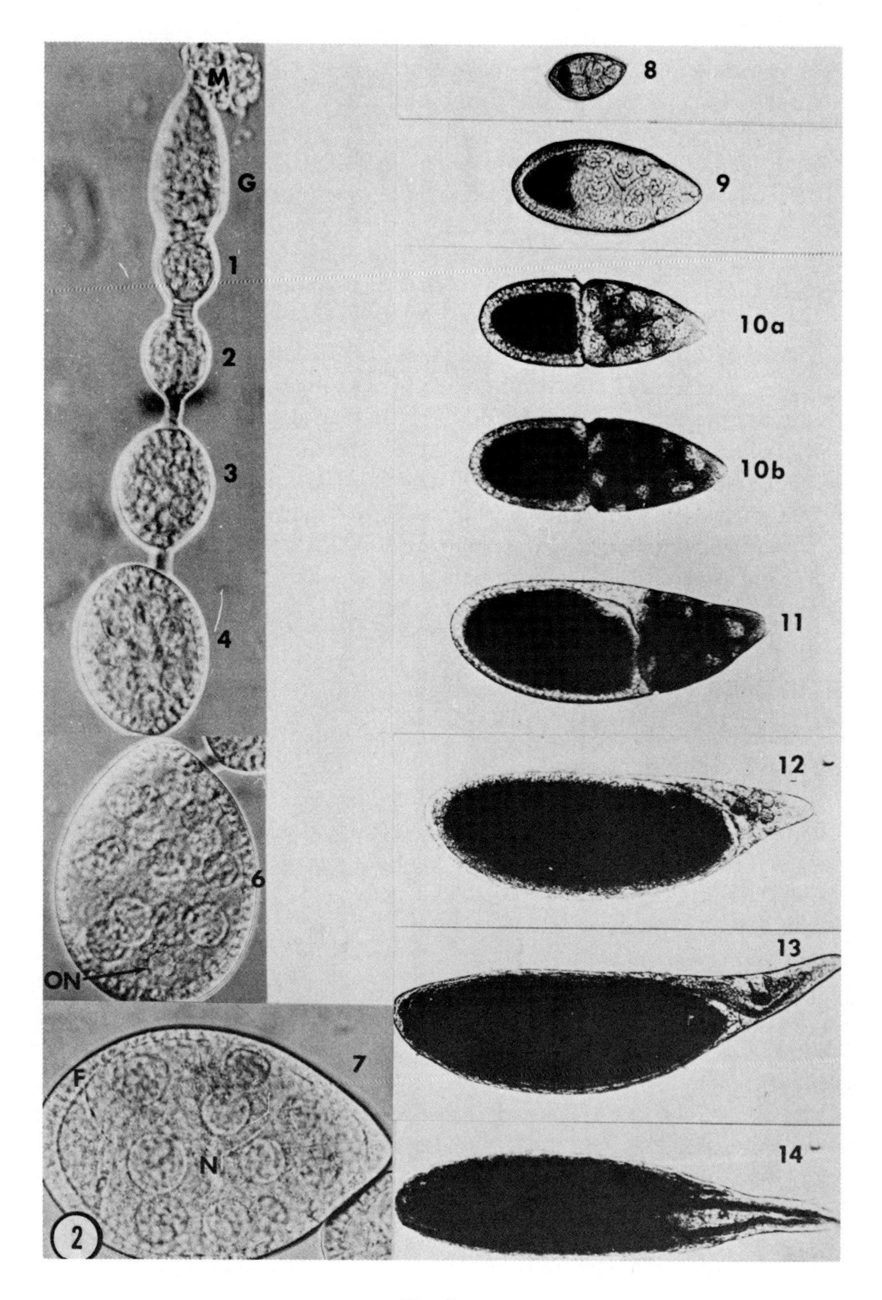

Fig. 2.

While class II phenotypes have been thought to be due to follicle cell dysfunction [Gans et al., 1975], this is certainly not necessarily true [cf., Wieschaus et al., 1978, 1981].

Finally, class III mutants are believed to be the most significant for the analysis of the informational content of eggs. Since the egg is produced with intact envelopes, follicle cell formation has been thought to be normal. In the few instances where this has been directly tested, this has been true.

CHIMERIC OVARIES AND THEIR IMPORTANCE

Not all mutations which affect follicle cell products (e.g., the chorion) produce their phenotype because of changes in the follicle cells. While *ocelliless* clearly affects choriogenesis via the follicle cells [Spradling and Mahowald, 1981; Underwood and Mahowald, 1980], *fs(1)K10* [Wieschaus et al., 1978] is dependent only upon the germ line for its defect even though the resulting phenotype is a grossly abnormal anterior chorion. Apparently, the mutant germ-line tissue causes a change in follicle cell orientation. This result suggests the important conclusion that the surface of the oocyte (on the assumption that the oocyte and follicle cells interact at their surface) determines the location and pattern of follicle cell function.

While pole cell transplants successfully produce chimeric follicles with germ-line components of a different genotype than the oocyte, the procedure is time consuming and not convenient for routine work. X-ray-induced somatic mosaics [Wieschaus et al., 1981] are simple and provide equivalent results. Recently, Perrimon and Gans [1983] have shown that the dominant female sterile, *fs(1)1237*, can be readily used to screen X-chromosomal sterility mutations for germ-line-specific mutations. In addition, *short eggs (seg),* which is follicle-cell specific, can be used to characterize somat-

---

Fig. 2. Light microscopic pictures of the stages of oogenesis in *Drosophila melanogaster*. In the germarium (G) one to three germ-line stem cells continually produce cytoblasts which divide synchronously four times to produce a cyst of 16 cells interconnected by intercellular bridges. Fifteen of these cells become the polyploid nurse cells (N) (stages 2–10) and one becomes the oocyte (ON). Follicle cells (F) surround the 16-cell cyst. The opacity of the oocyte in stages 8–14 is due to the accumulation of yolk. At stage 11 there is a dramatic change in the structure of the follicle. The cytoplasm and most of the karyoplasm of the nurse cells rapidly flows into the oocyte. Most of the follicle cells move around the anterior end of the oocyte where they synthesize the highly structured anterior end of the chorion. The duration of the stages in rapidly laying flies at 25°C has been determined [David and Merle, 1968]: 48 hours from stage 2 to the start of vitellogenesis; 16 hours for stages 8 through 10; stage 11 is only 25 minutes; finally, stages 12 through 14 constitute choriogenesis and require 4.7 hours. Figure from Mahowald and Kambysellis [1980]. Used with permission of Academic Press.

ically dependent mutations. When similar mutations are available for each chromosome arm, it should be routine to determine for every mutation the cellular basis for the female sterility. Because of the surprises already encountered with *fs(1)K10*, it is obvious that these tests are needed for careful interpretation of oogenesis.

SPECIFIC MUTATIONS AFFECTING POLE CELL FORMATION (CLASS III C)

Illmensee et al. [1976] showed conclusively that germ cell formation is dependent upon the presence of a characteristic germ plasm located at the posterior tip of the egg. Furthermore, they showed conclusively that the posterior polar plasm could function autonomously in ectopic locations (anterior or midventral) [Illmensee and Mahowald, 1974, 1976]. It was not surprising, then, to find that mutations which affect germ cell formation are maternal effect mutants. These have been called *grandchildless (gs)* [Spurway, 1948] because in the absence of germ cells the F1 progeny are sterile. Most of the *gs* mutations [cf., Mariol, 1981] found so far are pleiotropic and produce considerable embryonic lethality in addition to the failure to produce pole cells. Two mutations do not produce significant decreases in embryonic survival. The *grandchildless* mutation found in *D. subobscura* has the most specific polar plasm defect [Mahowald et al., 1979a]. During the terminal period of oogenesis localized regions of the posterior and possibly the anterior tip become segregated and the polar granules, the organelles unique to the germ plasm [Mahowald, 1968], within these regions disappear. Subsequently during cleavage nuclei fail to migrate into the posterior polar plasm properly, leading to no pole cells. The nature of this deficiency is still unknown. The mutant *agametic* in *D. melanogaster* [Engstrom et al., 1982] does not affect the formation of pole cells but it surprisingly results in pole cell death at the time when the embryonic gonads first form and the germ cells begin cell division. This cell death is autonomous for the mutant germ line and not rescued by the presence of wild-type germ line or somatic tissue.

Possibly the most interesting mutation is *tudor* [Wieschaus and Nüsslein-Volhard, 1983; Boswell and Mahowald, 1983]. Six alleles have been discovered with different degrees of expressivity for two phenotypes: (1) The amount of polar granule material varies from none in the extreme allele to approximately one third the amount found in wild-type eggs in the most moderate allele. This latter mutation is incompletely penetrant for the grandchildless phenotype as a homozygote but is completely agametic in combination with an extreme allele. (2) While flies homozygous for any of the *tudor* alleles produce embryos of which 30%–40% are lethal at late embryonic stages, the degree of abnormality correlates with the amount of polar granules: Extreme alleles with no polar granules produce lethal embryos with extreme segment deficiencies; moderate alleles with small polar granules produce lethal embryos with no segmental abnormalities.

We do not know the basis for this interesting connection between segmental organization and germ plasm. Since the germ plasm can function autonomously in the absence of a normal posterior blastoderm [Rice and Garen, 1975] or following transplantation [Illmensee and Mahowald, 1974], the interdependence of polar granules and segmentation is probably concerned with the establishment of the proper egg structure. For example, the mechanism for localizing polar granules may also function in establishing segmentation. Or the proper amount of polar granules (polar plasm) is needed for establishing the segmental pattern. At the least, these grandchildlesslike mutants clearly show that even a clear example of a localized determinant may be interdependent upon the remainder of the egg.

The mutant *tudor* is also important for displaying another correlation between the process of pole cell formation and the presence of germ plasm. Previously a detailed ultrastructural study of the original *grandchildless* in *D. subobscura* had correlated the failure to produce pole cells with a severe diminution of polar granules [Mahowald et al., 1979a]. In the mutant *tudor,* the amount of polar granule material correlates with the ability to produce pole cells. The nature of the mechanism responsible for this interrelation between polar granules and pole cell formation has not yet been determined. This relationship is consistent with the hypothesis that polar granules function in storing at the posterior tip the specific maternal mRNAs needed to produce pole cells [Mahowald, 1968].

MATERNAL EFFECT LETHAL MUTATIONS

Much current interest in the mutational analysis of oogenesis concentrates on a subset of female steriles (approximately 10% of the total) in which homozygous mutant females produce normal blastoderm stages but the embryos subsequently die. While many screens for these mutations have been carried out, only recently have attempts been made to exhaust the genome for these classes of female steriles. One interesting finding is that there are a number of different loci that produce equivalent developmental defects (cf. below). In the near future, detailed analyses will be possible on sets of mutations which can reasonably be assessed to affect common processes. We have recently begun investigations of specific mutations which belong to sets of loci, all of which produce either identical or very similar phenotypes.

Gastrulation-Defective Mutations (Class III B)

Zalokar et al. [1975] described the phenotype of a maternal effect mutation (A573) in which the embryos reach the blastoderm normally, but subsequently development became abnormal. Mohler [1977] found six alleles of A573 and we have found six more alleles [Konrad et al., 1983]. Flies, homozygous for extreme alleles or heterozygous for any allele and a deletion of the locus, produce eggs which fail to produce any of the normal gastrulation movements (at 3½ hours at 25°C) for forming mesodermal and endodermal structures. In addition, ventral and lateral cuticular structures fail to form although neuroblasts are present and produce neural structures. Because of these abnormalities we have called the locus *gastrulation-defective (gd)*.

One of the alleles (gd^{10}) has provided the opportunity to determine the temperature-sensitive period (TSP) for the action of the gene product [Konrad et al., 1983]. The most favorable combinations have been to take either the original A573 (gd^{1}) or Mohler's 4955 (gd^{3}) as a heterozygote with gd^{10}. In the first instance, the hatching rate for eggs laid by gd^{1}/gd^{10} heterozygotes at 25°C is 40% but drops to zero at 29°C. If eggs are shifted from 25° to 29°C during the first 90 minutes of development, the hatching rate dramatically decreases. However, if the switch from 25° to 29°C is delayed until 90 minutes after egg laying, there is no reduction in hatching rate. Similar shifts from 29° to 25°C or short periods at 29°C have supported the finding that the *gastrulation-defective* gene product is needed during the first 60–90 minutes after fertilization.

While we do not know the nature of the gene product, Goralski et al. (unpublished results) have been able to map it to 11A2-4 by means of deficiencies. It is hoped that molecular cloning of this region will provide the possibility to determine directly the structure of the gene and its role in development.

There are a number of other loci that produce comparable phenotypes. Nüsslein-Volhard [1979] discovered *dorsal* on the second chromosome which produces a phenotype similar to that found with extreme alleles of *gastrulation-defective*. Recently, in their comprehensive screen of the third chromosome for female-sterile mutations additional mutations have been obtained with a *gd*-like phenotype (Nüsslein-Volhard et al., personal communication). It is an exciting prospect that shortly we will have available for analysis mutations of sets of loci whose gene products interact to produce specific developmental defects. With the advent of molecular cloning, two-dimensional (2-D) gel analysis, and hybridomas, it is reasonable to hope that it will be possible to discover how these gene products interact to produce a specific developmental phenotype.

Recent mutation screens are certain to increase our repertoire of maternal effect lethals affecting gastrulation. For example, in a screen of 2,500 X-chromosomes, Engstrom et al. (unpublished results) found two loci that produce an embryo lacking all structures posterior to the seventh abdominal segment. Another X-chromosomal mutation *(fs(1)1502)* affects anterior development [Komorowska, 1980]. We have analyzed an allele of this locus which shows a preferential loss of only the clypeolabrum (Fig. 3) leading to a defective pharyngeal apparatus and subsequent lethality.

Finally, Rice and Garen [1975] have analyzed a set of third chromosomal loci which produce abnormal development at the blastoderm. Interestingly, one locus, mat(3)6, fails to produce a complete cellular blastoderm, and yet the pole cells move around the dorsal side of the embryo in a pseudoposterior midgut pouch [Rickoll and Counce, 1981], thus indicating that some of the gastrulation movements occur independently of the cellular blastoderm. These authors [Rickoll and Counce, 1981] have suggested that contractile filaments in the yolk sac may be at least partially responsible for these cellular movements of gastrulation. When the results of the exhaustive screen of chromosomes 2 and 3 are analyzed, it should be possible to make important generalizations about the types of developmental events dependent upon oogenetic information.

Rescuable Maternal Effect Mutation (Class III A)

Finally, a class of loci exists in which sufficient gene products can be produced for normal embryonic development either by the maternal or the

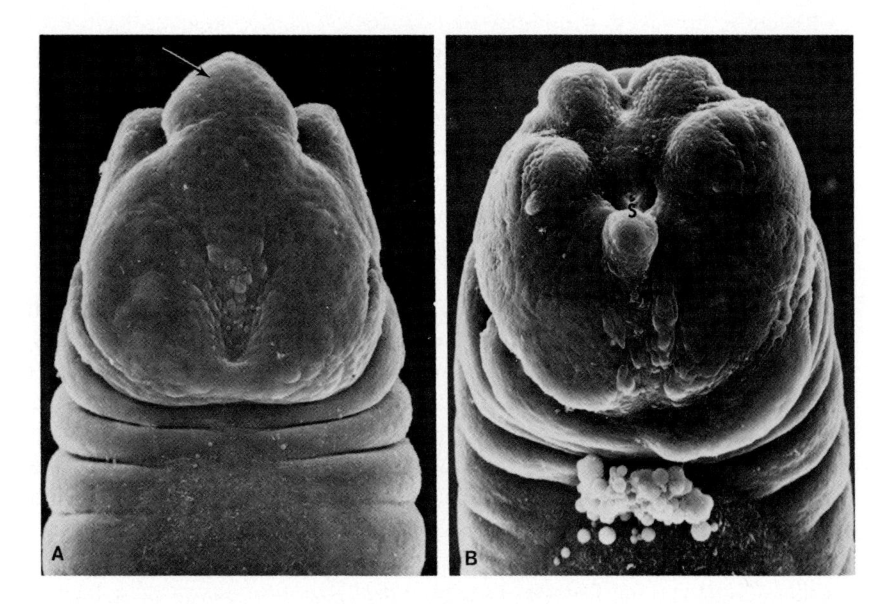

Fig. 3. Scanning electron micrograph of wild-type (A) and mutant (B) [allele of *fs (1) 15021*] embryos at approximately ten hours, showing the absence of the clypeolabrum (arrow) in the mutant embryo. The stomodeum (S) is clearly visible (Stephenson, Turner, and Mahowald, unpublished observations).

zygotic genome. On one hand, embryos produced by mothers heterozygous for the wild-type allele develop normally, even though the embryo lacks the wild-type allele. On the other hand, embryos produced by homozygous mutant mothers only survive if the fertilizing sperm carries the wild-type allele [Counce, 1956a–c]. The model for this type of mutation has been *rudimentary,* which has been shown to be the genetic locus for the first three enzymes of the pyrimidine biosynthetic pathway [Norby, 1970; Jarry and Falk, 1974; Rawls and Fristrom, 1975]. Significant rescue of embryonic development is produced by both diet and injection of pyrimidines [Okada et al., 1974]. Consequently, it is clear that embryonic lethality is due to insufficiencies of pyridimine stored in the embryo in the absence of the biosynthetic pathway in female flies. Rescue depends upon the early production of the new bases requiring the presence of the *rudimentary* gene products. In a similar manner, embryonic deficiencies in progeny of *deep orange* [Counce, 1956a] and *cinnamon* [Baker, 1973] females could be due to defects in pathways leading to some of the cyclic components of eye pigments. Searches for new loci affecting female sterility have shown little interest in the developmental aspects of these mutants [cf., Mohler, 1977; Gans et al., 1975; review by Mahowald and Kambysellis, 1980] because they apparently affect only metabolic pathways.

Recently, however, a number of new female sterile mutants have been discovered which suggest that this class may contain mutants of considerable developmental significance. Cuticle patterns of *fused* embryos show a distinct "gooseberry" pattern [cf., Nüsslein-Volhard and Wieschaus, 1980]—i.e., the abdominal ventral setal belts show a duplication in each segment of the anterior half of the segment in reverse symmetry. Thus the developmental defect probably concerns the pattern of development and not simply growth potential. Similarly, *almondex* [Shannon, 1972, 1973] has been shown to produce a hypertrophy of the central nervous system at the expense of the ventral ectodermal cells in a manner reminiscent of neurological zygotic lethals [Lehmann et al., 1981]. We have examined a new *almondex* allele as well as a new locus on the X-chromosome called *pecanex (pcx),* by both scanning electron microscopy and cuticular preparations. During early neurogenesis there is an increase in the number of ventral cells contributing to the nerve cord (Fig. 4). As a consequence, during dorsal closure, the dorsal ectoderm moves to the dorsal midline, resulting in a herniation of the ventral nerve cord. Extensive genetic analysis of *almondex* and six zygotic lethals with similar phenotypes is being carried out by Campos-Ortega and coworkers [Jimenez and Campos-Ortega, 1982]. On the basis of their interactions with each other, the loci fall into three groups: *amx,* which shows no interaction with the other loci; *mastermind, big brain, neuralizer,* and *Notch,* which show a maternal component and which are partially rescued by a

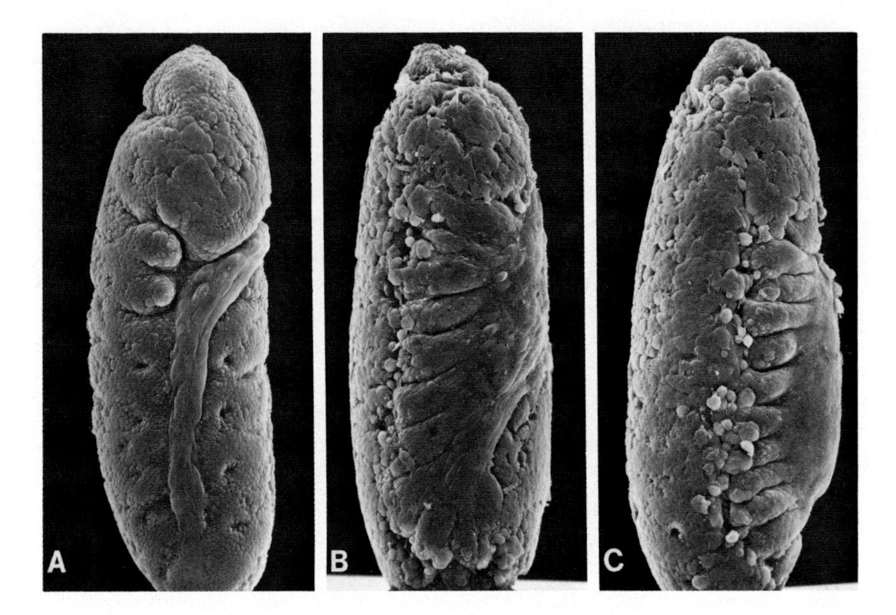

Fig. 4. Scanning electron micrographs of *almondex* embryos during segmentation and germ band shortening. At eight hours, thoracic and abdominal segmentation occurs normally (A). At nine to ten hours, as the posterior tip of the embryo moves posteriorly, the segments begin dorsal closure (B). However, in the absence of ventral hypoderm only small patches of dorsal hypoderm move to the dorsal midline and fuse (C). The absence of ventral cells is due to the hypertrophy of the ventral nervous system at the expense of the ventral hypoderm.

duplication containing *Notch*$^+$; and *Delta* and *Enhancer of split,* which show no maternal effect or interaction with *Notch* [Lehmann et al., 1981]. In all of these mutants, the developmental defect is similar and affects a small number of cells shortly after the blastoderm stage.

A fourth rescuable female sterile has been identified in my laboratory, temporarily called *fs(1)151*, which produces a transformation of a small region of the posterior ventral hypoderm to structures which appear to be anterior structures (Mahowald and Turner, unpublished results). While detailed description of the development of these embryos is out of place here, it is important to note that, analogous to *amx* and *pcx,* only a small portion of the blastoderm is altered in phenotype, and the change affects cellular determination. Thus, for these "rescuable" maternal effect mutations, in the absence of gene product from either oogenesis or transcription during the preblastoderm stage, specific cells of the blastoderm or early gastrula become transformed to a new cell type.

The nature of the gene product for these four mutations is not yet known, although cloning offers possibilities in the near future. Determination of cellular fates is the critical step during early development. It is difficult to envision the nature of the process which can be equally well established by information provided during oogenesis or by new gene activity produced during the first two hours of transcription [Zalokar, 1976; McKnight and Miller, 1976]. Both because of the rescuable nature of the genic activity and the possibility of cloning these genes, it should be possible to determine the nature of the control on development exercised by these gene products.

CONCLUSIONS

Our understanding of oogenesis and the role oogenetic gene products play on development has increased considerably since the reviews of only two years ago. The exciting prospect is that within the next few years nearly complete catalogs of female sterility mutations will be accumulated. Even now, apparently complete sets of loci are known which are essential for the production of specific embryological processes (e.g., neurulation, gastrulation). Because of the relative ease of identifying the genetic sequences associated with specific genes [e.g., Scallenghe et al., 1981], it should be possible to describe in detail the embryological defect for each mutation and combination of mutants, identify the gene and its product, demonstrate the rescue ability of the product during development [e.g., Illmensee and Mahowald, 1974], and finally provide a molecular understanding of key developmental events. Thus, as a result of the new tools of contemporary biology and the availability of mutants, our hope of fathoming the complexities of maternal information is becoming realized.

ACKNOWLEDGMENTS

I wish to thank Dr. F.R. Turner for the scanning electron micrographs, my colleagues at Indiana University, Drs. R. Boswell, L. Engstrom, T. Kaufman, K. Konrad, E. Stephenson, and J. Caulton, P. DiMario, T. Goralski, and C. Hallberg. This work has been supported by NIH-HD 7983 and HD 17607.

REFERENCES

Allis CD, Waring GL, Mahowald AP (1977): Mass isolation of pole cells from *Drosophila melanogaster*. Dev Biol 56:372–381.

Baker B (1973): The maternal and zygotic control of development by *cinnamon*, a new mutant in *Drosophila melanogaster*. Dev Biol 33:429–440.

Barnett T, Pachl C, Gergen JP, Wensink PC (1980): The isolation and characterization of Drosophila yolk protein genes. Cell 21:729–738.

Boswell RE, Mahowald AP (1983): A new grandchildless mutation in *Drosophila melanogaster* which affects both polar granules and segmentation. (in preparation).

Brennan MD, Weiner AJ, Goralski TJ, Mahowald AP (1982): The follicle cells are a major site of vitellogenin synthesis in *Drosophila melanogaster*. Dev Biol 89:225–236.

Counce SJ (1956a): Studies on female sterility genes in *Drosophila melanogaster*. I. The effects of gene *deep-orange* on embryonic development. Z Vererbungslehre 87:443–461.

Counce SJ (1956b): Studies on female sterility genes in *Drosophila melanogaster*. II. The effects of the gene *fused* on embryonic development. Z Vererbungslehre 87:462–481.

Counce SJ (1956c): Studies on female sterility genes in *Drosophila melanogaster*. III. The effects of the gene *rudimentary* on embryonic development. Z Vererbungslehre 87:482–492.

David J, Merle J (1968): A re-evaluation of the duration of egg chamber stages in oogenesis of *Drosophila melanogaster*. Drosophila Inform Serv 43:122–123.

Davidson E (1976): "Gene Activity in Early Development, 2nd Ed." New York: Academic Press, 452 pp.

Engstrom L, Caulton JH, Underwood EM, Mahowald AP (1982): Developmental lesion in the gametic mutant of *Drosophila melanogaster*. Dev Biol 91:163–170.

Fargnoli J, Waring GL (1982): Identification of vitelline membrane proteins in *Drosophila melanogaster*. Dev Biol 92:306–314.

Gans M, Audit C, Masson M (1975): Isolation and characterization of sex linked female sterile mutants in *Drosophila melanogaster*. Genetics 81:683–704.

Garen A, Gehring W (1972): Repair of the lethal developmental defect in *deep orange* embryos of *Drosophila* by injection of normal egg cytoplasm. Proc Natl Acad Sci USA 69:2982–2985.

Griffin-Shea R, Thireos G, Kafatos FC, Petri WH, Villa-Komaroff L (1980): Chorion cDNA clones of *D. melanogaster* and their use in studies of sequence homology and chromosomal location of chorion genes. Cell 19:915–922.

Illmensee K (1973): The potentialities of transplanted early gastrule nuclei of *Drosophila melanogaster*. Production of their descendants by germ-like transplantation. W Rouxs Arch 171:331–343.

Illmensee K, Mahowald AP (1974): Transplantation of posterior polar plasm in *Drosophila*. Induction of germ cells at the anterior pole of the egg. Proc Natl Acad Sci USA 71:1016–1020.

Illmensee K, Mahowald AP (1976): The autonomous function of germ plasm in a somatic region of the *Drosophila* egg. Exp Cell Res 97:127–140.

Illmensee K, Mahowald AP, Loomis M (1976): The ontogeny of germ plasm during oogenesis in *Drosophila*. Dev Biol 49:40–65.

Jarry B, Falk DR (1974): Functional diversity within the rudimentary locus of *Drosophila melanogaster*. Mol Gen Genet 135:113–122.

Jimenez F, Campos-Ortega JA (1982): Maternal effects of zygotic mutants affecting neurogenesis in *Drosophila*. W Rouxs Arch 191:191–201.

King RC, Bodenstein D (1965): The transplantation of ovaries between genetically sterile and wild type *Drosophila melanogaster*. Z Naturforsch 20/4:292–297.

King RC, Mohler JD (1975): The genetic analysis of oogenesis in *Drosophila melanogaster*. In King RC (ed): "Handbook of Genetics." New York: Plenum Press, Vol 3, pp 757–791.

Komorowska B (1980): L'effet d'une mutation de stérilité femelle sur la céphalogenése de *Drosophila melanogaster* pendant la vie embryonnaire. Thesis, University of P. and M. Curie.

Konrad KD, Goralski TJ, Turner FR, Mahowald AP (1982): Maternal effect mutation affecting gastrulation. J Cell Biol 95:159a.

Lehmann R, Dietrich U, Jimenez F, Campos-Ortega JA (1981): Mutations of early neurogenesis in *Drosophila.* W Rouxs Arch 190:226–229.

Mahowald AP (1968): Polar granules of *Drosophila.* II. Ultrastructural changes during early embryogenesis. J Exp Zool 167:237–262.

Mahowald AP (1979): Genetic control of oogenesis in *Drosophila.* In Ebert J, Okada T (eds): "Mechanisms of Cell Change." New York: Wiley & Sons, pp 101–117.

Mahowald AP, Tiefert M (1970): Fine structural changes in the *Drosophila* oocyte nucleus during a short period of RNA synthesis. W Rouxs Arch 165:8–25.

Mahowald AP, Caulton JH, Gehring WJ (1979a): Ultrastructural studies of oocytes and embryos derived from female flies carrying the grandchildless mutation in *Drosophila subobscura.* Dev Biol 69:118–132.

Mahowald AP, Caulton JH, Edwards MK, Floyd AD (1979b): Loss of centrioles and polyploidization in follicle cells of *Drosophila melanogaster.* Exp Cell Res 118:404–410.

Mahowald AP, Kambysellis MP (1980): Oogenesis. In Ashburner M, Wright T (eds): "Genetics and Biology of Drosophila." London: Academic Press, Vol 2, pp 141–224.

Mariol MC (1981): Genetic and developmental studies of a new *grandchildless* mutant of *Drosophila melanogaster.* Mol Gen Genet 181:505–511.

McKnight SL, Miller OL Jr (1976): Ultrastructural patterns of RNA synthesis during early embryogenesis of *Drosophila melanogaster.* Cell 8:305–319.

Mohler JD (1977): Developmental genetics of the Drosophila egg. I. Identification of 59 sex-linked cistrons with maternal effects on embryonic development. Genetics 85:259–272.

Norby S (1970): A specific requirement for pyrimidines in *rudimentary* mutants of *Drosophila melanogaster.* Hereditas 66:205–214.

Nüsslein-Volhard C (1979): Maternal effect mutations that alter the spatial coordinates of the embryo of *Drosophila melanogaster.* In Sultelny S, Konigberg I (eds): "Spatial Determinants of Development." New York: Academic Press, pp 185–211.

Nüsslein-Volhard C, Wieschaus E (1980): Mutations affecting segment number and polarity in *Drosophila.* Nature 287:795–801.

Okada M, Kleinman IA, Schneiderman HA (1974): Repair of a genetically-caused defect in oogenesis in *Drosophila melanogaster* by transplantation of cytoplasm from wild-type eggs and by injection of pyrimidine nucleotides. Dev Biol 37:55–62.

Perrimon N, Gans M (1983): Clonal analysis of tissue specificity of recessive female sterile mutations of *Drosophila melanogaster* using a dominant female sterile mutation, *FS(1)K1237* (submitted for publication).

Petri WH, Wyman AR, Kafatos FC (1976): Specific protein synthesis in cellular differentiation. III. The eggshell proteins of *Drosophila melanogaster* and their program of synthesis. Dev Biol 49:185–199.

Postlethwait JH, Jowett T (1980): Genetic analysis of the hormonally regulated yolk polypeptide genes in *D. melanogaster.* Cell 20:671–678.

Rawls JM, Fristrom JW (1975): A complex genetic locus that controls the first three steps of pyrimidine biosynthesis in *Drosophila.* Nature 255:138–740.

Rice TB (1973): Isolation and characterization of maternal-effect mutants: An approach to the study of early determination in *Drosophila melanogaster.* PhD Thesis, Yale University.

Rice TB, Garen A (1975): Localized defects of blastoderm formation in maternal effect mutants of *Drosophila.* Dev Biol 43:277–286.

Rickoll WL, Counce SJ (1981): Morphogenesis in the embryo of *Drosophila melanogaster.* Germ band extension in the maternal-effect lethal *mat*(3)6. W Rouxs Arch 190:245–251.

Scalenghe F, Turco E, Edstrom JE, Pirotta V, Melli M (1981): Microdissection and cloning of DNA from a specific region of *Drosophila melanogaster* polytene chromosomes. Chromosoma 82:205–216.

Shannon MP (1972): Characterization of the female-sterile mutant *almondex* of *Drosophila melanogaster*. Genetics 43:244–256.

Shannon MP (1973): The development of eggs produced by the female-sterility mutant almondex of *Drosophila melanogaster*. J Exp Zool 183:383–400.

Spradling AC, Mahowald AP (1979): Identification and genetic localization of mRNAs from ovarian follicle cells of Drosophila melanogaster. Cell 16:589–598.

Spradling AC, Digan ME, Mahowald AP, Scott MW, Craig EA (1980): Two clusters of genes for major chorion proteins of *Drosophila melanogaster*. Cell 19:905–914.

Spradling AC, Mahowald AP (1981): A chromosome inversion alters the pattern of specific DNA replication in Drosophila follicle cells. Cell 27:203–209.

Waring GL, Mahowald AP (1979): Identification and time of synthesis of chorion proteins in *Drosophila melanogaster*. Cell 16:599–607.

Wieschaus E, Nüsslein-Volhard C (1983): In preparation.

Wieschaus E, Marsh JL, Gehring WJ (1978): fs(1)K10, a germline dependent female sterile mutation causing abnormal chorion morphology in *Drosophila melanogaster*. W Rouxs Arch Entwicklungsmech Org 184:75–82.

Wieschaus E, Audit C, Masson M (1981): A clonal analysis of the roles of somatic cells and germ line during oogenesis in *Drosophila*. Dev Biol 88:92–103.

Zalokar M, Audit C, Erk I (1975): Developmental defects of female-sterile mutants of *Drosophila melanogaster*. Dev Biol 47:419–423.

Zalokar M (1976): Autoradiographic study of protein and RNA formation during early development of *Drosophila* eggs. Dev Biol 49:425–437.

Time, Space, and Pattern in Embryonic Development, pages 365–383

The Genetic Regulation of Segmentation in *Drosophila melanogaster*

Thomas C. Kaufman

Program in Genetics, Department of Biology, Indiana University, Bloomington, Indiana 47405

It is a reasonably widely held view that genes control the ontogenic processes which result in the production of the adult organism from the zygote. What is not clear, however, is the manner in which this regulation takes place. Put another way, we would like to understand how the two-dimensional information encoded in the DNA is translated into the three-dimensional structures of the organism. In this laboratory we have approached this problem through the paradigm of developmental genetics, utilizing *Drosophila melanogaster* as our experimental tool. More specifically, we have investigated the nature of the genes which regulate the basic pattern of segmentation in this particular insect. It has been shown that there are two basic ways in which mutations alter the pattern of segmentation in *Drosophila* [Kaufman and Wakimoto, 1982]. The first, in mutant individuals, is by changing the number and anterior-posterior polarity of the segments in the embryo [Nüsslein-Volhard and Weischaus, 1980]. The second is to cause or allow homoeotic transformations between and among segmental types [Garcia-Bellido, 1977]. This second class of mutation results in the replacement of one metamere and its derivatives with structures normally found in another locality. The segment mimicked by the transformed metamere, however, is not affected and develops normally. Since this class of mutation results in an altered pattern of development rather than a disruption of the ontogenic process, these genes have been hypothesized to be involved in the specification (determination) of cell fate in early development. Their role is viewed as a set of "selector" or "switch" genes which choose among a variety of cell fates [Kauffman, 1977]. The manner in which these genes function has been the subject of investigation in a number of laboratories, including ours.

Figure 1 shows the distribution of the various homoeotic genes in the genome of *Drosophila melanogaster* as well as some of the genes necessary for the normal pattern of segmentation. A brief survey of this map demonstrates that the class of homoeotic genes is not randomly distributed on the chromosome arms. Indeed, there are two distinct clusters of homoeotic mutations on the right arm of the third chromosome (see also Fig. 2). The more distal of the two is called the Bithorax Complex (BX-C) and has been the subject of genetic, developmental, and, more recently, molecular analysis in a number of laboratories [Lewis, 1978, 1981, 1982; Garcia-Bellido and Ripoll, 1978; Hogness, personal communication]. The complex is comprised of at least ten closely linked contiguous loci which, when mutant, cause segmental transformations in the posterior half of the organism. As in all insects, *Drosophila* is divided into a head, thorax, and abdomen. These three large regions are further divided into smaller segments. The abdomen has eight subterminal segments and the posterior, terminal caudal segment. There

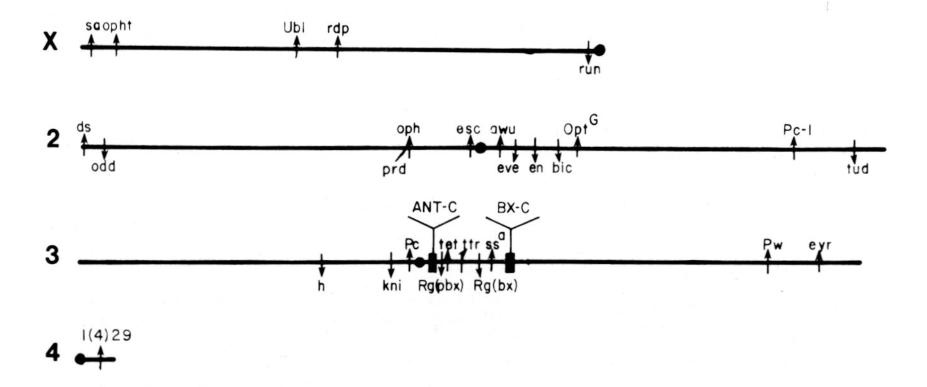

Fig. 1. Diagram of the genetic map of the *Drosophila melanogaster* genome showing the relative positions of the homoeotic loci (above the lines) and the segment number specifying genes (below the lines). X, 2, 3, and 4 indicate the respective chromosomes. The solid circles show the positions of the centromeres. *sa*, sparse arista; *opht*, opthalmoptera; *Ubl*, Ultrabithorax-like; *rdp*, reduplicated; *run*, runt; *ds*, dachsous; *odd*, odd-skipped; *prd*, paired; *oph*, opthalmopedia; *esc*, extra sex comb; *awu*, augenwulst; *eve*, evenskipped; *en*, engrailed; *bic*, bicaudal; Opt^G, Opthalmopteria; *Pc-1*, Polycomb-like; *tud*, tudor; *h*, hairy; *kni*, knirps; *Pc*, Polycomb; ANT-C, Antennapedia Complex; *Rg(pbx)*, Regulator of postbithorax; *tet*, tetraltera; *ttr*, tetraptera; *Rg(bx)*, Regulator of bithorax; ss^a, spineless aristapedia; BX-C, Bithorax Complex; *Pw*, Pointed wing; *eyr*, eyes reduced; *1(4)29*, lethal(4)29.

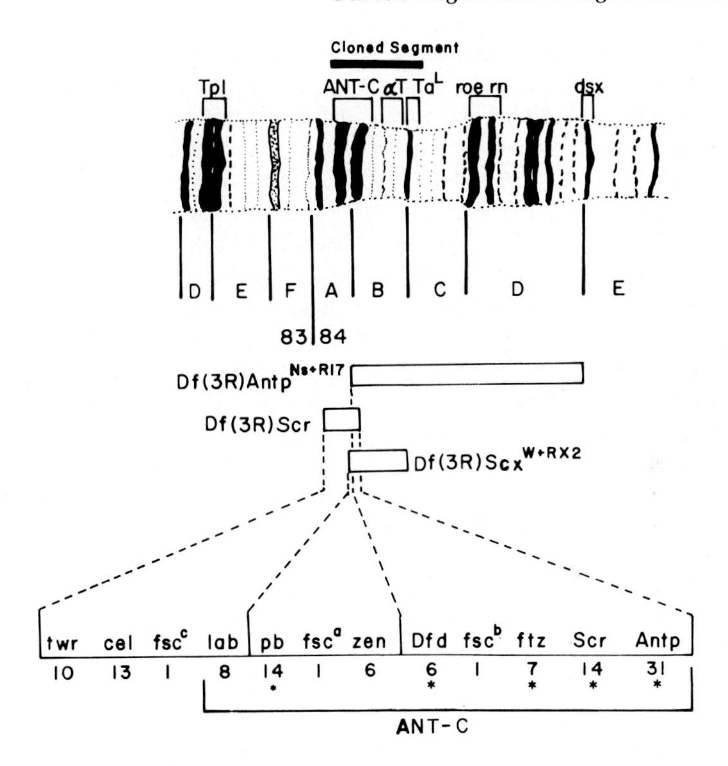

Fig. 2. Drawing of the portion of the polytene chromosome containing the locus of the ANT-C and showing the location of the cloned region of the genome. The open bars below the chromosome show the location and extent of three of the deletions recovered and utilized in the genetic and developmental studies carried out to date. The dashed lines emerging from the deletion boxes connect to a line which delineates the loci known to inhabit the 84A1,2-B1,2 interval. The loci in this interval thought to be members of the ANT-C are enclosed by the bracket below the line. The numbers below the line indicate the number of alleles we have identified at each locus. The asterisks indicate those loci at which at least one allelic variant is associated with a visible chromosome aberration. The cloned segment extends distally beyond the *Antp* locus; its extent proximal to the *Dfd* locus is presently not known. Abbreviations are as follows: *Tpl*, Triplo-lethal; αT, Alpha tubulin; Ta^L, Thickened arista-lethal; *roe*, roughened eye; *rn*, rotund; *dsx*, doublesex; *twr*, twisted bristles-rough eye; *cel*, cell-lethal; fsc^c, female sterile of Cain-c; *lab*, labial; *pb*, proboscipedia; fsc^a, female sterile of Cain-a; *zen*, zerknullt; *Dfd*, Deformed; fsc^b, female sterile of Cain-b; *ftz*, fushi tarazu; *Scr*, Sex combs reduced; *Antp*, Antennapedia.

are three thoracic segments, a pro-, meso-, and metathorax. In the adult fly, each of these segments bears a pair of walking legs. Dorsally, only the mesothorax is conspicuous. It forms a majority of the dorsal thorax and carries the single pair of wings found in the diptera. The dorsal pro- and metathoraces are reduced to a narrow band of cuticle, naked of any distinguishing morphological landmarks. These two structures lie anterior and posterior (respectively) to the mesothorax. Laterally the metathorax bears a pair of structures called halteres. In an evolutionary sense these can be viewed as vestigial wings—a point we will return to later. The segmental origin of the head in the higher diptera such as *Drosophila* is not entirely apparent in the adult or, for that matter, the larva. However, this segmentation is revealed in the early embryo. A cartoon of a lateral view of such an embryo is shown in Figure 3. There are three gnathocephalic segments—the mandibular, maxillary, and labial—which give rise to the larval and adult mouthparts (the mandibular lobe, however, is not thought to contribute to the adult mouthparts in *Drosophila*). The remainder of the head is comprised of the clypeolabral, procephalic, and optic lobes. These give rise to the remainder of the larval and adult head. With respect to the latter structure, this would correspond to the eye, antenna, and head capsule.

As stated earlier, mutations in the BX-C cause transformations in the posterior half of the animal. Thus the mutations *bithorax* (*bx*) and *postbithorax* (*pbx*) transform the anterior (*bx*) and posterior (*pbx*) metathorax into the corresponding portions of the mesothorax. This is evidenced in the adult fly by a second pair of wings developing in place of the normal halteres. The metathoracic or third pair of legs also are transformed to resemble the second or mesothoracic pair of legs in the same anterior-posterior fashion. If the double *bx pbx* mutant is contructed, the entire metathorax is transformed, resulting in an adult fly with two tandem mesothoracic segments. The *bithoraxoid* (*bxd*) mutation causes a similar transformation of the first abdominal segment to a thoracic identity. The dominant *Ultrabithorax* (*Ubx*) mutations fail to complement all three of the aforementioned recessive mutations and express all of the transformations observed in individuals carrying the recessive mutations. They differ by virtue of the fact that homozygous *Ubx* individuals do not reach the adult stage but die as larvae. These animals, however, do exhibit the transformations expected based on analyses of segment-specific morphologies present in the larvae. Therefore, the normal function of the *Ubx*-associated recessive mutations is necessary for the specification of proper segmental identity in the third thoracic (T_3) and first abdominal (A_1) metameres in the larva and adult. In the absence of these loci these two segments develop as second thoraces. As one moves in the BX-C more distally along the chromosome, more loci are found which specify segmental identity for more posterior segments. The spatial distribution in the embryo of these gene functions is shown diagramatically at the top of Figure 3.

Like the *Ubx* mutation, a deletion of the entire BX-C is lethal. These animals, however, exhibit a more extreme phenotype in that *all* of the abdominal segments as well as the third thoracic segment are transformed to

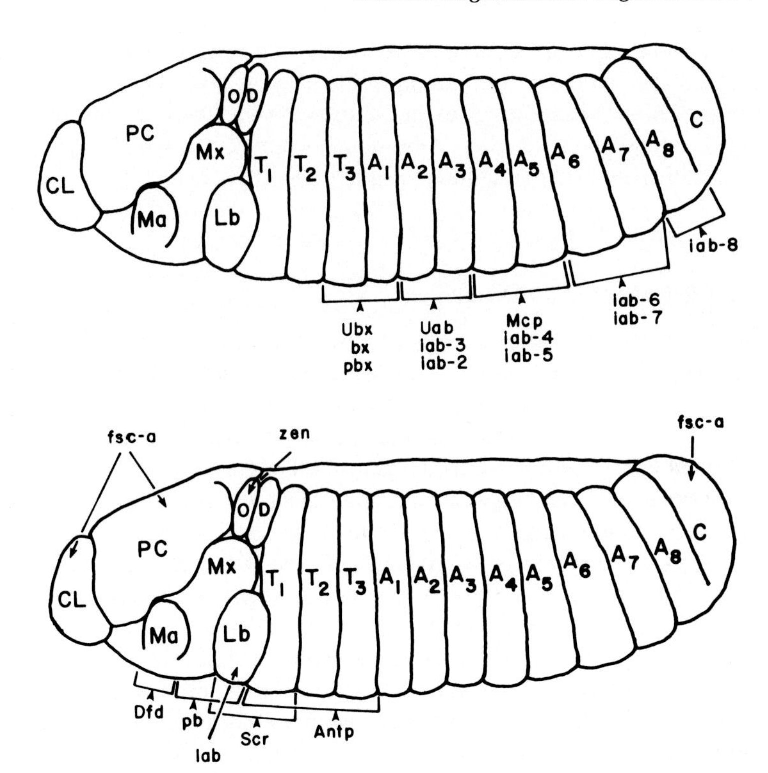

Fig. 3. Schematic drawing of the lateral view of an approximately 8-hour-old embryo showing the relative positions of all of the segments. The top animal shows the segments in which the various members of the bithorax complex are thought to be active in specifying segmental identity. Deletion of the various members of the complex results in the transformation of the more posterior abdominal segments into a second thoracic (T_2) identity. The lower animal shows the distribution of the segments affected by the member loci of the *Antennapedia Complex*. The nature of transformations and defects caused by mutation in the various loci are summarized in Table I. Abbreviations for the ANT-C loci are given in Figure 2; for BX-C, *Ubx*, *Ultrabithorax*; *pbx*, postbithorax; *Uab*, Ultraabdominal; *Mcp*, Miscadestral pigmentation; *iab*, infraabdominal. Abbreviations specifying the segments: CL, Clypeolabrum; PC, Procephalic; O, Optic; D, Dorsal Ridge; Ma, Mandibular; Mx, Maxillary; Lb, Labial; T_1-T_2-T_3, First-Second-Third thoraces; A_1–A_8, First-Eighth Abdominal; C, Caudal.

resemble a second thoracic segment. The identity of the more anterior segments and the caudal segment are apparently normal.

Developmental and genetic analyses of several of these BX-C mutations have revealed that these genes are active early in ontogeny at the cellular blastoderm stage, the time at which the inital determinative events take place in *Drosophila*. They also continue to be necessary to the maintenance of the determined state until differentiation takes place during metamorphosis in the pupal stage [for a review see Garcia-Bellido and Ripoll, 1978]. Moreover, these genes have been shown to be autonomous in their action and their activity is spatially restricted to only those metameres in which they play a role in establishing and maintaining segment identity.

THE ANTENNAPEDIA COMPLEX

The more proximal of the two groups of homoeotic mutations is the Antennapedia Complex (ANT-C) and has been the focus of investigation in this laboratory for several years [Duncan and Kaufman, 1975; Kaufman, 1978; Kaufman et al., 1980; Lewis et al., 1980a, b; Wakimoto and Kaufman, 1981; Denell et al., 1981]. We have localized this group of nine continguous loci to two adjacent polytene chromosome bands on the right arm of chromosome three. The location and order of these sites is shown in Figure 2. Four of the loci—*proboscipedia* (*pb*), *Deformed* (*Dfd*), *Reduced sex combs* (*Scr*), and *Antennapedia* (*Antp*)—are homoeotic and cause segmental transformations similar to those described for the BX-C, but in the anterior portions of the embryo and adult. The remaining members of the ANT-C are not ostensibly homoeotic in nature, but what is striking is that they are all restricted in action to the portions of the embryo not affected by the BX-C. This spatial division can be seen by comparing Figure 3B with 3A. We have obtained the information which allows the spatial (segmental domain of action) assignment of these nine loci by a variety of genetic and developmental analyses on a number of mutant alleles at each locus. These have included morphological observations of lethal animals (embryos, larva, and adults), determination of the temperature-sensitive periods of ts mutant alleles, and the analysis of somatic clones of mutant tissue created by x-ray-induced mitotic recombination. A summary of these results is presented in Table I. As can be readily seen, not all the identified loci have been investigated to an equal extent. Therefore, specific examples will be drawn from genes about which we have the firmest understanding and which in two cases bear a resemblance to loci of the BX-C. These criteria are best met by the *Scr* and *Antp* loci. Animals which are Scr^{-} are lethal and show a transformation of the embryonic and larval first or prothoracic segment into mesothorax as well as a transformation of the labial segment into a maxillary identity [Wakimoto and Kaufman, 1981]. Two newly recovered temperature-sensitive alleles show precisely the same transformations in adult flies. Somatic clones produced in the adult cuticle are normal in all parts of the body except the prothorax and labium where these structures are autonomously transformed into mesothorax and maxilla, respectively [Wakomoto, 1981; Struhl, personal communication]. Except for the fact that two adjacent segments are simultaneously transformed by a single mutation, the results obtained in our analysis of *Scr* lesions are directly analogous to those found with BX-C mutations. That is, the normal function of the *Scr* locus is to specify segmental identity in the labial and prothoracic (T_1) segments. Moreover, the gene is also necessary to the maintenance of that state throughout the larval and pupal stages until differentiation produces the final adult form after metamorphosis.

The *Antp* locus is genetically the most complex of the nine identified loci. There are three classes of striking dominant mutant alleles. The first is *Antp* itself. Flies heterozygous for *Antp* lesions show a transformation of their antennae into second or mesothoracic leg. Second is the *Cephalothorax* (*Ctx*) mutation which transforms the eye and head capsule into dorsal mesothorax. Lastly, there are the *Extra sex combs* (*Scx*) mutations in which the second and third legs are transformed into first legs. All of these dominant mutations are related by the fact that they also act as recessive lethals; i.e., *Antp*/*Antp* animals die. This recessive lethality is shared by all three types in that all heterozygous combinations of the three are also lethal, with death occurring in the late embryonic/early larval stages. Moreover, the phenotype of all of the lethal embryos and larvae is the same. The second and sometimes the third thoracic segments are transformed into the likeness of a first thorax, while the embryonic head is normal. Therefore, the recessive lethal phenotype resembles the dominant *Scx* lesions but not the *Antp* or *Ctx* mutations.

By examining individuals deleted for the *Antp* locus, it has been possible to demonstrate that all three dominant effects result from the improper activity of the locus and not its inactivation. That is, flies heterozygous for a deletion of *Antp* are normal morphologically. These same deletions, when homozygous, are lethal and yield embryos with the same morphology (T_2 + $T_3 \rightarrow T_1$) as those described above for the recessive lethality associated with the dominant lesions. We have also recovered and analyzed a number of recessive lethal mutations which are not deletions but do mimic the behavior of the deletions. Therefore, it is possible to obtain genetic lesions which exhibit all of the recessive properties of the dominant alleles but none of the dominant transformations.

Somatic clones of the above mutations have revealed some rather unexpected results. Clones homozygous for one of the recessive lethal mutations are normal in all parts of the adult cuticle (including the head and antennae) except the thoracic legs. The legs containing clones of $antp^-$ cells show a variety of defects and, if induced prior to 12 hours of embryogenesis (late gastrula), show a partial transformation of leg to antennae [Abbott, personal communication; Struhl, 1981]. The same is not true of clones of the *Antp* or *Scx* dominant mutations [Wakimoto, 1981], despite the fact that there is a shared recessive lethality among all three types. Clones of *Antp*/*Antp* or *Scx*/*Scx* cells express a more extreme form of the transformation seen in either

TABLE I. Summary of the Developmental Effects of Mutation in the ANT-C (Gene Symbols and Segment Designations as in Figures 2 and 3, Respectively)

Locus	*lab*	*pb*	*fsc*[a]	*zen*	*Dfd*	*fsc*[b]	*ftz*	*Scr*	*Antp*
No. of alleles	8	14	1	6	6	1	7	14	31
Lethal phase	Embryonic	Viable	Embryonic	Embryonic	Embryonic	Embryonic	Embryonic	Embryonic	Embryonic and pupal
Phenotype or nature of homoeotic transformation	Embryos are missing the derivatives of the labial segment	Labial palps transformed to first leg or antenna depending on allele and/or temperature; also the maxillary palps can be transformed to antennae by some alleles	Embryos are missing the derivatives of the Clypeolabral and caudal segments	All movements of gastrulation are blocked; optic lobe missing	Head involution blocked; derivatives of the mandibular lobe missing; antennal sense organs not well developed	Derivatives of the gnathocephalic segments missing; also gaps and fusions in segment pattern	Embryo has pair-wise association of segments giving 1/2 normal segment number; one dominant allele gives a pbx-like phenotype	*In embryo*, labial segment to maxillary and $T_1 + T_2$; *in adults*, labial palps to maxillary palps and $T_1 + T_2$; one dominant allele gives $T_2 + T_3 \rightarrow T_1$	*In lethals*, $T_2 + T_3 \rightarrow T_1$ *Antp + Ns* antenna → 2nd leg *Scx* $T_2 + T_3 \rightarrow T_1$ $T_1 \rightarrow T_2$ *Ctx* eye → T_2 *semilethals* 1st leg → antenna?

Temperature-sensitive period	?	Mid 2nd instar to early 3rd instar larva	?	Between 2 and 4 hr of embryogenesis	?	?	Between 2 and 4 hr of embryogenesis	Late 3rd instar to early pupa	3rd instar larva
Phenotype of somatic clones	?	Autonomous expression of homoeotic transformation	?	Clones made after 6 hr of embryogenesis are normal in adult cuticle	Clones made in the dorsal postgenal area of the head (mandibular?) show autonomous transformation of the region to ventral vibrissae (maxillary?)	?	Clones made after 6 hr of embryogenesis are normal in adult cuticle	Clones show autonomous expression of transformation $T_1 + T_2$ and labium → maxilla	*Lethals* $T_1, T_2 + T_3$ → antenna? *Antp* antenna → 2nd leg *Scx* $T_2 + T_3 \rightarrow T_1$ $T_1 \rightarrow T_2$ labial palps → maxillary palps

heterozygote (*Antp* = antenna → second leg and *Scx* = second and third leg → first leg). There is no evident transformation of leg to antenna.

The fact that *antp*$^-$ clones do not affect normal antennal development implies that this gene and its product(s) are not involved in the normal specification of that segmental derivative. However, both the defective legs and transformed embryonic thoraces indicate that *Antp*$^+$ is necessary for the specification of normal thoracic identity in an analogous fashion to *Scr* and *Ubx*. There is, however, an apparent critical and intriguing difference among the three. The *Antp* gene is causing two qualitatively different transformations, depending on the ontogenic stage being assayed. In the embryo, *Antp*$^+$ is necessary to specify what type of thoracic identity T_1 vs. T_2 or T_3 will be expressed. While in the adult it appears to function as a selector which, to some extent, specifies leg vs. antenna. With this latter information in hand, we can gain some insight into the nature of the dominant lesions we have described at the *Antp* locus. The *Scx* lesions apparently result from the misexpression of the *Antp* embryonic function (T_1 vs. T_2 or T_3) in the adult where it does not normally act. The *Antp* lesions result from the abnormal activation of the adult leg vs. antenna function in the antenna where it should not be active. Therefore, these dominant lesions can be seen as regulatory mutations causing the abnormal temporal (*Scx*) and spatial (*Antp*) expression of the *Antp*$^+$ gene.

Another example of our analyses of the ANT-C is presented by the *fushi tarazu* (*ftz*) locus. Mutations at this locus produce phenotypes which are unlike any lesions in the BX-C and are characteristic of those loci which regulate segment number rather than specifying identity, per se [Nüsslein-Volhard and Weischaus, 1980].

All of the recessive, nonconditional alleles of *ftz* die as late embryos and show a pair-wise association of normally adjacent segments producing an animal with one half the normal number of segments [Wakimoto and Kaufman, 1981] (Fig. 4). Temperature-conditional mutations show only partial paired associations when grown at their restrictive temperature. This partial expression has allowed us to deduce the frame of segmental associations which is: Mx + Lb; $T_1 + T_2$; $T_3 + A_1$; $A_2 + A_3$; $A_4 + A_5$; $A_6 + A_7$ (see

Fig. 4. Micrographs and tracings showing the gross morphological attributes of normal (G, H, I); *ftz*$^-$ (D, E, F); and *Rg(pbx)*$^-$ (A, B, C) embryos. A, D, and G are SEM preparations of 8-hour-old embryos of each genotype. B, E, and H are whole-mount embryo preparations of the same age stained for acetylcholinesterase activity, thus revealing the morphology of the brain and ventral nerve cord. C, F, and I are tracings of photographs of the cuticular patterns of the three genotypes at 24 hours. As can be seen, both the internal anatomy of the central nervous system and the external pattern of dentical belts is altered by the two segment disruptions. H, Head; T, Thorax; A, Abdomen.

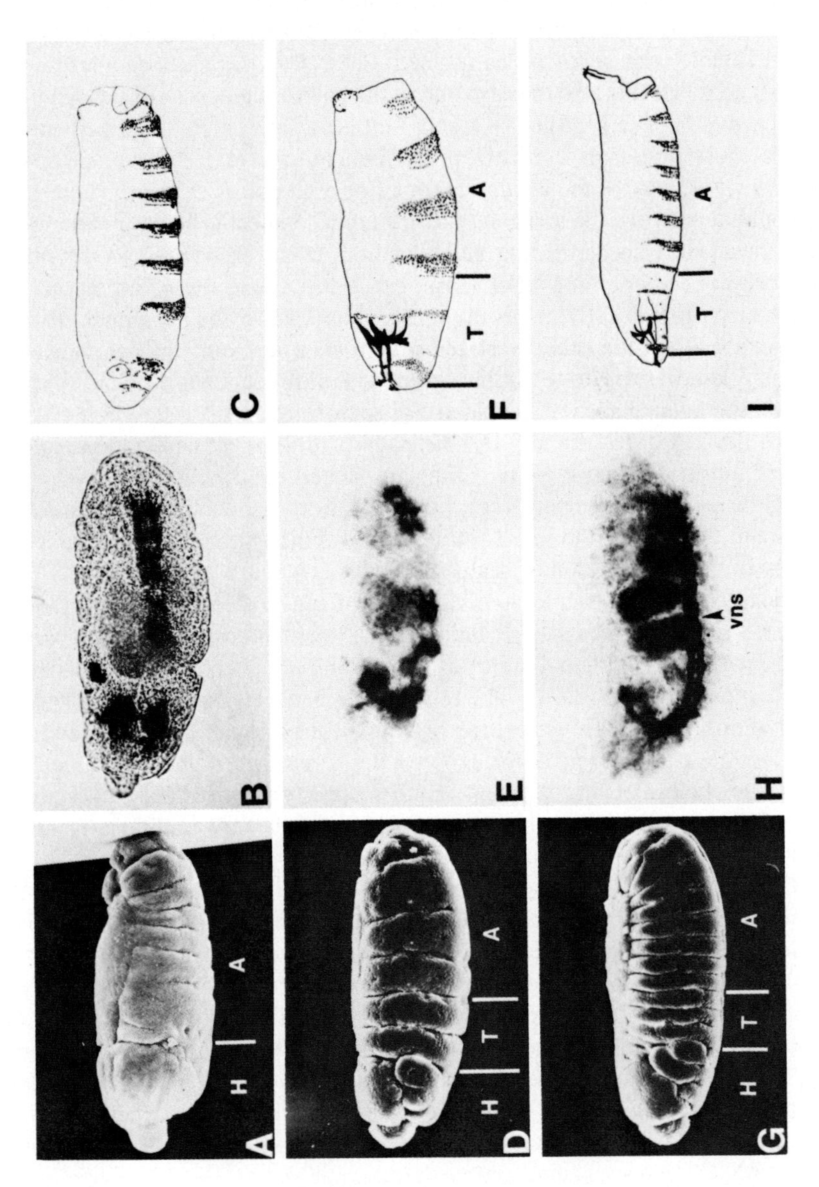

Fig. 4.

Fig. 3 for abbreviations). The most anterior and posterior segments are not affected. This failure in segmentation is apparent at the initial point of that process at about four hours of embryogenesis [Turner and Mahowald, 1977], where double-wide segments can be seen in the scanning electron microscope (SEM) (Fig. 4). It is also the case that in the paired segments of very different prospective fates (e.g., Mx + Lb and T_3 + A_1), no vestige of the derivatives of the more posterior segment can be found in the *ftz*$^-$ embryo. Only the segmental identity of the leading member of each pair is evident. Therefore, it would appear that the anlagen of every other segment is being incorporated into the identity of its leading neighbor and *ftz* can be viewed as a type of homoeotic gene—not in a strict sense, but it does cause a transformation.

The tsp of *ftz* is early in development, and clones made subsequent to this tsp do not affect the process of segmentation. Therefore, we can conclude that *ftz*$^+$ is necessary to the initial segmentation process but does not participate in the maintenance of the pattern of segmentation.

We have also determined that the pattern of segment associations is not dependent on segment identity. Embryos deleted for the BX-C and *ftz* show the same pattern of segmentation; however, all of the double-wide segments resemble the second thorax. Therefore, *ftz*$^+$ functions prior to or independently of the specification of segment identity.

Finally, there is a single dominant gain of function lesion at the *ftz* locus which shares the recessive lethality of *ftz*$^-$ mutations and shows partial segmental fusions in lethal embryos. The dominant phenotype is like that of *pbx*, a member of the BX-C. The relationship between the two phenotypes is at present not clear. However, the proximity of *ftz* to *Scr* and *Antp*, and the fact that this dominant *ftz* lesion shows a BX-C-associated phenotype, has led us to conclude that this segment number specifying locus may play some "regulatory" role in the proper expression of the more traditionally homoeotic loci of the ANT-C and the BX-C.

Our conclusion that the *ftz* locus represents a "regulator" of homoeotic genes is supported by the existence of another such segment number specification locus just distal to the ANT-C on the right arm of chromosome three. A dominant gain of function mutation exists at this locus and, like the *ftz* mutation described above, produces a homoeotic transformation of posterior haltere to wing similar to the *pbx* mutation in the BX-C. This dominant lesion has given the locus its name: *Regulator of postbithorax* [*Rg(pbx)*]. Developmental and genetic studies on this mutation have led to the conclusion that the gene functions early in development and acts as an activator of a portion of the BX-C [Capdevila and Garcia-Bellido, 1981].

We have analyzed the morphology of animals which carry apparent null alleles of *Rg(pbx)* as well as deletions of the locus. *Rg(pbx)*$^-$/+ animals do not show a pbx phenotype, supporting the conclusion that the dominant mutation results from a gain rather than a loss of function. Homozygous *Rg(pbx)*$^-$ individuals die at the end of embryogenesis and fail to hatch from the egg cases. As can be seen in Figure 4, these lethal individuals lack the elements normally derived from the anterior segments of the embryo (the mouthparts and the three thoracic segments). Instead, a single large abdomenlike segment is found in their place. Indeed, the only segments formed in these animals resemble abdominal metameres. It is also the case that the seventh and eighth abdominal segments are fused into one larger element. Using the SEM we have determined that like *ftz*$^-$ embryos this failure of normal segmentation is evident at the initial stages of that process. What is not clear, however, is the relationship of the defects observed in the deleted animals vis-a-vis the dominant homoeotic phenotype. Put another way, it is not clear how and at what level the two segment number specifying loci interact with the homoeotic genes. We do know, however, that the *Rg(pbx)* locus does not perform its function based on the identity of the anterior segments. By combining a deletion for both the *Rg(pbx)* locus and the BX-C, we have asked if thoracic identity is necessary for the failure of these elements to form in *Rg(pbx)*$^-$ embryos. An animal deleted for both *Rg(pbx)* and BX-C is shown in Figure 5 adjacent to an animal deleted for only the *Rg(pbx)* locus. As can be seen, both animals have the same number of segments, but the identity of those segments is quite different. The *Rg(pbx)*$^-$ animal has only abdominallike metameres, while in the double mutant only second thoracic (T_2) elements are seen. Therefore, animals deleted for the *Rg(pbx)* locus are capable of forming thoracic segments, but only in the posterior of the embryo. With respect to our original question, we can conclude that thoracic identity, as specified by the BX-C and the ANT-C, is not a prerequisite for *Rg(pbx)* action and, like *ftz*, this locus acts prior to or independently of the homoeotic loci. Based on the phenotypes of the two dominant gain of function lesions at the *ftz* and *Rg(pbx)* loci, we feel that independence of the two segment specifying systems (homoeotic and enumeration) is unlikely. Rather, we conclude that the segment number specification acts prior to and as a prerequisite for the normal spatial expression of the homoeotic loci. The actual manner of this interaction or sequential action of the two types is not at present known. However, we feel that the work at a molecular level, which will be described below, offers an opportunity for us to determine the manner in which the two types are integrated.

MOLECULAR ANALYSIS OF ANT-C

All of our conclusions and hypotheses about the functions of the various analyzed loci exist only in a formal sense. We have no direct proof or data as to the proximate role of these genes or their products in ontogeny. In order

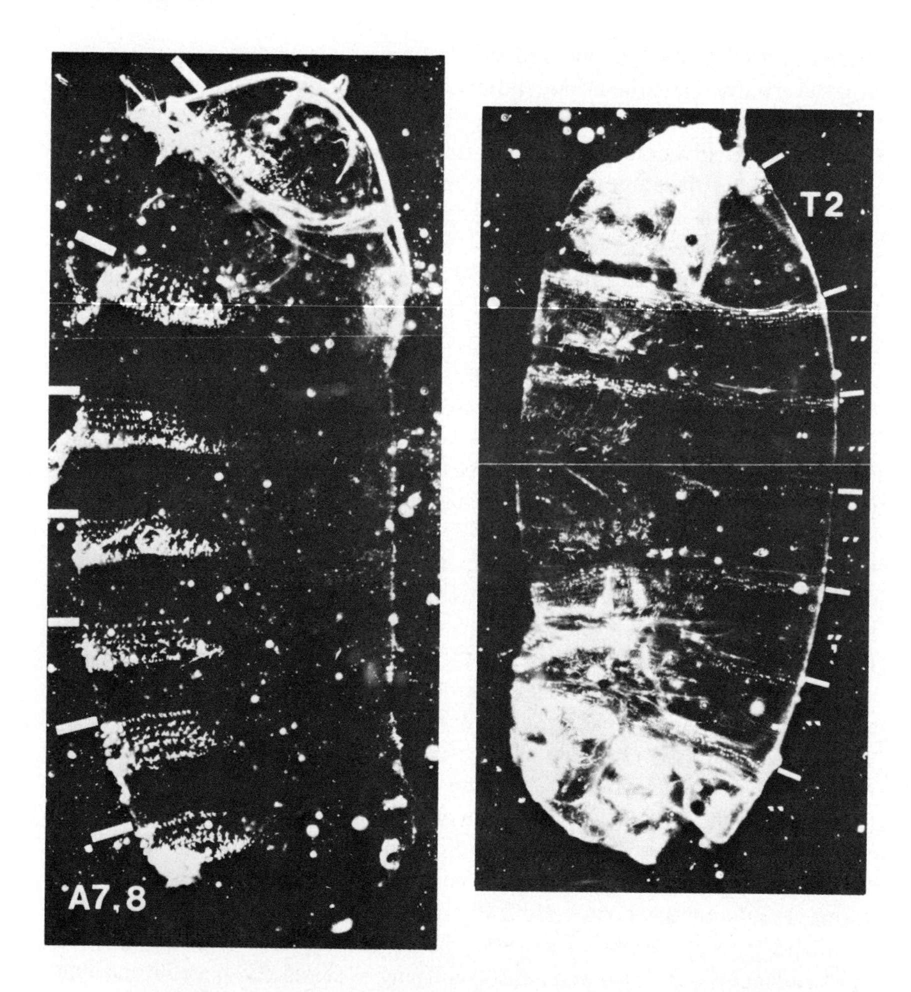

Fig. 5. Photomicrographs taken with darkfield lighting of *Rg(pbx)*$^-$ (left) and *Rg(pbx)*$^-$ BX-C$^-$ (right) embryos. Both animals are 24 hours old. Note that both animals have the same number of ventral denticle belts (indicated by white bars to the left and right of each animal) but that in the *Rg(pbx)* animal they have the morphology of the heavier abdominal type, while in the doubly mutant individual they resemble the pattern seen in the second thorax (T_2).

to gain more insight into this role we have embarked on a more molecular series of investigations involving the cloning of the DNA from the ANT-C.

We have succeeded in recovering DNA clones from the Antennapedia Complex by three separate approaches. The first involved the use of a chromosomal transposition of material from another site in the right arm of chromosome 3. Specifically, a segment of the genome represented on the polytene chromosome by section 86E10 through 87E1,2 was induced as and associated with a reversion of the *Deformed* mutation (Fig. 2). The distal end of this transposition brought into opposition the already cloned sequences in 87E1,2 [Bender et al., 1979] and the 84B1,2 doublet, the site of the ANT-C. We therefore constructed a "library" using *Drosophila* DNA derived from a stock of transposition-bearing flies and the λ vector 1059 [Maniatis et al., 1978]. This library was probed [Benton and Davis, 1975] using labeled cloned sequences from 87E1,2 (kindly supplied by D. Hogness and W. Bender). Inserts from the library were purified by standard procedures and were subjected to restriction enzyme analysis. Of nine such recovered clones, the restriction pattern of one could be seen to have DNA not normally associated with segment 87. This was confirmed by using the new 84-like fragments for in situ hybridization [Pardue and Gall, 1975]. They hybridized to bands in 84A,B as was expected. This jumper clone was then used to probe a library of wild-type *Drosophila* DNA [Maniatis et al., 1978], and inserts containing only DNA from bands 84A-B were recovered. The extent of these sequences was expanded by "walking" down the chromosome [Bender et al., 1979]. This entailed the recovery of new cloned sequences from the library which partially overlap the original unique probe. The first step of the walk was taken by probing two nitrocellulose lifts of the library with probes derived as restriction enzyme fragments from opposite ends of the first unique clone. By comparing the two filters, it was possible to identify plaques which had DNA homologous to one but not both ends of the original insert. Since the library was constructed from size-selected DNA, the *Drosophila* inserts are all of roughly the same size and approximately 16 kb in length. Therefore, inserts identified in the described manner should overlap one end of the original 16-kb fragment and extend to varying degrees to the left and right, respectively. We were able to specify the chromosomal direction of the walk by virtue of our knowledge of the proximal-distal order of the original transposition. This in turn could be correlated with the genetic map, and we had a good idea as to our location with respect to those genes of the ANT-C we were interested in cloning. Further steps in the walk were accomplished by using probes derived from the two previously isolated and overlapping phages. That is, we sought inserts which had homology to the distal- or proximalmost insert at the open end of the walk but did not have homology to the cloned sequence adjacent and in the internal portion of the walk.

The second entree into the ANT-C was obtained by walking proximally on the chromosome from a cloned α-tubulin gene in segment 84B3-5 (Fig. 2). This clone was supplied to us by Drs. M. Pardue and D. Mischke. Indeed, Dr. Mischke performed the initial portion of this walk [Mischke and Pardue, 1982].

The third source of cloned sequences came from Dr. V. Perrotta, who dissected the 84A-B interval from a polytene chromosome and cloned that DNA, which was sent to us in the form of plaques. The viruses from those plaques were grown up and purified and used to probe the wild-type library. The inserts obtained by Perrotta were very small (less than 1 kb), which lessened their utility, especially for in situ hybridization. Therefore, we obtained larger but homologous inserts from the probing of the library.

Approximately 40 new inserts were recovered, and their cross homology was ascertained by dotting each of the purified inserts onto 40 separate filters and hybridizing each of the labeled inserts separately to each of the filters containing 40 dots. The resulting matrix demonstrated only a few overlaps of sequence. This was confirmed by in situ hybridizations of the overlapping groups and characterization of the inserts by restriction enzyme analysis.

One of the groups defined by the above procedure was comprised of eight separate but overlapping inserts. This group hybridized in situ to 84B1,2, the residence of the *Antennapedia* locus. This group was therefore chosen as a start point for a third chromosome walk as described above.

We walked distally on the chromosome from the *Dfd* entry point, proximally from the α-tubulin, and in both directions from the Perrotta-derived clones. All three walks have been joined, and we now possess in overlapping phage inserts somewhat more than 440 kb of contiguous *Drosophila* DNA. Interestingly, in that interval we have discovered only one copialike intermediate repeat element. There are also some other moderate repeat elements in the interval, but they are less than 1 kb long and therefore were no hindrance to the walking procedures. A majority of the 440 kb has been mapped using six different restriction enzymes.

Many of the mutations that originally aroused our interest in the ANT-C are associated with chromosome aberrations. We have begun localizing these breakpoints on the DNA in order to correlate the molecular map with our previously defined genetic map. We have used two methods to locate the various breakpoints. First, ^{3}H-labeled probes were prepared by nick translation of the clones from the walk. These were then hybridized in situ to polytene chromosomes of the aberration-associated mutations. This allowed us to determine if the cloned fragment was distal or proximal to the breakpoint. Once the position of the breakpoint had been narrowed to only a few fragments, DNA was extracted from the mutant stocks and subjected to a "Southern blot" analysis [Southern, 1975; Wahl et al., 1979] using the wild-type cloned fragments as labeled probes. By these techniques we have been able to map several of the breakpoints on the DNA. The cloned fragments almost certainly contain those loci on the genetic map (Fig. 2) from *Dfd* distal to *Antp*. Whether or not the loci proximal to *Dfd* are represented is not known. Two striking results have been obtained from their initial mapping.

Lesions associated with the *Antp* locus span 100 kb of DNA. Our genetic studies have demonstrated that there are no other detectable functions in that interval. All of the detectable *ftz* lesions map to a small 3.3-kb HindIII fragment proximal to *Antp*. Three of these lesions were EMS-induced and apparently associated with the insertion of an extra 2 kb of sequence of unknown origin into the 3.3-kb element. One of these, ftz^{f47}, is ts.

Just distal and closely adjacent to the *ftz* mutations is the only localized *Scr* lesion, *In*(3R)*Msc*.

In this large amount of cloned sequence we wanted to find those regions which encode RNA. However, we also desired to avoid doing "Northern blots" [Alwine et al., 1977; Thomas, 1980] using all of the cloned fragments.

This was accomplished by making labeled cDNA to polyadenylated RNA using an oligo-dT primer. The RNA was extracted from unfertilized eggs and 1-, 4-, and 17-hour-old embryos and adults. This labeled cDNA was used to probe nitrocellulose blots of restriction enzyme-digested insert DNA purified from the overlapping set of clones defining a majority of the 440-kb cloned segment. To date we have found by this method 16 cloned regions (there are 17 if one includes the α-tubulin RNA) apparently associated with transcripts.

We are currently investigating the various transcripts in more detail in mutant and normal individuals. It is our hope that this analysis will reveal the validity of our models of *Antp*, *Scr*, and *ftz* gene action. One thing that is already apparent is that the genetic and developmental complexity of the *Antp* locus is directly reflected in a like molecular intricacy.

ACKNOWLEDGMENTS

The author would like to acknowledge the contributions of all of the members of the laboratory, past and present. The developmental and genetic work was performed by Drs. R. Turner, R. Lewis, B. Wakimoto, and T. Hazelrigg, and Mr. M. Abbott. The molecular efforts were carried out by Dr. M. Scott and Ms. A. Weiner. Additionally, the technical help of P. Fornili, D. Otteson, and R. Laymon has been essential to the research. Supported by PHS-GM24299 and PHS-GM29709.

REFERENCES

Alwine JC, Kemp DJ, Stark GR (1977): Method for detection of specific RNAs in agarose gels by transfer to diazobenzyloxymethyl-paper and hybridization with DNA probes. Proc Natl Acad Sci USA 74:5350–5354.

Bender W, Spierer P, Hogness D (1979): Gene isolation by chromosomal walking. J Supramol Struct [Suppl] 3:32 (Abstract).

Benton WD, Davis RW (1975): Screening λgt recombinant clones by hybridization to single plaques in situ. Science 196:180–182.

Capdevila MP, Garcia-Bellido A (1981): Genes involved in the activation of the Bithorax Complex of *Drosophila*. W Rouxs Arch 190:339–350.

Denell RE, Hummels KR, Wakimoto BT, Kaufman TC (1981): Developmental studies of lethality associated with the Antennapedia gene complex in *Drosophila melanogaster*. Dev Biol 81:43–50.

Duncan IW, Kaufman TC (1975): Cytogenetic analysis of chromosome 3 in *Drosophila melanogaster*: Mapping of the proximal portion of the right arm. Genetics 80:733–752.

Garcia-Bellido A (1977): Homoeotic and atavic mutations in insects. Am Zool 17:613–630.

Garcia-Bellido A, Ripoll P (1978): Cell lineage and differentiation in *Drosophila*. In Gehring WJ (ed): "Results and Problems in Cell Differentiation." Berlin: Springer-Verlag, Vol 9, pp 119–156.

Kauffman SA (1977): Characteristic waves, compartments, and binary decision in *Drosophila* development. Am Zool 17:631–648.

Kaufman TC, Lewis RA, Wakimoto BT (1980): Cytogenetic analysis of chromosome 3 in *Drosophila melanogaster*: The homoeotic gene complex in polytene chromosome interval 48A-B. Genetics 94:115–133.

Kaufman TC, Wakimoto BT (1982): Genes that control high-level developmental switches. In Bonner JT (Ed): "Evolution and Development." Dahlem Konferenzen, pp 189–205.

Lewis EB (1978): A gene complex controlling segmentation in *Drosophila*. Nature 276:565–570.

Lewis EB (1981): Developmental genetics of the Bithorax Complex in *Drosophila*. In Brown DD, Fox CF (Eds): "Developmental Biology Using Purified Genes." ICN-UCLA Symposium on Molecular and Cellular Biology. New York: Academic Press, Vol 23.

Lewis EB (1982): Control of body segment differentiation in *Drosophila* by the Bithorax gene complex. In Burger MM (Ed): "Embryonic Development: Genes and Cells." Proceedings of the IX Congress of the International Society of Developmental Biologists. New York: Alan R. Liss, Inc., pp 269–288.

Lewis RA, Kaufman TC, Denell RE, Tallerico P (1980a): Genetic analysis of the Antennapedia gene complex (ANT-C) and adjacent chromosomal regions of *Drosophila melanogaster*. I. Polytene chromosome segments 84B-D. Genetics 95:367–381.

Lewis R, Wakimoto B, Denell R, Kaufman T (1980b): Genetic analysis of the Antennapedia gene complex (ANT-C) and adjacent chromosomal regions of *Drosophila melanogaster*. II. Polytene chromosome segments 84A-84B1, 2. Genetics 95:383–397.

Maniatis T, Hardison RC, Lacy E, Lauer J, O'Connell C, Quon D, Sim GK, Efstradiatis A (1978): The isolation of structural genes from libraries of eucaryotic DNA. Cell 15:687–701.

Mischke D, Pardue ML (1982): Organization and expression of α-tubulin genes in *Drosophila melanogaster*. J Mol Biol 156:449–466.

Nüsslein-Volhard C, Weischaus E (1980): Mutations affecting segment number and polarity in *Drosophila*. Nature 287:795–801.

Pardue ML, Gall JG (1975): Nucleic acid hybridization to the DNA of cytological preparations. In Prescott D (Ed): "Methods in Cell Biology." New York: Academic Press, Vol 10, pp 1–16.

Southern EM (1975): Detection of specific sequences among DNA fragments separated by gel electrophoresis. J Mol Biol 98:503–517.

Struhl G (1981): A homoeotic mutation transforming leg to antenna in *Drosophila*. Nature 292:635–638.

Thomas PS (1980): Hybridization of denatured RNA and small DNA fragments transferred to nitrocellulose. Proc Natl Acad Sci USA 77:5201.

Turner F, Mahowald A (1977): Scanning electron microscopy of *Drosophila melanogaster* embryogenesis. II. Gastrulation and segmentation. Dev Biol 57:403–416.

Wahl GM, Stern M, Stark GR (1979): Efficient transfer of large DNA fragments from agarose gels to diazobenzyloxymethyl-paper and rapid hybridization by using dextran sulfate. Proc Natl Acad Sci USA 76:3683–3687.

Wakimoto BT (1981): Genetic and developmental studies of the loci in the polytene interval 84A-84B1, 2 of *Drosophila melanogaster*. PhD Thesis, Department of Biology, Indiana University.

Wakimoto BT, Kaufman TC (1981): Analysis of larval segmentation in lethal genotypes associated with the Antennapedia gene complex in *Drosophila melanogaster*. Dev Biol 81:51–64.

Index